Fractal Models in the Earth Sciences

Fractal Models in the Earth Sciences

G. Korvin

Department of Geology and Geophysics
University of Adelaide
P.O. Box 498, Adelaide, S.A. 5001, Australia

1992

ELSEVIER
AMSTERDAM • LONDON • NEW YORK • TOKYO

ELSEVIER SCIENCE PUBLISHERS B.V.
Sara Burgerhartstraat 25
P.O. Box 211, 1000 AE Amsterdam, The Netherlands

Library of Congress Cataloging-in-Publication Data

Korvin, Gabor.
Fractal models in the earth sciences / Gabor Korvin.
p. cm.
Includes bibliographical references and index.
ISBN 0-444-88907-8 (acid-free paper)
1. Fractals. 2. Earth sciences--Mathematics. I. Title.
QE33.2.F73K67 1992
550'.1'51474--dc20 92-19046
CIP

ISBN: 0 444 88907 8

This book is printed on acid-free paper.

Printed in The Netherlands.

It is significant that in the greatest religious poem existent, the Book of Job, the argument which convinces the infidel is not (as has been represented by the merely rational religionism of the eighteenth century) a picture of the ordered beneficence of the Creation, but, on the contrary, a picture of the huge and undecipherable unreason of it. "Hast Thou sent the rain upon the desert where no man is?" This simple sense of wonder at the shapes of things, and at their exuberant independence of our intellectual standards and our trivial definitions, is the basis of spirituality as it is the basis of nonsense.

— G.K. CHESTERTON: *A Defence of Nonsense*

Preface

Australian surfing authority, Thor Svenson, once said that "choosing a (surf) board is a very individual thing, like choosing a wife or girl-friend, almost". I never fail telling my students that choosing the first technical book to be read from cover to cover on a long vacation is as important as choosing a surfboard.

Almost thirty years ago, as a fourth year Applied Maths student I was recommended by a physicist friend (Gabor Domokos, now Professor at John Hopkins) to read the "Introduction to the Theory of Quantized Fields" (Bogolyubov and Shirkov, 1959) "from cover to cover". Reading the book made me forget my pre-occupation with Graph Theory, and I decided to be a mathematical physicist. After graduation I joined the Hungarian Geophysical Institute to devote twenty years to seismic exploration. A well-thumbed copy of Dobrin's (1960) "Introduction to Geophysical Prospecting" saw me through the formative years of my career. When, by chance, I came across Chernov's (1960) "Wave Propagation in Random Media", I recognized I have found *the* research topic: seismic and ultrasonic waves in inhomogeneous geologic structures. Some ten years, a dozen research papers and a Ph.D. dissertation later I realized from Ziman's (1979) "Models of Disorder" that disorder, indicated by incoherent wave scattering, is a basic aspect of Nature, and I went through the books of Norbert Wiener, Landau and Lifshitz's (1981) "Statistical Physics" and books on turbulence to learn the mathematics of disorder. An article in "Scientific American" called my attention to Mandelbrot's work; his "The Fractal Geometry of Nature" (1982) has become my faithful companion (and one among the few books I could take when leaving Hungary, in 1986, with a 20 kg baggage to teach Geophysics at the University of Adelaide). Feder's delightful "Fractals" (1988) just arrived in time to cheer me up during my recovery from a surgery, and it gave me strength to start writing the present work. To all the authors of these wonderful books, most of whom I have never met, I offer my sincere thanks for teaching and guidance.

The present book is meant to be a detailed treatment of the fractal models proposed in various fields of Earth Sciences, by no means an introductory text on the Theory of Fractals. Those unfamiliar with some basic concepts are urged to look them up in a new edition of Mandelbrot's "The Fractal Geometry of Nature" or Feder's "Fractals" (indeed, throughout this book Mandelbrot will be cited almost hundred times, Feder some fifty times). An important application, viscous fingering (with a possible role in secondary oil recovery) will only be mentioned in passing; the interested Reader can find the details in Feder's book and the technical literature. Sedimentologists and civil engineers intending to use fractals for fine particle analysis will find additional information in Kaye's (1989) "A Random Walk Through

Fractal Dimensions". Aspects of computer graphics (generation of fractal landscapes) and the fractal coding of natural forms would have required a significantly higher level of mathematics than the rest of the book — and expertise in programming — so that I have reluctantly decided to refer the interested to Peitgen et al.'s "The Science of Fractal Images" (1988) and Barnsley's "Fractals Everywhere" (1989).

I have tried to give an unbiased coverage of all geological-geophysical applications of fractals that have come to my attention and have been published to the end of 1989. If particular topics, results or references are missing this only implies my ignorance and certainly not a value judgement. Preprints, conference abstracts and the like are slow to reach us here "down under": if anyone find his/her contribution omitted, please send me the reprints (or other comments) for a possible next edition. I am especially anxious to get further materials from Russian, French, Japanese and Chinese Colleagues, as these parts of the world are obviously under-represented in my collection.

The expected readership of the book includes geophysicists, geologists, geochemists, geostatisticians, geomorphologists, geographers, oceanographers, oceanologists and meteorologists involved in describing and analysing irregular spatial data.

The book will also be useful to applied mathematicians, physicists and computer scientists looking for new fields of research. A group of these people have devoted a life-time to deconvolve, unwrap, filter, simulate, krige, predict, up- and downward continue, to cross-plot and transform and prewhiten; in general: to apply the latest what mathematics, physics and electronics have to offer to improve data quality and build better geological models. Their endeavours are often met with suspicion or hostility: the bitter words of John Dowds — from his 1969 paper, cited in Chapter 2 — are still valid:

> ...in the report, information theory was mentioned, but I decided it best to avoid unfamiliar words such as entropy, ergodic, Markovian, etc., as these words can cause antagonisms.

The present book is dedicated to these fine and brave people: to the mathematicians, physicists and engineers who have become earth scientists.

It is my pleasant duty to acknowledge indebtedness to all who helped. Special thanks are due to Dr Robert L. Goodman (New York) and Dr Martin Tanke (Amsterdam) of Elsevier Science Publishers who encouraged me to write the book; and to those colleagues at the University of Adelaide who have indicated interest in the project: Professor David Boyd, Professor Larry Frakes, Drs Yvonne Bone, Peter Brooker, John Cooper; Jane Francis, Vic Gostin, John Jones, Nick Lemmon, Robin Oliver, Bill Stuart, Peter Tingate and Rawley Twidale. I gratefully acknowledge the help and expertise of Mr Rick Barrett, Mrs Sherry Proferes, Mr Geoff Trevelyan and Mr John Willoughby in preparing the artwork and in other technical tasks. The

University of Adelaide generously supported these technical tasks by a grant-in-aid in 1991.

My postgraduate students, Dr Robert O'Dowd and Dr Shaou Zhou helped me with the computations leading to the diagrams in Figures 2.20, 2.21 and 2.32. The computations in connection with Figures 2.20 and 2.21 are based on the SEDPAK program package, developed by Professor Kendall and his group at the University of South Carolina. Thanks are due to Prof. Christopher Kendall for his generosity in kindly sending the SEDPAK program code for teaching and research.

The photo of the limestone cliffs in Figure 1.33, also displayed on the cover, shows an *actual* formation of the Nullarbor karst, Australia, though it looks like a computer-generated "fractal" landscape. I am honoured and delighted that Drs Yvonne Bone (Geology Department, University of Adelaide, South Australia) and Noel P. James (Geology Department, Queens University, Canada) gave their permission to use the slide.

As there are over three hundred illustrations — mostly taken from works of other researchers, several tables and a great number of literary quotes, a legion of acknowledgements should go to the Authors and Publishers who kindly consented to the use of their materials. As there will be a separate List of Credits, here I would only like to register my thanks and indebtedness to all these people by quoting the words of Bishr the Barefoot's sister, as recorded in "The Book of the Thousand Nights and One Night" (translated by J.C. Mardus and P. Mathers; Routledge and Kegan Paul Ltd., London, 1941; vol. i, p. 581):

> Bishr the Barefoot's sister came one day to the Imaam Ahmad ibn Hanbal, and said: "Make a thing plain to me, O holy Imaam of the Faith. It is my custom to sit upon our terrace at night and spin thread by the light of the torches which pass in the street, for we have no lamp in our house. Tell me if it is lawful so to use light which does not belong to me."

Adelaide, South Australia, 15 September, 1991
GABOR KORVIN

Acknowledgements for reproductions of previously published material

Thanks are extended to the following Publishers, Learned Societies, Institutions, Editors and Authors who permitted to use previously published figures, tables and short extracts. Figure and table numbers in parentheses are as they appeared in the original publications. The complete reference to "X. Y. Author (3: 1978)" can be found in the References to Chapter 3.

Academic Press Inc. (London) Ltd.:

Fig. 2.41b (1) of G.W. Barlow (2: 1974) from Vol. 22 of Anim. Behav.

Fig. 4.3 (1) of M. Sernetz et al. (4: 1985) from Vol. 117 of J. Theor. Biol.

Fig. 3.7C (10.20) and two sentences of D.E. Grady and M.E. Kipp (3: 1987) *in:* B.A. Atkinson (Editor): Fracture Mechanics of Rock. Academic Press, London.

Academic Press, Orlando, Fla:

Fig. 3.9b (3) of F. Capaccioni et al. (3: 1986) from Vol. 66; Figs. 3.7B, 3.9a (2, 6) of A. Fujiwara et al. (3: 1977) from Vol. 31; Table 3.3 (Tables I, II, modified) of W.K. Hartmann (3: 1969) from Vol. 10. of Icarus;

Fig. 4.5 (4, 6) of D. Avnir et al. (4: 1985) from Vol. 103; Table 4.1 (1) of M. Ben Ohoud et al. (4: 1988) from Vol. 124; Figs. 4.48d, 4.49a–b (1, 2a, 3a) and Table 4.5 (part of Table I) of W.I. Friesen and and R.J. Mikula (4: 1987) from Vol. 120; Fig. 4.4 (4) of H. Van Damme et al. (4: 1988) from Vol. 122 of J. Colloid Interf. Sci.

One sentence of A.J. Chorin (2: 1982) from Vol. 46 of J. Comp. Phys.

Allied Books Ltd.:

Fig. 4.32 (Figs. 1, 2, 3 on pp. 245–246; the original of the El Greco portrait is in the Metropolitan Museum of Art, N.Y., the Modigliani in the Museum of Modern Art, N.Y.) of J. Schillinger (1948): The Mathematical Basis of the Arts. Philosophical Library, N.Y.

American Association for the Advancement of Science:

Fig. 1.30 (2) of R. Dolan and J.C. Ferm (1: 1968) from Vol. 159 (© 1968 by ASSS); Figs. 4.1.c, 4.2 (1, 2) of H. Elias and D. Schwartz (4: 1969) from Vol. 166 (© 1969 by ASSS); Fig. 3.20a (l.h.s. of 1) of G.R. Heath (3: 1979) from Vol. 205 (© 1979 by ASSS); Fig. 1.47 (1) of S. Lovejoy (1: 1982) from Vol. 216 (© 1982 by ASSS); Fig. 2.33e (1) from R. Willden and D.R. Mabey (2: 1961) from Vol. 133 (© 1961 by ASSS) of Science.

American Association for State and Local History:

Figs. 2.33, 2.36b (detail of Fig. 78, and diagrams on p. 120) from C.K. Keck (1965): A Handbook on the Care of Paintings. Amer. Assoc. State Local Hist.

American Association of Petroleum Geologists:

Fig. 1.1 (6) of J.C. Lorenz et al. (1: 1985) from Vol. 69; Fig. 2.33 (2.B) of R.A. Nelson (2: 1979) from Vol. 63 of Bull. AAPG.

American Chemical Society:

Figs. 4.8, 4.10a–b (1, 10.a, 11.a) of D. Farin et al. (4: 1985) from Vol. 1 of Langmuir. © 1985 by American Chemical Society.

American Geological Institute:

Fig. 2.35 (1) of T.M. Chowns (2: 1973) from Vol. 18 of Geotimes.

American Geophysical Union:

Figs. 2.56, 2.57 (1, 4) of C.H. Scholz and C.A. Aviles (2: 1986) from S. Das et al. (Editors): Earthquake Source Mechanisms. Geophys. Monogr., 37. © 1986 by AGU.

Figs. 3.18a–b (1, 2) of D.K. Smith and T.H. Jordan (3: 1987) from Vol. 14 of Geophys. Res. Lett. © 1987 by AGU.

Fig. 4.36b (9) from M. Auberger and J.S. Rinehart (4: 1961) from Vol. 66 (© 1961 by AGU); Fig. 4.21a (3) of S.R. Brown and C.H. Scholz (4: 1985) from Vol. 90B (© 1985 by AGU); Table 4.4 (after Tables 1, 2) of C.E. Krohn (4: 1988a) from Vol. 93B (© 1988 by AGU); Fig. 2.53 (6) of P. G. Okubo and K. Aki (2: 1987) from Vol. 92B (© 1987 by AGU); Figs. 4.19a–b, 4.20, 4.21c (3, 8, 2, 6) of D.A. Rothrock and A.S. Thorndike (4: 1980) from Vol. 85C (© 1980 by AGU); Figs. 2.41a, 2.47b–d, 3.8a (6, 4, 2, 7) of D.A. Rothrock and A.S. Thorndike (2: 1984) from Vol. 89C (© 1984 by AGU); Figs. 3.12, 3.15a–c (3, 4, 5, 7) and Table 3.3 (1, modified) of D.L. Turcotte (3: 1986) from Vol. 91B (© 1986 by AGU); Fig. 4.24 (1) of D.L. Turcotte (4: 1987) from Vol. 92B (© 1987 by AGU); Fig. 1.15 (1.a) of J.B. Walsh and W.F. Brace (1: 1984) from Vol. 89B (© 1984 by AGU) of J. Geophys. Res.

Figs. 4.21b, 4.23a–b (4, 7, 10) of C.G. Fox and D.E. Hayes (4: 1985) from Vol. 23 of Rev. Geophys. © 1985 by AGU.

Four sentences of V. Klemes (5: 1974) from Vol. 4 (© 1974 by AGU); Fig. 5.2b (11) of B.B. Mandelbrot and J.R. Wallis (5: 1969) from Vol. 5 (© 1969 by AGU) of Water Resour. Res.

American Institute of Physics:

Figs. 4.34a–b (2.a–b) of D.W. Schaefer et al. (4: 1987); Fig. 4.51 (4.b) and one sentence of P.-Z. Wong (4: 1987) from J.R. Banavar, J. Koplik and K.W. Winkler (Editors): Physics and Chemistry of Porous Media II. AIP Conf. Proc., 154, New York.

Fig. 4.9 (1) of D. Avnir et al. (4: 1983) from Vol. 79 of J. Chem. Phys.

Figs. 4.10c–d (1, 2) of F. Varnier et al. (4: 1988) from Vol. A6 of J. Vac. Sci. Technol.

Fig. 4.37 (1) of D.L. Koch (4: 1987) from Vol. 30 of Phys. Fluids. Tables 5.1, 5.3 (I [simplified], VI) of B.M. Smirnov (5: 1986) from Vol. 29 of Sov. Phys. Uspekhi.

American Journal of Science:

Fig. 1.2b (inset of 4) of C.W. Carlston (1: 1965) from Vol. 263 of Am. J. Sci.

American Meteorological Society:

Fig. 3.20d (3) from R.E. Lopez (3: 1977) from Vol. 105 of Monthly Weather Rev.

American Physical Society:

Figs. 2.51a–b (1, 2.c) of Y. Sawada et al. (2: 1982) from Vol. 26 of Phys. Rev. A.

Fig. 4.27 (1.b, 6) (the original, kindly provided by Dr. M. Cloitre) of C. Allain and M. Cloitre (4: 1986) from Vol. 33; Fig. 2.14a (1) of R. Bruinsma and P. Bak (2: 1983) from Vol. 27; Figs. 4.48e–f (1.a–b) from G. Grinstein and Sh.K. Ma (4: 1983) from Vol. 27 of Phys. Rev. B.

Fig. 4.33 (1) of H.D. Bale and P.W. Schmidt (4: 1984) from Vol. 53; Fig. 4.53a (1.b) of G. Daccord et al. (4: 1986) from Vol. 56; Fig. 4.41 (1) and Table 4.3 (I) of A.J. Katz and A.H. Thompson (4: 1985) from Vol. 54; Fig. 4.30 (top row of Fig. 1) of M. Kolb and H.J. Hermann (4: 1987) from Vol. 59; Figs. 4.53b–c (2.a–b) of J. Nittmann et al. (4: 1987) from Vol. 58; one sentence of J.G. Powles and N. Quirke (5: 1984) from Vol. 52; Fig. 1.48 (2.a) of F.S. Rys and A. Waldvogel (1: 1986) from Vol. 56; Fig. 5.13iv. (2) of T. Vicsek and A.S. Szalay (3: 1987) from Vol. 58; one sentence of T.A. Witten (4: 1987b) from Vol. 59 of Phys. Rev. Lett.

American Society of Mechanical Engineers:

One sentence of W. Weibull (3: 1951) from Vol. 18 of J. Appl. Mech.

Annual Reviews Inc.:

Fig. 5.17 (1) of O.E. Lanford (5: 1982), and Fig. 3.2b (1) of A.C. Payatakes (3: 1982) from Vol. 14 of Annu. Rev. Fluid Mech. © 1982 by Annual Reviews Inc.

Architectural Book Publ. Co., N.Y.:

Fig. 1.41b (Fig. 1-8 on p. 21) of Christopher Williams (1981): Origins of Form. Architectural Book Publ. Co, N.Y.

Association of Exploration Geophysicists of India:

Fig. 2.31a (1) of D.V. Subba Rao et al. (2: 1987) from Vol. 8 of J. Assoc. Explor. Geophys. (Hyderabad).

Bailliere Tindall:

Figs. 5.13i., 5.18a-1. (50, 51) from F.D. Adams (1938): The Birth and Development of the Geological Sciences. Bailliere, Tindall and Cox, London.

Australian Petroleum Exploration Association Ltd., Sydney:

Fig. 1.16 (4) of I.R. Campbell and D.N. Smith (1: 1982) from Vol. 22 of The APEA Journal.

Bergakademie Freiberg:

Fig. 4.36a after H. Militzer and R. Stoll (4: 1968) from Vol. 20 of Bergakademie.

Bestseller Publications, Ltd.:

Fig. 2.44 (black and white reproduction of Plate 87) and two sentences from O. Jones (1867): The Grammar of Chinese Ornaments. S. and T. Gilbert, London (reprint edition: Studio Editions, Ltd., London, 1987).

Galerie Beyeler, Basle:

Fig. 4.26 (black and white reproduction of Paul Klee's (1929) "Fleeting appearances on the water"), from: P. Klee Notebooks, Vol. 2, p. 71. Lund Humphries Publ. Ltd., London, 1973.

Birkhauser Verlag AG:

Fig. 5.7 (1.a–c) of D.J. Crossley and O.G. Jensen (5: 1989) from Vol. 131; Fig. 5.2c-2 (1) of R.H. Fluegeman, Jr. and R.S. Snow (5: 1989) from Vol. 131; Figs. 2.50, 2.58 (2.a-b, 3, 5.a-b) of T. Hirata (2: 1989) from Vol. 131; Figs. 2.33, 3.21 (1.a-d, 3) of G. Korvin (2: 1989) from Vol. 131; Figs. 2.61a-b (4, 5) of O. Nishizawa et al. (2: 1984/85) from Vol. 122; Figs. 3.13a-b (19, 20) of C.G. Sammis et al. (3: 1986) from Vol. 124; Figs. 5.12a-b (6 [top], 9) of J.P. Todoeschuck and O.G. Jensen (5: 1989) from Vol. 131; Fig. 3.11 (2) and Table 3.3 (1, modified) of D.L. Turcotte (3: 1989) from Vol. 131 of PAGEOPH. (Fig. 3.21 [= Fig. 3 in G. Korvin (2: 1989)] is detail of a map from the Ph.D. Thesis of Dr H.M. Hammouda [3: 1986], © 1986 by H.M. Hammouda and The University of Cairo, Giza, Egypt.)

Blackwell Publishers, Oxford:

Four sentences from E.A. Abbott (1932): Flatland. A Romance of Many Dimensions. 4th, rev. edition, Basil Blackwell, Oxford.

Blackwell Scientific Publications Ltd., Oxford:

Figs. 2.60a-c (2.a-c) of Y.Y. Kagan and L. Knopoff (2: 1980) from Vol. 62 of Geophys. J. R. Astron. Soc.

Figs. 2.31b, 2.32 (2.a, 3) of G. Korvin et al. (2: 1990) from Vol. 100 of Geophys. J. Int.

Two sentences of N.A. Anstey and P. Newman (5: 1966) from Vol. 14 of Geophys. Prospect.

Figs. 4.15-4.17, 4.43-4.45 (6, 5, 7, 12, 1 [based on Fig. 1 of J.P. Rigaut, 1984, J. Microsc., 133: 41–54], 14) of H. Pape et al. (4: 1987a) from Vol. 148 of J. Microsc.

Figs. 1.36, 4.10 (8, and the quartz particle 2f from Fig. 9) of J.D. Orford and W.B. Whalley (1: 1983) from Vol. 30 of Sedimentology.

Gebrüder Bornträger Verlagsbuchhandlung, Stuttgart:

Fig. 1.29 (1) of R. Dolan et al. (1: 1974) from Vol. 18 of Z. Geomorphol. N.F.

Marion Boyars Publishers Ltd, London:

Sixteen lines from p. 137 of J. Cage (1975): A Year from Monday. Marion Boyars, London.

Cambridge University Press:

Fig. 5.18c-1 (1) of L.-S. Young (5: 1982) from Vol. 2 of Ergodic Theor. Dyn. Syst.

Fig. 1.2a (inset of 4) of M.R. Leeder (1: 1973) from Vol. 110; Fig. 2.41c (2) of I.J. Smalley (2: 1966) from Vol. 103 of Geol. Mag.

Figs. 1.50, 5.19 (3, 2) and two sentences of K.R. Sreenivasan and C. Menevau (1: 1986) from Vol. 173 of J. Fluid. Mech.

One sentence from Thompson, D'Arcy W. (1942): On Growth and Form. Cambridge Univ. Press, Cambridge.

Table 1.1 (part of 9.1) from J.M. Ziman (1979): Models of Disorder. Cambridge Univ. Press, Cambridge.

Caravan Books:

Fig. 4.6 (F273) from F. Ragette (1980): Architecture in Lebanon. Caravan Books, Delmar, N.Y.

Centrale des Revues, Dunod:

Fig. 5.18b-2 (1.a–c) of J. Bébien et al. (5: 1987) from Vol. 305, Ser. II.; Fig. 5.14 (5) of A. Thomas (5: 1986) from Vol. 303, Ser. II. of C. R. Acad. Sci. Paris

Chapman and Hall Ltd., London:

Figs. 1.53, 1.54 (1, 4) of C.S. Pande et al. (1: 1987a) from Vol. 6 of J. Mater. Sci. Lett.

CNRS Publications, Paris:

Figs. 2.49a–c (3.I., 4.b, 4.a) of P.G. Thomas and P. Masson (2: 1983) from Vol. 1 of Ann. Geophys.

Collins/Angus and Robertson Publishers Pty Ltd:

Four sentences from Mark Warren (1988): Atlas of Australian Surfing. Angus and Robertson, N.Z.

Colorado School of Mines Press:
One sentence of J.P. Dowds (2: 1969) from Vol. 64 of Col. Sch. Mines Quart.

Czechoslovakian Academy of Sciences Publ. House:
Figs. 2.25, 2.26 (2, 1) of B. Sprušil and F. Hnilica (2: 1985) from Vol. 35 of Czech. J. Phys.

J.M. Dent and Sons, Ltd.:
An extract (11 lines) of "A Defence of Nonsense", from G.K. Chesterton's Stories, Essays and Poems. J.M. Dent and Sons, Ltd., London, 1957.

Dept. of Fisheries and Oceans, Canada, Scientific Publications:
Figs. 1.37a–b; 1.46 (1, 4, 5); Table 3.1 (3) and four sentences of C. Kent and J. Wong (1: 1982) from Vol. 39 of Can. J. Fish. Aquat. Sci.

Dover Publications, Inc.:
Fig. 1.32 (Plate CXIV) of J.P. Richter (Compiler and Editor): The Notebooks of Leonardo da Vinci. Vol. II. Dover, N.Y., 1970.

Les Éditions de Physique:
Fig. 5.16B (3.b) of E. Louis and F. Guinea (5: 1987) from Vol. 3 of Europhys. Lett. Fig. 3.9 (1) of X. Campi (3: 1988) from Vol. 50C of J. Phys.

Elsevier Science Publishers:
One sentence of P.R. Vogt (3:1974) from Vol. 21 of Earth Planet. Sci. Lett.
Fig. 4.25 (7.A) of F.P. Agterberg (4: 1982) from Vol. 2 of Geo-Processing.
Fig. 2.16 (1) of P.M. Sadler (2: 1980) from Vol. 35 of Mar. Geol.
Fig. 1.21 (the original micrograph of the gold film was kindly provided by Drs. M. and. P. Gadenne) of M. Gadenne et al. (1: 1989) from Vol. 71 of Optics Commun.
Fig. 2.54 (8) of Y. Fukao and M. Furumuto (2: 1985) from Vol. 37; Fig. 3.7D (1) of G. Lucido (3: 1985) from Vol. 39; Figs. 2.63–2.65, 2.66a–c (1, 2, 3b, 4, 6a–b) of T. Ouchi and T. Uekawa (2: 1986) from Vol. 44 of Phys. Earth Planet. Int.
Figs. 4.1a–b (1, 2) of M. Mendes France (4: 1984) from Vol. 103; one sentence of K.G. Wilson and J. Kogut (3: 1974) from Vol. 12 of Phys. Rep.
Fig. 4.50c (1) of I.J. Smalley (4: 1967) from Vol. 8 of Sedimentology. Two sentences of P. Pfeifer et al. (4: 1983) from Vol. 126 of Surf. Sci.
Fig. 2.31e (1) of D.A. Weitz and J.S. Huang (2: 1984) from F. Family and D.P. Landau (Editors): Aggregation Gelation. North-Holland, Amsterdam, 1984.
Figs. 2.52a–b (3, 5) of H. Takayasu (2: 1986) from L. Pietronero and E. Tosatti (Editors): Fractals in Physics. Elsevier, Amsterdam–N.Y., 1986.
Fig. 5.1 (5.2.b) of D.C. Boes (5: 1988) from J.N. Srivastava (Editor): Probability and Statistics. Essays in Honor of F.A. Graybill. Elsevier, Amsterdam–N.Y., 1988.

Elsevier Sequioa S.A., Lausanne:
Fig. A2.2 (10) of E.E. Underwood and K. Banerji (2: 1986) from Vol. 80 of Mater. Sci. Eng.

Energy Publications, Mt. Hawthorn, Western Australia:
Fig. 1.41c (advertisement on p. 21) of Oil and Gas Australia, October, 1988.

The English Universities Press, Ltd., London:
Figs. 4.50a–b (6.15, 6.16) of H.G. Flegg (1974): From Geometry to Topology. The English Universities Press, Ltd., London.

Eötvös Lorand Geophysical Institute, Budapest:
Fig. 4.36 (5.1, 5.4) of G. Korvin (4: 1977) from Vol. 24, Suppl. 2; Figs. 1.22–1.24 (8, 10, 9) of G. Korvin (1: 1992) from Vol. 39 of Geophys. Trans.

Faber and Faber Ltd. Publishers:
Four lines of "The Maze" from W.H. Auden: Collected Poems (Edward Mendelson, Editor). Faber and Faber, London, 1976.

W.H.Freeman and Co., N.Y.:
Fig. 1.4. (5.4) of L.B. Leopold, M.G. Wolman and J.P. Miller (1964): Fluvial Processes in Geomorphology. Freeman and Co., San Fransico.
Figs. 1.13, 2.47a, 3.1a–b (Plates 13 bottom, 170b, 117b, 119b) of B.B. Mandelbrot (1982): The Fractal Geometry of Nature. Freeman, N.Y.
Fig. 2.14c (22-4) of Ch.F. Richter (1958): Elementary Seismology. Freeman and Co., San Francisco.

Geological Society of America:
Fig. 1.3 (3) of L.B. Leopold and M.G. Wolman (1: 1960) from Vol. 71 of Bull. Geol. Soc. Am.
Fig. 2.36a (1) of H. Lachenbruch (2: 1962) from Geol. Soc. Am. Spec. Pap. No. 70.
Fig. 5.18c-2 (7) of K.T. Pickering et al. (5: 1988) from Vol. 16 of Geology.

Geological Society of South Africa:
Figs. 2.48a–d (6, 2, 7.a–b) of A.R. Newton (2: 1987) from Vol. 90 of S. Afr. J. Geol.

Gordon and Breach Science Publishers:
Figs. 5.5, 5.11 and three sentences of W.H. Press (5: 1978) from Vol. 7 of Comments Astrophys.

Hamish Hamilton Ltd., London and Alfred A. Knopf Inc., N.Y.:
Seven sentences from Pavič, M. (1988): The Dictionary of the Khazars. Male Edition. (Transl. by Ch. Pribičevič-Zorič.) Hamish Hamilton, London, 1988.

Hamlyn Publ. Group Ltd.:
Six sentences from "Sylvie and Bruno Concluded" and 2 lines from "The Hunting of the Snark" from R.L. Green (Editor): The Works of Lewis Carroll. Hamlyn, London, 1965.

Hodder and Stoughton Ltd.:
Two sentences from G.K. Chesterton (1943): St. Thomas Aquinas. Hodder and Stoughton, London, 1943.

Holt, Rinehart and Winston, Inc.:
Fig. 3.7A (82) of M.E. Bevlin (1984): Design Through Discovery. 4th edition, Holt, Rinehart and Winston, N.Y.

IEEE Inc.:
Fig. 4.52 (5.a–b) and three sentences of A.P. Pentland (4: 1984) from Vol. PAMI-6 of IEEE Trans. Pattern Anal. Mach. Intell. © 1984 by IEEE.

Institute of Geosciences, University Tsukuba, Japan:
Figs. 1.38a–c, 1.39 (3.A, 3.A-4, 2.A, 5) of T. Nakano (1: 1983) from Vol. 9; Fig. 1.40 (2) of T. Nakano (1: 1984) from Vol. 10 of Annu. Rep. Inst. Geosci. Univ. Tsukuba. (The originals were kindly provided by Dr. T. Nakano.)

Institute of Physics Publ., Ltd., Bristol:
Figs. 4.39, 4.40 (3, 5) of M.V. Berry (4: 1972) from Vol. 5; Fig. 4.38 (1) of M.V. Berry and T.M. Blackwell (4: 1981) from Vol. 14; Fig. 2.31d (1.a) of S.R. Forrest and T.A. Witten, Jr. (2: 1979) from Vol. 12; Fig. 3.3 (1) of J. Hoshen et al. (3: 1979) from Vol. 12; Fig. 5.13iv. (2) of T. Vicsek (5: 1983) from Vol. 16 of J. Phys. A: Math. Gen.
Fig. 4.31 (2) of D.F.R. Mildner and P.L. Hall (4: 1986) from Vol. 19 of J. Phys. D: Appl. Phys.
Figs. 4.28, 4.29 (9, 11) of S.H. Liu (4: 1986) from Vol. 39 of Solid State Phys.

International Society for Educational Information, Tokyo:
Fig. 4.18b (Hokusai's "Great Wave of Kanazawa")

International Society for the Systems Sciences:
Fig. 1.26 (17) and the Table in Chapter 1.4.1 (Table IX) of L.F. Richardson (1: 1961) from Vol. 6 of Gen. Sys.

Israel Physical Society:
Fig. 1.18 (1) from R. Zallen (1: 1983) *in*: G. Deutscher, R. Zallen and J. Adler (Editors): Percolation Structures and Processes. Ann. Isr. Phys. Soc., Vol. 5.

Japanese Journal of Applied Physics:
Fig. 2.51c (1) of S.Fujimori (2: 1985) from Vol. 24 of Jap. J. Appl. Phys.

Kluwer Academic Publishers, Dordrecht:
Fig. 3.20c (1) of W.K. Brown et al. (3: 1983) from Vol. 94 of Astrophys. Space Sci. Two sentences of H.E. Stanley (2: 1986) *in*: H.E. Stanley and N. Ostrowsky (Editors): On Growth and Form. Fractal and Non-fractal Patterns in Physics. Martinus Nijhoff, Dordrecht, 1986.

Alfred A. Knopf, Inc., N.Y. and Octopus Publishing Group Library, England:
Four sentences from Thomas Mann (1943): Doctor Faustus. The Life of the German Composer Adrian Leverkühn as Told by a Friend. (Transl. by H. T. Lowe-Porter.) Secker and Warburg, London.

Prof. G. Lucido (Ist. di Miner. Petr. e Geochim., Univ. di Palermo):
Fig. 5.18a-2 (1) of G. Lucido (5: 1984) from Vol. 28 of Miner. Petrol. Acta.

Macmillan Magazines Ltd.:
Figs. 3.14, 3.15 (2.a–b) of C.J. Allègre et al. (3: 1983) from Vol. 297 (© 1983 Macmillan Magazines Ltd.); Table 4.2 (parts of Table 1) of P.A. Burrough (4: 1981) from Vol. 294 (© 1981 Macmillan Magazines Ltd.) ; Fig. 5.2.c1 (1.b) of P.Grassberger (5: 1986) from Vol. 323 (© 1986 Macmillan Magazines Ltd.); one sentence of S. Lovejoy et al. (2: 1986) from Vol. 319 (© 1986 Macmillan Magazines Ltd.); Fig. 1.52 (1) of B.B. Mandelbrot et al. (1: 1984) from Vol. 308 (© 1984 Macmillan Magazines Ltd.); Fig. 4.7 (1.b) of D.R. Morse et al. (4: 1985) from Vol. 314 (© 1985 Macmillan Magazines Ltd.); one sentence of Jenny Nelson (4: 1989) from Vol. 339 (© 1989 Macmillan Magazines Ltd.); Figs. 2.40a–c (4.a–c) of D.C. Pieri (2: 1981) from Vol. 289 (© 1981 Macmillan Magazines Ltd.); Fig. 4.22 (1) of R.S. Sayles and T.R. Thomas (4: 1978b) from Vol. 273 (© 1978 Macmillan Magazines Ltd.); Fig. 5.10 (1) of A.M. Schroeder (5: 1987) from Vol. 325 (© 1987 Macmillan Magazines Ltd.); Figs. 5.15, 5.16A (1, 2) of Y. Termonia and P. Meakin (5: 1986) from Vol. 320) (© 1986 Macmillan Magazines Ltd.); one sentence of D.L. Turcotte et al. (3: 1985) from Vol. 313 (© 1985 Macmillan Magazines Ltd.) of Nature.

Methuen London, Ltd.:
Two sentences from Graham Chapman et al.: Monty Python and the Holy Grail (Methuen, London, no year)

Munksgaard International Publishers Ltd., Copenhagen:
Figs. 1.57a–b, 1.58 (2a–b, 1) of J.R. Krummel et al. (1: 1987) from Vol. 48 of OIKOS.

W. W. Norton and Company, Inc.:
One sentence from "Gulliver's Travels", Pt. 2, Chap. 5; and Lines 337–340 from "On Poetry. A Rhapsody" from: The Writings of Jonathan Swift (R.A. Greenberg and W.B. Piper, Editors). W. W. Norton and Co., Inc., N.Y., 1973.

Octopus Publishing Goup Library, England:
Three sentences from Somerset Maugham (1936): Cosmopolitans. Very Short Stories. William Heinemann Ltd, London.

Off-Shore Technology Conference, Richardson, Texas:
Fig. 3.20b (9) of A.H. Magnuson (3: 1981) from OTC Paper No. 4133.

Oxford University Press:
Figs. 2.34a, c (1.a–b) of W.E. Le Gros Clark (2: 1947) *in*: W.E. Le Gros Clark and P.B. Medawar (Editors): Essays on Growth and Form. Clarendon Press, Oxford, 1947.
Four lines of Alfred Lord Tennyson's "In Memoriam A.H.H." from: Alfred Tennyson Poetical Works. Geoffrey Cumberlege, Oxford Univ. Press, London, 1953.
Fig. 1.43 (map of Utopia) from: The Utopia of Sir Thomas More, edited by J.H. Lupton. Oxford, at the Clarendon Press, M.DCCC.XCV.
Fig. 1.41a (map of "Brobdingnag" on p. 106) and Fig. 3.2a (islands visited by Gulliver, p. 208) from: Gulliver's Travels by Jonathan Swift. London, First Edition Club, 1926. Printed by John Johnson, Univ. Press, Oxford.

Palaeontological Association, Dundee, Scotland:
Figs. 5.13ii, iii (1, 2) of D. Hamilton (5: 1965) from Vol. 4, Pt.3 of Palaeontology.

Paleontological Society, Jacksonville, N.Y.:
Table 5.2 (1, modified) of J.M. Garcia-Ruiz et al. (5: 1990) from Vol. 16 of Paleobiology.

Dr. Hansgeorg Pape (Dept. Geophys., Univ. of Clausthal, Germany):
Figs. 4.47, 4.48 (7, 6) of H. Pape et al. (4: 1984) (3rd Eur. Symp. Particle Char., Preprint); Fig. 4.46 (10) of H. Pape et al. (4: 1985b) (4th Int. Symp. on Agglom., Preprint).

Penguin Group, England:
Three sentences from R.D. Laing (1965): The Divided Self. An Existential Study in Sanity and Madness. Penguin, Baltimore.

Pergamon Press PLC.:

Fig. 1.55 (4) of C.S. Pande et al. (1: 1987b) from Vol. 35 of Acta Metall. © 1987 Pergamon Press PLC.

Fig 1.11 (4) of Jay Kapraff (1: 1986) from Vol. 12B of Comput. Math. Appl. © 1986 Pergamon Press PLC.

Petroleum Exploration Society of Australia Ltd.:

Fig. 4.14a (Plate 1/c) of J.P. Schulz-Rojahn and S. Phillips (4: 1989) *in*: B.J. O'Neill (Editor): The Cooper and Eromanga Basins, Australia. P.E.S.A., Adelaide, South Australia, 1989.

Physical Society of Japan:

Figs. 3.8b, 3.10 (1, 2) of M. Matsushita (3: 1985) from Vol. 54 of J. Phys. Soc. Jap.

Plenum Publishing Corporation:

Two sentences of B.B. Mandelbrot (2: 1984) from Vol. 34; two sentences of E.W. Montroll and M.F. Shlesinger (3: 1983) from Vol. 32 of J. Stat. Phys..

Figs. 2.28, 2.29 (5, 6.b) of J.P. Chilès (2: 1988) from Vol. 20; Figs. 3.4, 3.5 (1, 2) of R.L. Curl (3: 1986) from Vol. 18; Fig. 1.56 and Table 3.2 (based on Tables 1, 2) of M.F. Goodchild (1: 1988) from Vol. 20; Figs. 4.35a–b (4, 5) of P.-Z. Wong and J.-S. Lin (4: 1988) from Vol. 20 of Math. Geol.

Figs. 1.27 (2.8), 1.45 (12.1), 2.5 (5.1), 2.6 (5.2), 2.7 (2.15), 2.8 (2.16), 2.10 (6.1), 2.11 (6.2), 2.22 (9.10), 2.31f (2.14), 2.55 (2.17), 4.11 (13.1), 4.13 (2.4) and one sentence (from p. 236) of J. Feder (1988): Fractals. Plenum Press, N.Y. and London.

Progress of Theoretical Physics, Kyoto:

Fig. 2.52c (4) of H. Takayasu (2: 1985b) from Vol. 74 of Progr. Theor. Phys.

Random House, Inc., New York:

One line of John Donne's "An Anatomie of the World" from: The Complete Poetry and Selected Prose of John Donne (edited by Ch.M. Coffin). The Modern Library, New York, 1952.

Routledge (International Thomson Publishing Services Ltd., England):

Three sentences from J.C. Mardus and P. Mathers (Translators): The Book of the Thousand Nights and One Night. Routledge and Kevan Paul Ltd., London, 1941; vol. I, p.581.

The Royal Society of London:

Fig. 2.33f (2) of J.J. Jarrige et al. (2: 1986) from Vol. 317A of Phil. Trans. R. Soc. London. Fig. 5.18b-1 (2, 3, 4) of P.G. Saffman and G.I. Taylor (5: 1958) from Vol. 245 of Proc. R. Soc. London.

Scientific American, Inc.:
Fig. 2.30 (illustration on p. 85 by Andrew Christie) of L.M. Sander (2: 1987) from the January 1987, international edition of Sci. Am. (© 1987 by Scientific American, Inc. All rights reserved); Fig. 2.37 (illustration on p. 181 by Michael Goodman) of J. Walker (2: 1986) from the October, 1986, international edition of Sci. Am. (© 1986 by Scientific American, Inc. All rights reserved). Fig. 2.43 [photo *d* on p. 181] by courtesy of Dr. Jearl Walker, Physics Dept., Cleveland State University.)

Scolar Press Academic Publishing Group:
Fig. 1.42 (map of Utopia) from St. Thomas More's "Utopia" (1516). A Scolar Press Facsimile. Leeds, England, 1966.

Dr. E.R. Segnit (Div. Building Res., CSIRO, Melbourne):
Original slide of Fig. 2.33a (7) of E.R. Segnit et al. (2: 1970) from Vol. 1 of Search.

Seismological Society of America:
Fig. 2.62. (5) of R.F. Smalley, Jr. et al. (2: 1987) from Vol. 77 of Bull. Seismol. Soc. Am.

Serindia Publications, London:
Figs. 1.49, 4.18a ("clouds" on p. 117, and "flames" on p. 159) from David P. Jackson and Janice A. Jackson (1988): Tibetan Thangka Painting. Methods and Materials. 2nd rev. edition. Serindia Publ., London.

Les Éditions du SIRAC à Paris:
Fig. 4.14c (illustration from "Voyage au centre de la terre") from Simone Vièrne (1973): Jules Verne et le roman initiatique. Ed. SIRAC, Paris.

Society of Exploration Geophysicists, Tulsa, Okla.:
Fig. 2.14b (3.c–e, 4) of D.H. Rothman (2: 1986) from Vol. 51; Fig. 5.12c (4) of J.P. Todoeschuck et al. (5: 1990) from Vol. 55 of Geophysics.

Society of Petroleum Engineers, Richardson, Tex.:
Fig. 5.2d-3 (C-1) of A.S. Emanuel et al. (5: 1987) from SPE Paper No. 16971 (© 1987 by SPE); Figs. 5.2d-1–d-2, 5.3, 5.4a–b (5, 8, 15, 16, 18) of T.A. Hewett (5: 1986) from SPE Paper No. 15386 (© 1986 by SPE); Fig. 1.17 (1) of J.W. Neasham (1: 1977) from SPE Paper No. 6858 (© 1977 by SPE).

Society of Professional Well Log Analysts, Inc., Houston, Tex.:
Fig. 4.17b (1) of H. Pape et al. (4: 1985c) from Trans. SPWLA 26th Annu. Logging Symp.

Society for Sedimentary Geology, Tulsa, Okla.:
Figs. 2.33d, 2.34b of P.S. Plummer and V.A. Gostin (2: 1981) from Vol. 51 of J. Sediment. Petrol.

Society of Systematic Biologists, Washington, D.C.:
Fig. 2.13b (5) of L. Dingus and P. M. Sadler (2: 1982) from Vol. 31 of Syst. Zool.

Springer-Verlag, Heidelberg:
Figs. 2.27a–d (3, 4, 5, 6) of M. Hénon (2: 1976) from Vol. 50 of Commun. Math. Phys.
Figs. 2.38, 2.46 (24, 25, 17) and one sentence of D. Weaire and N. Rivier (2: 1984) from Vol. 25 of Contemp. Phys.
Two sentences of M. McGuire (4: 1988) and Figs. 5.8, 5.9 (1.11–12) of R.F. Voss (5: 1988) *in*: H.-O. Peitgen and O. Saupe (Editors): The Science of Fractal Images. Springer, Berlin–Heidelberg, 1988.

Stereol. Iugosl. (c/o Inst. of Histology and Embriology, Ljubljana):
Figs. 4.17c–d (4, 5) of H. Pape et al. (4: 1981a) from Vol. 3, Suppl. 1 of Stereol. Iugosl.

Stevens and Sons, Ltd.:
Two sentences from Richardson, L.F. (1960): Arms and Insecurity (posthumous edition, by N. Rashevsky and N. Truco). Atlantic Books, London.

Systematics Association, London:
Fig. 5.20 (10) of J. Wiedmann and J. Kullmann (5: 1981) *in*: M.R. House and J.R. Senior (Editors): The Ammonoidea. Syst. Assoc. Spec. Vol. No. 18. Academic Press, London–N.Y., 1981.

Tate Museum (c/o. Dept. of Geol. and Geophys., Univ. of Adelaide, South Australia):
Permission to photograph and publish the "landscape marble" (Fig. 1.44) from their collection. (Courtesy of the Curators, Ms Mima Stojanovic and Dr Fran Parker.)

Taylor and Francis, Ltd.:
Figs. A2.3, A2.4, 4.42 (2, 3, 17) of A.H. Thompson et al. (2: 1987) from Vol. 36 of Adv. Phys.
Fig. 3.7E (2) of R. Englman et al. (3: 1984) from Vol. 50 of Phil. Mag. B.

United States Geological Survey:
Fig. 1.6 after J. T. Hack (1: 1957) from USGS Prof. Pap. No. 294-B; Figs. 1.8, 1.9 (8, 9) of L.B. Leopold and W.B. Langbein (1: 1961) from USGS Prof. Pap. No. 500-A.

University of California Press:
Eight lines from “The Picket Fence” from Christian Morgenstern’s “Galgenlieder” (Gallows Songs) (Transl./ed. by Max E. Knight). Univ. of Calif. Press, Berkeley and Los Angeles, 1966. (© 1963 Max E. Knight.)

University of Chicago Press:
Fig. 1.14 (1) of M.A. Carson and M.F. Lapointe (1: 1983) of Vol. 91; Figs. 2.12, 2.13a (2, 5) and Table 2.3 (part of 1) of R.E. Plotnick (2: 1986) from Vol. 94; Figs. 2.1, 2.2, 2.17 (1, 8, 10) and one sentence of P.M. Sadler (2: 1981) from Vol. 89; Fig. 2.15 (1) and one sentence of A. Yamaji (2: 1988) from Vol. 96 of J. Geol.

VCH Verlagsgesellschaft, Weinheim, Germany:
Four lines of a poem from B.H. Kaye (1989): A Random Walk Through Fractal Dimensions. VCH Verlagsgesellschaft, Weinheim, Germany.

Verlag Freies Geistesleben GmbH, Stuttgart:
Fig. 1.12 (illustr. on p. 74) of Theodor Schwenk (1980): Das Sensible Chaos. Verlag Freies Geistesleben, Stuttgart.

The Wallace Collection, London:
Fig. 2.19 (black and white reproduction of Fragonard’s “The Swing”). Reproduced by permission of the Trustees.

Wayland Publishers, Ltd., London:
Fig. 1.25 (3.9) of Diana Hindley and Geoffrey Hindley (1972): Advertising in Victorian England 1837–1901. Wayland, London.

Westview Press, Boulder, Colo.:
One sentence of R.A. Randolph (2: 1988) *in*: R.A. Randolph, D.M. Schneider and M.N. Diaz (Editors): Dialectics and Gender. Anthropological Approaches. Westview Press, Boulder and London, 1988.

John Wiley and Sons, Inc., N.Y.:
Two sentences from D.V. Ager (1981): The Nature of the Stratigraphic Record. 2nd edition, Halsted, N.Y.
Six sentences from N. Wiener (1948): Cybernetics, or Control and Communication in the Animal and the Machine. The Technology Press, N.Y.

John Wiley and Sons., Ltd., Sussex, England:
Two sentences of W.E.H. Culling and M. Datko (4: 1987) and Fig. 3.6 (1) of M. Laverty (3: 1987) from Vol. 12 of Earth. Surf. Proc. Landforms. © 1987 by John Wiley and Sons, Ltd.

World Scientific Publishing Co. Pte. Ltd., Singapore:
Fig. A2.1 (9) of R. Jullien and R. Botet (1987): Aggregation and Fractal Aggregates. World Scientific, Singapore.
Fig. 4.14b (5, bottom) of Y. Kantor (1989): Properties of tethered surfaces. *In*: D. Nelson et al. (Editors): Statistical Mechanics of Membranes and Surfaces. Vol. 5. World Scientific, Singapore. © 1989 by World Scientific, Singapore.

Wydawnictwo Naukowe PWN, Warsaw:
Three sentences of H. Steinhaus (1: 1954) from Vol. 3 of Colloq. Math.

The Yale J. Biol. Med., Inc., New Haven, Conn.:
Figs. 3.17, 5.6 (parts of 5, 8.A–B) of A.L. Goldberger and B.J. West (3: 1987) from Vol. 60 of Yale J. Biol. Med. © 1987 by Yale J. Biol. Med., Inc.

Yale University Press:
Description of "Utopia" (lines 7–18, p. 11, from Book II, English translation) from: The Complete Works of St. Thomas More, vol. iv (E. Surtz, S.J. and J.H. Hexter, Editors). Yale University Press, New Haven and London, 1965.

* * *

Due to the well-known problems of communication in these difficult times, I could not get replies from the following Publishers and Authors, from the (former) Soviet Union.

Izdatel'stvo "Nauka", Moscow:
Figs. 2.3, 2.4, 2.9 (20, 21, 32) of N.Ya. Vilenkin (1968): Stories About Sets. Academic Press, New York–London. © 1968 by Izdatel'stvo "Nauka", Moscow. Permission requested.

Mir Publishers, Moscow:
Fig. 1.51 (39, 40) of A.L. Efros (1986): Physics and Geometry of Disorder. Percolation Theory. Mir Publishers, Moscow. © 1986 by Mir Publishers. Permission requested.

Okeonologiya (c/o. Shirshov Inst. of Oceonology of the Academy of Sciences, Moscow):
Figs. 2.59a–c (1, 3) of G.I. Barenlatt et al. (2: 1984) from Vol. 24 of Okeonologiya. © 1984 by Okeonologiya. Permission requested.

* * *

All Bible quotes (Gen. 7: 11–12; Gen. 7: 17–20; Gen. 8: 3–13; Gen. 41: 29–30; Eccles. 11: 9; Luke 3: 5) are from (different revisions) of the Authorized (King

James) Bible. The excerpt from G. Cantor (2: 1883–84) in Chapter 2.1.2 is in my free translation. Heaviside (in Mathematical Appendix to Chapter 1) is quoted from his (1893) "Electromagnetic Theory". Two lines from Molière is from C.C. Clarke (1865): "Molière Characters" (William P. Nimmo, Edinburgh).

I could not find the source of the D'Alambert motto (Chapter 1). One sentence by Jaroslav Hašek was previously used by Zel'dovich and Sokolov (4: 1985). One sentence (in English) of Korčak (3: 1940) was previously used in Curl (3: 1986). Two sentences of Lloyd (5: 1967) were taken from Mandelbrot (1: 1982). One sentence from Macaulay (3: 1922) was previously quoted by Mandelbrot (3: 1963) and Brock (3: 1971).

Contents

1 What on earth fractal?

Human understanding progresses naturally from individuals to species, from species to families, from closely related families to those more remote, and at each level it constructs a science.
— D'ALAMBERT

1.1 The self-similarity of rivers

Monsieur Jourdain: Only think! Here is more than forty years I've been talking prose without knowing it.
— MOLIÈRE

Daisetz Suzuki once remarked that the best book ever written by a European on Zen Buddhism is the "Zen in the Art of Archery" by Herrigel which, except in its title, does not even mention the word Zen. One of the best case histories on the application of self-similarity and fractals in petroleum geology, the remarkable "Determination of widths of meander-belt sandstone reservoirs from vertical downhole data, Mesaverde Group, Piceance Creek Basin, Colorado" (1985) by J.C. Lorenz and co-workers of the Sandia National Laboratories, Albuquerque, does not explicitly refer to the concept of self-similarity, let alone to fractals. In this study the authors analysed data of three wells drilled through the Upper Cretaceous Mesaverde Group in the east-central part of the Piceance Creek Basin in northwestern Colorado, in the depth range 1,220–2,500 m (4,000–8,200 ft). The upper 610 m (2,100 ft) of the Mesaverde Group was deposited in an environment of meandering streams of high sinuosity. The aim of the study was to estimate the maximum width of the meanders of the ancient fluvial system (that is, the present day reservoir extension) using only the depths of the palaeostreams which are measurable in wells. Geomorphologists have found empirical relations connecting the parameters characterizing fluvial systems. For rivers of sinuosity greater than 1.7, Leeder (1973) found that bankfull width is related to bankfull depth as:

$$W_c = 6.8\,h^{1.54} \tag{1.1.1}$$

where W_c is channel width and h is channel depth (both in m). Similar relations have been observed between meander width and channel width by Leopold and Wolman (1960) and Carlston (1965), whose combined data can be approximated by the empirical rule:

$$W_m = 7.44\,W_c^{1.01} \qquad (W_m \text{ and } W_c \text{ in feet}) \tag{1.1.2}$$

(see Figure 1.1, from Lorenz et al., 1985).

Bankfull depth h and width W_c may be broadly defined as the maximum val-

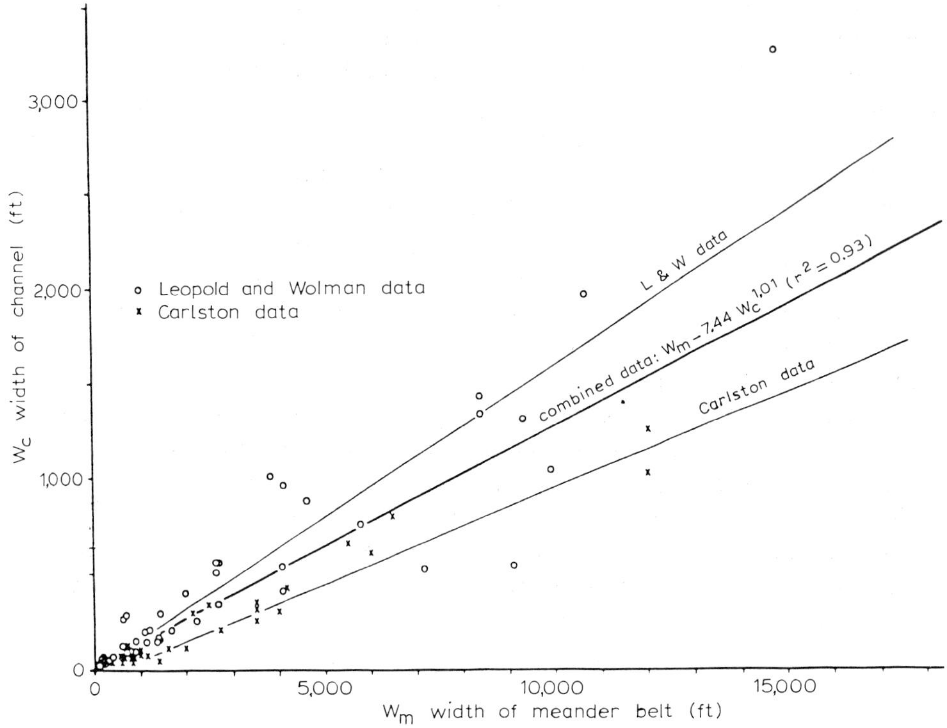

Fig. 1.1. Empirical relation between channel widths W_c and meander-belt widths W_m (from Lorenz et al., 1985).

ues of the depth and width attained in a stream channel (Figure 1.2a), the width W_m, amplitude A and wavelength L are defined in Figure 1.2b. The *sinuosity* of a river (Leopold and Wolman, 1960) is the ratio of the length measured along the channel centre line from one point of the inflection to the next, to half the meander wavelength. In their study, Lorenz et al. (1985) first identified altogether 49 fining-upward individual sand bodies from the meandering fluvial environment of

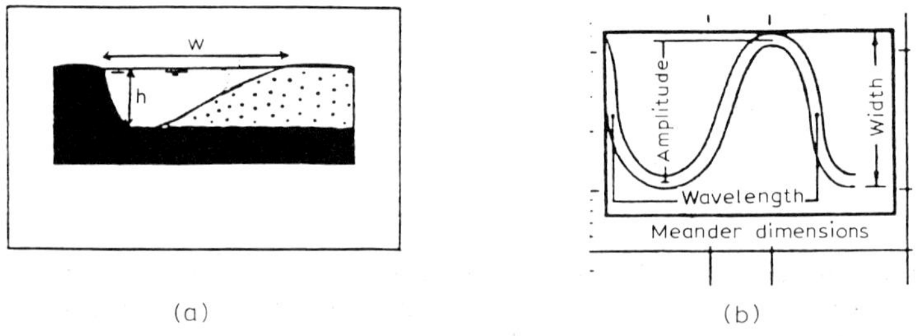

Fig. 1.2. Meander characteristics (a: after Leeder, 1973, b: after Carlston, 1965)

the Mesaverde Group. For this, they have only relied on the core data and geophysical logs of a single well. From the histogram of the preserved thicknesses of the sand bodies they estimated the range of the maximum palaeochannel depth as 3–4 m (10–13 ft). To take into account postdepositional compaction of sand into sandstone, they increased the sandstone body thicknesses by 11% (cf. Ethridge and Schumm, 1978); this converted the above channel depths to 3.3–4.4 m (11–14 ft). Application of Leeder's equation (1.1.1) yielded palaeochannel widths between 45–67 m (150–220 ft) which, according to the empirical correlation (1.1.2), implies a predicted meander belt width of 350–520 m (1,100–1,700 ft). The result has been corroborated by three other techniques: a probabilistic method based on well-to-well correlation, a palaeohydrologic reconstruction, and a study of the continuity of the sand bodies in outcrop.

The basic assumption in the previous study has been that rivers of very different size behave in an essentially similar manner, and their geometric properties can be summarized in simple equations like (1.1.1) or (1.1.2). Among the rivers used to derive the empirical rule (1.1.2), for example, there were small creeks in Wyoming of 10–15 ft average width as well as the Mississippi (5000 ft wide near Wynnburg, Tenn.) — the relation is reasonably well satisfied even by the 1–3 ft wide model rivers of the U.S. Waterways Experimental Station (Leopold and Wolman, 1960)!

It is a common experience that all rivers look similar on the maps (Figure 1.3,

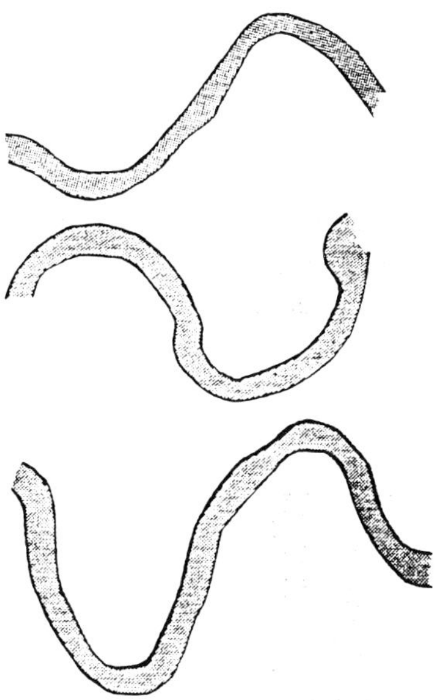

Fig. 1.3. Planimetric maps of three rivers: a. Mississippi at Point Breeze, Louisiana, average width 3,400 ft; b. New Fork River near Pinadale, Wyoming, average width 60 ft; c. Duck Creek near Cora, Wyoming, average width 8 ft (redrawn from Leopold and Wolman, 1960).

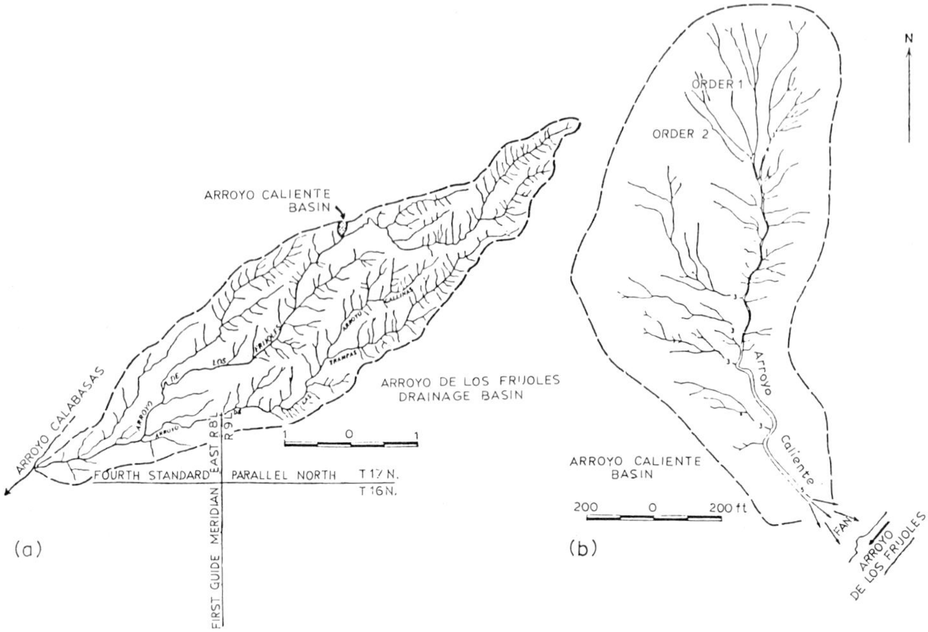

Fig. 1.4. a. Drainage network of the Arroyos near Santa Fe, New Mexico. b. Small tributaries of the Arroyo Caliente basin. (From Leopold et al., 1964.)

after Leopold and Wolman). In fact, without first glancing at the map scale, it is not immediately obvious whether a river is large or small. Small details of a whole drainage network, when properly magnified, can also be surprisingly similar to the entire net (Figure 1.4, from Leopold et al., 1964). The network of the smallest ephemeral channels and rills reveals the same degree of complexity as the system of the master streams and their tributaries.

Equations (1.1.1) and (1.1.2) share the common form:

$$y = ax^p \tag{1.1.3}$$

where y and x are two characteristic dimensions of the stream, a is a coefficient which depends on the system of units in which x and y are measured, and p is a generally non-integer, positive or negative exponent. All the other equations describing the correlation among the further properties of meandering streams (between meander width and wavelength, wavelength and radius of curvature, wavelength and discharge, etc.; see Leopold et al., 1964; Carlston, 1965; Bruinsma, 1990) have the same general form as equation (1.1.3).

It is easy to show that the power-function like relationship of the form (1.1.3) between the different characteristics of a meandering river is a simple consequence of the similarity of all meandering rivers, or the different parts of the same river.

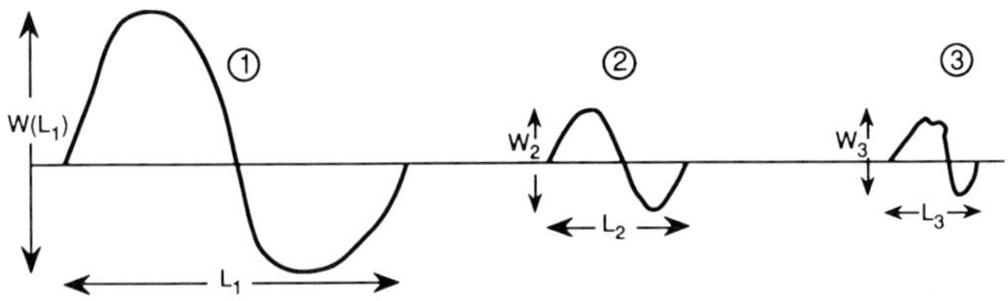

Fig. 1.5. To the derivation of equation (1.1.3).

Let us consider for example the wavelengths and widths of meanders of different size (Figure 1.5) and assume that there exists a unique analytical expression:

$$W_{\mathrm{m}} = W(L_{\mathrm{m}}) \tag{1.1.4}$$

relating width and wavelength. The three meanders in Figure 1.5 are of different size but, if different scales were chosen as to make their wavelengths equal on the printed page, their amplitudes would also be approximately equal. Thus, the ratio of meander widths — say of W_1 and W_2 — only depends on the ratio of wavelengths:

$$\frac{W(L_1)}{W(L_2)} = f\left(\frac{L_1}{L_2}\right) \tag{1.1.5}$$

where f is an as yet unknown function. To find the functional form of f, observe that, in Figure 1.5, to make the third meander look as large as the first, we either immediately magnify it to L_1/L_3 times its original size, or we can make this in two steps, once by a factor L_2/L_3 which transforms it to the second meander and then once more, by a factor L_1/L_2.

By equation (1.1.5) we have:

$$W_1 = W_3 f\left(\frac{L_1}{L_3}\right) \tag{1.1.6}$$

in the first case, and:

$$W_1 = W_2 f\left(\frac{L_1}{L_2}\right) = W_3 f\left(\frac{L_1}{L_2}\right) f\left(\frac{L_2}{L_3}\right) \tag{1.1.7}$$

for the two-step magnification. Comparing equations (1.1.6) and (1.1.7) we find that:

$$f\left(\frac{L_1}{L_3}\right) = f\left(\frac{L_1}{L_2}\right) f\left(\frac{L_2}{L_3}\right) \tag{1.1.8}$$

that is, using the notations $L_1/L_2 = a$, $L_2/L_3 = b$, we have $L_1/L_3 = ab$ and the unknown function $f(x)$ should satisfy the functional relation:

$$f(ab) = f(a)f(b) \qquad \text{for all } a, b > 0 \tag{1.1.9}$$

Equation (1.1.9) is the celebrated Cauchy's functional equation whose only continuous solution (apart from the trivial $f(x) = 0$ for all x) is:

$$f(x) = x^p \tag{1.1.10}$$

for some real number p (see Aczel, 1966, and the Mathematical Appendix to Chapter 1). If we plug equation (1.1.10) into (1.1.5), assume that $L_2 = 1$, and simply write L for L_1, we find that the meander width–meander wavelength relation has the general power-function like form:

$$W = wL^p \tag{1.1.11}$$

as stated. The constant w in equation (1.1.11) is the width of the meander having unit wavelength, the exponent p can not be determined from principles of self-similarity. (Obviously, p is positive if there is a positive correlation, and negative if there is an inverse correlation between the quantities in question.)

In some cases, the empirical value of exponent p can be derived from physical considerations.

A well known example for power function type laws in hydrology is Hack's law for the relation between the length of the longest stream L in a drainage region, and the area of that region, A (Hack, 1957; Leopold and Langbein, 1962; Leopold et al., 1964, chapters 5 and 10; Stevens, 1974; Mandelbrot, 1977; Green and Moore, 1982). In 1957 the American hydrologist J.T. Hack made an extensive study of the geometry of the drainage regions for several streams of the Shenandoah Valley and adjacent mountains in Virginia. When he measured successive distances from the headwaters along a watercourse and the total contributing drainage area at each of these points, he found that the progressive increase in drainage area with increasing channel length can be mathematically described as:

$$L = 1.4A_{\mathrm{d}}^{0.6} \tag{1.1.12}$$

where L is length of the longest stream, in miles, in the drainage region and A_{d} is the area of this region in square miles (Figure 1.6A). He later extended these studies to other parts of the United States and to various large rivers of the world (Figure 1.6B) and found that the general form of equation (1.1.12) has retained its validity though there was a slight variation in the coefficient (1.0–2.5) and the value of the exponent (0.6–0.7).

Hack's law is one of the very few examples where the value of the exponent p in the power-law relation (1.1.3) can be predicted from physical principles.

Because of the self-similarity of the drainage network we can assume, on the basis of equation (1.1.3) proved above, that the different parameters of a smaller subregion are interrelated by power-law relations $x \propto y$, where the relation "$\propto$" denotes proportionality. Thus, if we relate everything to the channel width W at the given locality, we can write:

$$X \propto W^{p_X} \qquad Y \propto W^{p_Y} \qquad L \propto W^{p_L} \tag{1.1.13}$$

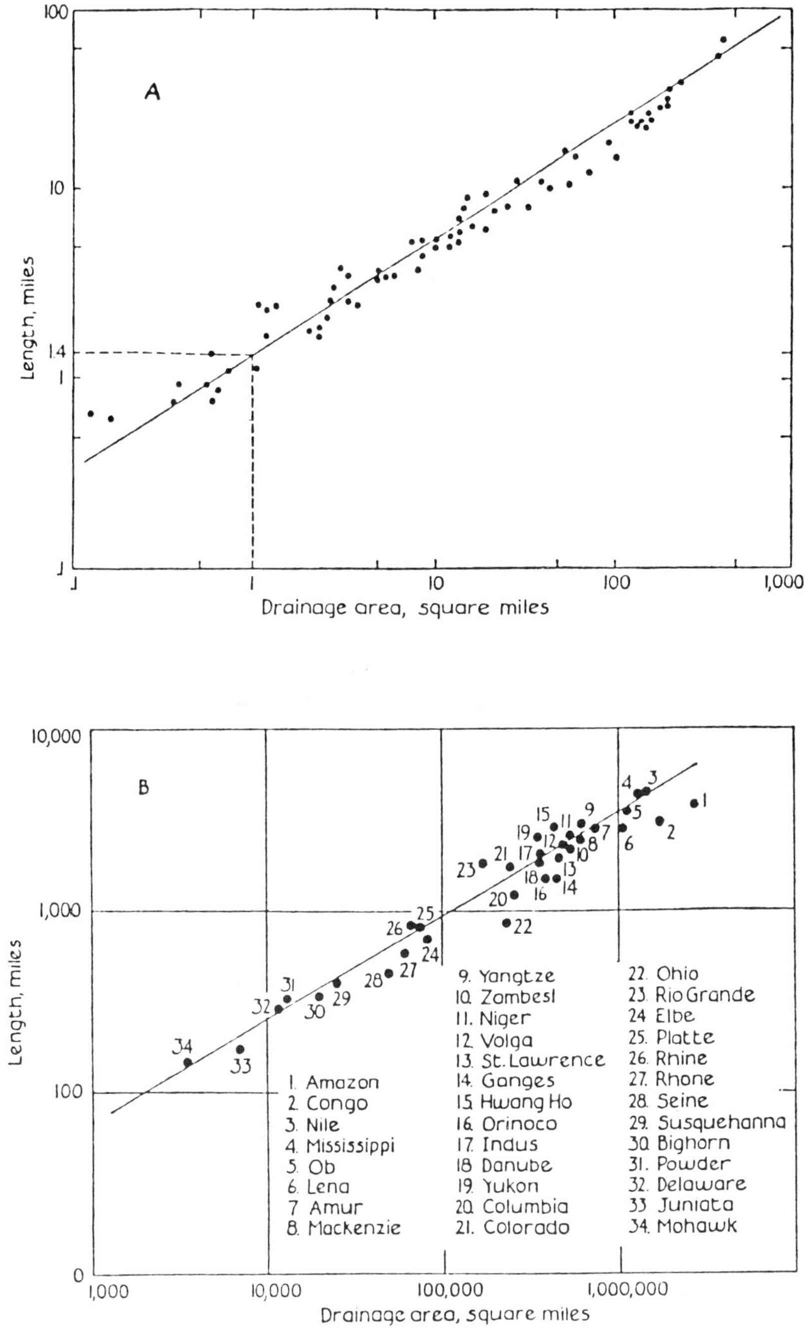

Fig. 1.6. Relation of the length of the longest stream to drainage area: A. for rivers of the Shenandoah Valley, Virginia; B. for various rivers of the world (after Hack, 1957).

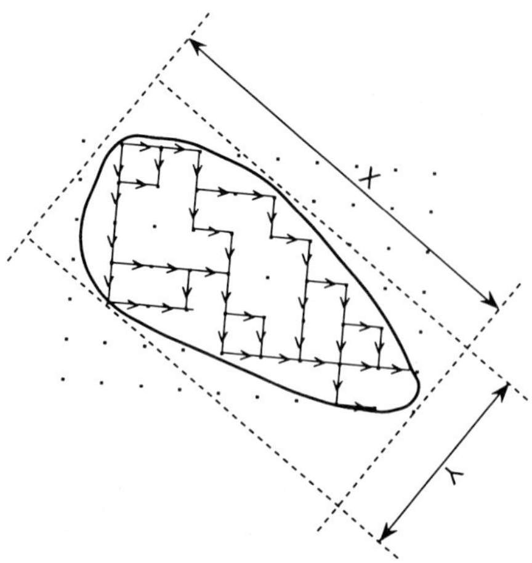

Fig. 1.7. To the derivation of the stream-length–drainage-area relationship.

where X is the size of a typical drainage region corresponding to channel width W in the preferential direction of flow, Y is the width of the region in the perpendicular direction, and L is the length along the longest stream from the given locality to the drainage divide (Figure 1.7). Because of the strong asymmetry, the exponents p_X and p_Y are not equal but $p_X > p_Y$. Also, as X is measured in the overall direction of flow, the exponent p_L should be closer to p_X than to p_Y.

As the drainage area A_{d} is proportional to XY, from equations (1.1.13):

$$A_{\mathrm{d}} \propto XY \propto W^{p_X + p_Y} \tag{1.1.14}$$

and we obtain Hack's law (1.1.12) in the form:

$$L \propto A_{\mathrm{d}}^{D} \quad \text{where the exponent } D = \frac{p_L}{p_X + p_Y} \tag{1.1.15}$$

According to recent results of the statistical physical theory of directed percolation (Green and Moore, 1982; Redner and Yang, 1982; Takayasu and Nishikawa, 1986), the exponent p_X is typically about 1.6 times larger than p_Y, that is if we assume that p_L is close to p_X, D can be estimated as:

$$D = \left(\frac{p_L}{p_X}\right)\left\{\frac{1}{(1 + p_Y/p_X)}\right\} = \frac{1}{1 + 0.625} = 0.62 \tag{1.1.16}$$

in good agreement with the value of the exponent, $D = 0.6$, in Hack's law (1.1.12). It is rather unusual that the stream length L in Hack's law (1.1.1) is proportional to the 0.64th power of drainage area A_{d}. Simple dimensional considerations suggest that length must be proportional to the square root, that is the 0.5th power of the area. As we have seen in the derivation of equation (1.1.15), the fact that the exponent D is not 0.5 but 0.6–0.7 is a consequence of the asymmetrical shape of the drainage basins. With increasing size, drainage basins elongate — similarly to the irrigation systems in leaves where the "large elongated leaf of the banana stands in contrast to the small rounded leaf of the blueberry" (Stevens, 1974, p. 110). In 1962 Leopold and Langbein (see also Leopold et al., 1964) simulated the development of drainage networks by a random walk on a rectangular grid paper. In their model

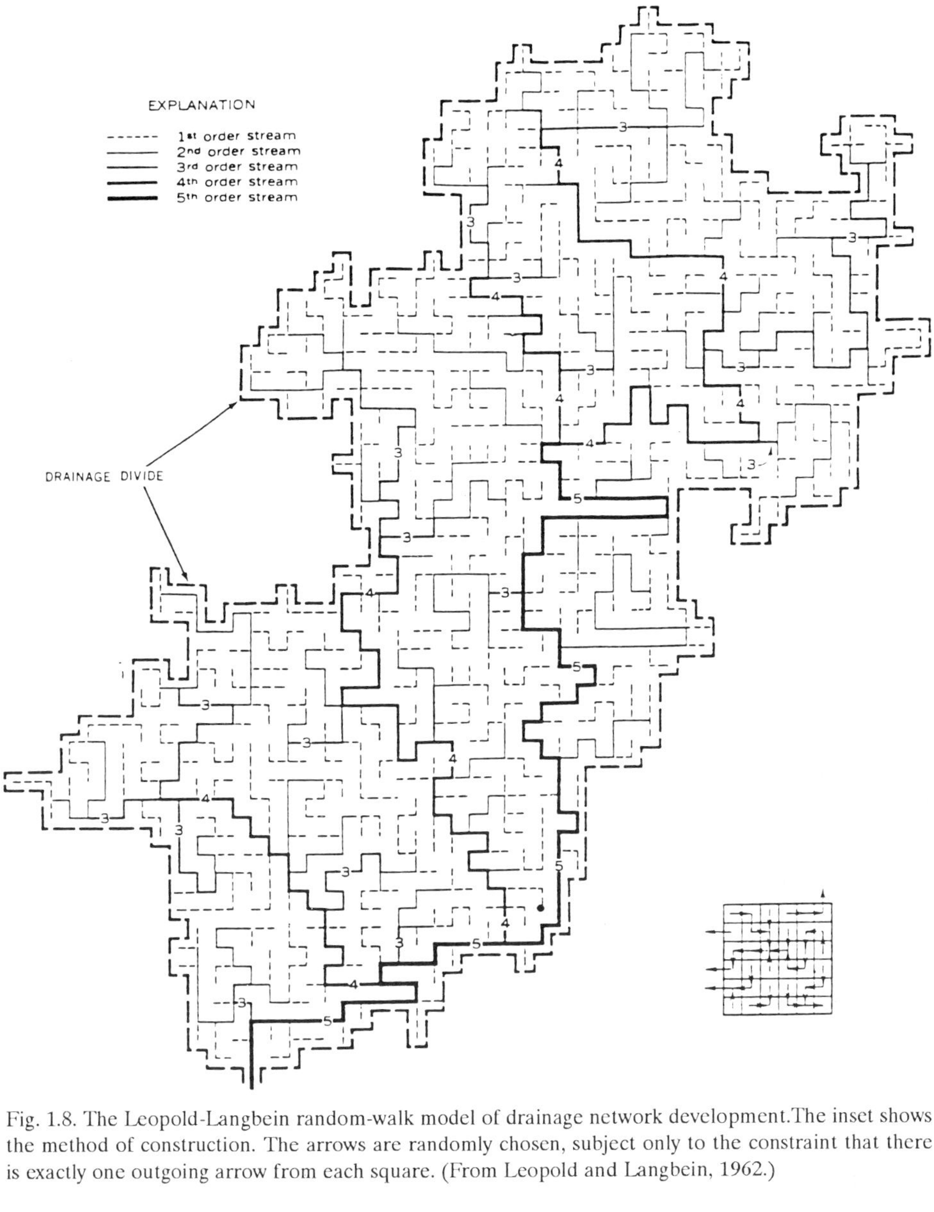

Fig. 1.8. The Leopold-Langbein random-walk model of drainage network development.The inset shows the method of construction. The arrows are randomly chosen, subject only to the constraint that there is exactly one outgoing arrow from each square. (From Leopold and Langbein, 1962.)

each square represented a unit area in the developing drainage net and it was assumed that each square was to be drained. From each square there were four equally possible directions of drainage, to one of the four adjacent squares. The actual directions of the flows (represented by arrows in the inset of Figure 1.8) were randomly chosen, subject only to the condition that no internal sinks could develop. Leopold and Langbein found that the resulting networks were strikingly similar to the real stream networks, "divides" had developed, and the arrows joined as to form "rivers" and "drainage areas" of various size. The relation between drainage area and stream length (Figure 1.9) satisfied Hack's law with an exponent $D = 0.64$. The main difference between the way we have derived equations (1.1.12) and (1.1.15) and the Leopold-Langbein (L&L) random-walk model (Figure 1.8) is that in the L&L model there is no preferential direction of flow, the "drainage basins" are symmetrical, that is their length and width increase as the same power $p_X = p_Y$ of the increasing stream length. According to equation (1.1.15), in the symmetrical L&L model the empirically obtained $D = 0.64$ implies $p_L/p_X = 1.28$, that is, *the*

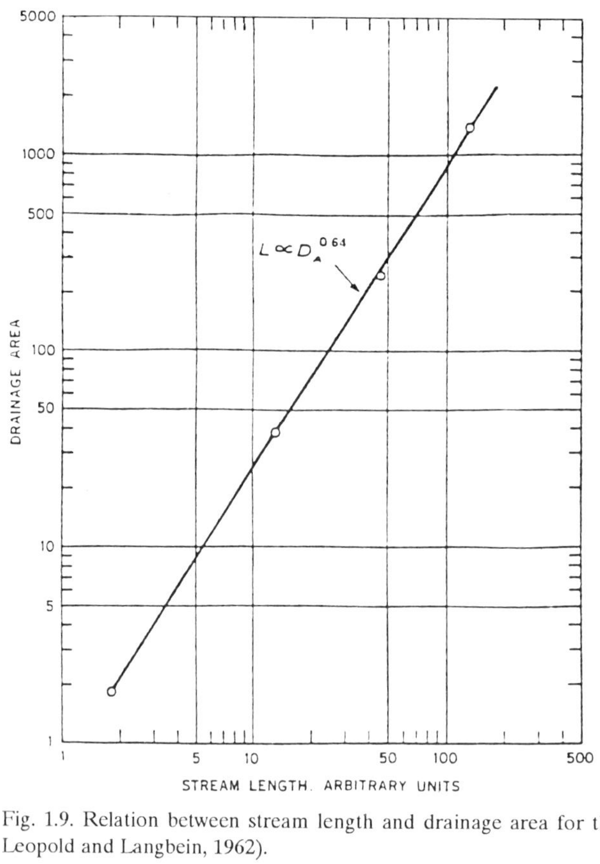

Fig. 1.9. Relation between stream length and drainage area for the L&L random walk model (from Leopold and Langbein, 1962).

length of the longest "stream" in the L&L random walk model increases as the 1.28th power of the linear size of the drainage area. In other words — as we shall see in the next section — the stream in the L&L model is 1.28-dimensional.

1.2 How long is the Vistula River?

> "What do you consider the *largest* map that would be really useful?"
> "About six inches to the mile."
> "Only *six inches*!" exclaimed Mein Herr."We very soon got to six *yards* to the mile. Then we tried a *hundred* yards to the mile. And then came the grandest idea of all! We actually made a map of the country, on the scale of *a mile to a mile*!"
> "Have you used it much?" I enquired.
> "It has never been spread out, yet," said Mein Herr: "the farmers objected: they said it would cover the whole country, and shut out sunlight. So now we use the country itself as its own map."
> — LEWIS CARROLL: *Sylvie and Bruno Concluded*, Chapter XI

There are many ways to measure the length of an irregular curve, for example we can drive a measuring wheel along its sinuosities, or pace along them with a compass.

In 1930 the Polish mathematician H. Steinhaus found an interesting new technique to measure length, based on a result of geometrical probability.

As is well known, and intuitively clear, if we have some planar figure F of arbitrary shape and finite extent — say inside a square of unit side — then the probability that a point randomly thrown onto the unit square would "hit" F will be equal to the area $A(F)$ of F.

Letting angular brackets denote expectation (here, and throughout the book), and defining a function $f(X)$ of the spatial point X as $f(X) = 1$ inside F and 0 outside it, we have:

$$\langle f(X)\rangle = A(F) \tag{1.2.1}$$

or, in an actual experiment:

$$\frac{\text{Number of hits}}{\text{Number of throws}} \to A(F) \tag{1.2.2}$$

It is less well known that the length of an arbitrary curve C can be similarly found (Crofton, 1868): apart from a constant factor the length of C is equal to the expected number of intersections of C with a randomly drawn line e:

$$l(C) \propto \langle \text{number of intersections of } C \text{ and } e\rangle \tag{1.2.3}$$

where the constant of proportionality in equation (1.2.3) does not depend on the form of C.

A convenient way to parametrize random straight lines is by introducing a fixed point O and some fixed direction $\overrightarrow{OQ}$, then given any angle $(0 \leq \theta \leq \pi)$ and any real number p we can find a direction $\overrightarrow{OT}$ such that the angle $QOT = \theta$, and a point P on $\overline{OT}$ so as to have $\overline{OP} = p$. The straight line e is uniquely defined by the parameters (θ,p). Denoting the number of intersections of the line $e(\theta,p)$ with the curve C by $a(\theta,p)$, Crofton's theorem can be stated (and proved, see Blaschke, 1936; Steinhaus, 1954) in the form:

$$\text{length of } C = l(C) = \frac{1}{2}\int_{-\infty}^{\infty}\int_{0}^{\pi} a(\theta,p)\,\mathrm{d}\theta\,\mathrm{d}p \tag{1.2.4}$$

Steinhaus (1938, 1954) devised an ingenious technique for the practical application of equation (1.2.4) to measure the length of a river or a highway traced on a map. He prepared a transparent sheet containing a family of parallel straight lines e_i ($i = \ldots,-2,-1,0,1,2,\ldots$) spaced a distance d apart. If we place the sheet in a random position over the curve C to be measured, C will intersect e_i in a_i points, and $s_0 = \sum a_i$ is the total number of intersections. Then we turn the sheet through an angle $\pi k/m$ ($k = 0,1,\ldots,m-1$) and count the number of intersections s_k in all the m directions. The expression:

$$\frac{d\pi}{2m}\cdot\sum_{k=0}^{m-1} s_k \tag{1.2.5}$$

which is an approximation of the integral (1.2.4), tends to the length of C for $d \to 0, m \to \infty$ (according to Steinhaus, in most practical applications $d = 2$ mm and $m = 6$ different rotations of the sheet give sufficient accuracy).

As a first application of the method, Steinhaus measured the length of the left bank of the Vistula, first on a school map of Poland and then on a map of 1:200,000. And he made a striking discovery: *the length read off the school map was appreciably smaller than that on the 1:200,000 map* — "... the left bank of the Vistula,when measured with increased precision would furnish lengths ten, hundred and even thousand times as great as the length read off the school map... This is ... contrary to the belief that not rectifiable arcs are inventions of mathematicians and that natural arcs are rectifiable; it is the opposite that is true." (Steinhaus, 1954).

Steinhaus attributed the divergence of expression (1.2.5) to the unboundedness of the integrand $a(\theta,p)$ in equation (1.1.4), for the irregular curves of Nature: with increasing magnification there will be straight lines intersecting them arbitrarily many times. Alas, he has not analysed the increase of the measured length as a function of map scale. This task has been left to the British physicist and meteorologist, Lewis Richardson, who in the late 1930's carried out similar studies (posthumously published in 1961) related to the length of coast-lines and state frontiers; and to the French-American mathematician Benoit Mandelbrot whose history-making paper "How long is the Coast of Britain? Statistical Self-Similarity and Fractional Dimension" (1967) marks the birth of the science of *fractals*. (Coastlines will be discussed in Section 1.4.)

Let us return to the question posed in Section 1.1: What is the meaning (if any) of an allegedly 1.28-dimensional river? We show first that if one measures the length of a self-similar river (or any other naturally occurring self-similar, irregular curve) on maps of different scales, the length L_r measured with an r times larger accuracy will increase as a power of r:

$$L_r \propto r^{D-1} \tag{1.2.6}$$

where D is the exponent figuring in the scaling law:

$$L(A,B) \propto \overline{AB}^D \tag{1.2.7}$$

Next we shall make it plausible that the exponent D in equation (1.2.7) is a generalisation of our usual concept of dimension and — as it is usually non-integer — can be properly called the *fractal dimension* of the curve.

Consider a curve C, connecting points A and B, which has a self-similar structure over a wide range of scales (Figure 1.10). We can assume that $\overline{AB} = 1$.

Instead of magnifying the curve, we shall carry out a series of measurements on the same curve, using "measuring rods" of gradually decreasing size $\varepsilon = 1/r, r \to \infty$ (Figure 1.11).

If we approximate the curve C by straight line segments of length ε ($\varepsilon \ll 1$), their total number will be a function of ε, $N(\varepsilon)$. To determine $N(\varepsilon)$ observe that the endpoints of the segments divide the curve C into $N(\varepsilon)$ pieces. Because of the assumed self-similarity of C, its length between the endpoints of any of these segments is related to the length of the segment as:

$$L(X_i, X_{i+1}) = a\overline{X_i X_{i+1}}^D = a\varepsilon^D \tag{1.2.8}$$

where the constant a is the length of the curve between any two points a unit distance apart, say between A and B. As:

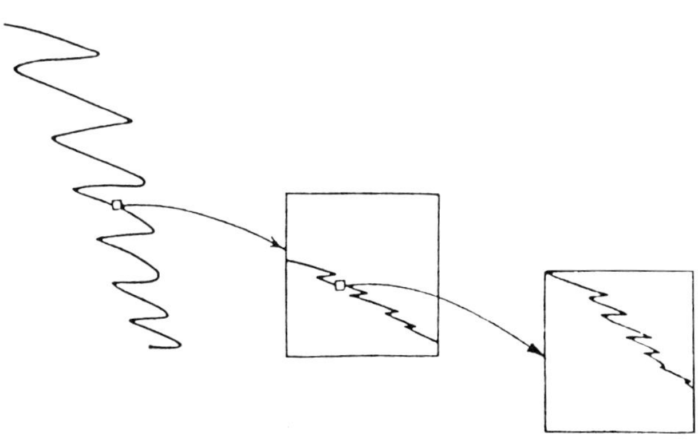

Fig. 1.10. A fractal curve has a self-similar structure over a wide range of scales.

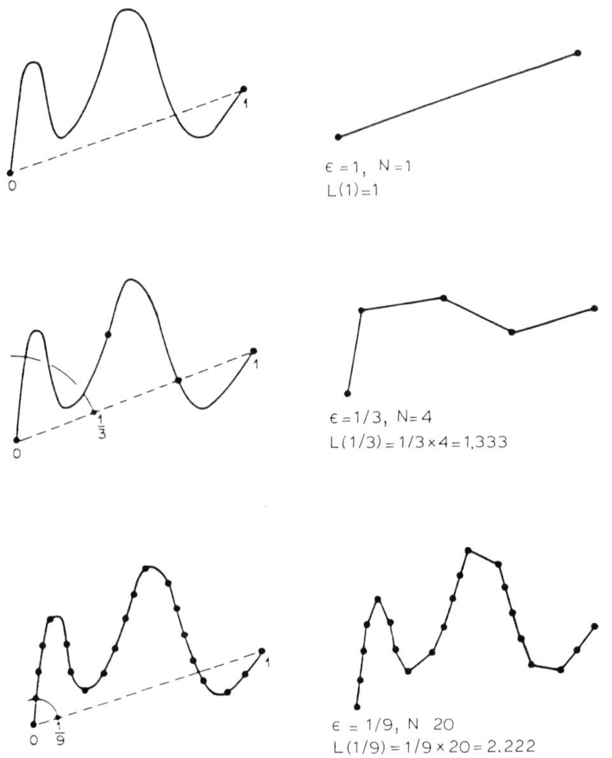

Fig. 1.11. Measurement of the length of a fractal curve with increasing precision (from Kapraff, 1986).

$$L(X_0, X_{N(\varepsilon)}) = L(\overline{AB}) = a \sum_{i=0}^{N(\varepsilon)-1} L(X_i, X_{i+1}) \tag{1.2.9}$$

from equations (1.2.8) and (1.2.9) we have:

$$N(\varepsilon) \propto \frac{1}{\varepsilon^D} \tag{1.2.10}$$

and the length of C measured with a precision ε becomes:

$$L(\varepsilon) \propto \frac{\varepsilon}{\varepsilon^D} = \varepsilon^{1-D} \tag{1.2.11}$$

As $D \geq 1$, the number of segments and the estimated length diverge (as ε^{-D} and ε^{1-D}, respectively) with increasing accuracy $\varepsilon \rightarrow 0$, in conformity with Steinhaus's empirical findings.

To understand why is it reasonable to call a self-similar planar curve *D-dimensional* if the length $L(\overline{XY})$ of any of its parts scales as the D-th power of the distance between its endpoints, observe that the line of thought which has been used to derive (1.2.10) can be easily modified to estimate the number of squares of a fine

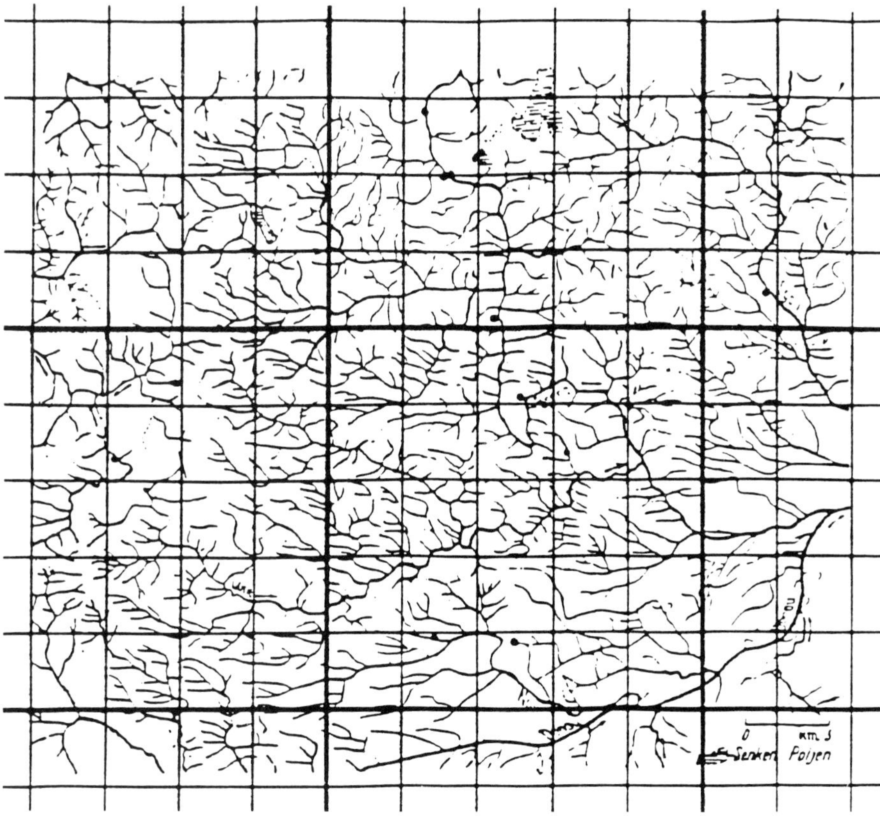

Fig. 1.12. Drainage network (from Schwenk, 1980) superimposed on a square grid.

grid containing points of our object (Figure 1.12). If the curve in question is inside a square of unit sides then, when we divide the unit square to r^2 small squares of side $1/r$, the number of small squares covering the curve would increase as r^D. On the other hand, if we take an Euclidean figure, say a circle of radius 1/2 inscribed in the unit square, its *one-dimensional perimeter* will occupy about $2r$ small squares (2 in each row, 1 or 2 in the lowermost and uppermost ones, depending on whether r is odd or even). The *area* of the circle contains about $r^2\pi/4$ small squares or, using a more exact estimation: $\pi/4\{r^2 + O(r^{2/3})\}$ small squares (Sierpinski's theorem, see Fricker, 1982). That is, the number of squares covering a *one-dimensional* object increases as r^1, the number of squares covering a *two-dimensional* object behaves asymptotically as r^2, and by analogy it is reasonable to call a curve occupying $\propto r^D$ small squares as $r \to \infty$ a *D-dimensional curve.*

The above method of determining the fractal dimension is called *box counting* (Forrest and Witten, 1979); it will be frequently used in this book.

A well known example for a statistically self-similar curve is the Brownian motion

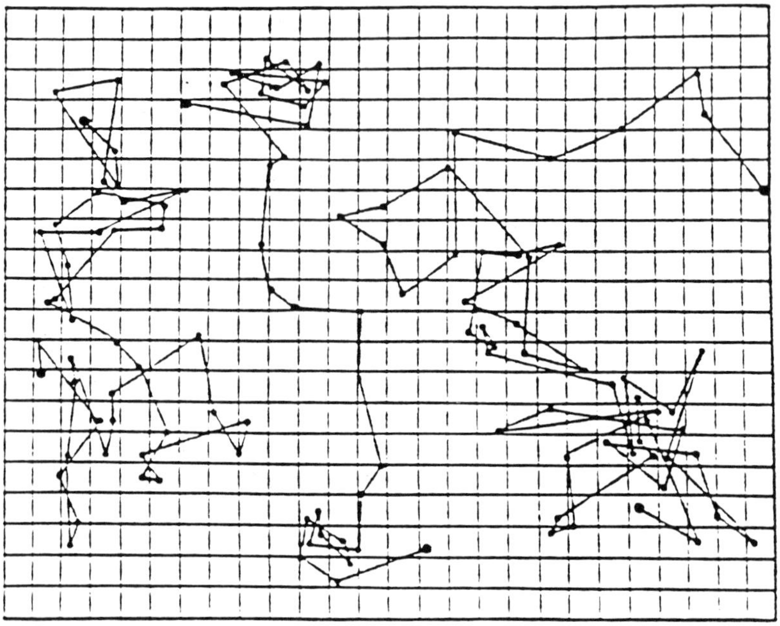

Fig. 1.13. Perrin's observation of the Brownian motion of a small colloidal particle. Grid size: 3.2 μm; successive positions (dots) observed every 30 seconds. (From Mandelbrot, 1982.)

of the very fine particles in a liquid. Figure 1.13 (from Mandelbrot, 1982) reproduces the motion of a colloidal particle of radius 0.53 μm as observed by the French physicist Jean Perrin under the microscope. For the Brownian motion the change of the position $X(t)$ of the particle over time, $\Delta X(t) = X(t + \Delta t) - X(t)$ obeys a Gaussian distribution with mean $\langle \Delta X(t) \rangle = 0$ and variance $\langle [\Delta X(t)]^2 \rangle^{1/2} \propto \Delta t$.

Thus, assuming a constant average velocity, during the time Δt the particle would describe a path of length proportional to Δt, while the distance between the endpoints $A = X(t)$ and $B = X(t + \Delta t)$ of this path is only of the order of $(\Delta t)^{1/2}$. Thus, $\overline{AB} \propto (\Delta t)^{1/2}$, $L(\overline{AB}) \propto \Delta t$, that is: $L(\overline{AB}) \propto \overline{AB}^2$ and the *Brownian motion is two-dimensional.*

A scrupulous philosopher (like Zenon) would argue that for a turtle (or the ant, the favourite pet of Percolation Theorists) a stroll along a 1.28-dimensional river takes an infinite time. Richardson (1961) had a ready answer to this objection: "When a man says that he 'walked 10 miles along the coast', he usually means that he walked 10 miles *near* the coast."

Rivers can also flow in even the most irregular channels (Figure 1.14), though they do slow down and the laminar part of the flow obviously does not follow the small-scale irregularities of the river bank. For high fractal dimensions however, the tortuosity of the channel becomes infinite, obstructing the flow. (The stream becomes a "wat'ry labyrinth" as Milton described Lethe, the "River of Oblivion"

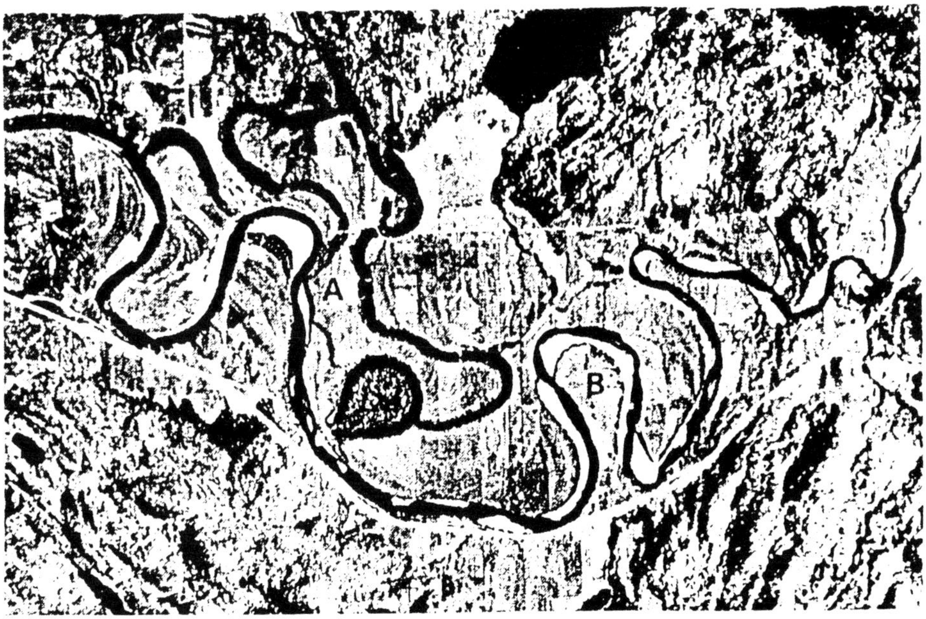

Fig. 1.14. Asymmetric meander style of high tortuosity. The Rouge River valley, Quebec. River flow is from left to right. (From Carson and Lapointe, 1983.)

in the "Paradise Lost", Book ii, l. 582.) As we shall see in the next section, this is exactly what happens in certain sandstones which have a reasonable porosity (4–5%) but zero permeability. The fluid can not flow through such a rock because of the *infinite tortuosity* of the pore network.

1.3 The paradox of tortuosity: permeability of kaolinite-bearing sandstones

> All Mathematics would suggest
> A steady straight line as the best,
> But left and right alternately
> Is consonant with History. — W.H. AUDEN: *The Maze*

1.3.1 *The permeability of shaly sandstones*

The permeability of porous rocks can be expressed as:

$$k = \frac{1}{b}\Phi^3\left(\frac{V}{A_S}\right)^2\frac{1}{\tau^2} \tag{1.3.1.1}$$

where k is permeability, A_S/V is the surface area per unit volume, τ is tortuosity of

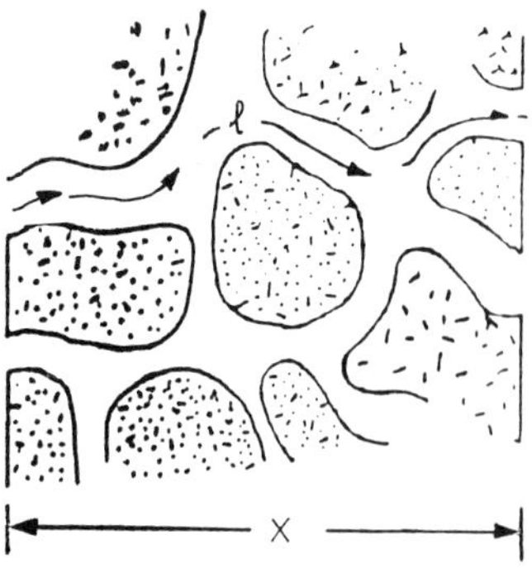

Fig. 1.15. The definition of tortuosity: $\tau = l/x$. (From Walsh and Brace, 1984.)

the flow path (Figure 1.15), Φ is porosity, and the constant b is equal to 2 for circular tubes and equal to 3 for cracks. An equivalent expression is:

$$k = \frac{R^2_{\text{HYD}}}{b} \Phi \frac{1}{\tau^2} \tag{1.3.1.2}$$

where R_{HYD} is the hydraulic radius, defined as the ratio of the pore volume to the wetted area (Walsh and Brace, 1984). By definition, a porous material has a permeability of 1 darcy if a pressure difference of 1 atm produces a flow rate of 1 cm^3/s of fluid with 1 cP viscosity through a cube having sides 1 cm in length (Dullien, 1979). It is easy to check that 1 darcy=0.987 μm^2 , that is, if we express R_{HYD} in equation (1.3.1.2) in mm and k in millidarcies, the equation becomes:

$$k = \frac{R^2_{\text{HYD}}}{b} \Phi \frac{1}{\tau^2} \tag{1.3.1.3}$$

The tortuosity in equations (1.3.1.1)–(1.3.1.3) is between 2–4 in clean sands (Volarovich et al., 1968); its role has generally been neglected in permeability studies. Equations (1.3.1.1) and (1.3.1.2) correctly describe the empirical facts (Martin and Hamilton, 1981) that permeability generally increases with porosity, and with the square of the grain size. (To see the latter, consider a simple rock model consisting of spherical grains of radius r. Then a volume of rock of porosity Φ will contain $N = V(1 - \Phi)/(4r^2\pi/3)$ grains of total surface area $A_S = 4r^3\pi N$, that is $V/A_S \propto r$).

It has recently been realized that the permeability of *claybearing* sandstones can not be described by simple equations like (1.3.1.1) or (1.3.1.2) in a way universally valid for all values of the clay volume content and for all clay morphologies. Any theoretical model attempting to describe fluid flow in shaly sands must conform with the experimental finding that the permeability of shaly sandstones rapidly decreases with increasing clay content and becomes almost zero — even for non-zero porosities — if the clay content is greater than about 15% (Hanin, 1951; Denson et al., 1968). This permeability cutoff around 15% clay content is readily observed in the

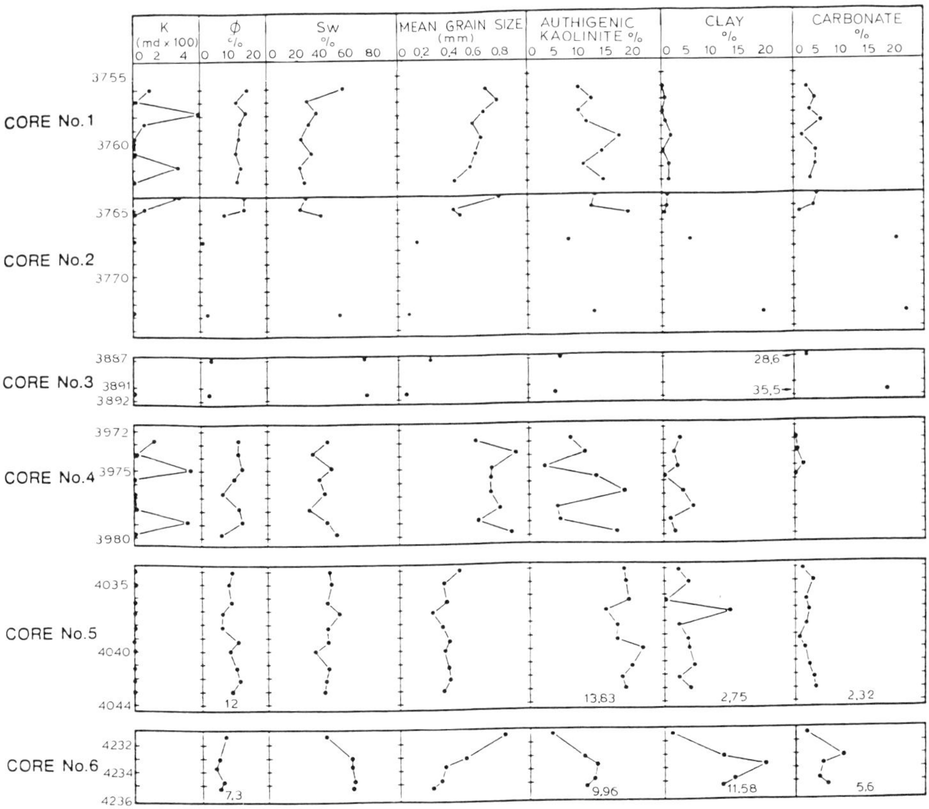

Fig. 1.16. Core analysis and petrographic properties of the Late Triassic Mungaroo Formation sandstones, offshore W.Australia. Dominant clay mineral is kaolinite. (From Campbell and Smith, 1982.)

data of Campbell and Smith (1982) if we compare the permeability, porosity and kaolinite-fraction curves (Figure 1.16).

The relation between porosity and permeability also depends on clay morphology. Amaefule et al. (1988) found different trends in the permeability versus porosity crossplots for reservoir sands, depending on whether the dominant clay minerals were of the "pore bridging" (illite), "pore lining" (chlorite) or "discrete particle" (kaolinite) type — using Neasham's (1977) terminology (Figure 1.17).

As by equation (1.3.1.1) permeability is inversely proportional to the square of the specific surface of the rock, a possible way to introduce the effect of shaliness to the permeability equations is to assume that in shaly samples the V/A_S ratio is affected, or even dominated, by the enormous specific surface of the clay particles [Michaels and Lin (1954), Goode and Sen (1988) and Van Olphen and Fripiat (1979) quote 46 m^2/g specific surface for montmorillonite, 8–13 m^2/g for kaolinite, 100 m^2/g for illite). Goode and Sen (1988) have developed a model based on this assumption, and described permeability in terms of porosity and the internal surface

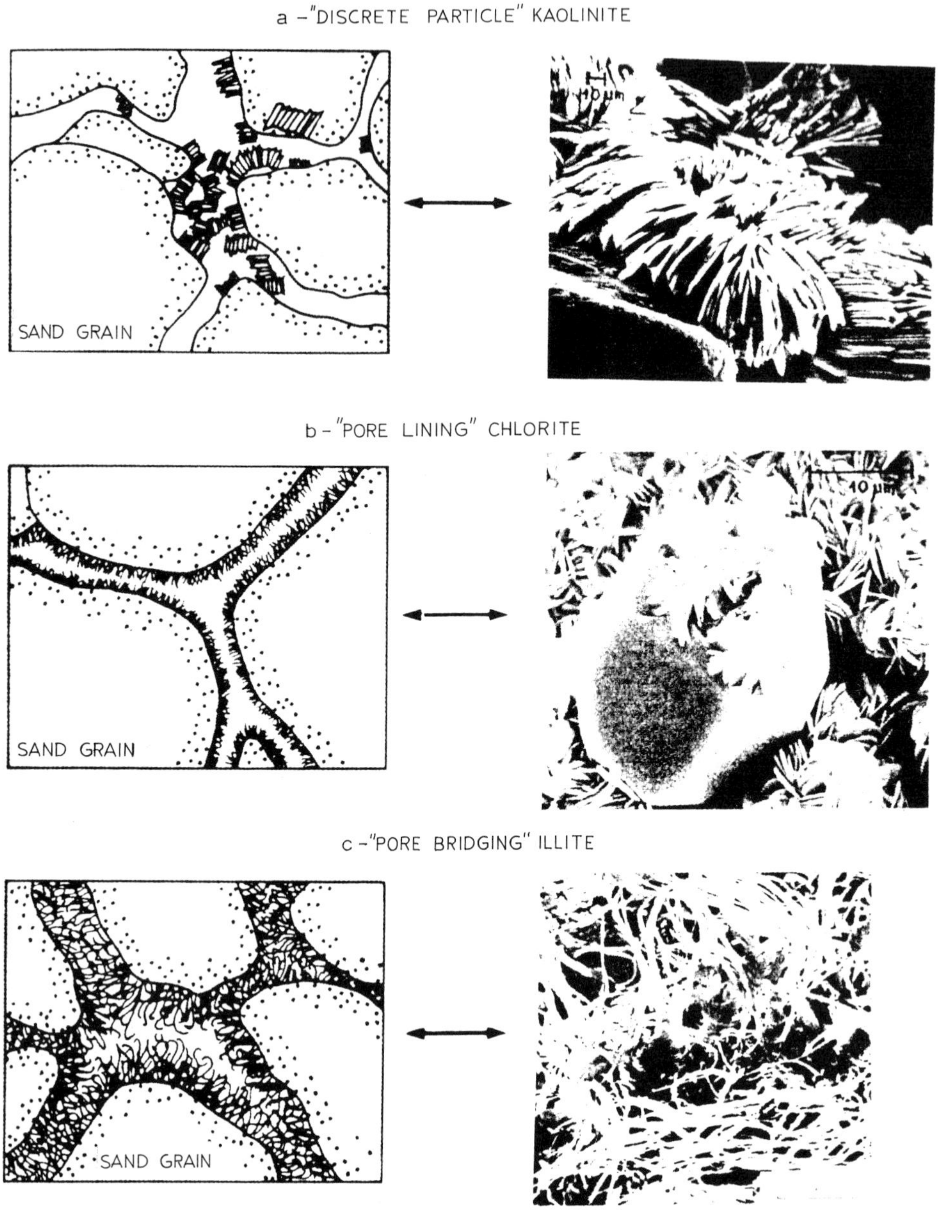

Fig. 1.17. Basic types of dispersed clay in sandstone reservoirs (from Neasham, 1977).

of the *clay particles*. They proposed that the latter could be found from the measured cation exchange capacity (CEC), on the basis of the strong correlation between CEC and the specific surface of clay minerals (Patchett, 1975).

This "specific surface" approach, however, can not answer some basic questions:

– In Darcy's law (Dullien, 1979) we are only concerned with that part of the internal surface which actually becomes wetted. In case of pore-lining (chlorite) or discrete particle (kaolinite) clays only a small fraction of the total clay surface will be exposed to the fluid flow.

– The permeability of shaly sands is very sensitive to the net confining pressure (Amaefule et al., 1988) while the total internal surface of the clay particles is obviously unaffected.

– The specific surface model can not explain the abrupt disappearance of the permeability at some critical clay fraction.

In a recent study (Korvin, 1992) based on the analysis of 229 kaolinite-bearing sandstone samples from Jurassic to Early Cretaceous reservoirs from the Eromanga Basin, South Australia, I could prove that the disappearance of the permeability below some critical clay volume fraction is due to the divergence of tortuosity ($\tau \rightarrow \infty$) in expressions (1.3.1.1)–(1.3.1.3), as a consequence of the intricated, *fractal* shape of the average fluid path.

To make this result plausible, we should have recourse to the Percolation Theory of Statistical Physics [Dullien (1979), Essam (1972), Ziman (1979), Efros (1986) and Feder (1988) are good general introductions].

1.3.2 *Basic concepts of percolation theory*

Historically, the very first published problem in percolation theory was a question related to the design of impermeable gas masks. It was raised by S.R. Broadbent — in abstract mathematical form — at a Symposium of the Royal Statistical Society on Monte Carlo Methods (Broadbent, 1954; Hammersley, 1983). At that time Broadbent was working at the British Coal Utilization Research Association on the design of gas masks for use in coal mines. The masks contained porous carbon granules into which the gas could penetrate. Broadbent found that *if the pores were large enough and sufficiently well connected*, the gas could permeate the interior of the granules; but if the pores were too small or inadequately connected, the gas would not get beyond the granules' surface. There was a critical porosity and pore interconnectedness, above which the mask worked well and below which it was ineffective. Thresholds of this kind are typical of percolation processes.

The basic result of Percolation Theory is represented in Figure 1.18 (from Zallen, 1983). In the (bond-)percolation problem we assume that a fraction $1 - p$ ($0 \leq p \leq 1$) of the bonds of a regular grid are randomly cut and a fraction p are left uncut. Then there exists a critical fraction p_c (called *percolation threshold*) such that for $p < p_c$ there is no continuous connection along the bonds of the network between the opposite faces, while there exists a connection with probability one if $p > p_c$. The lattice is assumed infinitely extended.

For the two-dimensional square lattice (Figure 1.18) we shall soon make it plausible that the percolation threshold is 0.5. In the general case the percolation thresh-

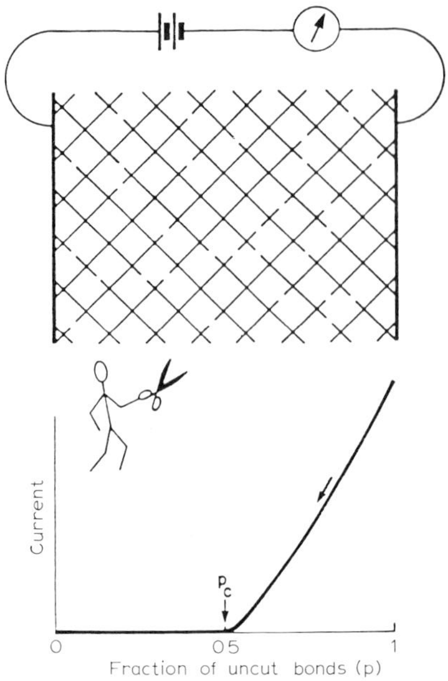

Fig. 1.18. Randomly cut network as an example of percolation. The current represents the continuity of connection along the bonds. (From Zallen, 1983.)

old depends on the dimensionality of the network, d, and on its coordination number Z (where the coordination number is the average number of bonds connected to any node of the network), but it is independent of the detailed geometry of the network. Table 1.1 (from Ziman, 1979) lists coordination numbers and percolation thresholds for some regular lattices. It was first observed by Vyssotsky et al. (1961) that the percolation thresholds of Table 1.1 conform quite closely to the simple empirical rule:

TABLE 1.1

Bond percolation thresholds

Network	Dimension, d	Coordination no., Z	p_c
Honeycomb	2	3	0.6527
Square	2	4	0.5
Triangular	2	6	0.3473
Tetrahedral (diamond)	2	4	0.39
Simple cubic	3	6	0.25
Body centred cubic	3	8	0.18
Face centred cubic	3	12	0.12
Hexagonal close packing	3	12	0.12

Source: Ziman (1979).

$$Zp_c = \frac{d}{d-1} \tag{1.3.2.1}$$

For a three-dimensional network $d/(d-1) = 1.5$, that is percolation only occurs if there are on the average more than 1.5 links to any node.

Close to the percolation threshold ($p \approx p_c$) the nodes which are connected with each other by continuous paths form large clusters of average size ξ; ξ is called the *correlation distance* or *connectedness length*. The correlation distance diverges for $p \rightarrow p_c$ as:

$$\xi \propto (p - p_c)^{-\nu} \tag{1.3.2.2}$$

where for three-dimensional networks $\nu = 0.83$, independently of the coordination number (Fisch and Harris, 1978).

Before proceeding further, it will be instructive to consider the simple case of two-dimensional bond-percolation on a square lattice (Figure 1.18). We shall show, heuristically that:

(a) the percolation network is self-similar at the percolation threshold;
(b) that $p_c \approx 0.5$;
(c) that the correlation distance diverges according to equation (1.3.2.2); and
(d) we shall also find a reasonable estimation for the value of the exponent ν.

In order to decide whether a given network is self-similar, we have to consider it at different reductions. A simple way of scaling (introduced by Madden, 1983) is when one substitutes four small squares of the original lattice by a single square which is twice as large:

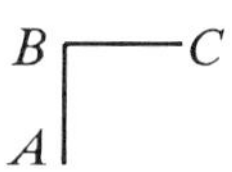

will correspond to

(1.3.2.3a)

The horizontal bond BC will be present if and only if there is a horizontal continuity across the four small squares: B——C if

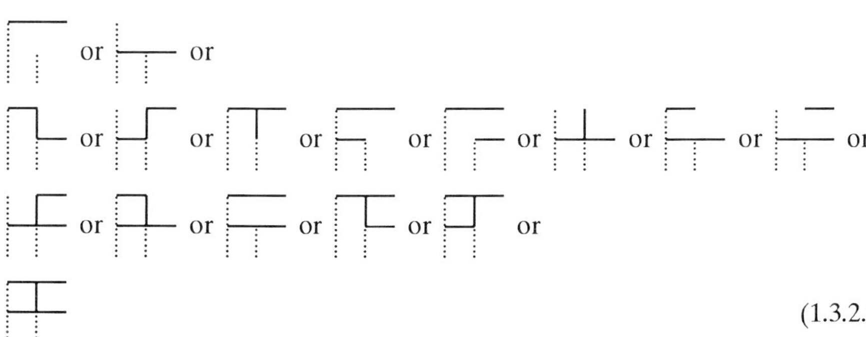

(1.3.2.3b)

where a solid line represents the presence of a bond. The bonds denoted by dashed

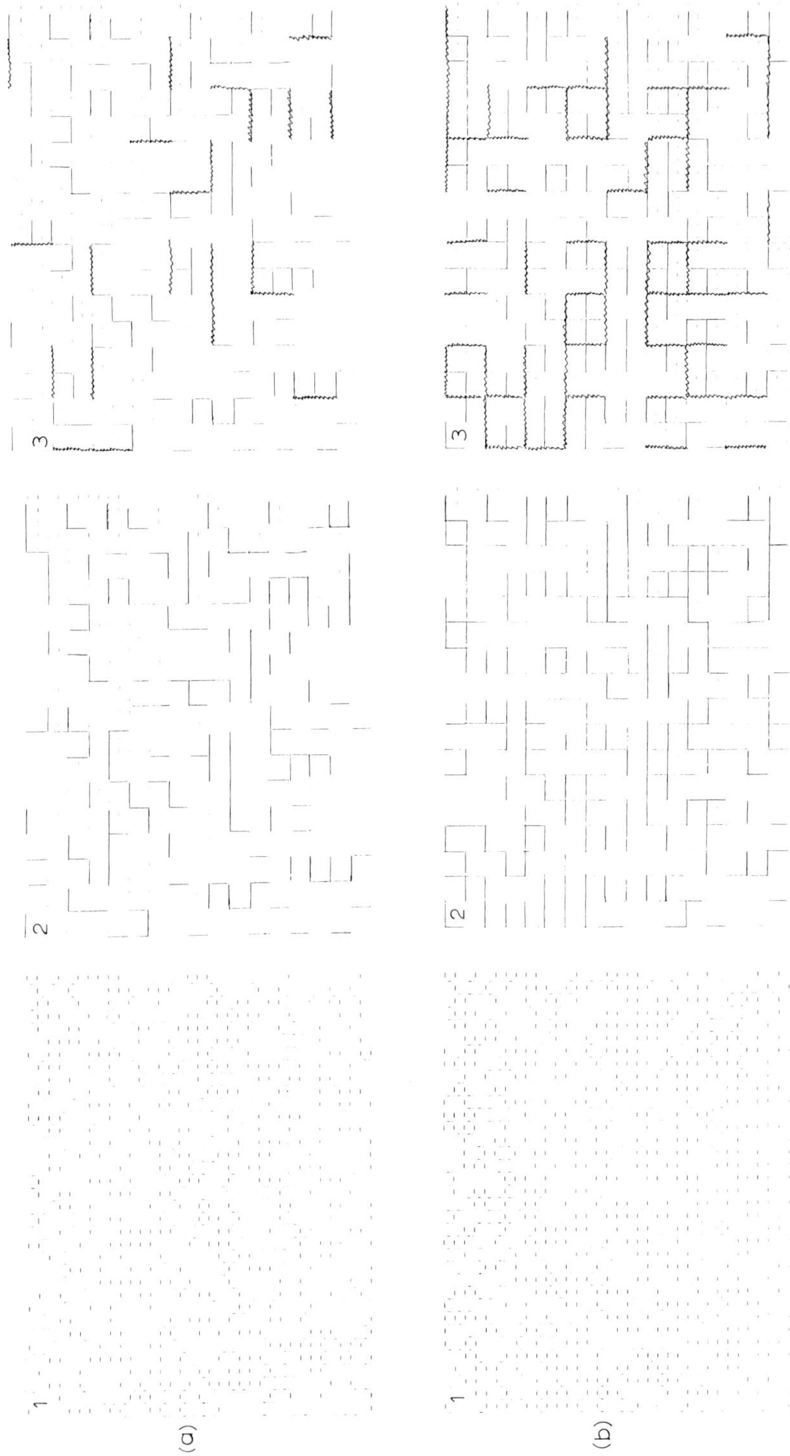

(a)
1
2
3
(b)
1
2
3

Fig. 1.19. Random percolation networks and their scaling according to the rule (1.3.2.3). a1: $p = 0.4$, a2: first scaling of a1, a3: second scaling of a1; b1: $p = 0.51$, b2: first scaling of b1, b3: second scaling of b1; c1: $p = 0.65$, c2: first scaling of c1, c3: second scaling of c1.

lines are irrelevant in the definition of the bond BC of the scaled network. The vertical bond AB is similarly defined. This operation (called "coarse graining" in Renormalization Group Theory, cf. Wilson and Kogut, 1974) can be repeated several times.

Figures 1.19a1–c3 show three randomly generated percolation networks (a1, b1 and c1) corresponding to the probabilities $p = 0.4$, $p = 0.51$ and $p = 0.65$, respectively, and their first two scalings in accordance with rule (1.3.2.3). For $p = 0.4$ the percolation clusters become smaller after repeated applications of (1.3.2.3): the lattice becomes ever more disconnected. For $p = 0.65$ the size of the clusters increases as we iterate scaling: after the second scaling lattice c3 is almost without gaps. On the other hand, when $p = 0.51$ scaling does not cause any difference: the lattices b1, b2 and b3 show the same statistical properties, that is, *the lattice b1 is self-similar.*

Intuitively, we feel that for the lattice a1 ($p = 0.4$) repeated scaling would have eventually lead to an empty lattice with no bonds at all ($p = 0$); repeated scaling of lattice c1 (with $p = 0.65$) would have converged to the regular lattice ($p = 1$); while the lattice corresponding to $p = 0.51$ would have retained its structure after an arbitrary large number of scalings.

To prove this mathematically, let us express the probability p' of the event that there is a bond between nodes B and C of the scaled lattice. Denoting $1 - p$ by q, and observing that the dashed lines do not enter into the calculation of probabilities, we find from (1.3.2.3b):

$$p' = 2p^2q^3 + 8p^3q^2 + 5p^4q + p^5 \qquad (q = 1 - p) \tag{1.3.2.4}$$

The graph of p' vs. p is shown in Fig 1.20. For $p < 0.5$ the scaling operation decreases the probability of the presence of bonds: $p' < p$.

For $p > 0.5$ we have $p' > p$, that is the probability of bonds increases.

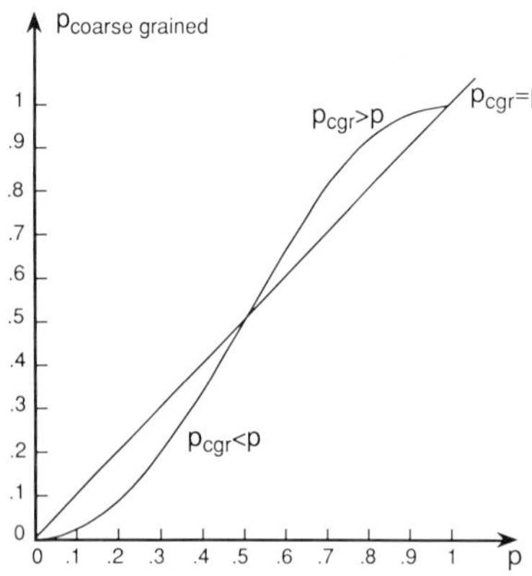

Fig. 1.20. The $p'(p)$ function.

For $p = 0.5$ the percolation structure is statistically invariant to scaling: $p' = p$. The value $p_c = 0.5$ satisfies the equation:

$$p = 2p^2q^3 + 8p^3q^2 + 5p^4q + p^5 \qquad (q = 1 - p) \tag{1.3.2.5}$$

and yields an *approximation* of the critical percolation probability p_c. The *exact* critical probability for the square lattice happens to be also 1/2 (Sykes and Essam, 1964), but this should be considered as fortuitous. In general we can not expect such an accuracy from any scaling technique. A less fortunate scaling rule, like the $(2)^{1/2}$-times scaling:

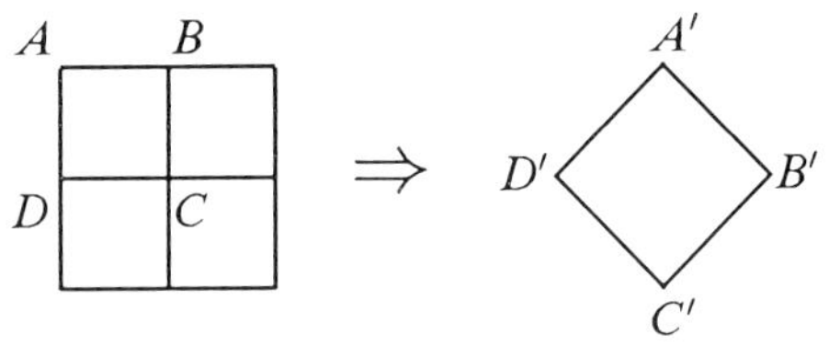

used by Young and Stinchcombe (1975) would yield $p_c = 0.618$.

As for $p = p_c$ there will be arbitrary large clusters, the correlation length ξ diverges for $p \to p_c$, where ξ is defined as the average root mean square distance between nodes that belong to the same cluster. To find the functional form of the dependence of ξ on $(p - p_c)$ for $p \to p_c$, we apply the usual arguments of renormalization group theory (Wilson and Kogut, 1974; Feder, 1988).

Denote the scaling factor in rule (1.3.2.3a) by k, and develop the function $p'(p)$ in equation (1.3.2.4) into Taylor series around $p = p_c$:

$$p'(p_c) = p_c + \lambda(p - p_c) + \text{higher order terms} \tag{1.3.2.6}$$

where:

$$k = 2 \qquad p_c = \frac{1}{2} \qquad \lambda = \frac{26}{16} = 1.625 \tag{1.3.2.7}$$

As the percolation network and its scaled version are similar around p_c (apart from the scale factor k):

$$\xi' = \xi(p') = \frac{\xi(p)}{k} \tag{1.3.2.8}$$

Substituting (1.3.2.6) into (1.3.2.8):

$$\xi[p_c + \lambda(p - p_c)] = \frac{1}{k}\xi(p) = \frac{1}{k}\xi[p_c + (p - p_c)] \tag{1.3.2.9}$$

Introducing the new functions:

$$f(x) = \xi[p_c + x] \qquad g(\lambda) = k^{-1} \tag{1.3.2.10}$$

by equation (1.3.2.9) we find that $f(x)$ satisfies Pexider's functional equation:

$$f(\lambda x) = g(\lambda) f(x) \tag{1.3.2.11}$$

whose only solution is:

$$f(x) = ax^{\nu} \qquad g(\lambda) = \lambda^{\nu} \tag{1.3.2.12}$$

where a and ν are arbitrary real numbers. (Pexider's equations and other functional equations characterizing homogeneity and self-similarity will be discussed in the Mathematical Appendix to Chapter 1.)

Using equations (1.3.2.8)–(1.3.2.12) we obtain:

$$\xi(p - p_c) \propto |p - p_c|^{-\nu} \qquad \text{with } \lambda^{-\nu} = \frac{1}{k} \tag{1.3.2.13}$$

that is, by (1.3.2.7) the value of ν becomes:

$$\nu = -\frac{\log k}{\log \lambda} = -\frac{\log 2}{\log 1.625} = 1.428 \tag{1.3.2.14}$$

which is a fair approximation of the theoretical value $\nu = 4/3$ (Den Nijs, 1979).

The percolation between two nodes of a cluster, a distance ξ apart, takes place along tortuous zig-zag paths. Near the percolation threshold the length $L(\xi)$ of a typical zig-zag path spanning a distance ξ also grows as a power of $(p - p_c)$:

$$L(\xi) \propto (p - p_c)^{-\beta} \qquad \text{for } p \to p_c \tag{1.3.2.15}$$

or, using equation (1.3.2.2):

$$L(\xi) \propto \xi^{\alpha} \qquad \text{for } p \to p_c \tag{1.3.2.16}$$

For three-dimensional lattices $\beta = \nu\alpha = 0.83\alpha$. As the correlation length is the natural length scale in percolation problems, we shall follow Ritzenberger and Cohen (1984) and define the tortuosity τ of the percolation path as:

$$\tau = \frac{L(\xi)}{\xi} \propto \xi^{\alpha - 1} \tag{1.3.2.17}$$

The exponents describing the divergences of the length and tortuosity, respectively, of the percolating paths are compiled in Table 1.2 for different percolation models. Figure 1.21 (Gadenne et al., 1989) is the most beautiful published example for anomalous tortuosity, it shows the dc conductive path in a thin gold film near the percolation threshold.

The exponent α in equation (1.3.2.17) has a simple physical meaning: for distances X smaller than ξ, α is the fractal dimension of the fluid paths between two nodes a distance X apart [see Ritzenberger and Cohen (1984) or recall our definition (1.2.7)].

TABLE 1.2

Percolation exponents

Definition of the path	Exponent [a]			Reference [b]	Note [c]
	α	β	γ		
Correlation length	1	0.83	0	a	1
Minimum path	1.3	1.08	0.25	b, c	1
Conductive path	1.35	1.12	0.29	d	2
Self-avoiding random walk on uncut bonds	1.7	1.41	0.58	e	1
Brownian motion in 3-D	2	1.66	0.83	f	
Brownian walk on a d_f-dimensional fractal	$(3/2)\,d_f$	0.83	$0.83\,(\alpha - 1)$	c	3
Brownian walk on a 3-D fractal	4.5	3.74	2.91		4

[a] The percolation exponents describe the divergence of the path length L and of its tortuosity τ: $L \approx \xi^{\alpha}$; $L \approx (p - p_c)^{-\beta}$; $\tau \approx (p - p_c)^{-\gamma}$.
[b] a: Fisch and Harris (1978); b: Ritzenberger and Cohen (1984); c: Stanley (1986); d: Lubensky (1977); e: Le Guillou and Zinn-Justin (1977); f: Mosolov and Dinaryev (1987); g: Wong (1988).
[c] 1: 3-D percolation; 2: 3-D percolation–conduction; 3: $\alpha = (3/2)\,d_f$ is called the "Alexander-Orbach conjecture" (reference c); 4: the pore space of certain sandstones forms an almost three-dimensional fractal (reference g).

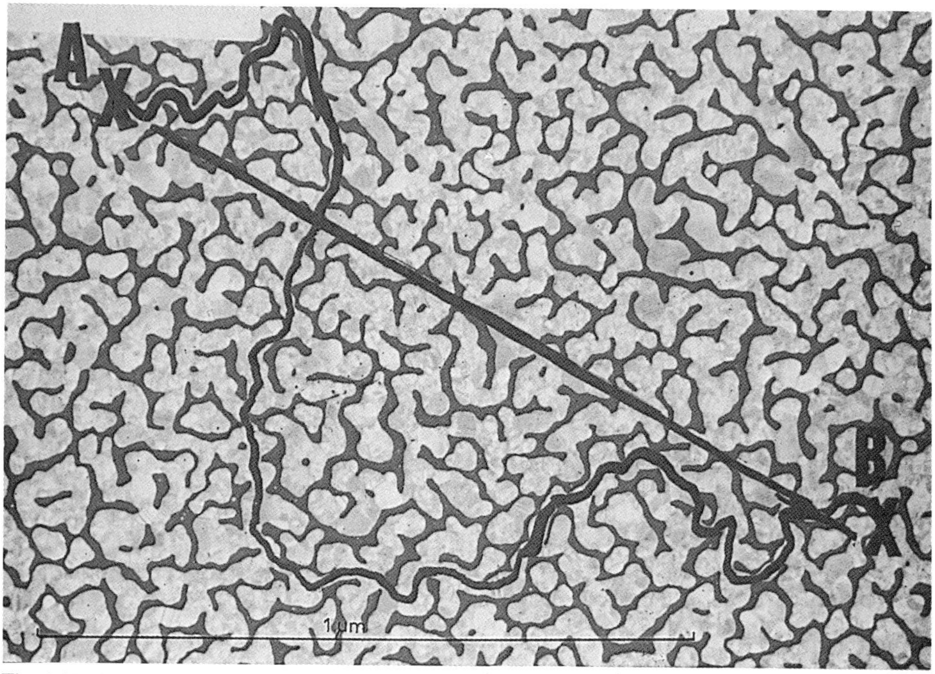

Fig. 1.21. Conductive path in a thin gold film near the percolation threshold (from Gadenne et al., 1989).

1.3.3 *Percolation models of rock permeability*

The pore geometry of a sedimentary rock can be converted to a discrete lattice model by letting the pores correspond to nodes and the throats to bonds (Figure 1.22, from Korvin, 1992). The coordination number of the pore system is defined as the average number of throats which connect each pore; it is a measure of connectivity of the network of pores and can be experimentally determined by serial sectioning (Dullien, 1979; Wardlaw and McKellar, 1981).

Recent theoretical work in continuum percolation (Elam et al., 1984; Halperin et al., 1985) has proved the general applicability of discrete lattice models in simulating continuous problems, though the percolation transport exponents for electric conductivity and hydraulic permeability have been found larger than their discrete lattice counterparts.

Early applications of percolation theory have centred around qualitative problems of oil recovery (Dullien, 1979) and mercury porosimetry (Wardlaw and McKellar, 1981). Recent quantitative results are reviewed by Thompson et al. (1987) and Wong (1988). In 1985 Halperin and his group at the Harvard University introduced a "Swiss cheese" permeability model in which the holes play the role of sand grains

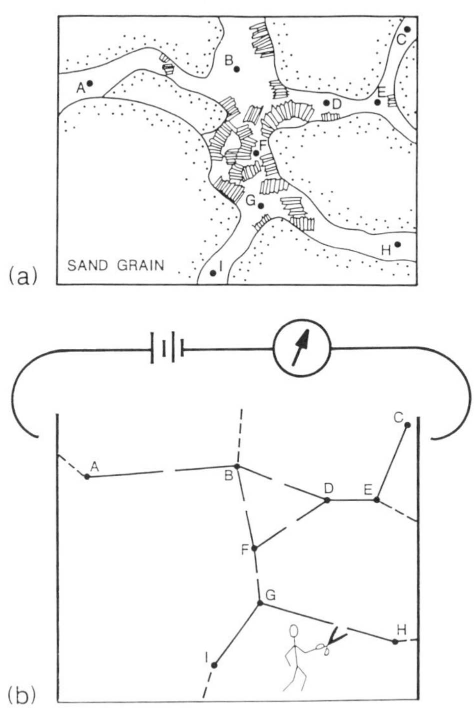

Fig. 1.22. Fluid transfer through a kaolinite-bearing sandstone (a) and the corresponding percolation model (b). The symbolic "current" can be an arbitrary transfer process. (From Korvin, 1992.)

and the cheese is the fluid content. They found that if one makes more and more holes there is a critical fraction of cheese $\Phi_c = 0.03$–0.04 at which electric conductivity vanishes as $(\Phi - \Phi_c)^t$ and hydraulic permeability vanishes as $(\Phi - \Phi_c)^e$, with $t = 2.4$ and $e = 4.4$.

In an important paper Katz and Thompson (1986) of Exxon Production Research, Houston, assumed that only throats wider than a given characteristic length l_c can significantly contribute to permeability and then applied percolation arguments to derive permeability in the form:

$$k = \beta \, \Phi \, l_{max}^2 \, [p(l_{max}) - p(l_c)]^t \tag{1.3.3.1}$$

with $\beta = 1/32$; for t they took the percolation conductivity exponent $t = 1.9$ (Fisch and Harris, 1978). In equation (1.3.3.1) $p(l)$ means the probability that a throat is wider than l; l_c is a critical width such that the throats wider than l_c still form a connected net across the rock; l_{max} is an other size parameter. The critical width l_c can be experimentally determined using mercury intrusion.

Recently (Korvin, 1992) I have developed another percolation theoretical model to describe the permeability of the kaolinite-bearing sandstones from oil reservoirs of the Eromanga Basin, South Australia. I could prove that percolation transition occurs at a certain critical kaolinite volume-fraction and that at the percolation threshold the tortuosity τ of the flow path [figuring in equations (1.3.1.1)–(1.3.1.3)] has a power-function divergence of the form (1.3.2.17).

In an attempt to explain the experimental data (viz. the very low or zero permeabilities above a certain clay content, reported by Gravestock and Alexander, 1986, 1989) I transformed the continuous Darcy flow to a lattice percolation problem (Figure 1.22).

Pores of the rock correspond to the nodes of the discrete lattice, throats correspond to the bonds, a given bond is considered as "cut" or "uncut" depending on whether the corresponding throat is blocked by the kaolinite particles, independently of the actual radius of the throat. The coordination number Z of the lattice depends on the original packing of the quartz grains and on the subsequent compaction and diagenesis history (Taylor, 1950) — observe the missing bond between nodes A and F in Figure 1.22b because of the concavo-convex contact between the adjacent grains. Generally, Z is less than 6 for sandstones. Using the empirical rule (1.3.2.1) and assuming a three-dimensional lattice, the bond percolation threshold becomes:

$$p_c = \frac{1.5}{Z} \tag{1.3.3.2}$$

Finally, after tedious calculations (full details will be found in Korvin, 1992) I found an expression for the tortuosity near the percolation threshold, substituted this into equation (1.3.1.2) and arrived at the final result:

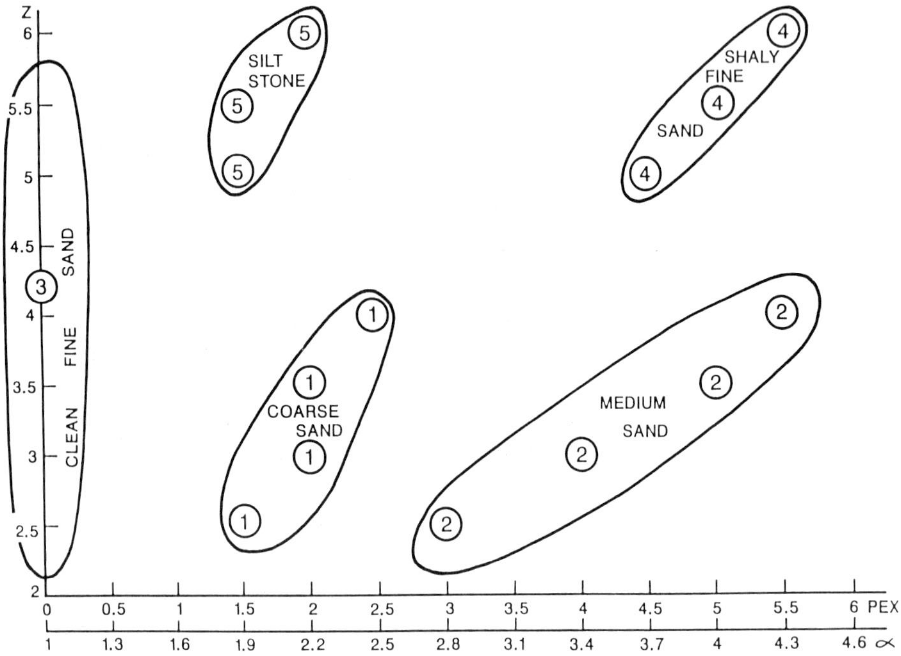

Fig. 1.23. Optimal percolation parameters Z and PEX for the different lithologies. [Z = coordination number; PEX = percolation exponent in equation (1.3.3.3); α = fractal dimension of the tortuous fluid path.] (From Korvin, 1992.)

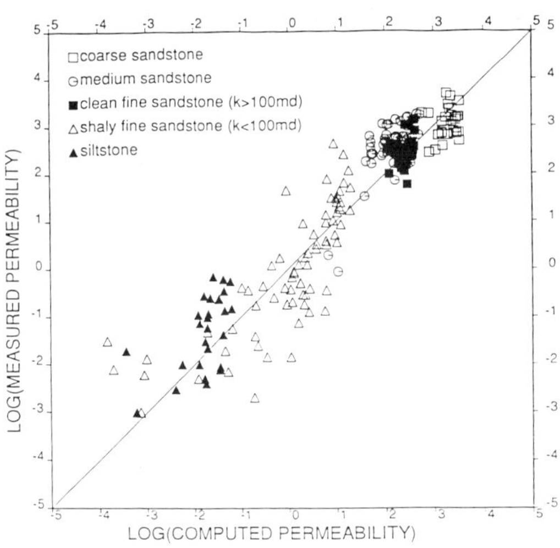

Fig. 1.24. Crossplot of measured and computed permeabilities (from Korvin, 1992).

$$
\begin{aligned}
k &= \frac{R_{\mathrm{HYD}}^2}{b\tau_0^2}\,\Phi 10^9\,\frac{(p-p_{\mathrm{c}})^{\mathrm{PEX}}}{(1-p_{\mathrm{c}})^{\mathrm{PEX}}} && \text{if } p \geq p_{\mathrm{c}} \\
&= 0 && \text{if } p < p_{\mathrm{c}}
\end{aligned}
\tag{1.3.3.3}
$$

with $b = 2$ and $\tau_0 = 4$; R_{HYD} and p can be expressed in terms of the grain size, porosity and kaolinite content. There are only two adjustable parameters, the coordination number Z, and the percolation exponent PEX. The latter expresses the divergence of the tortuosity at the critical kaolinite content, and it is closely related to the *fractal dimension of the fluid paths*.

The parameters Z and PEX had been optimized as to minimise the deviation between measured and computed permeabilities. The optimal values are shown in Figure 1.23 for the different lithologies. Application of equation (1.3.3.3) with these optimal parameters has resulted in a reasonable agreement between computed and measured permeabilities over more than seven orders of magnitudes (Figure 1.24).

History, alas, also proceeds along circuitous lines. In a clever Victorian advertisement the artist discovered a meaningful message in Lord Robert's route (Figure 1.25). More recently, in 1989 — before the demolishing of the Wall — tens of thousands of East Berliners selected extremely tortuous paths via Leipzig–Prague–Brno–Budapest–Wien–Passau–Frankfurt to reach West Berlin. Which is a further proof that the divergence of tortuosity is a precursor of phase transition.

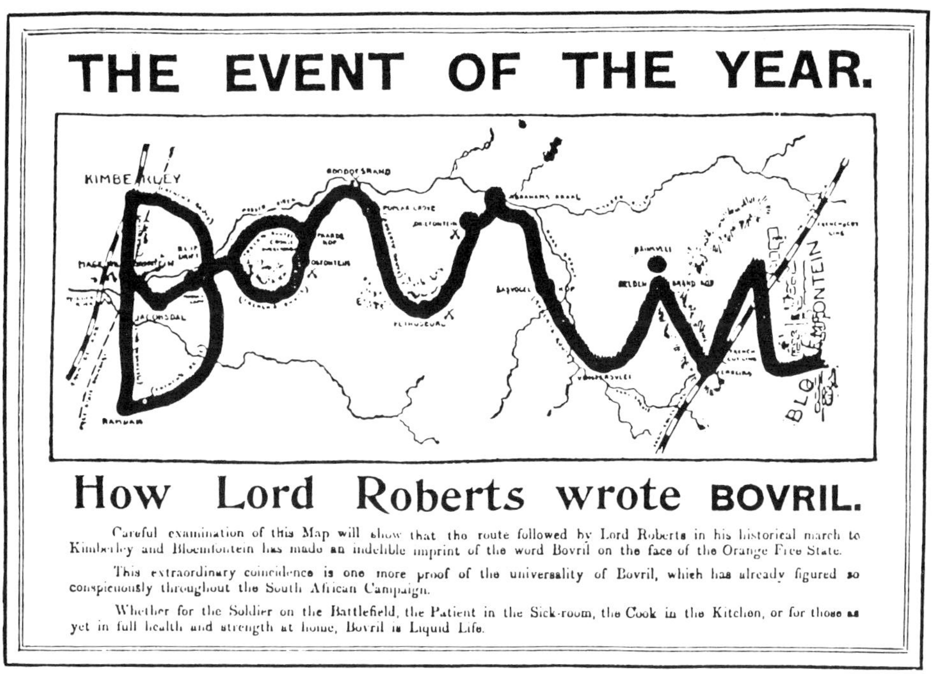

Fig. 1.25. Victorian advertisement (from D. and G. Hindley, 1972).

1.4 Deadly quarrels and coastlines

How long is our coastline? Measuring the length of any given stretch of coastline is not as easy as you may think. If you trace the outline of the continent and measure the line, the coast seems to be about 15,000 km long. But if you measure all the twists and turns in the coast created by headlands, inlets and beaches the effective length is much greater, perhaps as much as 50,000 km. And that's more than the circumference of the earth.

— MARK WARREN: *Atlas of Australian Surfing* (1988)

1.4.1 *The coastline of Britain*

"Deadly quarrels" in this chapter will have nothing to do with the heated (though not deadly) discussions whether multifractals are fractal. The term comes from the title "Statistics of Deadly Quarrels", which was Lewis F. Richardson's life work on the mathematical theory of wars (posthumously published in 1961).

In the late 1930's the British physicist and meteorologist, L.F. Richardson decided to devote his time to research on the causes of war. "Someone told me", wrote Richardson in his Diary, "that Helmholtz had been a medical doctor before he became a physicist. It thereupon occurred to me that Helmholtz had eaten the meal of life in the wrong order, and that I would like to spend the first half of my life under the strict discipline of physics, and afterwards to apply that training to researches on living things". He retired from the Paisley Technical College and School of Art, where he had been Principal, to do research on wars and eddy diffusion.

In 1939 an American journal received for publication a paper from Richardson. In his letter of transmittal Richardson urged that the paper be published immediately because it might avert an impending war. (The editors of the journal not only did not rush the paper but rejected it — see the Preface to Richardson, 1960).

The famous "coastline of Britain" paradox appeared in an appendix to the "Statistics of Deadly Quarrels" (1961). Richardson applied statistical methods to decide whether two nations with a common boundary are more likely to go to war with one another than two nations without a common boundary, and if yes, whether the incidence frequencies of "deadly quarrels" correlate with the length of the political boundary.

During the practical execution of this plan he encountered an unexpected difficulty: there were disagreements in the official statements about the length of the common land-frontier (see Table 1.3).

To resolve this inconsistency, Richardson set out to measure himself the length of the frontier between Spain and Portugal, then the German land-frontier..., then different coastlines, among others the west coast of Britain from "Land's End to Duncansby Head... as an example of a coast that looks more irregular than most other coasts in an atlas of the world". His method has been to walk a pair of dividers along the map of the frontiers so as to count the number of equal sides of a polygon

TABLE 1.3

Land-frontier between:	Kilometres as stated by:	
	the former country	the latter country
Spain and Portugal	987	1214
Netherlands and Belgium	380	449.5
U.S.S.R. and Finland	1590	1566
U.S.S.R. and Roumania	742	812
U.S.S.R. and Latvia	269	351
Estonia and Latvia	356	375
Yugoslavia and Greece	262.1	236.8

All data from the Armaments Year-Book 1938 as quoted by Richardson.

the corners of which lie on the frontier. And — as in the case of Steinhaus's Vistula River, see Section 1.2 — Richardson found that the length L of the frontiers and coastlines varies inversely with a power of the side l of the polygon, for the coast of Britain, for example:

$$L(l) \propto l^{-0.25} \tag{1.4.1.1}$$

or, in the general case:

$$L(l) \propto l^{-\alpha} \tag{1.4.1.2}$$

where α is "expected to have some positive correlation with one's immediate visual perception of the irregularity of the frontier" (Figure 1.26). Interestingly, Richardson has not tried to explain these findings, though he has had at his disposal all the necessary tools from his studies on the self-similarity of turbulence. (Recall his famous doggerel: "Big whorls have little whorls,/Which feed on their velocity;/And little whorls have lesser whorls,/And so on to viscosity.")

Richardson's and Steinhaus's work has been followed up in 1967 by the French-American mathematician Benoit B. Mandelbrot in his history making paper "How long is the coast of Britain? Statistical selfsimilarity and fractional dimension". He has shown that the divergence of the length $L(l)$ for $l \rightarrow 0$ is due to the fact that the curve has a non-integer *fractal* dimension, proved that the fractal dimensionality is intimately connected with the *self-similarity* of the curve, and identified the exponent "$-\alpha$" in Richardson's equation (1.4.1.2) with the exponent $1 - D$ of equation (1.2.11) expressing the length–resolution relation for a D-dimensional fractal curve. Thus, Richardson's data yield $D = 1.25$ for the west coast of Britain, $D = 1.15$ for the land frontier of Germany (in 1900), $D = 1.14$ for the land frontier between Spain and Portugal, $D = 1.13$ for the Australian coast. More recently Nakano (1983, 1984) reported fractal dimensions between 1.12–1.39 for ria shorelines in Japan, and Feder (1988) has found that the southern coast of Norway is 1.52-dimensional. The highest fractal dimension reported has come from Queensland, Australia: Bradbury

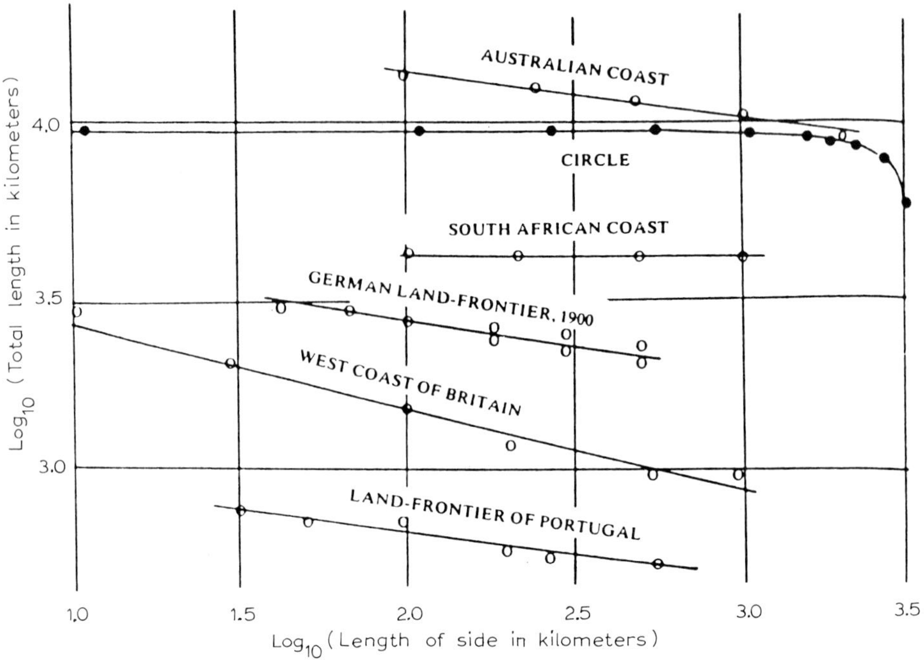

Fig. 1.26. Richardson's data concerning the measurements of geographical curves with increasing precision. The circle had the same plane area on the map as the Australian continent. (From Richardson, 1961.)

and Reichelt (1983) measured the contour of a reef slope at 2 m water depth with specially constructed dividers and proved that it was fractal over three orders of magnitude, with $D = 1.9$–2.

To show how easy to construct curves with an arbitrary fractal dimension $1 \leq D \leq 2$, Mandelbrot (1967) also described some variants of Von Koch's (1906) continuous, non-differentiable, non-rectifiable curves, hitherto only known to Pure Math. students as tricky counterexamples and major tools of torture.

A simple case of the Von Koch construction is shown in Figure 1.27. We start out from an *initiator*, which is a simple polygon or broken line consisting of straight line segments of equal sides. (It is a straight line segment in Figure 1.27.) Also, we prescribe a *generator* (the second broken line in Figure 1.27). The Von Koch construction proceeds recursively: The initiator is the 0-th generation of the curve; in the 1st generation we replace each straight line section of the initiator by a scaled-down version of the generator;..., in the $(k + 1)$-st generation we replace all segments of the k-th generation by scaled down versions of the generator. The Von Koch curve is the limit of this procedure for $k \to \infty$. If the length of the initiator line segment in Figure 1.27 is $L(1) = 1$, the first generation will consist of 4 smaller segments of length 1/3, that is $L(1/3) = 4/3$ where

Fig. 1.27. Construction of the Von Koch curves (from Feder, 1988).

$L(x)$ denotes the length measured with resolution x. By mathematical induction, the k-th generation will consist of 4^k segments of length $1/3^k$; that is $L(1/3^k) = (4/3)^k$.

Putting $\varepsilon = 1/3^k$, we have $k = -\log\varepsilon/\log 3$ and the length becomes:

$$L(\varepsilon) = \varepsilon^{1-D} \qquad \text{with } D = \frac{\log 4}{\log 3} = 1.2628 \tag{1.4.1.3}$$

In the more general case, when any generation consists of N parts deducible from the whole by a similarity of ratio r, the fractal dimension becomes:

$$D = -\frac{\log N}{\log r} \tag{1.4.1.4}$$

(Mandelbrot, 1967).

In a second paper devoted to the subject, Mandelbrot (1975) proposed a random geomorphological process responsible for the fractal nature of coastlines. Rather than dealing directly with coastlines, he considered them as solutions of the equation $B(x,y) = 0$, where $B(x,y)$ describes the topography over a flat Earth with coordinates x and y. If one could prove that $B(x,y)$ is a *fractal surface* with dimension D ($2 \leq D \leq 3$), it would be easy to deduce that the contour consisting of the points (x,y) for which $B(x,y) = 0$ is also fractal, with dimension $D - 1$. Mandelbrot proposed the following model for a random relief of the Earth: we start out from a flat Earth of zero altitude, break it along a succession of randomly placed and randomly oriented straight faults, and in each case displace the two sides vertically by a random amount to form a cliff. The curvature of the Earth and the roles of isostasy and erosion (Takayasu and Inaoka, 1992) have been neglected.

More exactly, Mandelbrot has made the following assumptions: (a) the directions of the fault lines are isotropic, uniformly distributed between 0°–180°; (b) their spacing follows a Poisson distribution with density λ; and (c) the cliff heights are normalised so as to decrease as $\lambda^{1/2}$ with increasing λ's (that is the short-wavelength features of the topography have smaller amplitude). Then, by the central limit theorem, for $\lambda \to \infty$ the surface will tend to a *Brownian surface* $B(x,y)$ (Paul Lévy, 1965) which is a generalization of the Brownian curve.

The characteristic property of a Brownian surface is that for any two points $P = (x,y)$, $Q = (x',y')$ the deviation $B(P) - B(Q)$ is a zero mean Gaussian random variable of variance proportional to $\overline{PQ}$:

$$\left\langle [B(P) - B(Q)]^2 \right\rangle \propto \overline{PQ}$$

Let us now see a plausibility proof that the dimension of the Brownian surface is $D = 5/2$, and that its contour-lines $B(x,y) = \text{const}$ are 1.5-dimensional.

Consider a part of the surface $B(x,y)$ inside the unit cube ($0 \leq x,y,z \leq 1$), and divide the unit cube to $1/(\Delta x)^3$ small cubes of side Δx. Estimate the number of small cubes which have points common with the surface B. The unit square ($0 \leq x,y \leq 1$) is covered by $1/(\Delta x)^2$ small squares of side Δx. As the greatest distance between any two points of a small square is $2^{1/2}\Delta x$, the expected variation of $B(x,y)$ over the small square will be:

$$B_{\max} - B_{\min} \propto \left\langle (\Delta B)^2 \right\rangle^{1/2} \propto (\Delta x)^{1/2}$$

This total variation can be covered by $\propto (\Delta x)^{1/2}/(\Delta x) = (\Delta x)^{-1/2}$ small cubes. As the unit square consists of $1/(\Delta x)^2$ small squares, the covering of the whole surface requires $(\Delta x)^{-2}(\Delta x)^{-1/2} = (\Delta x)^{-5/2}$ small cubes which shows that B is

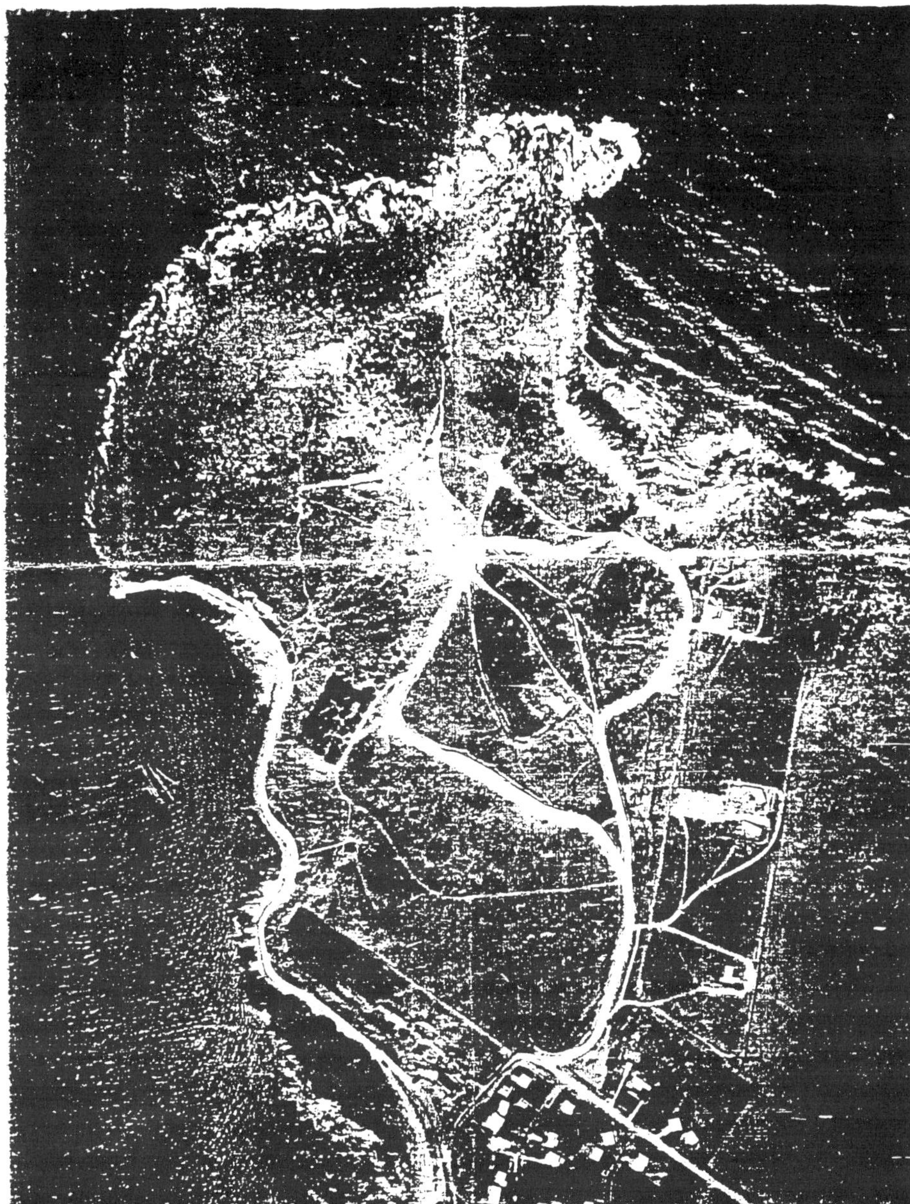

Fig. 1.28. Bifractal coastline (Rosetta Head, Victor Harbour, South Australia).

5/2-dimensional. To find the fractal dimension of a contour-line $B(x,y)$ = constant, observe that the unit cube consists of $1/(\Delta x)$ layers, that is in an average layer the contour-line of B occupies $\approx (\Delta x)^{5/2}/(\Delta x) = (\Delta x)^{-3/2}$ cubes, implying that the contour-line is 3/2-dimensional.

While the computer realizations of Mandelbrot's Brownian surface model have lead to the much publicized real-looking landscapes that "never were", the fractal dimension $D = 1.5$ that has resulted from the Brownian surface model is much higher than the experimentally found dimensions of coastlines. (The single exception is Feder's (1988) Norwegian coastline with $D \approx 1.5$. The almost two-dimensional reef — as pointed out by Bradbury and Reichelt (1983) themselves — should be considered as a living ecosystem maximising its contact with the water, hence its large fractal dimension.) To solve this difficulty, Mandelbrot (1975) has generalized the concept of Brownian surface, and defined the *fractional Brownian surface* $B_H(X,Y)$ by the property that the variance satisfies:

$$\left\langle [B(P) - B(Q)]^2 \right\rangle \propto \overline{PQ}^{2H} \quad \text{with } 0 < H < \frac{1}{2} \text{ or } \frac{1}{2} < H < 1 \tag{1.4.1.5}$$

(For $H = 1/2$, B_H is the ordinary Brownian surface.) Mandelbrot has shown that H and the dimension D of the contour-lines are related by $H = 2 - D$, that is it suffices to assume $H = 0.7$ in order to get a "coastline" which is 1.3-dimensional. However, more recent studies of Kent and Wong (1982) on the shoreline of lakes, and Nakano's (1983, 1984) analysis of ria coastlines in Japan have proved that in certain cases it is impossible to describe the divergence of $L(\varepsilon)$ as $\varepsilon \to 0$ by assuming a single fractal dimension: *A different dimension belongs to the zero contour-line of the original surface, and an other one to the coastline that has been modified (smoothed) by subsequent erosional processes* (Figure 1.28). Nakano (1983) could also prove, for a ria coast in northeastern Japan, that the cross-over point between the two fractal processes is changing with geological time. Before analysing Nakano's (1983, 1984) findings, we shall develop a fractal model of coastal erosion in order to explain the observed low fractal dimensions of coastlines.

1.4.2 *A fractal model of coastal erosion*

> Every valley shall be filled, and every mountain and hill shall be brought low; and the crooked shall be made straight, and the rough ways shall be made smooth. — LUKE 3, 5

Remarkable patterns of almost evenly spaced crescentic embayments separated by seaward projections are common features of most shorelines. Among the smallest of these forms is the *beach cusplet*, measuring about 1.5 metres between tips of the crescent. These are associated with swash-zones and appear and disappear with the turn of the tide. Beach cusplets belong to a complete hierarchy of crescentic shoreline forms (Figure 1.29), ranging from the cusplets to major

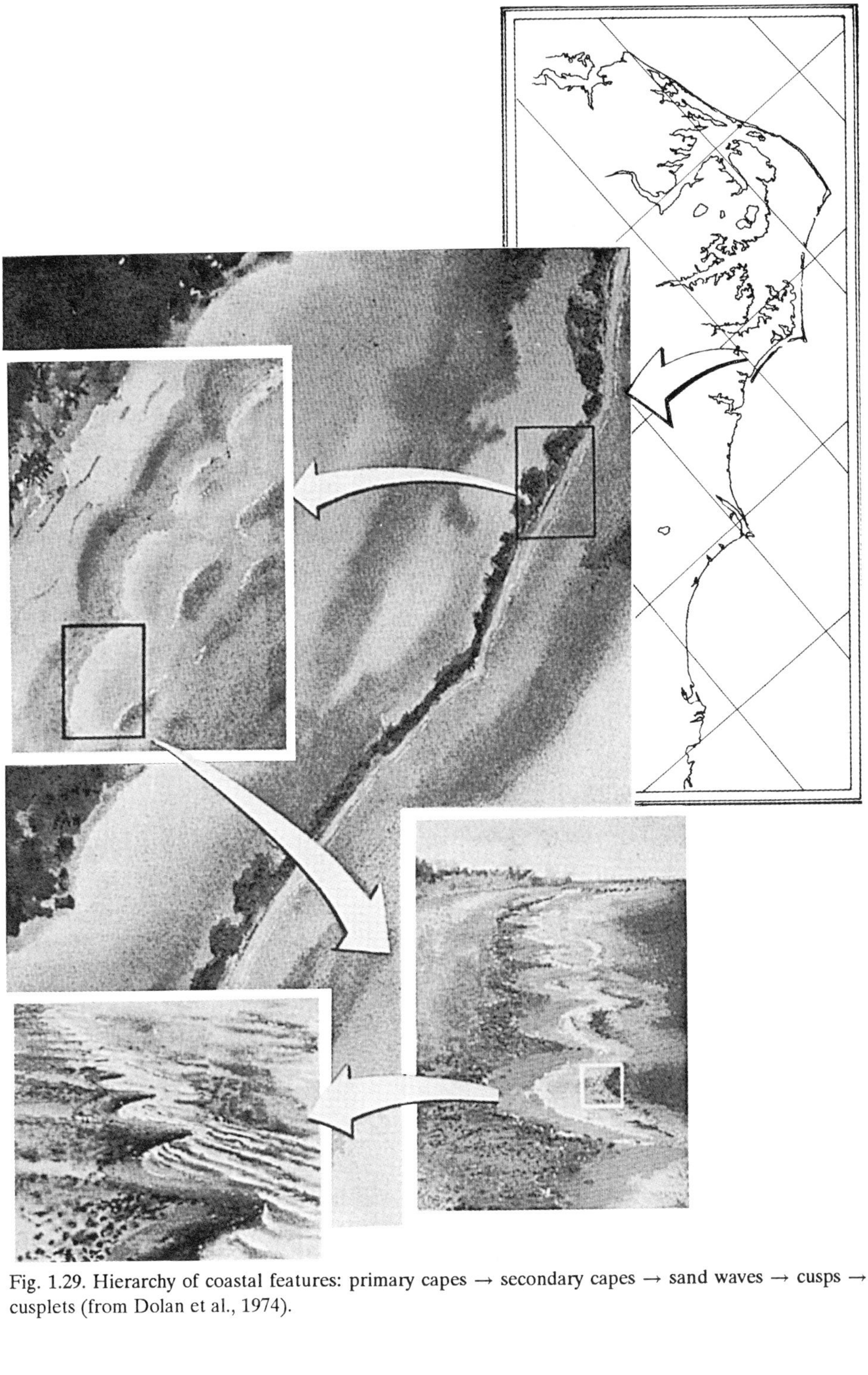

Fig. 1.29. Hierarchy of coastal features: primary capes → secondary capes → sand waves → cusps → cusplets (from Dolan et al., 1974).

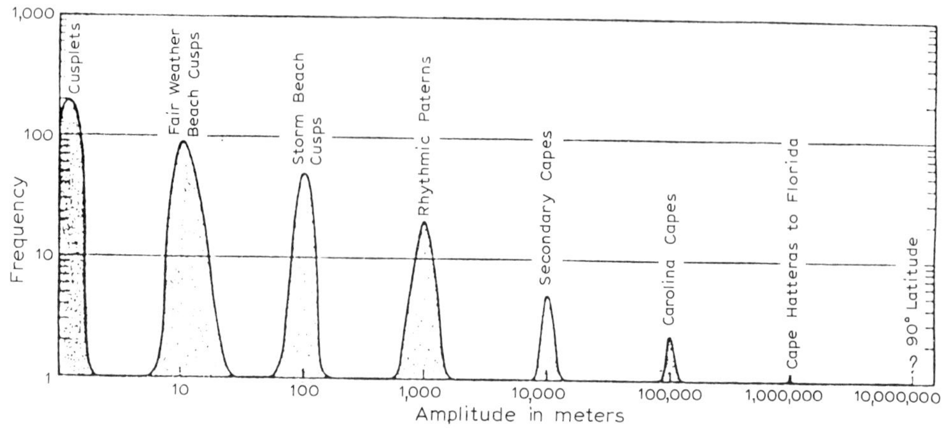

Fig. 1.30. Frequency vs. amplitude plot of crescentic forms observed along the middle of the Atlantic coast of the United States (from Dolan and Ferm, 1968).

cuspate indentations of the coastline of hundred kilometers of amplitude. The hierarchy of these features has been first recognised by Dolan and Ferm (1968; see also Dolan et al., 1974 and the recent review of Hallet, 1987), in the sequence of crescentic landforms along the Atlantic Coast of the United States. In the next stage of the hierarchy we find the *beach cusps* which range from about 8 to 25 m. Still larger crescentic forms are the shoreline *rhythms* or *sand waves* (100 m–3 km) and the large cuspate sedimentary *capes*, subdivided to *primary capes* ($\gg$200 km) and *secondary capes* ($\approx$100 km). Dolan and Ferm (1968) plotted the size–frequency distribution of 850 cuspate features along the central Atlantic coast of the U.S.A. (Figure 1.30) and they have found that the modes of the hierarchical groups are logarithmically spaced. The frequency is linearly decreasing on the double-logarithmic plot indicating the self-similar distribution of these features.

Contemporaneously with Dolan and Ferm, R. McLean (1967) has described the form of 88 beaches on the east coast of New Zealand in terms of their radius of curvature, of their orientation, and the angle subtended by the radii from the ends of the beach. He has drawn the following conclusions:

– the outline of the beaches fitted a circular arc;

– of the beaches 45 had values of the radius less than 1.6 km; 13 had values between 1.8–3.2 km; 12 had values between 3.4–8 km; 7 had values between 8.15–14.4 km; 9 beaches had radii of curvature >14.4 km;

– the majority of beaches were facing the direction from which the dominant swells come;

– the subtended angles ranged between 20–110° with a mean of 64°, rather than the theoretical 14.3° = 0.25 radians corresponding to the equilibrium beach form (cf. Hoyle and King, 1958).

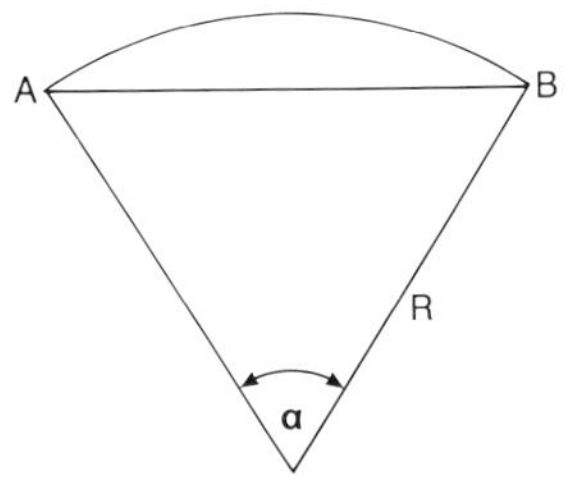

Fig. 1.31. To the derivation of equations (1.4.2.1) and (1.4.2.2).

Based on the findings of Dolan and Ferm(1968), Dolan et al. (1974) and McLean (1967), we can build up a simple recursive model of coastal erosion.

The basic step of the process consists of the replacement of a line segment $\overline{AB}$ by the circular arc $\overset{\frown}{AB}$ (Figure 1.31). As $\overline{AB} = 2R\sin(\alpha/2)$ and $\overset{\frown}{AB} = R\alpha$ (if α is in radians), we have:

$$\frac{\overset{\frown}{AB}}{\overline{AB}} = \frac{(\alpha/2)}{\sin(\alpha/2)} > 1 \qquad (1.4.2.1)$$

Introducing the notation:

$$\frac{\overset{\frown}{AB}}{\overline{AB}} = 1 + \delta \qquad (1.4.2.2)$$

we see that a single replacement of a line segment $\overline{AB}$ by the circular arc $\overset{\frown}{AB}$ increases the total length by an amount $\Delta L = \delta \overline{AB}$ where δ only depends on the suspended angle α. Let us now consider a decreasing sequence of characteristic sizes $r_k \to 0$, and assume that r_k represents the size (i.e. the distance between the tips) of the crescentic waveforms in the k-th iteration of the process. The sequence r_k is not arbitrary; we assume that the hierarchic set of the crescentic forms is self-similar. Thus, if we consider any two numbers of the sequence, say r_i and r_k $(i < k)$, the same magnification that transforms r_i to r_k should transform r_{i+1} to r_{k+1}; wherefrom $r_i/r_k = r_{i+1}/r_{k+1}$, that is:

$$r_k \propto \frac{1}{R^k} \qquad \text{for some } R > 1 \qquad (1.4.2.3)$$

Suppose that we are also given an increasing sequence of real numbers $0 < \lambda_1 < \lambda_2 < \ldots$ to be specified later. After these preparations, the iterative construction will proceed as follows. The initial curve C_0 is arbitrary, of length L_0. We shall only assume that C_0 can be reasonably well approximated by a polygon P_1 consisting of sides of equal length r_1. We then randomly select $N_1 = \lambda_1 L_0$ points on the curve C_0 according to a Poisson process of density λ_1, and replace those sides of the polygon P_1 which contain projections of the randomly selected points by small circular arcs of length $r_1(1 + \delta)$, see equation (1.4.2.2).

The total length of the curve will increase, on the average, by $L_0\lambda_1 r_1\delta$, that is the length of the "decorated" curve becomes:

$$L_1 = L_0 + L_0\lambda_1 r_1\delta = L_0(1 + \lambda_1 r_1\delta) \tag{1.4.2.4}$$

Generally, in the k-th generation of the process, straight line segments of length r_k approximating the curve C_{k-1} will be randomly selected according to a Poisson process of density λ_k, and replaced by $(1 + \delta)$-times longer circular arcs. The length of C_k will be:

$$L_k = L_{k-1}(1 + \delta\lambda_k x_k) \tag{1.4.2.5}$$

A typical step of the process can be well observed on Leonardo da Vinci's map of the sea shore between Rome and Naples (Figure 1.32), while the Tertiary limestone cliffs in Figure 1.33 represent an advanced stage. To assure self-similarity, we still have to relate λ_k to r_k. As the average spacing between crescentic features of the amplitude r_k is $1/\lambda_k$; a magnification transforming r_k to r_l should also transform $1/\lambda_k$ to $1/\lambda_l$; that is for any k, l we have:

$$\lambda_k r_k = \lambda_l r_l = \Lambda \tag{1.4.2.6}$$

where λ is a dimensionless positive constant. Substituting equation (1.4.2.6) into (1.4.2.5) yields:

$$L_k = L_0(1 + \Lambda\delta)^k \tag{1.4.2.7}$$

Putting $\varepsilon = r_k$, from equation (1.4.2.3) we have $k = -\log\varepsilon/\log R$, and — after the usual manipulations — $L(\varepsilon)$ can be written as:

$$L(\varepsilon) \approx \varepsilon^{1-\frac{\log\{R(1+\Lambda\delta)\}}{\log R}} \tag{1.4.2.8}$$

where we recognize that:

$$D = \frac{\log\{R(1 + \Lambda\delta)\}}{\log R} \tag{1.4.2.9}$$

is the fractal dimension of the limiting "coastline" $C = \lim_{k\to\infty} C_k$. The recursive construction of the curves $C_0, C_1, C_2 \ldots$ is similar to the way we have defined the Von Koch curves in the previous section (Figure 1.27). For the Von Koch curves we have $R = 3$; $\lambda = 1/3$ (because, in the first step, one out of the three smaller segments has been replaced by a $\wedge$-shaped "decoration", and $\delta = 1$ (because the decoration doubles the length of the original segment). Substituting these values into equation (1.4.2.9) we get $D = \log 4/\log 3$ in agreement with equation (1.4.1.3).

According to equation (1.4.2.9) the fractal dimension D tends to $D = 1$ for large R's:

$$\lim_{R\to\infty} D = 1 \tag{1.4.2.10}$$

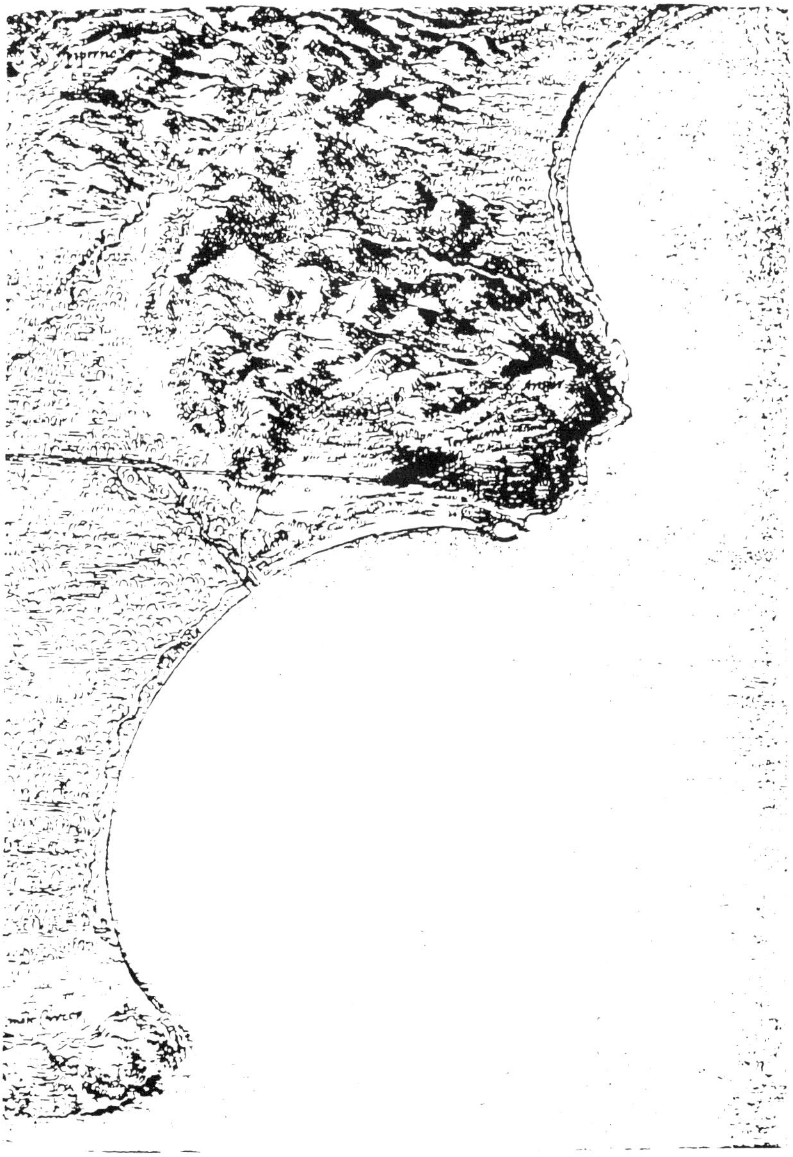

Fig. 1.32. Leonardo da Vinci's map of the sea shore between Rome and Naples (detail, from "The Notebook of Leonardo da Vinci", 1970).

To estimate the fractal dimension of the coastline resulting from this erosional process we assume $R = 10$ in accordance with Dolan and Ferm's (1968) data (Figure 1.30). In the simplest geometry (the "pigeon hole" model of Pape et al., 1981) the crescentic forms belonging to the same order of hierarchy follow each

Fig. 1.33. Tertiary limestone cliffs, the Nullarbor karst, Australia (courtesy Dr. Yvonne Bone and Dr. Noel P. James).

other in an uninterrupted sequence (as the large capes Hatteras, Lookout, Fear, Romain,. . . , along the Atlantic coast of the United States).

Thus $\lambda_k = 1/r_k$ and we have $\lambda = 1$ by equation (1.4.2.6). The factor δ in equation (1.4.2.9) has been defined in equations (1.4.2.1) and (1.4.2.2); it depends on the central angle of the arc (Figure 1.31). The values of the fractal dimension D are tabulated below for $R = 10$, $\lambda = 1$ and a few representative angles α:

Angle (α)*	Fractal dimension (D)
14.3°	1.001
60°	1.02
90°	1.05
120°	1.08
180°	1.2

*$\alpha = 14.3°$ is the theoretical equilibrium beach angle, cf. Hoyle and King (1958); $\alpha \approx 60°$ is the average angle found by McLean (1967) for New Zealand beaches.

Observe that all these values are very close to 1, only the model based on semicircular "decorations" ($\alpha = 180°$) has lead to a larger value $D = 1.2$. According to equation (1.4.2.10), this is due to the large ($\approx$1 decade) differences between the size of the "decorations" in the successive steps of the coastal erosion process (Figure 1.30). Which, in its turn, might very well be connected with the extreme range of the oceanic processes causing the erosion, including both planetary currents and the breaking waves in the surf zone. Which, in turn,... big fleas have even bigger fleas..., and there are people who continue the chain of causes, if not *ad infinitum*, at least to the "astronomical long distance controls" (Steiner, 1967).

1.4.3 *Bifractal coastlines*

The construction of the Von Koch curves and the fractal erosion of coastlines can be formally described as follows (Suzuki, 1984): we start out from an initial figure F_0, transform it to F_1 by a transformation T, and so on. That is, we have:

$$TF_0 = F_1 \qquad TF_1 = F_2 \ldots TF_n = F_{n+1} \tag{1.4.3.1}$$

The fractal figure is defined by the limiting figure F^*:

$$\lim_{n\to\infty} F_n = F^* \tag{1.4.3.2}$$

For F^* one has $TF^* = F^*$, that is, F^* is a *fixed point* of the transformation T. The resolution of measurement of length for F is ε_n.

Then, by Mandelbrot's (1982) definition, the fractal dimension of F^* is given by:

$$D = \lim_{n\to\infty} \frac{\log N(F_n)}{\log (1/\varepsilon_n)} \tag{1.4.3.3}$$

where $N(F_n)$ is the number of straight line segments of size ε_n approximating F_n, see equation (1.2.10). There are, however, many systems where the similarity is only valid in a certain range of the scale factor ε_n, that is, the slope of the $\log N(F_n)/\log (1/\varepsilon_n)$ plot depends on n (Figure 1.34). In this case a convenient definition of the fractal dimensionality is (Mandelbrot, 1982; Orford and Whalley, 1983; Suzuki, 1984):

$$D = \frac{\log N(F_{n+1})/N(F_n)}{\log (\varepsilon_{n+1}/\varepsilon_n)} \tag{1.4.3.4}$$

These fractal dimensionalities $\{D_n\}$ are functions of the scales ε_n and of a certain characteristic scale factor ε^*:

$$D_n = D(\varepsilon_n, \varepsilon^*) \tag{1.4.3.5}$$

Suzuki (1984) lists the following possibilities:

(a) D_n is independent of ε_n and ε^*, we observe a single "ideal" fractal dimensionality.

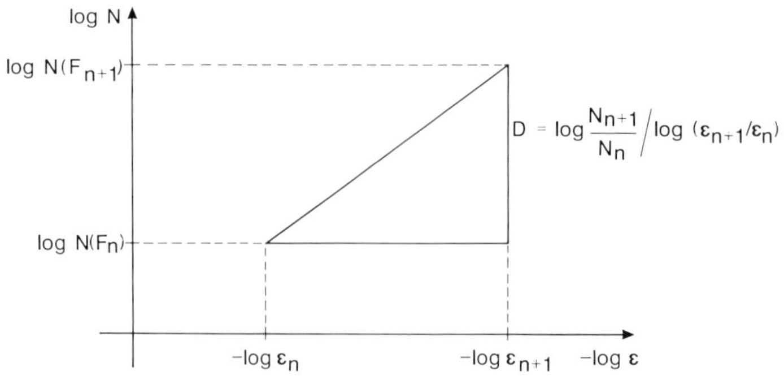

Fig. 1.34. To the derivation of equation (1.4.3.3).

(b) There occurs a cross-over effect around $\varepsilon_n \approx \varepsilon^*$, so that we have different fractal dimensions for $\varepsilon_n \ll \varepsilon^*$ and for $\varepsilon^* \ll \varepsilon_n$.

(c) There are several ranges in which $\{D_n\}$ are approximately constant, these ranges are separated by the characteristic scales $\varepsilon_1^*, \varepsilon_2^*, \ldots$

(d) $\{D_n\}$ is a gradually changing continuous function of ε_n.

In an influential paper, Orford and Whalley (1983) proposed to use the fractal dimension of the particle outline for a quantitative characterisation of the irregularity of sedimentary particles. In the majority of cases (see e.g. Figure 1.35) they recognised two distinct fractal dimensions:

D_1 related to the smallest step lengths, describing the micro-scale edge textural effects; and

D_2 corresponding to the larger step lengths and related to the particle shape.

Orford and Whalley adopted the terminology used in powder technology (Kaye,

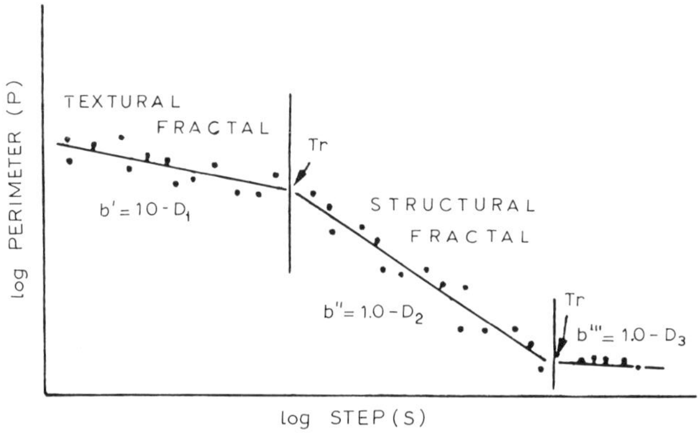

Fig. 1.35. Bifractality of the contour of a quartz grain. T_r marks the cross-over points between the textural and structural fractal regimes. There is a second crossover point beyond which the shape is Euclidian. (From Orford and Whalley, 1983.)

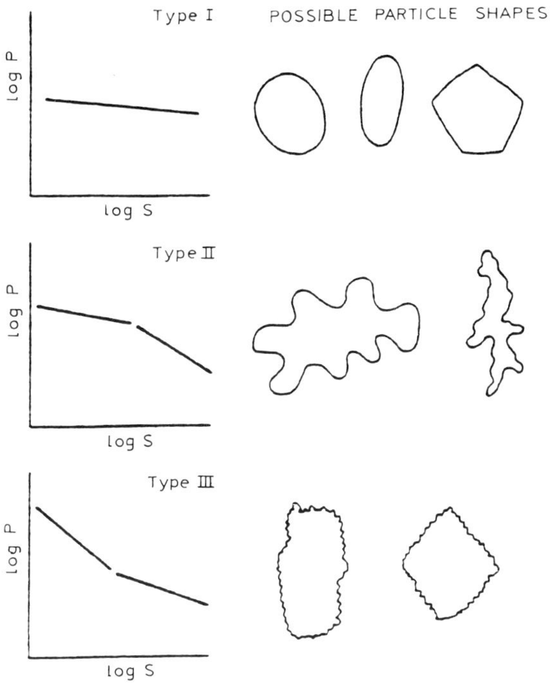

Fig. 1.36. Three basic types of fractality of the outline of sedimentary particles (from Orford and Whalley, 1983).

1978; Flook, 1979) and referred to these fractal regimes as *textural fractals* and *structural fractals*, respectively.

(Observe that in Figure 1.35 there is also a third regime; the non-fractal "Euclidian" part of the plot with $D_3 \approx 1$.)

Orford and Whalley described three basic combinations of textural and structural fractals (Figure 1.36). Type I is the ideal, single fractal population, the particle shape dominates the outline with very little edge roughness. Type II corresponds to the case when the particle shape is very irregular, the edge texture is smoother (hence the greater slope of the second line). Type III corresponds to a very irregular edge texture relative to a more regular structural outline.

Orford and Whalley's classification can also be applied to the much larger scales of the closed and open forms of geomorphology. Kent and Wong (1982) analysed the shoreline of six Canadian shield lakes in the Muskoka and Haliburton regions, Ontario, and found that all six were bifractal, of type II according to Orford and Whalley's classification (Figure 1.37). They recognized that the bifractal nature of the $\log N(\varepsilon)$ vs. $\log \varepsilon$ plot is due to the presence of "...at least two different geomorphological fractal processes. Each process modifies the shoreline, but at different scales... If we have two processes working at two different scales, then the

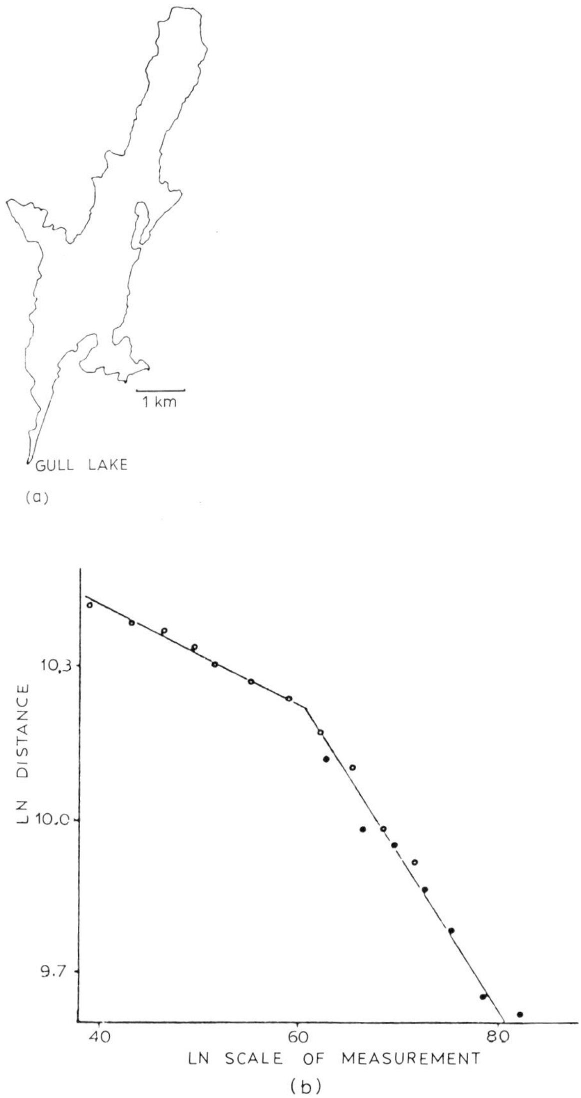

Fig. 1.37. The Gull Lake, Ontario, Canada (a), and the fractal analysis of its shoreline (b) (from Kent and Wong, 1982).

resulting curve would have two parts. These two line segments would be joined at the characteristic scale or breakpoint separating the processes" (Kent and Wong, 1982, p. 850).

In an important study, T. Nakano (1983) of the Tsukuba University, Japan, analysed six ria coastlines from different parts of Japan and found that, irrespective of their location, tectonic division and the age of the rocks constituting them, *all six*

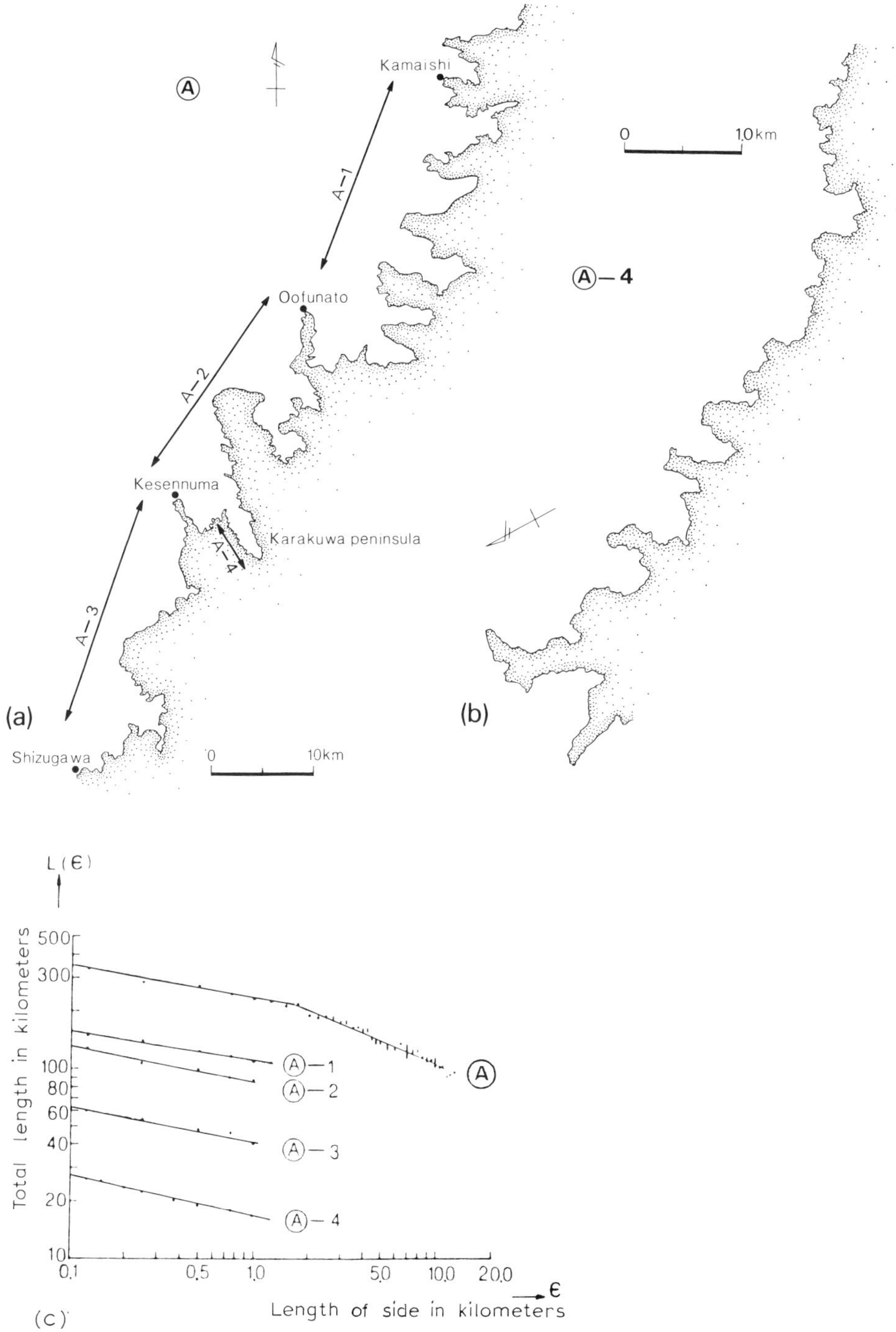

Fig. 1.38. a. Ria coastline in northeastern Japan. b. Magnified version of its smaller part (*A–4*). c. Fractal analysis of the whole coastline and of its parts. The break in line *A* defines the *cross-over* point. (From Nakano, 1983.)

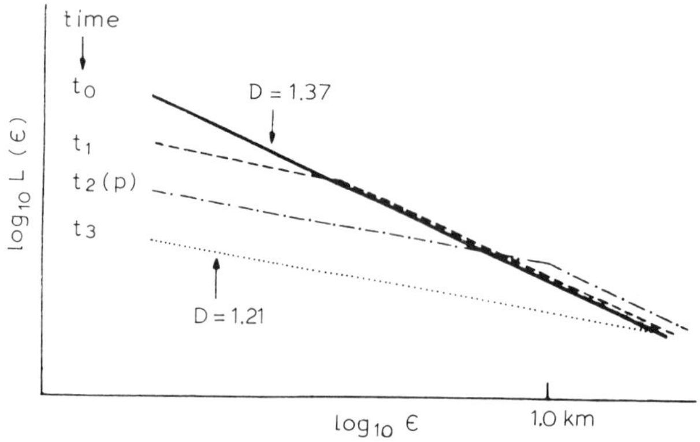

Fig. 1.39. Hypothetical model for the change of the $L(\varepsilon)$ vs. ε relation with geological time for ria coastlines in Japan. t_0 = past, t_1 = intermediate, t_2 = present, t_3 = future. (From Nakano, 1983.)

were bifractal of type II (Figure 1.38). In most of the cases the "structural" dimension and the "textural" dimension were about 1.37 and 1.21, respectively. Nakuno identified the smaller fractal dimensional form as a result of the smoothing effect of erosion, and suggested that the *cross-over point between the two fractal domains has continuously moved from $D = 1.37$ towards $D = 1.21$ during geological time* (Figure 1.39). In a further study, Nakano (1984) has extended the fractal analysis to the +10, +30 and +50 m contour-lines of the near-shore topography. As expected,

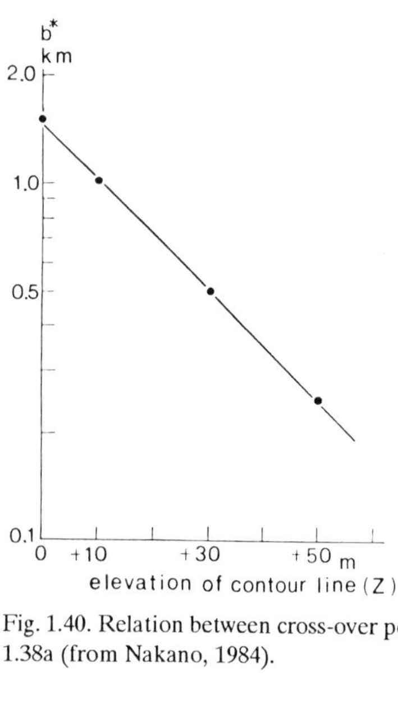

Fig. 1.40. Relation between cross-over point (b^*) and elevation of the contourline for the coast of Figure 1.38a (from Nakano, 1984).

the cross-over point separating the two fractal regimes continuously increased with increasing elevation of the contour-lines (Figure 1.40).

The Reader might recall how displeased Captain Gulliver had been with the sight of the Brobdingnagian ladies:

> Their Skins appeared so coarse and uneven, so variously coloured when I saw them near, with a Mole here and there as broad as a Trencher, and Hairs hanging from it thicker than Pack-threads... — (Part two, Chapter five)

However, he has not noticed any coarseness or unevenness in Brobdingnag's coastline (Figure 1.41) or topography, as these fractal features are independent of scale.

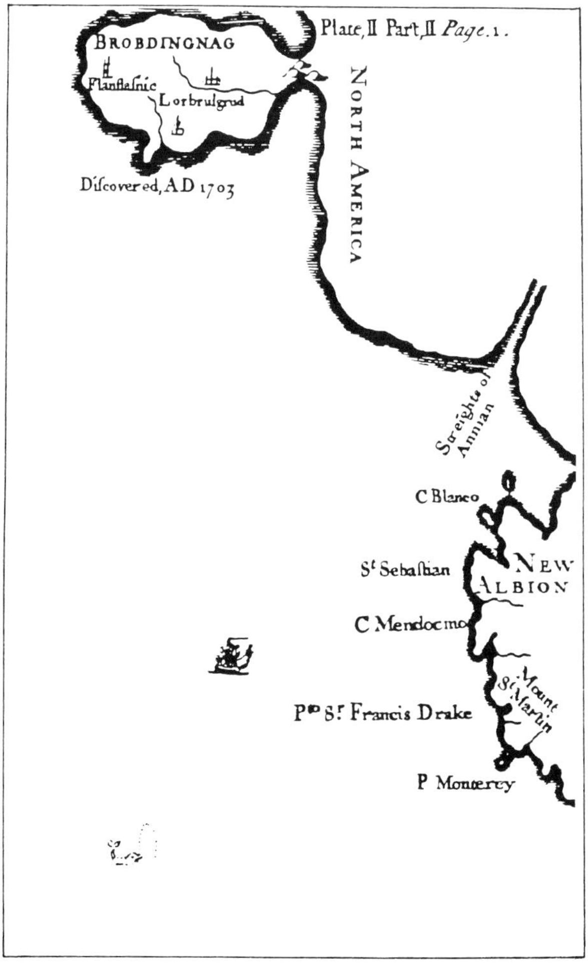

(a)

Fig. 1.41. a. Map of Brobdingnag, the second land visited by Gulliver.

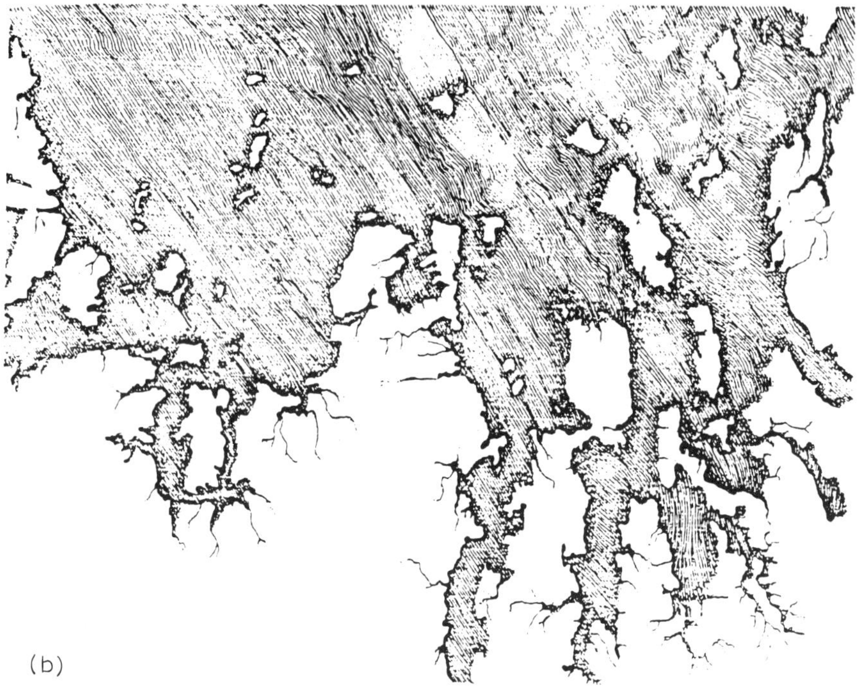

(b)

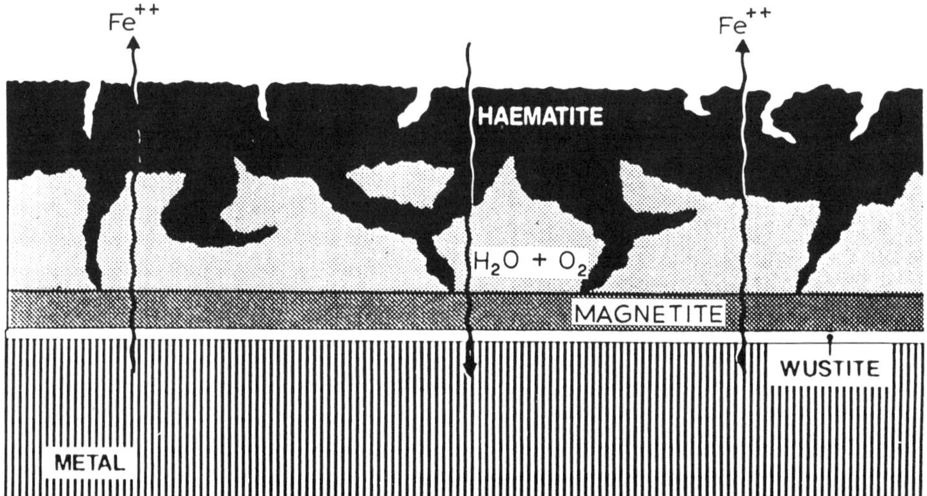

(c)

Fig. 1.41 (continued). b. The northern coastline of the United States demonstrating the effect of the erosion processes of the sea (from Williams, 1981). c. Cross-section of a corroding steel (Advertisement, Oil and Gas Australia, Oct. 1988).

Coastlines are not only self-similar, they also resemble *each other*, over an enormous range of scale. The dendritic configuration in Figure 1.41b, which is an unusual representation of the northeastern coastline of the United States, could be easily confused with a corroding alloy under magnification (Figure 1.41c).

1.5 The perimeter–area rule of Mandelbrot

1.5.1 *Islands and lakes*

> The island of the Utopians extends in the centre (where it is broadest) for two hundred miles and is not much narrower for the greater part of the island, but toward both ends it begins gradually to taper. These ends form a circle five hundred miles in circumference and so make the island look like a new moon, the horns of which are divided by straits about eleven miles across. The straits then unfold into a wide expanse...
> The mouth of this bay is rendered perilous here by shallows and there by reefs. Almost in the centre of the gap stands one great crag which, being visible, is not dangerous. A tower built on it is occupied by a garrison.

Sir Thomas More's Utopia is probably the most famous of all invented islands, even among those who have never heard of its author. Unlike many later works set in imaginary lands, it provides a clear description of the fictitious topography. The first edition of the Utopia, published at Louvain in 1516 is illustrated with an anonymous woodcut showing a sketch of the island (Figure 1.42). In the third edition (in Basle, 1518) this sketch has been replaced by a woodcut attributed to Ambrosius Holbein, brother of Hans, which is far more elegant but far less conformable to the text (Figure 1.43).

Straightforward as More's description might appear, both illustrators had had a great difficulty in preparing the sketches, because Utopia *can not be mapped.* According to a remarkable geometrical property of curves of constant breadth: all curves of constant breadth having the same breadth b have the same perimeter $b\pi$, which is the circumference of the circle of diameter b (Rademacher and Töplitz, 1957). That is, if the island is 200 miles wide for its greater part, its perimeter should be of the order of 600 miles and a circumference of 500 miles is clearly impossible.

Similarly, the *area* (A) and *perimeter* (P) of an island are also interconnected and cannot be arbitrarily chosen. For any closed planar figure the circle has the least possible perimeter for a given area (Rademacher and Töplitz, 1957), that is for any closed planar figure:

$$\frac{P}{(4\pi A)^{1/2}} \geq 1 \tag{1.5.1.1}$$

On the other hand the perimeter P in inequality (1.5.1.1) can be arbitrarily large.

Fig. 1.42. Utopia (unknown artist, 1516).

As large values of the ratio $P/(4\pi A)^{1/2}$ indicate more convoluted figures, this ratio can be used to characterize the complexity of the shorelines of islands or lakes.

In case of lakes the $P/(4\pi A)^{1/2}$ ratio is called the *shoreline development index* (Kent and Wong, 1982) and it is of considerable interest for biologists as it reflects the potential for greater development of the littoral communities in proportion to the volume of the lake. However, it had already been recognized by Hutchinson (1957) that the shoreline length depends "on the fineness of detail of the map and for this reason the indexes obtained on maps of different scale may be somewhat different". [The "somewhat different" is an understatement: Kent and Wong (1982)

Fig. 1.43. Utopia (A. Holbein, 1518).

reported that some of the shoreline lengths were as much as 100% greater than the published values when re-measured on detailed maps.] If we have a collection of *similar figures* (lakes, islands) of different sizes and with *non-fractal boundaries*, the ratio:

$$\rho = \frac{P}{A^{1/2}} \tag{1.5.1.2}$$

will be the same for all objects, independently of their size, that is one has:

$$P \propto A^{1/2} \tag{1.5.1.3}$$

Let us now assume that we are given a collection of *similar* closed planar figures $F_1, F_2, F_3, \ldots$ with *self-similar fractal boundaries of dimension D*. Let us select a yardstick of length $\varepsilon \to 0$, and measure the perimeter $P_i(\varepsilon)$ and area $A_i(\varepsilon^2)$ of each figure F_i in units of ε and ε^2, respectively. Mandelbrot (1977) has shown that for the collection of fractal figures a modified form of equation (1.5.1.3) holds:

$$P_i \propto \left(A_i^{1/2}\right)^D = A_i^{D/2} \tag{1.5.1.4}$$

where the constant of proportionality is the same for all figures, but it depends on the selected resolution ε. Equation (1.5.1.4) is the celebrated *perimeter–area relation* of Mandelbrot.

Before proceeding to prove equation (1.5.1.4), let us see a familiar example where the fractal dimension can be easily determined using this rule. The dendritic growths of the Bristol landscape marble of Figure 1.44 are obviously fractal, and also similar to each other. To find their fractal dimension, we simply copy them onto a sufficiently fine grid and for each object count how many small squares the whole object occupies, and its boundary occupies, respectively.

If the objects are similar and self-similar, then by equation (1.5.1.4) the points $\{P_i, A_i\}$ plot along a straight line on a doubly logarithmic paper. The slope of the straight line will be $D/2$ where D is the fractal dimension.

The proof of equation (1.5.1.4) will follow Hentschel and Procaccia (1984).

Let us consider an object of the collection, F, having an area A. As F has a fractal boundary, if we measure its perimeter with a yardstick of length $\varepsilon (\varepsilon \ll 1)$, we shall need:

$$N(A, \varepsilon) = N(A)\varepsilon^{-D} \tag{1.5.1.5}$$

segments to approximate it by a polygon, that is the length of the perimeter will be found at this resolution as:

$$P(A, \varepsilon) = N(A)\varepsilon^{-D}\varepsilon = N(A)\varepsilon^{1-D} \tag{1.5.1.6}$$

[cf. equations (1.2.10) and (1.2.11)]. In equations (1.5.1.5) and (1.5.1.6) the coefficient $N(A)$ depends on the area of the object, but it cannot depend on its shape as all objects have been assumed similar.

To find the function $N(A)$ let us consider two objects of the collection, F_1 and F_2, such that the first has a unit area, and the second an area A. If we take a large number M ($M \gg 1$), the object F_1 can be covered by M^2 squares of side $1/M$. By a similarity transformation the covering of F_1 goes over to a covering of F_2 by M^2

Fig. 1.44. Dendritic growth in a "landscape marble"

Fig. 1.45. To the derivation of equation (1.5.1.7) (after Feder, 1988).

squares of side $A^{1/2}/M$. Because of the similarity of the two objects, in both cases the same number of squares would contain boundary points (Figure 1.45):

$$N(1, \frac{1}{M}) = N(A, \frac{A^{1/2}}{M}) \tag{1.5.1.7}$$

and by (1.5.1.5):

$$N(1)M^D = N(A)\frac{M^D}{(A^{1/2})^D}$$

Consequently:

$$N(A) = N(1)(A^{1/2})^D \tag{1.5.1.8}$$

and by equation (1.5.1.6) we obtain Mandelbrot's perimeter–area relation in the form:

$$P(A, \varepsilon) = N(1)(A^{1/2})^D \varepsilon^{1-D} \tag{1.5.1.9}$$

where $N(1)$ is a constant, depending on the shape of the objects.

Because of the factor ε^{1-D} in equation (1.5.1.9) the P–A relation depends on the resolution ε.

It frequently occurs in practice that we have a mixed set of data obtained from maps of different resolution. Writing out equation (1.5.1.9) for two different resolutions r and s:

$$P(A,r) = N(1)A^{D/2}r^{1-D} \qquad P(A,s) = N(1)A^{D/2}s^{1-D}$$

we find that:

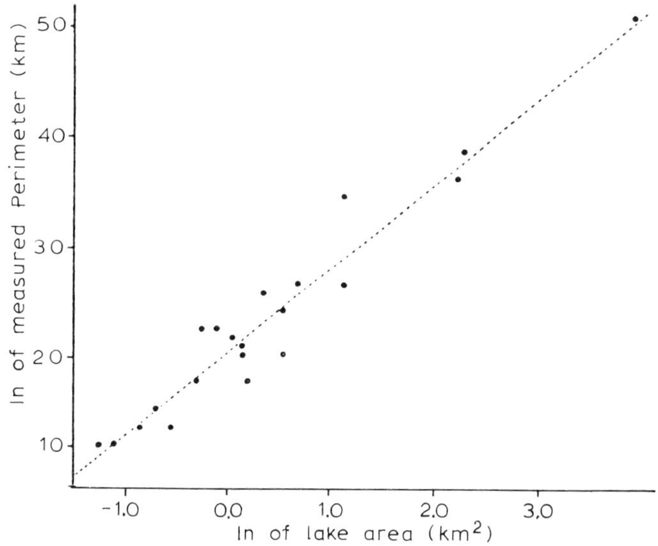

Fig. 1.46. Perimeter vs. area plot for 21 lakes in Ontario, Canada (from Kent and Wong, 1982).

$$P(A,s)\left[\frac{r}{s}\right]^{1-D} = N(1)A^{D/2}r^{1-D} \qquad (1.5.1.10)$$

that is, the perimeters measured with accuracy s should be corrected by the factor $(r/s)^{1-D}$. (Otherwise they plot on a different, though parallel, line on the P–A crossplot.) In the derivation of (1.5.1.10) it has been assumed that both r and s are sufficiently fine so that the area does not depend on the resolution. A practical example will be provided in the next section: if we use radar data sampled on a grid of 1 × 1 km resolution, *and* satellite data of 4.8 × 4.8 km resolution for a P–A analysis, the perimeters of clouds measured on the satellite picture must be increased by the factor $(1/4.8)^{1-D}$ (Lovejoy, 1982).

In 1982 Kent and Wong checked the validity of the perimeter–area rule on 21 lakes in the Muskoka and Haliburton regions of Ontario, Canada (Figure 1.46). As the shoreline length determinations were all taken from maps drawn at the same scale, no corrections of the form (1.5.1.10) had been necessary. The value of D was estimated to be 1.50 ± 0.18.

On the other hand, the length–resolution analysis of the same lakes has shown that their shorelines are *bifractal* (Figure 1.37) with the dimensions of the large scale features clustering around a mean value of 1.44 ± 0.09, while the dimensions of the finer scales clustered around 1.14 ± 0.02. Kent and Wong observed that the P–A analysis has provided a unique D value which is close to the dimension characterising the *large-scale behaviour* of the bifractal boundaries.

1.5.2 *The fractal shape of clouds*

Hamlet: Do you see yonder cloud that's almost in shape of a camel?
Polonius: By th'mass, and 'tis like a camel indeed.
Hamlet: Methinks it is like a weasel.
Polonius: It is backed like a weasel.
Hamlet: Or like a whale?
Polonius: Very like a whale. — SHAKESPEARE: *Hamlet*, iii. 3

In 1982 Lovejoy investigated the geometry of satellite and radar determined cloud and rain areas over six orders of magnitudes of areas from 1 km^2 to $1.2^6 \times 10$ km^2 and found that the areas and perimeters of the clouds satisfied Mandelbrot's relation $P \propto A^{D/2}$ with $D = 1.35 \pm 0.05$ (Figure 1.47).

The rain areas in Lovejoy's study were analysed by digitizing radar pictures at a resolution of 1×1 km. The radar senses the reflected microwave radiation primarily from the large rain drops. Rain areas were defined as regions for which the rain rate exceeded 0.2 mm per hour. The curvature of the earth limited the radar analysis to rain areas less than 40,000 km. In order to get data at larger scales Lovejoy also

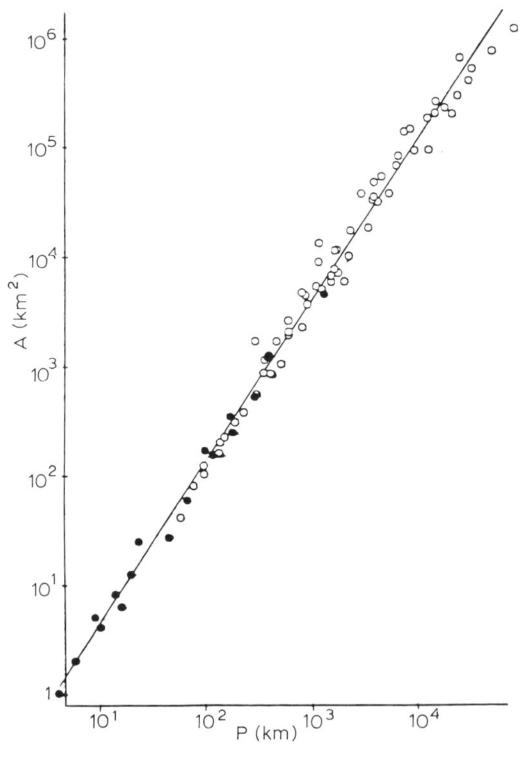

Fig. 1.47. Area vs. perimeter plot for rain and cloud areas determined from radar (•) and satellite (o) data (from Lovejoy, 1982).

used the infrared pictures of the geostationary operational environment satellite (GOES) sampled on a grid of 4.8 × 4.8 km. For clouds more than 300 m thick, the infrared channel of the GOES yields an estimate of the cloud-top temperature. In Lovejoy's study pixels corresponding to temperatures below −10°C were considered as belonging to clouds.

Because of the different resolutions of the two data sets, the perimeters determined on the satellite pictures have been corrected by the factor $(1/4.8)^{1-D}$ in agreement with equation (1.5.1.10).

The most striking aspect of Lovejoy's data (Figure 1.47) is that all points lie on the same line on the area–perimeter plot, over more than six orders of magnitude in area, without any apparent change of slope. This finding has generally been interpreted as evidence for the absence of any characteristic length scales in the turbulent processes involved in the formation of clouds, in the range 1 km to 10,000 km (Lovejoy, 1982; Hentschel and Procaccia, 1984; Henderson-Sellers, 1986; Beech, 1987; Feder, 1988). In an important paper Hentschel and Procaccia (1984) have proposed a theoretical explanation for the fractal shape of clouds based on the theory of turbulent diffusion, and calculated the fractal dimension of the cloud perimeters to be between $1.37 < D < 1.41$, in agreement with the measurements of Lovejoy (1982).

Lovejoy's study has been repeated by Henderson-Sellers (1986) for 11 clouds of the planet Mars in the range of perimeters 10–1000 km, using images of the Viking orbiter of unspecified resolution. In spite of the many differences between the Martian atmosphere and that of the Earth (smaller gravitational force, surface air pressure one hundredth of the Earth's, dominant atmospheric gas is the electromagnetically active CO_2 rather than the inert N_2 found in the Earth's atmosphere) Henderson-Sellers found that the Martian clouds also satisfy Mandelbrot's perimeter–area relationship (that is they are fractal) with dimension $D = 1.60$.

Not surprisingly, the galactic dark nebulae have also proved to be fractal (Beech, 1987) — as a matter of fact, it is exactly because of their "cloudlike" appearance that these molecular and dust clusters had been called "clouds". The perimeter–area analysis of 24 opacity class 5 and 6 clouds has yielded a fractal dimension $D = 1.4$, which might be in connection with turbulent processes operating in these objects.

In 1986 Rys and Waldvogel extended Lovejoy's investigations to clouds of smaller perimeters ($P = 1$–100 km). They analysed radar echo data of 0.3 × 0.3 km resolution from severe convective clouds (Figure 1.48), and found that there was a characteristic perimeter $P_0 \approx 9$ km (shown by dotted line in the figure) such that clouds with $P > P_0$ had a fractal boundary with dimension $D = 1.36 \pm 0.1$ in agreement with Lovejoy's (1982) data; while clouds with perimeter $P < P_0$ had an almost non-fractal boundary ($D = 1.0 \pm 0.1$). As the position of the cross-over point P_0 seemed to depend on the strength of the vertical winds, Rys and Waldvogel proposed that the non-fractal nature of the smaller contours is due to the smoothing

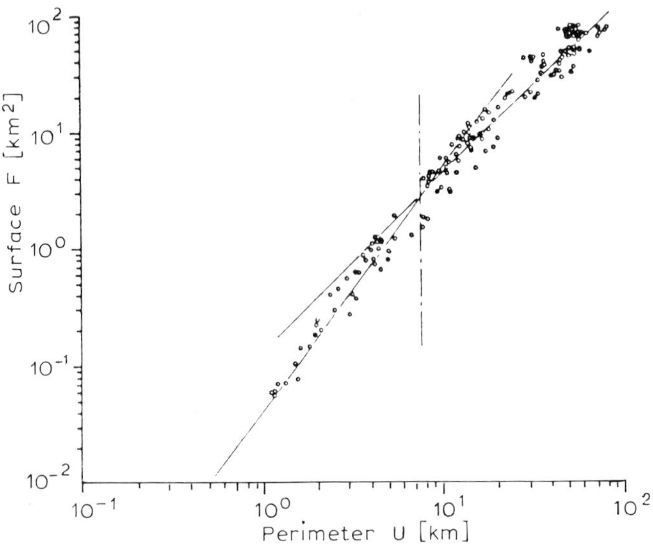

Fig. 1.48. Perimeter–area relation for a typical hail storm. Every point corresponds to a particular time during the evolution of the storm. The full lines are fits of 24 different hail storms. The dotted vertical line shows the cross-over point. (From Rys and Waldvogel, 1986.)

effect of the strong vertical winds which are known to be present near the centre of severe convective clouds.

Recent experiments of Sreenivasan and Meneveau (1986) have raised some doubts concerning the usual interpretation of Lovejoy's results.

They observed that the cloud areas and perimeters measured by Lovejoy (1982) are generally thought of as if they were the areas and perimeters of clouds intersected with a horizontal plane, while in reality "Lovejoy was looking at the boundary of the *projection of a cloud* onto a horizontal plane… However, a projection can be thought of as a superposition of a large number of sections." According to Sreenivasan and Meneveau (1986), this superposition increases the boundary roughness, leading to an increased fractal dimension. (This phenomenon can be very well observed on the cloud representation of the Tibetan paintings, Figure 1.49.) To prove their point, Sreenivasan and Meneveau marked a turbulent boundary layer developing on a flat plate by smoke, and illuminated it by a thin sheet of light (Figure 1.50). When they increased the thickness of the light-sheet from η to about 5η (where the Kolmogorov scale η is the smallest dynamical scale of the turbulence), the fractal dimension of the boundary increased from 1.37 to 1.43.

As we have seen in the previous Section, in connection with the *P–A* analysis of lakes, the *P–A* analysis can provide a unique fractal dimension even in cases when the boundary is bifractal, and this dimension will be close to the largest fractal dimension observed on the boundary.

Fig. 1.49. Cloud representations in Tibetan paintings (R.Beer's drawings from D.P. and J.A. Jackson, 1988).

Fig. 1.50. Smoke photograph of a turbulent boundary layer developing on a flat plane. The thickness of the intersecting light sheet is of the order of the Kolmogorov scale η. (From Sreenivasan and Meneveau, 1986.)

According to Rys and Waldvogel's results (1986) there is a cross-over between smooth behaviour and fractal behaviour which can be detected on the P–A plot of the smaller ($P < 10$ km) clouds. The experiments of Sreenivasan and Meneveau (1986) indicate that there might be a further cross-over separating the "real" convolutedness of the clouds' perimeter from the fractal decorations arising from the superposition of the great number of independent fractal outlines contained in the projection of the cloud onto the horizontal plane. The presence or absence of this further cross-over could only be ascertained by a resolution–perimeter analysis over the entire length scale.

1.5.3 *The "slit island" analysis of fracture surfaces*

> And the flood was forty days upon the earth; and the waters increased, and bare up the ark, and it was lift above the earth.
> And the waters prevailed, and were increased greatly upon the earth; and the ark went upon the face of the waters.
> And the waters prevailed exceeding upon the earth; and all the high hills, that were under the whole heaven, were covered.
> Fifteen cubits upward did the waters prevail; and the mountains were covered.
>
> — *Genesis*, 7, 17–20

This majestic vision of the Deluge in the Bible is followed a few verses later (Gen. 8, 3–13), by the description of the painfully slow subsiding of waters: "the waters asswaged" and then "returned from off the earth continually". Figure 1.51 schematically illustrates the withdrawal of waters: first the highest peaks of the mountains break to the surface and form islands (Figure 1.51B-c), then the zone of alpine meadows is liberated (Figure 1.51B-b) and, finally, the water drops to the foothills to form lakes (Figure 1.51B-a).

If the original surface of the mountains had been fractal, of dimension D ($2 < D < 3$) then at any water level the islands (or lakes) would also be fractal planar figures of statistically similar shape. It is generally assumed — though actually never proved — that for any water level the dimension D' of the lakes or islands is the same and that it is related to the dimension D of the surface as $D' = D - 1$. The dimension D' of the lakes or islands can be easily determined using Mandelbrot's perimeter–area relation $P \propto A^{D'/2}$, and this provides us with an estimate of the otherwise intractable surface fractal dimension $D = D' + 1$.

In 1984 Mandelbrot et al. introduced a new technique, based on the perimeter–area analysis, to quantitatively characterize the rough and irregular fracture surfaces of metals. They studied fractured samples of 300 Grade Maraging steel by plating the fracture surfaces with nickel. The specimens were polished parallel to the plane of the fracture. "Islands" of steel surrounded by nickel appeared under the microscope which, on subsequent polishing, grew and merged (as in the subsiding of the water, Figure 1.51B).

Mandelbrot et al. (1984) called these objects *slit islands*, and the technique *slit island analysis*.

The perimeter–area plot of these islands (Figure 1.52) gave:

$$A \propto P^{2/D'} \qquad \text{with } D' = 1.28$$

that is, the fracture surface was estimated to be 2.28-dimensional.

Fig. 1.51. Map of a mountain system flooded with water; the submerged regions are shaded (after Efros, 1986).

Since the pioneering study of Mandelbrot et al. (1984) the slit island analysis has been widely used in fractography (Pande et al., 1987a, b, for titanium specimens; Mu and Lung, 1988, for steel; Stupak and Donovan, 1988, for rubber wear surfaces; Mecholsky et al., 1988, for polycrystalline and single-crystal ceramics). In all experiments reported: (a) the fracture surfaces had fractal cross sections (Figure 1.53); (b) the slit islands were fractal planar objects (Figure 1.54); and (c) the perimeter–area relation plotted as a straight line on a double logarithmic grid. However, the dimensions determined from the perimeter–area analysis of the slit islands were sometimes larger than the dimension obtained from the resolution–length analysis of the vertical sections through the fracture surface (Pande et al., 1987a). In some cases the fractured surfaces did not have a simple fractal structure according to the length–resolution analysis, while their slit-island analysis has still resulted in a straight line in the $\log P$ vs. $\log A$ plot (Pande et al., 1987b).

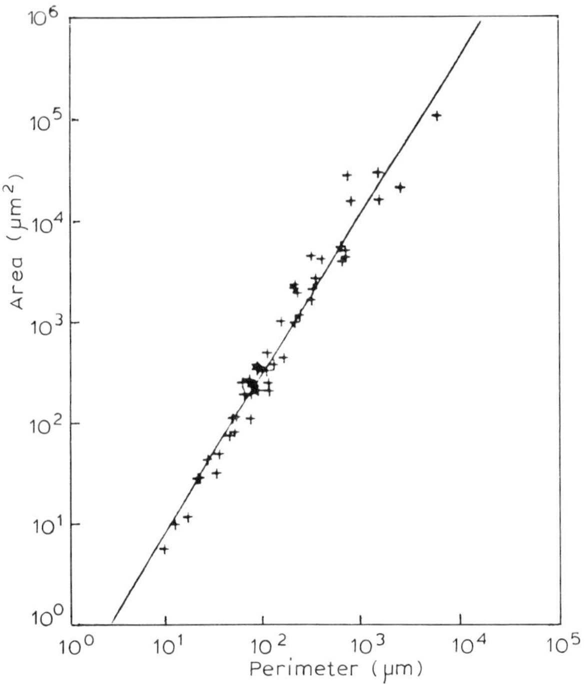

Fig. 1.52. Area–perimeter relation for the slit islands of the fracture surface of a 300 Grade Maraging steel specimen (from Mandelbrot et al., 1984).

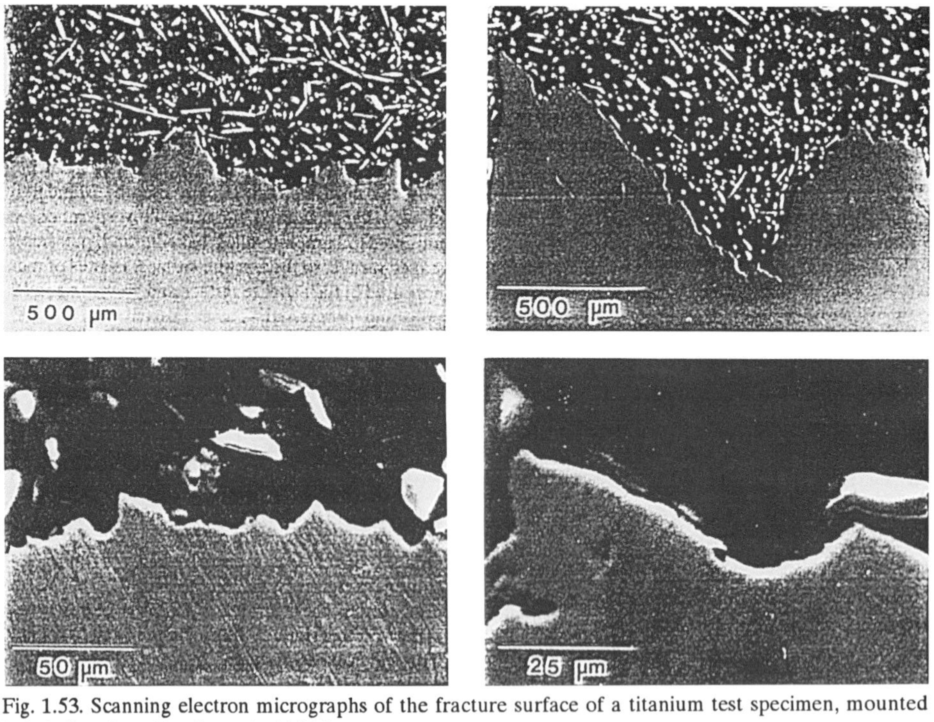

Fig. 1.53. Scanning electron micrographs of the fracture surface of a titanium test specimen, mounted in bakelite (from Pande et al., 1987a).

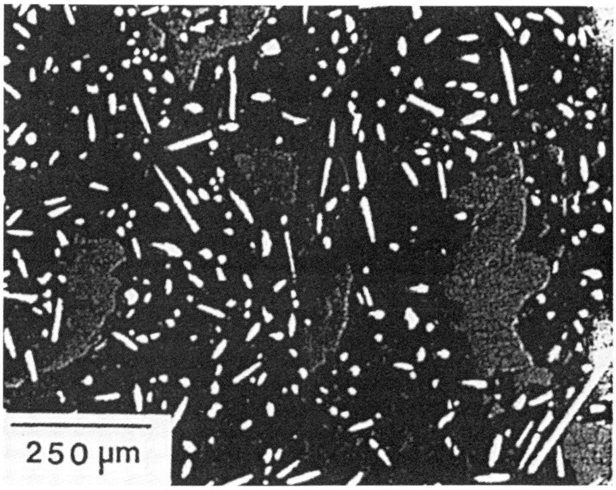

Fig. 1.54. Scanning electron micrograph showing the "slit islands" (irregular grey shapes) of the fracture surface of a titanium specimen in a bakelite mount (from Pande et al., 1987a).

A possible pitfall of the slit-island technique has been recognized by Pande et al. (1987b). As it has been an essential assumption in the derivation of Mandelbrot's perimeter–area relation (1.5.1.9) that the objects are geometrically similar to each other (see p. 60), a systematic failure of this constraint would lead to spurious values for D' in $A \propto P^{2/D'}$. Pande et al. (1987b) have found that the shortcomings of the slit-island technique had partly been due to the fact that the area of the slit islands depended on the length/breadth ratio of these objects (Figure 1.55).

I have not come across any references to application of the slit island method to the investigation of the fracture surfaces in rocks, though there have been several

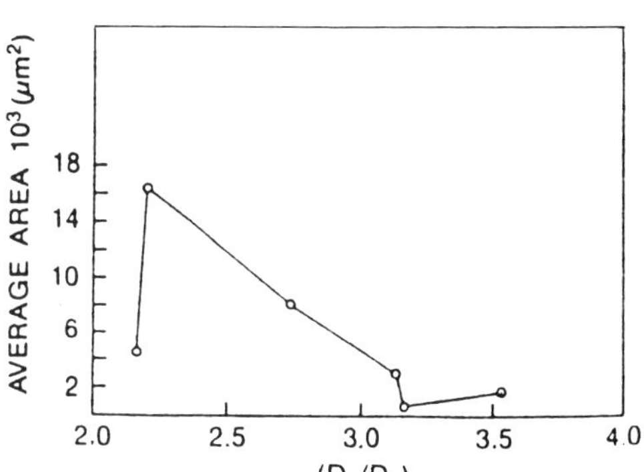

Fig. 1.55. Average area of the "slit islands" in a fracture surface of a titanium alloy specimen, as function of the form factor D_1/D_2, where D_1 is length, D_2 is breadth of the islands (from Pande et al., 1987b).

important studies on the fractal nature of cracks, fractures and faults, based on resolution–length analysis (Aviles et al., 1987; Okubo and Aki, 1987; Pape et al., 1987; Termonia and Meakin, 1986); on Fourier analysis (Brown and Scholz, 1985; Scholz and Aviles, 1986), on boxcounting techniques (Barton and Larsen, 1985; Chilès, 1988), on spatial correlation analysis (Hirata et al., 1987) and on statistical analysis of the distribution of fault lengths (Korvin, 1989; Segall and Pollard, 1983; Villemin and Sunwoo, 1987). Some of these studies will be reviewed in forthcoming chapters.

Goodchild (1988) generated random fractal surfaces by realizing a stochastic process known as fractional Brownian motion, flooded them by water to obtain lakes (Figure 1.56) and checked whether the areas and perimeters of these lakes satisfy $A \propto P^{2/D'}$ with $D' = D - 1$, where D is the fractal dimension of the surface.

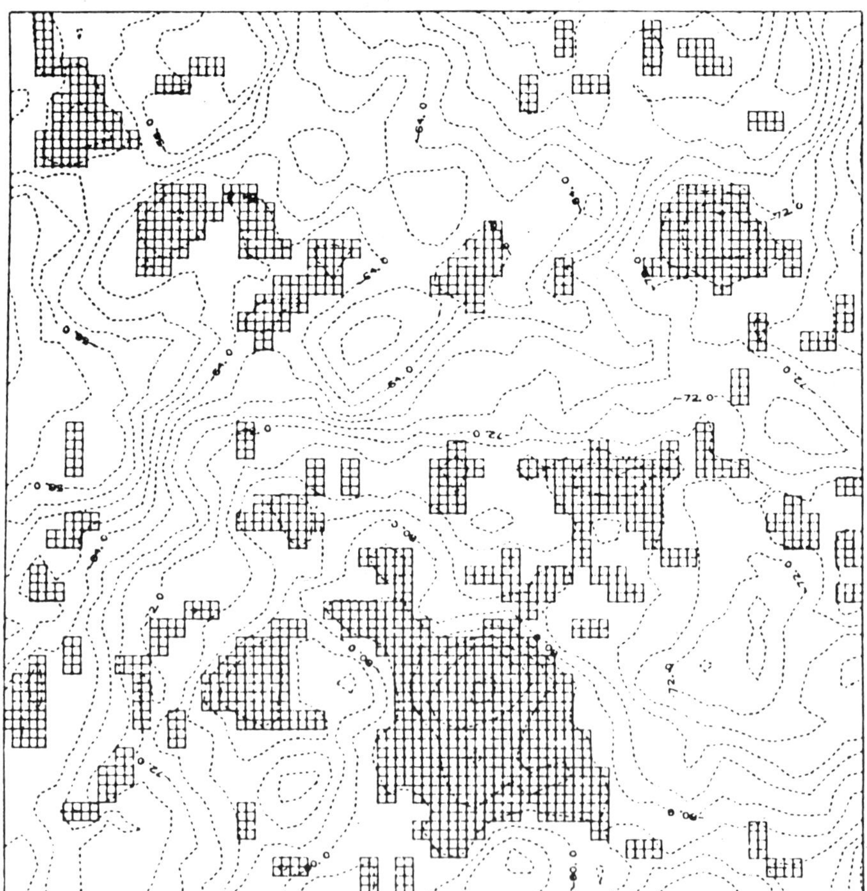

Fig. 1.56. A computer-generated 2.4-dimensional random fractal surface flooded by water to show lakes (from Goodchild, 1988).

His results are tabulated below:

D'	$D-1$
1.44	1.7
1.38	1.6
1.34	1.5
1.28	1.4
1.28	1.3
1.28	1.3
1.28	1.3

Goodchild concluded that D' is different from (as a matter of fact, less than) the theoretical value $D-1$ at 0.05 level of significance.

Finally, we should mention an interesting ecological application of the perimeter–area method. Krummel et al. (1987) analysed deciduous forest patterns in the U.S. Geological Survey 1:250,000 Natchez Quadrangle. They found significant differences between the shape of small forest patches (Figure 1.57a) and larger forested areas (Figure 1.57b). The fractal dimensions obtained from perimeter–area analysis were also different for small as compared to large forest patches (Figure 1.58).

The most interesting aspect of Krummel et al.'s results is that the large forested area in Figure 1.57b is similar to a *percolation cluster*, and the sudden change in the fractal dimension of the forests at a certain critical area is characteristic to a *phase transition*.

Comparing Figures 1.57 and 1.51B, the large forest of Figure 1.57b corresponds to the continuous waterway of Figure 1.51B-b. In course of the continuous transition

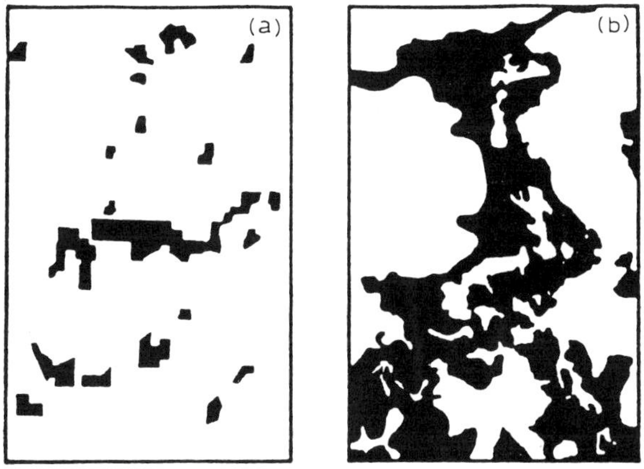

Fig. 1.57. Computer plots of deciduous forest in two different areas of the U.S.A. a. Small forest patches in the Mississippi flood plain. b. A larger forested area on the eastern edge of the flood plain. (From Krummel et al., 1987.)

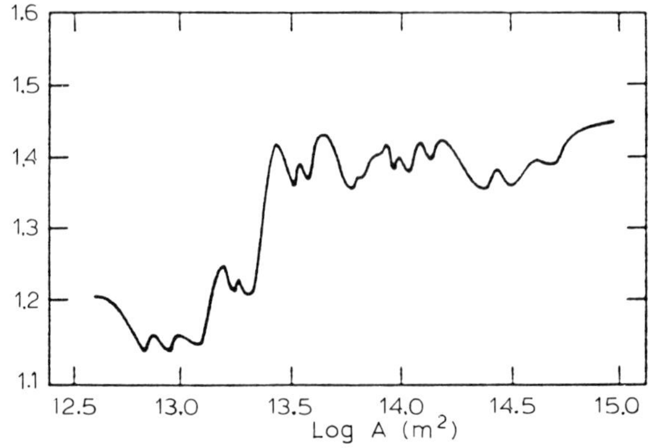

Fig. 1.58. Fractal dimension D determined by P–A analysis for deciduous forest patterns, as functions of the forest area (from Krummel et al., 1987).

from Figure 1.51B-c to 1.51B-a there is a critical water level for which the connected waterway suddenly disappears. It is easy to prove by symmetry arguments that for Gaussian random surfaces the *percolation transition* occurs for that water level for which in the plane section 50% of the total area is covered by water (Efros, 1986).

At the percolation transition the connected clusters of water (or forest) are D^*-dimensional selfsimilar fractal objects, with $D^* < 2$. [For two-dimensional site percolation, for example, $D^* = 1.895$; see Stauffer (1985), Feder (1988).]

If the random surface is a D-dimensional fractal, the fractal dimension of the perimeter of the clusters at the percolation threshold will be $D' = D - 1$. That is, for a cluster of characteristic size X its perimeter and area, respectively, will scale as:

$$P \propto X^{D'} \qquad A \propto X^{D^*} \tag{1.5.3.1}$$

From the perimeter–area relation we have:

$$X^{D'} \propto A^{D(\text{estimated})/2} \tag{1.5.3.2}$$

that is, from equations (1.5.3.1) and (1.5.3.2):

$$D(\text{estimated}) = \frac{2}{D^*} D' > D' \tag{1.5.3.3}$$

because $D^* < 2$. Consequently, around the percolation threshold (Figures 1.51B-b and 1.57b) the fractal dimension estimated from the perimeter–area relation will be larger than the real one. Far from the threshold, the area of both the islands and the lakes (Figs. 1.51B-c or 1.51B-a, 1.57a) scales as X^2 (Feder, 1988), that is, the P–A analysis yields unbiased estimate of D'.

A1 Mathematical appendix

Three of the pernicious results of overmuch rigour may be mentioned here. First, its enfeebling action on the mind. Secondly, it leads to the omission from mathematical works of the most interesting parts of the subject, because the authors cannot furnish rigorous proofs. Thirdly, it leads to an inability to recognise the good that may be in other men's work, should it be unconventional, and be devoid of rigorous pretence.

— HEAVISIDE: *Electromagnetic Theory* (1893)

A1.1 *The functional equations of similarity*

THE CAUCHY EQUATIONS

In 1821 Cauchy considered the functional equations:

$$f(x+y) = f(x) + f(y) \tag{A1.1.1}$$

$$f(x+y) = f(x)f(y) \tag{A1.1.2}$$

$$f(xy) = f(x) + f(y) \tag{A1.1.3}$$

$$f(xy) = f(x)f(y) \tag{A1.1.4}$$

and proved that:

THEOREM 1. *If (A1.1.1) is satisfied for all real* x, y, *and* $f(x)$ *is continuous everywhere then*:

$$f(x) = cx \qquad \text{for some real constant } c \tag{A1.1.5}$$

THEOREM 2. *The continuous solutions of (A1.1.2) are*:

$$f(x) = e^{cx} \quad \text{or} \quad f(x) = 0 \qquad \text{for all real } x,y \tag{A1.1.6}$$

THEOREM 3. *The continuous solution of (A1.1.3) for positive* x *and* y *is*:

$$f(x) = c \log x \tag{A1.1.7}$$

THEOREM 4. *The continuous solutions of (A1.1.4) for positive* x *and* y *are*:

$$f(x) = x^c \quad \text{or} \quad f(x) = 0 \tag{A1.1.8}$$

PROOF: The proof will be essentially that given by Cauchy (1821).

(1) From (A1.1.1) we have, by induction:

$$f(x+y+z+u+\ldots) = f(x) + f(y) + f(z) + f(u) + \ldots$$

In particular, for m equal terms $x = y = z = \ldots = a$ one finds $f(ma) = mf(a)$. To extend the validity of the last equation for the positive rational numbers m/n, let $b = (m/n)a$, that is:

$$nb = ma$$

$$nf(b) = mf(a)$$

$$f(b) = f\left(\frac{m}{n}a\right) = \frac{m}{n}f(a)$$

If r is any *positive real* number, we can select a sequence of positive rational numbers $m_i/n_i \to r$, and, because of the assumed continuity of $f(x)$ we get $f(ra) = rf(a)$ for any real $r > 0$ and a. In particular, for $a = 1$ we have $f(r) = rf(1)$ for all positive values of r, and — because of the continuity of $f(x)$ — we also have $f(0) = 0$. On the other hand, by putting $x = 0$, $y = -r$ in equation (A1.1.1):

$$f(-r) = f(0) - f(r) = -rf(1)$$

that is, $f(r) = rf(1)$ is generally valid for positive and negative real numbers and for zero. Equation (A1.1.5) follows by putting $c = f(1)$. Q.E.D.

(2) The proof of THEOREM 2 follows similar lines. From (A1.1.2) we first observe that for any real x one has:

$$f(x) = f\left(\frac{x}{2} + \frac{x}{2}\right) = \left[f\left(\frac{x}{2}\right)\right]^2 \geq 0$$

It is easy to prove that either $f(x) > 0$ for all values of x or $f(x) = 0$ everywhere. Indeed, suppose that for some value x_0 we have $f(x_0) = 0$. Then for any x:

$$f(x) = f(x - x_0 + x_0) = f(x - x_0)\, f(x_0) = 0$$

We can disregard the trivial solution $f(x) = 0$ in what follows and assume that $f(x) > 0$. By induction, for any number of variables $f(x + y + z + \ldots) = f(x) \cdot f(y) \cdot f(z) \cdot \ldots$; in particular for any *integer* $m > 0$ and any real number a we have $f(ma) = [f(a)]^m$. Again, if b is a positive rational multiple of a, say $b = (m/n)a$, then:

$$nb = ma$$

$$f(b)^n = f(a)^m$$

$$f(b) = f\left(\frac{m}{n}a\right) = [f(a)]^{m/n}$$

where the extraction of the n-th root is meaningful because of the non-negativity of $f(x)$. If r is a positive real number and m_i/n_i is a sequence of positive rational numbers approaching r, we find in the limit $f(ra) = [f(a)]^r$ because of the continuity of $f(x)$. In particular, $f(r) = f(r \cdot 1) = [f(1)]^r$, and, letting $r \to 0$, $f(0) = 1$.

On the other hand, for $r < 0$, from equation (A1.1.2):

$$f(-r) = f(0 - r) = \frac{f(0)}{f(r)} = [f(1)]^{-r}$$

that is, we have $f(x) = [f(1)]^x$ for all *real* values x. Equation (A1.1.6) is obtained if we choose c as to satisfy $\exp c = f(1)$. Q.E.D.

(3) The remaining two Cauchy equations (A1.1.3) and (A1.1.4) can be solved in a similar manner, but they can also be reduced to the first two equations very easily (see e.g. Aczel, 1966).

The substitutions:

$$x = e^u \qquad y = e^v \qquad f(e^u) = g(u) \tag{A1.1.9}$$

transform the equations:

$$f(xy) = f(x) + f(y) \tag{A1.1.3}$$

$$f(xy) = f(x)\, f(y) \tag{A1.1.4}$$

into:

$$g(u + v) = g(u) + g(v) \quad \text{and} \quad g(u + v) = g(u)\, g(v)$$

respectively, which were treated above. That is, the continuous solutions of (A1.1.3) and (A1.1.4) for positive x, y are $f(x) = c \log x$ for equation (A1.1.3), and $f(x) = \exp(c \log x) = x^c$ or $f(x) = 0$ for equation (A1.1.4). Q.E.D.

PEXIDER'S EQUATIONS

J.V. Pexider (1903, see Aczel, 1966) studied the following generalizations of Cauchy's functional equations:

$$f(x + y) = g(x) + h(y) \tag{A1.1.10}$$

$$f(x + y) = g(x)\, h(y) \tag{A1.1.11}$$

$$f(xy) = g(x) + h(y) \tag{A1.1.12}$$

$$f(xy) = g(x)\, h(y) \tag{A1.1.13}$$

and proved the following theorems:

THEOREM 5. *The most general continuous system of solutions of (A1.1.10) is:*

$$f(x) = cx + a + b \qquad g(x) = cx + a \qquad h(x) = cx + b \tag{A1.1.14}$$

THEOREM 6. *The most general continuous system of solutions of (A1.1.11) is:*

$$f(x) = ab \exp(cx) \qquad g(x) = a \exp(cx) \qquad h(x) = b \exp(cx) \tag{A1.1.15}$$

(apart from the trivial solutions $f(x) = g(x) = 0$, $h(x)$ arbitrary; $f(x) = h(x) = 0$, $g(x)$ arbitrary).

THEOREM 7. *The most general continuous system of solutions of (A1.1.12) for positive x,y is:*

$$f(x) = c\log(abx) \qquad g(x) = c\log(ax) \qquad h(x) = c\log(bx) \tag{A1.1.16}$$

with $a,b > 0$.

THEOREM 8. *The most general continuous system of solutions of (A1.1.13) for positive x,y is (apart from the trivial solutions):*

$$f(x) = abx^c \qquad g(x) = ax^c \qquad h(x) = bx^c \tag{A1.1.17}$$

PROOF: We only sketch here the proof of THEOREM 8, the other proofs proceed along similar lines.

Let us write out $f(xy) = g(x)h(y)$ first for $x = 1, y = t$, then for $x = t, y = 1$:

$$f(t) = g(1)\,h(t) \qquad h(t) = \frac{f(t)}{a} \quad \text{with} \quad a = g(1)$$

$$f(t) = g(t)\,h(1) \qquad g(t) = \frac{f(t)}{b} \quad \text{with} \quad b = h(1)$$

$$f(xy) = \frac{f(x)\,f(y)}{ab} \qquad (ab \neq 0)$$

Let $F(t) = f(t)/ab$; then $F(t)$ satisfies $F(xy) = F(x)F(y)$; that is, by Cauchy's fourth functional equation $F(t) = t^c$ and we have:

$$f(x) = abx^c \qquad g(x) = ax^c \qquad h(x) = bx^c$$

[The trivial solutions occur in the cases $g(1) = 0$ or $h(1) = 0$, these are: $f(x) = 0$, $g(x) = 0$, $h(x)$ arbitrary; or $f(x) = 0$, $g(x)$ arbitrary, $h(x) = 0$.] Q.E.D.

EULER'S EQUATION FOR HOMOGENEOUS FUNCTIONS

THEOREM 9 (Euler, 1755; see Aczel, 1966). *If the function $F(x,y)$ satisfies the equation of k-th order homogeneity ($k \neq 0$):*

$$F(rx,ry) = r^k F(x,y) \qquad (x \neq 0,\ r \neq 0) \tag{A1.1.18}$$

then:

$$F(x,y) = x^k f\left(\frac{y}{x}\right) \tag{A1.1.19}$$

PROOF:

$$F(x,y) = F\left(x\cdot 1, x\frac{y}{x}\right) = x^k F\left(1,\frac{y}{x}\right) = x^k f\left(\frac{y}{x}\right)$$

Q.E.D. (A careful discussion of the cases $x = 0, y \neq 0$ and $x = 0, y = 0$ as well as $k = 0$ can be found in Aczel, 1966.)

A1.2 *Fractal curves are hot*

In this paragraph C will denote a planar curve. The length of C is represented by $[C]$, it is infinite for non-rectifiable curves.

Let H and ∂H respectively be the convex hull of C and its boundary (Figure A1.1). The lengths $[C]$ and $[\partial H]$ satisfy the inequality:

$$2[C] \geq [\partial H] \tag{A1.2.1}$$

where equality holds if and only if C is a straight segment. Steinhaus's result (1.2.5) can be stated as the following:

THEOREM 10. *The average number of intersection points of C with a random straight line L is:*

$$\sum_{n=0}^{\infty} n p_n = \frac{2[C]}{[\partial H]} \tag{A1.2.2}$$

where p_n is the probability that a random straight line L intersects C in n points, $\sum_0^\infty p_n = 1$.

[The definition of "random" straight lines was given in Section 1.2. Equation (A1.2.2) is due to Steinhaus (1954); its proof can also be found in Santalo (1976), p. 31.]

M. Mendes France (1984, 1989; DuPain et al., 1986) used Steinhaus's formula (A1.2.2) to develop the thermodynamics of irregular planar curves.

Mendes France defined the *entropy* of a finite length curve C as:

$$S(C) = \sum_{n=0}^{\infty} p_n \log p_n \tag{A1.2.3}$$

and proved the following:

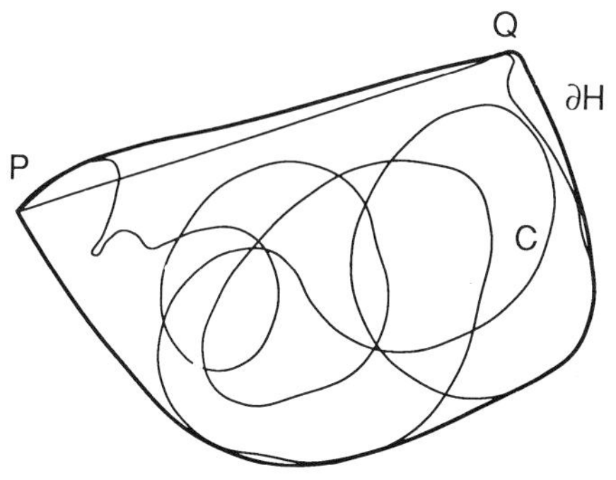

Fig. A1.1. Curve C connecting points P and Q, and the boundary of its convex hull, ∂H.

THEOREM 11.

$$S(C) \leq S_t(C) = \log \frac{2[C]}{[\partial H]} + \frac{\beta}{\exp(\beta) - 1} \tag{A1.2.4}$$

where:

$$\beta = \log \frac{2[C]}{2[C] - [\partial H]} > 0 \tag{A1.2.5}$$

PROOF (Mendes France, 1989; DuPain et al., 1986): We use the Lagrange technique to maximize the entropy $S(C) = S(p_1, p_2, \ldots)$ subject to the two conditions:

$$\sum_0^\infty p_n = 1 \qquad \text{and} \qquad \sum_0^\infty n p_n = \frac{2[C]}{[\partial H]}$$

The maximum is attained for the Gibbs distribution:

$$p_n^* = a \exp(-\beta n) \tag{A1.2.6}$$

where the constants a and β are:

$$a = \exp(\beta) - 1 \qquad \text{and} \qquad \beta = \log \frac{2[C]}{2[C] - [\partial H]}$$

Hence:

$$p_n^* = \frac{[\partial H]}{2[C] - [\partial H]} \left\{1 - \frac{[\partial H]}{2[C]}\right\}^n \tag{A1.2.7}$$

and it is easy to check that $S(p_1^*, p_2^*, \ldots) = S_t(C)$. Q.E.D.

Mendes France called $S_t(C)$ the *topological entropy* of the curve C.

In statistical physics one usually identifies the exponent β figuring in the Gibbs distribution with the inverse of the absolute temperature (in a system of units where the Boltzmann constant is 1):

$$T = \left\{\log \frac{2[C]}{2[C] - [\partial H]}\right\}^{-1} \tag{A1.2.8}$$

Observe that $T \geq 0$ as it should and, by (A1.2.1), $T = 0$ if and only if $2[C] = [\partial H]$. In other words *at zero temperature the curves "freeze" to straight segments*. To pursue the parallel with thermodynamics, Mendes France has defined the "volume" V of the curve C as its length: $V = [C]$, and the "pressure" P as the ratio $1/[\partial H]$. With these notations equation (A1.2.8) becomes the equation of state:

$$PV = \frac{1}{2}\left\{1 - \exp\left(-\frac{1}{T}\right)\right\}^{-1} \tag{A1.2.9}$$

For $T \to 0$, the product PV tends to 1/2; that is $[C]/[\partial H] = 1/2$, and the curve C collapses to a straight line. On the other hand, for high temperatures $T \gg 1$ equation (A1.2.9) reduces to "Boyle's law":

$$PV \approx \frac{T}{2} \tag{A1.2.10}$$

Let us now apply Mendes Frances' formalism to study the temperature of *fractal* curves.

First we observe from "Boyle's law" (A1.2.10) that for $1 < D \leq 2$ any D-dimensional fractal curve C which is entirely contained within a finite (non-fractal) convex hull H has infinitely high temperature.

Indeed, if we measure in equation (A1.2.10) the length of C and of ∂H, respectively, by a yardstick $\varepsilon \to 0$, we get $[C] \approx \varepsilon^{1-D} \to \infty$ while $[\partial H]$ remains independent of ε, that is $T = \infty$.

The temperature is not necessarily uniform along a curve. DuPain et al. (1986) defined the local lower temperature $T_*(A)$, the local upper temperature $T^*(A)$ and the local temperature $T(A)$ of a curve C at the point A as follows:

$$T_*(A) = \liminf_{M,N\to A} T(\widehat{MN}) \qquad T^*(A) = \limsup_{M,N\to A} T(\widehat{MN}) \tag{A1.2.11}$$

where $A, M, N \in C$. If $T_* = T^* = T(A)$, this is called the local temperature at A.

We shall need a further function defined on the curve C:

$$d(A) = \limsup_{M,N\to A; M,N,A\in C} \left(\frac{[\widehat{MN}]}{[\overline{MN}]} \right) \tag{A1.2.12}$$

($\widehat{MN}$ refers to the arc, $\overline{MN}$ to the straight chord, between M and N.)

DuPain et al. (1986) proved the following:

THEOREM 12. *$T(A)^* = 0$ if and only if $d(A) = 1$. If $d(A) = \infty$ then $T(A)^* \geq 1/\log 2 = 1.442695$.*

PROOF: We observe that:

$$2[\overline{MN}] \leq [\partial H(\widehat{MN})] \leq [\overline{MN}] + [\widehat{MN}] \tag{A1.2.13}$$

(see Figure A1.1, where there are P and Q instead of M and N.) Then:

$$\frac{2[\widehat{MN}]}{[\widehat{MN}] + [\overline{MN}]} \leq \frac{2[\widehat{MN}]}{[H(\widehat{MN})]} \leq \frac{[\widehat{MN}]}{[\overline{MN}]} \tag{A1.2.14}$$

and simple algebra gives:

$$\left\{\log \frac{2d(A)}{d(A)-1}\right\}^{-1} \leq T^*(A) \leq \left\{\log \frac{d(A)}{d(A)-1}\right\}^{-1} \tag{A1.2.15}$$

which proves the theorem.

Recalling from the theory of real functions that for any rectifiable curve $d(A) = 1$ *almost everywhere* (Riesz and Sz.-Nagy, 1956, p. 27), THEOREM 12 implies that the local temperature of a rectifiable curve is almost everywhere zero, though the curve might contain "hot" points.

DuPain et al.(1986) have shown, for example, that the local temperature of the exponential spiral $\rho = \exp(-\theta)$ vanishes everywhere except in the origin where it is $T(0) = 0.5296$; while the local temperature of the spiral:

$$\rho = \theta^{-\lambda} \qquad (\lambda > 1)$$

is zero everywhere and infinite at its centre (as in the centre of the galaxies!).

THEOREM 13. *For all points of a fractal curve* $T = \infty$.

PROOF: If S is a planar point set, let diam (S) denote its diameter:

$$\text{diam}\,(S) = \limsup_{X,Y \in S}\,[\overline{XY}]$$

For any finite curve $\overset{\frown}{AB}$ we have the inequality:

$$2\,\text{diam}\,(\overset{\frown}{AB}) \leq [\partial H(\overset{\frown}{AB})] \leq \pi\,\text{diam}\,(\overset{\frown}{AB}) \tag{A1.2.16}$$

Let C be a fractal curve of dimension $D > 1$; let $M, N, A \in C$; $M, N \rightarrow A$. Let us select a "yardstick" $\varepsilon \rightarrow 0$, and for any arc $\overset{\frown}{MN}$ let $[\overset{\frown}{MN}_\varepsilon]$ denote the arclength measured with resolution ε.

Consider the set $\overset{\frown}{MN} \cup \overline{MN}$ and let X denote its diameter: $X = \text{diam}\,(\overset{\frown}{MN} \cup \overline{MN})$. Because of the fractality of C, we have $[\overset{\frown}{MN}_\varepsilon] \propto X^D$, that is:

$$X = k[\overset{\frown}{MN}_\varepsilon]^{1/D} \tag{A1.2.17}$$

where the constant $k = k(\varepsilon)$ is independent of A, M and N. By inequality (A1.2.16) the length of the convex hull $[\partial H(\overset{\frown}{MN} \cup \overline{MN})]$ satisfies:

$$2k[\overset{\frown}{MN}_\varepsilon]^{1/D} \leq [\partial H(\overset{\frown}{MN} \cup \overline{MN})] \leq k\pi[\overset{\frown}{MN}_\varepsilon]^{1/D} \tag{A1.2.18}$$

and considerations similar to those in the proof of THEOREM 12 yield:

$$\left\{\log \frac{2P}{2P - 2kP^{1/D}}\right\}^{-1} \geq T^*(A) \geq T_*(A) \geq \left\{\log \frac{2P}{2P - kP^{1/D}}\right\}^{-1} \tag{A1.2.19}$$

where for brevity we have put $P = [\overset{\frown}{MN}_\varepsilon]$.

Letting $\varepsilon \rightarrow 0$, we have $P \rightarrow \infty$ and $T_*(A) = T^*(A) = T(A) = \infty$. Q.E.D.

References

Aczel, J. (1966). Lectures on Functional Equations and Their Applications. Academic Press, New York and London.

Amaefule, J.O., Kersey, D.G., Marschall, D.M., Powell, J.D., Valencia, L.E. and Keeland, D.K. (1988). Reservoir description: a practical synergistic engineering and geological approach based on analysis of core data. SPE Paper No. 18167.

Aviles, C.A., Scholz, C.H. and Boatwright, J. (1987). Fractal analysis applied to characteristic segments of the San Andreas fault. J. Geophys. Res., 92, B1: 331–344.

Barton, C.C. and Larsen, E.F. (1985). Fractal geometry of two-dimensional fracture networks at Yucca Mountain, Southwestern Nevada. Proc. Int. Symp. on Fundamentals of Rock Joints, Bjorkliden, pp. 77–84.

Beech, M. (1987). Are Lynds dark clouds fractals? Astrophys. Space Sci., 133: 193–195.

Blaschke, W. (1936). Vorlesungen über Integralgeometrie. Leipzig.

Bradbury, R.H. and Reichelt, R.E. (1983). Fractal dimension of a coral reef at ecological scales. Mar. Ecol. Progr. Ser. 10: 169–171.

Broadbent, S.R. (1954). Discussion on Symposium on Monte Carlo Methods. J. R. Stat. Soc., B, 68.

Brown, S.R. and Scholz, C.H. (1985). Broad bandwidth study of the topography of natural rock surfaces. J. Geophys. Res., 90, B14: 12575–12582.

Bruinsma, R. (1990). The statistical mechanics of meandering. J. Phys. France, 51: 829–845.

Campbell, I.R. and Smith, D.N. (1982). Gorgon 1 — Southernmost Rankin platform gas discovery. The A.P.E.A. Journal, 22, Pt.1: 102–111.

Carlston, C.W. (1965). The relation of free meander geometry to stream discharge and its geomorphic implications. Am. J. Sci., 263: 864–885.

Carson, M.A. and Lapointe, M.F. (1983). The inherent asymmetry of river meander platform. J. Geol., 91: 41–55.

Cauchy, A.L. (1821). Course d'Analyse de l'Ecole Polytechnique. Vol.1. Analyse Algebraïque. *In*: Oeuvres, Ser. 2., Vol.3, pp. 98–113. Gauthier-Villars et Fils, Paris, 1897.

Chilès, J.P. (1988). Fractal and geostatistical methods for modelling of a fracture network. Math. Geol., 20(6): 631–654.

Crofton, M.W. (1868). On the theory of local probability, applied to straight lines drawn at random in a plane. Phil. Trans. R. Soc., 158: 181–199.

Denson, K.H., Shindala, A.S. and Fenn, C.D. (1968). Permeability of sand with dispersed clay particles. Water Resource Res., 4(6): 1275–1276.

Dolan, R. and Ferm, J.C. (1968). Crescentic landforms along the Atlantic coast of the United States. Science, 159: 627–629.

Dolan, R., Vincent, L. and Hayden, B. (1974). Crescentic coastal landforms. Z. Geomorph. N. F., 18(1): 1–12.

Dullien, F.A.L. (1979). Porous Media. Fluid Transport and Pore Structure. Academic Press, New York.

DuPain, Y., Kamae, T. and Mendes France, M. (1986). Can one measure the temperature of a curve? Arch. Rat. Mech. Anal., 94(2): 155–163.

Efros, A.L. (1986). Physics and Geometry of Disorder. Percolation Theory. Mir Publishers, Moscow.

Elam, W.T., Kerstein, A.R. and Rehr, J.J. (1984). Critical properties of the void percolation problem for spheres. Phys. Rev. Lett., 52(17): 1516–1519.

Essam, J.W. (1972). Percolation and cluster size. *In*: C. Domb and M.S. Green (Editors), Phase Transitions and Critical Phenomena , Vol.2, Academic Press, New York, pp. 197–270.

Ethridge, F.G. and Schumm, S.A., (1978). Reconstructing paleochannel morphologic and flow characteristics: methodology, limitations, and assessment. *In*: A.D. Miall (Editor), Fluvial Sedimentology , Can. Soc. Petr. Geol. Mem., 5: 703–721.

Euler, L. (1755). Institutiones Calculi Differentialis. §222. *In*: Opera Omnia, Ser.1, Vol.10, p. 150, Leipzig, 1913.

Feder, J. (1988). Fractals. Plenum Press, New York and London.

Fisch, R. and Harris, A.B. (1978). Critical behavior of random resistor networks near the percolation threshold. Phys. Rev. B., 18(1): 416–420.

Flook, A.G. (1979). The characterization of textural and structural profiles by the automated measurement of their fractal dimension. 2nd European Symp. Particle Characterization, pp. 591–599.

Forrest, S.R. and Witten, T.A., Jr. (1979). Long-range correlations in smoke particle aggregates. J. Phys. A. (Math. Gen.), 12: L109–L117.

Fricker, F. (1982). Einführung in die Gitterpunktlehre. Birkhäuser Verlag, Basel.

Gadenne, M., Lafait, J. and Gadenne, P. (1989). Infrared optical absorption and fractal structure of granular media at percolation. Optics Comm., 71(5): 273–278.

Goodchild, M.F. (1988). Lakes on fractal surfaces: a null hypothesis for lake-rich landscapes. Math. Geol., 20(6): 615–630.

Goode, P. and Sen, P.N. (1988). Charge density and permeability in clay-bearing sandstones. Geophysics, 53(12): 1610–1612.

Gravestock, D.I. and Alexander, E.M. (1986). Porosity and permeability of reservoirs and caprocks in the Eromanga Basin, South Australia. Austr. Petr. Expl. Assoc. J., 26: 202–213.

Gravestock, D.I. and Alexander, E.M. (1989). Petrophysics of oil reservoirs in the Eromanga Basin, South Australia. Proc. Cooper and Eromanga Basins Conf., Adelaide, pp. 629–645.

Green, J.E. and Moore, M.A. (1982). Application of directed lattice animal theory to river networks. J. Phys. A. (Math. Gen.), 15: L597–L599.

Hack, J.T. (1957). Studies of longitudinal stream profiles in Virginia and Maryland. U.S. Geol. Surv. Prof. Pap., 294-B: 46–94.

Hallet, B. (1987). On geomorphic patterns with a focus on stone circles viewed as a free convection phenomenon. *In*: C. and G. Nicolis (Editors), Irreversible Phenomena and Dynamical Systems Analysis in Geosciences. D. Reidel Publ. Co., Dordrecht–Boston–Lancaster–Tokyo.

Halperin, B.I., Feng, S. and Sen, P.N. (1985). Differences between lattice and continuum percolation transport exponents. Phys. Rev. Lett., 54(22): 2391–2394.

Hammersley, J.M. (1983). Origins of percolation theory. *In*: G. Deutscher, R. Zallen and J. Adler (Editors), Percolation Structures and Processes. Ann. Isr. Phys. Soc., 5: 48–57.

Hanin, A.A. (1951). On the effect of the cement content of sandstones on the decrease of permeability and porosity. Tr. VNIIGAZ, No. 232 (in Russian).

Henderson-sellers, A. (1986). Are Martian clouds fractal? Q. J. R. Astron. Soc., 27: 90–93.

Hentschel, H.G.E. and Procaccia, I. (1984). Relative diffusion in turbulent media; the fractal dimension of clouds. Phys. Rev. A., 29(3): 1461–1470.

Hindley, D. and Hindley, G. (1972). Advertising in Victorian England 1837–1901. Wayland Publishers, London.

Hirata, T., Satoh, T. and Ito, K. (1987). Fractal structure of spatial distribution of microfracturing in rock. Geophys. J. R. Astron. Soc., 90: 369–374.

Hoyle, J.W. and King, G.T. (1958). Origin and stability of beaches. Proc. 6th Coastal Eng. Conf., pp. 281–301.

Hutchinson, G.E. (1957). A Treatise on Limnology. Vol. 1. Wiley, New York.

Jackson, D.P. and Jackson, J.A. (1988). Tibetan Thangka Painting. Methods and Materials. 2nd Revised Edition. Serindia Publs., London.

Kapraff, J. (1986). The geometry of coastlines: A study in fractals. Comp. Maths. Appl., 12B(3/4): 655–671.

Katz, A.J. and Thompson, A.H. (1986). Quantitative prediction of permeability in porous rock. Phys. Rev. B., 34: 8179–8181.

Kaye, B.H. (1978). Specification of the ruggedness and/or structure of a fine particle profile by its fractal dimension. Powder Technol., 21: 1–16.

Kent, C. and Wong, J. (1982). An index of littoral zone complexity and its measurement. Can. J. Fish. Aquat. Sci., 39: 847–853.

Korvin, G. (1989). Fractured but not fractal: fragmentation of the Gulf of Suez Basement. PAGEOPH, 131(1/2): 289–305.

Korvin, G. (1992). A percolation model for the permeability of kaolinite-bearing sandstones. Geophys. Trans. (in press).

Krummel, J.R., Gardner, R.H., Sugihara, G., O'Neill, R.V. and Coleman, P.R. (1987). Landscape patterns in a disturbed environment. OIKOS, 48: 321–324.

Leeder, M.R. (1973). Fluviatile fining-upwards cycles and the magnitude of paleochannels. Geol. Mag., 110(3): 265–276.

Le Guillou, J.C. and Zinn-Justin, J. (1977). Critical exponents for the n-vector model in three dimensions from field theory. Phys. Rev. Lett., 39: 95–98.

Leonardo Da Vinci: The Notebooks (Editor: J.P. Richter). 2 Vols. Dover Publs..Inc., New York, 1970.

Leopold, L.B. and Langbein, W.B. (1962). The concept of entropy in landscape evolution. U.S. Geol. Surv. Prof. Pap., 500-A, 20 pp.

Leopold, L.B. and Wolman, M.G. (1960). River meanders. Bull. Geol. Soc. Am., 71: 769–794.

Leopold, L.B., Wolman, M.G. and Miller, J.P. (1964). Fluvial Processes in Geomorphology. Freeman and Co., San Francisco.

Levy, P. (1965). Processus Stochastiques et Mouvement Brownien. Gauthiers Villars, Paris, 2nd Edition.

Lorenz, J.C., Heinze, D.M., Clark, J.A. and Searls, C.A. (1985). Determination of widths of meander-belt sandstone reservoirs from vertical downhole data, Mesaverde Group, Piceance Creek Basin, Colorado. Bull. AAPG, 69(5): 710–721.

Lovejoy, S. (1982). Area–perimeter relation for rain and cloud areas. Science, 216: 185–187.

Lubensky, T.C. (1977). Scaling theory of phase transitions in diluted systems near the percolation threshold. Phys. Rev. B., 15(1): 311–314.

Madden, T.R. (1983). Microcrack connectivity in rocks: a renormalization group approach to the critical phenomena of conduction and failure in crystalline rocks. J. Geophys. Res., 88, B1: 585–592.

Mandelbrot, B. (1967). How long is the coast of Britain? Statistical self-similarity and fractional dimension. Science, 165: 636–638.

Mandelbrot, B.B. (1975). Stochastic models for the Earth's relief, the shape and the fractal dimension of the coastlines, and the number–area rule for islands. Proc. Nat. Acad. Sci. USA, 72(10): 3825–3828.

Mandelbrot, B.B. (1977). Fractals: Form, Chance, and Dimension. Freeman, San Francisco.

Mandelbrot, B.B. (1982). The Fractal Geometry of Nature. Freeman, New York.

Mandelbrot, B.B., Passoja, D.E. and Paullay, A.J. (1984). Fractal character of fracture surfaces of metals. Nature, 308: 721–722.

Martin, K.R. and Hamilton, N.J. (1981). Diagenesis and reservoir quality, Toolache formation, Cooper Basin. Austr. Petr. Explor. Assoc. J., 21, Pt.1: 143–154.

Mclean, R. (1967). Plan shape and orientation of beaches along the east coast, South Island. N. Z. Geogr., 23(1): 16–22.

Mecholsky, J.J., Mackin, T.J. and Passoja, D.E. (1988). Self-similar crack propagation in brittle materials. *In*: V.D. Frechette and J.R. Varner (Editors), Fractography of Glasses and Ceramics. Advances in Ceramics, 22: 127–134.

Mendes France, M. (1984). Folding paper and thermodynamics. Phys. Rep., 103(1–4): 161–172.

Mendes France, M. (1989). Chaos implies confusion. *In*: M.M. Dodson and J.A.G. Vickers (Editors), Number Theory and Dynamical Systems. Lond. Math. Soc. Lect. Note Ser. 134, Cambridge Univ. Press, Cambridge, pp. 137–152.

Michaels, A.S. and Lin, C.S. (1954). Permeability of kaolinite. Ind. Eng. Chem., 46(6): 1239–1246.

Mosolov, A.B. and Dinaryev, O.Yu. (1987). Fractal models of porous media. J. Tech. Phys., 57(9): 1679–1685 (in Russian).

Mu, Z.Q. and Lung, C.W. (1988). Studies on the fractal dimension and fracture toughness of steel. J. Phys. D, Appl. Phys., 21: 848–850.

Nakano, T. (1983). A "fractal" study of some rias coastlines in Japan. Ann. Rep. Inst. Geosci., Univ. Tsukuba, 9: 75–80.

Nakano, T. (1984). A systematics of "transient fractals" of rias coastline; an example of rias coast from Kamaishi to Shizugawa, Northeastern Japan. Ann. Rep. Inst. Geosci., Univ. Tsukuba, 10: 66–68.

Neasham, J.W. (1977). The morphology of dispersed clay in sandstone reservoirs and its effects on sandstone shaliness, pore space and fluid flow properties. SPE Paper No. 6858.

Okubo, P.G. and Aki, K. (1987). Fractal geometry in the San Andreas Fault System. J. Geophys. Res., 92, B1: 345–355.

Orford, J.D. and Whalley, W.B. (1983). The use of fractal dimension to quantify the morphology of irregular-shaped particles. Sedimentology, 30: 655–668.

Pande, C.S., Richards, L.R. and Smith, S. (1987a). Fractal characteristics of fractured surfaces. J. Mater. Sci. Lett., 6: 295–297.

Pande, C.S., Richards, L.E., Louat, N., Dempsey, B.D. and Schwoeble, A.J. (1987b). Fractal characterization of fractured surfaces. Acta Metall., 35(7): 1633–1637.

Pape, H., Riepe, L. and Schopper, J.R. (1981). A solution attempt to the "coastline of Britain problem". Stereol. Iugosl., 3/Suppl.1: 331–336.

Pape, H., Riepe, L. and Schopper, J.R. (1987). Theory of self-similar network structures in sedimentary and igneous rocks and their investigation with microscopical and physical methods. J. Microsc., 148, Pt. 2: 121–147.

Patchett, J.G. (1975). An investigation of shale conductivity. SPWLA 16th Ann. Log. Symp. Trans., Paper U.

Pexider, J.V. (1903). Notiz über Funktionaltheoreme. Monatsh. Math. Phys., 14: 293–301.

Rademacher, H. and Töplitz, O. (1957). The Enjoyment of Mathematics. Princeton Univ. Press, Princeton, NJ.

Redner, S. and Yang, Z.R. (1982). Size and shape of directed lattice animals. J. Phys. A, Math. Gen., 15: L177–L187.

Richardson, L.F. (1960). Arms and Insecurity. A Mathematical Study of the Causes and Origins of War. (Posthumous edition, edited by N. Rashevsky and E. Trucco.) Atlantic Books, Stevens and Sons, Ltd., London.

Richardson, L.F. (1961). The problem of contiguity. An appendix to the "Statistics of Deadly Quarrels". Gen. Sys., 6: 139–187.

Riesz, F. and Sz.-Nagy, B. (1956). Functional Analysis. Blackie and Son, Ltd., London and Glasgow.

Ritzenberger, A.L. and Cohen, R.J. (1984). First passage percolation: scaling and critical exponents. Phys. Rev. B, 30(7): 4038–4040.

Rys, F.S. and Waldvogel, A. (1986). Fractal shape of hail clouds. Phys. Rev. Lett., 56(7): 784–787.

Santalo, L.A. (1976). Integral Geometry and Geometric Probability. Encycl. Math. Appl., Vol.1. Addison-Wesley Publ.Co., Reading, Mass.

Scholz, C.H. and Aviles, C.A. (1986). The fractal geometry of faults and faulting. *In*: S. Das, J. Boatwrigh, and C.H. Scholz (Editors) Earthquake Source Mechanics. Geophys.Monogr., 37, Maurice Ewing Vol.6, A.G.U., Washington, D.C., pp. 147–155.

Schwenk, T. (1980). Das Sensible Chaos. Strömendes Formenschaffen in Wasser und Luft. Verlag Freies Geistesleben, Stuttgart.

Segall, P. and Pollard, D.D. (1983). Joint formation in granitic rock of the Sierra Nevada. Bull. Geol. Soc. Am., 94: 563–575.

Sreenivasan, K.R. and Meneveau, C. (1986). The fractal facets of turbulence. J. Fluid Mech., 173: 357–386.

Stanley, H.E. (1986). Form: an introduction to self-similarity and fractal behaviour. *In*: H.E. Stanley and N. Ostrowsky (Editors), On Growth and Form. Fractal and Non-Fractal Patterns in Physics. Martinus Nijhoff Publ., Dordrecht, pp. 21–53.

Stauffer, D. (1985). Introduction to Percolation Theory. Taylor and Francis, London.

Steiner, J. (1967). The sequence of geological events and the dynamics of the Milky Way Galaxy — the present cosmic year; a preliminary study. J. Geol. Soc. Austr., 14: 99–132.

Steinhaus, H. (1938). Kalejdoskop Matematyczy. English translation: Mathematical Snapshots. G.E. Stechert and Co., New York, Leipzig, London, Paris, 1944

Steinhaus, H. (1954). Length, shape, area. Colloq. Math., 3(1): 1–13.

Stevens, P.S. (1974). Patterns in Nature. An Atlantic Monthly Press Book. Little, Brown and Co., Boston and Toronto.

Stupak, P.R. and Donovan, J.A. (1988). Fractal analysis of rubber wear surfaces and debris. J. Mater. Sci., 23: 2230–2242.

Sykes, M.F. and Essam, J.W. (1964). Exact critical percolation probabilities for site and bond problems in two dimensions. J. Math. Phys., 5(8): 1117–1127.

Suzuki, M. (1984). Finite-size scaling for transient similarity and fractals. Prog. Theor. Phys., 71(6): 1397–1400.

Takayasu, H. and Inaoka, H. (1992). A new type of self-organized criticality in a model of erosion. Phys. Rev. Lett. (in press).

Takayasu, H. and Nishikawa, I. (1986). Directed dendritic fractals. *In*: S. Ishizaka (General Editor), Science on Form: Proc. First Int. Symp. KTK Scientific Publishers, Tokyo.

Taylor, J.M. (1950). Pore space reduction of sandstone. Bull. AAPG, 34(4): 701–716.

Termonia, Y. and Meakin, P. (1986). Formation of fractal cracks in a kinetic fracture model. Nature, 320: 429–431.

Thompson, A.H., Katz, A.J. and Krohn, C.E. (1987). The microgeometry and transport properties of sedimentary rock. Adv. Phys., 36(5): 625–694.

Van Olphen, H. and Fripiat, J.J. (1979). Data Handbook for Clay Materials and Other Non Metallic Minerals. Pergamon Press, New York.

Villemin, T. and Sunwoo, C. (1987). Self-similarity and logarithmic distribution of fault-offsets and fault-lengths in the Lorraine coal mines. C. R. Acad. Sci. Paris, 305, Ser.II: 1309–1312 (in French).

Volarovich, M.P., Marmorshtein, L.M. and Mekler, Yu.B. (1968). Variation in the structure of the pore space in sandstones under pressure. Izv. Earth Phys., 6: 5–19.

Von Koch, H. (1906). Une méthode géometrique élémentaire pour l'étude de certaines questions de la théorie des courbes planes. Acta Math., 30: 145–174.

Vyssotsky, V.A., Gordon, S.B., Frisch, H.L. and Hammersley, J.M. (1961). Critical percolation probabilities (bond problem). Phys. Rev., 123(5): 1566–1567.

Walsh, J.B. and Brace, W.F. (1984). The effect of pressure on porosity and the transport properties of rock. J. Geophys. Res., 89, B11: 9425–9431.

Wardlaw, N.C. and Mckellar, M. (1981). Mercury porosimetry and the interpretation of pore geometry in sedimentary and artificial models. Powder Technol., 29: 127–143.

Williams, Ch. (1981). Origins of Form. Architectural Book Publ. Co., New York.

Wilson, K.G. and Kogut, J. (1974). The renormalization group and the ε-expansion. Phys.Rep., 12(2): 75–200.

Wong, P.-Z. (1988). The statistical physics of sedimentary rock. Physics Today, December 1988, pp. 23–32.

Young, A.P. and Stinchcombe, R.B. (1975). A renormalization group theory for percolation problems. J. Phys. C, Solid State Phys., 8: L535–L540.

Zallen, R. (1983). Introduction to percolation: A model for all seasons. *In*: G. Deutscher, R. Zallen and J. Adler (Editors), Percolation Structures and Processes. Ann. Isr. Phys. Soc., 5: 4–16.
Ziman, J.M. (1979). Models of Disorder. Cambridge Univ. Press, Cambridge.

2 Fractals in Flatland: a romance of <2 dimensions

If my Readers have followed me with any attention up to this point, they will not be surprised to hear that life is somewhat dull in Flatland. I do not, of course, mean that there are no battles, conspiracies, tumults, factions, and all those other phenomena which are supposed to make History interesting. — E.A. ABBOTT: *Flatland. A Romance of Many Dimensions*

2.1 The paradox of sedimentation rate

2.1.1 *Stratigraphic hiatuses and sedimentation rate*

Overall, he became a handsome and educated young man, and only occasionally did he exhibit barely noticeable signs that he was unlike others. For example, on Monday evenings he could take a different day from his future and use it the following morning, in place of Tuesday. When he came to the day he had taken, he would use the skipped Tuesday in its place, thereby adjusting the total. Under these conditions, of course, the connecting seams of the days could not fit together properly, and cracks appeared in time...

— M. PAVIČ: *Dictionary of the Khazars. The Red Book*
(translated by Ch. Pribičevič-Zorič, 1988)

To say that Professor Ager does not like the concept of "continuous sedimentation" would be an understatement. Connoisseurs have savoured his remark (in "The Nature of the Stratigraphic Record", 1981): "The late unlamented Field Marshall Goering once said that when he heard the word 'Culture', he reached for his gun; I feel rather the same about the phrase 'continuous sedimentation' ".

In the Chapter "More Gaps than Record" of his book Ager recalls the child's definition of a net ("a lot of holes tied together with string") to state that "the stratigraphical record is a lot of holes tied together with sediment".

More recent studies (Sadler, 1980, 1981; Schindel, 1980; Dingus and Sadler, 1982; Plotnick, 1986) have indicated that even the fishnet analogy is too generous: while the stratigraphic gaps (unconformities and diastems) exist on all scales, the preserved epochs of sedimentation only form a negligibly small fraction (as Plotnick, 1986, suggested, a *less than one-dimensional fractal "dust"*) of the total time.

Unfortunately, little is known about the distribution of the temporal length or spacing of the hiatuses, mainly because of the technical difficulties involved in measuring their absolute duration. There is however, a large volume of published estimates of the sedimentation rates measured over different time spans (Schwab, 1976; Sadler, 1980, 1981; Schindel, 1980; Dingus and Sadler, 1982) which allow an indirect estimation of the temporal distribution of stratigraphic hiatuses. According to

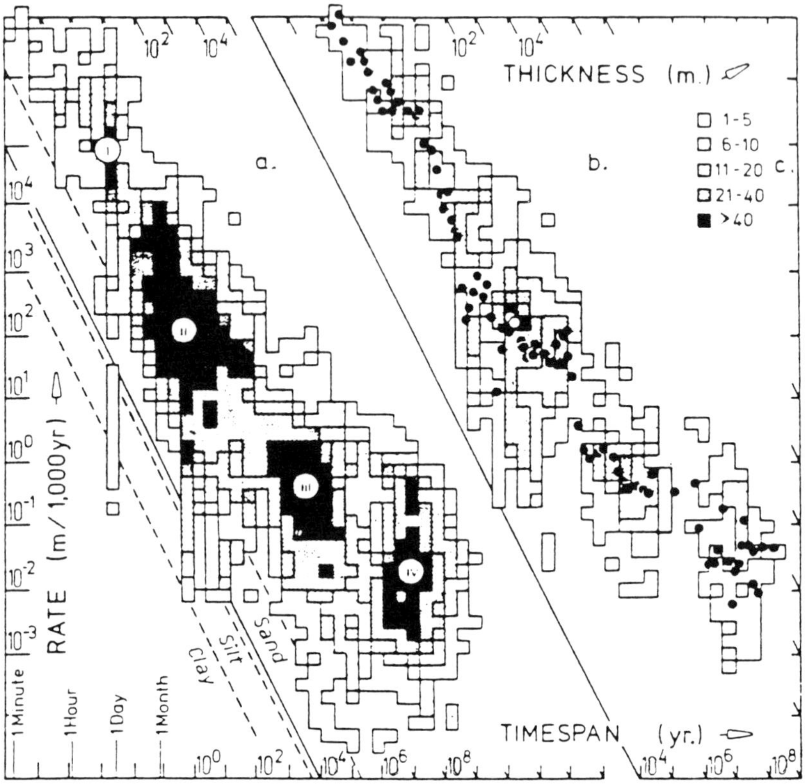

Fig. 2.1. Cluster density of sediment accumulation rates plotted against the time span over which they were determined (compiled on the basis of more than 25,000 data from over 700 references). All depositional environments are included. The key to density scale shows number of rates in a counting square of the same size as the key boxes. The modes *i* to *iv* correspond to the most popular ways of rate estimation: *i* = continuous observation; *ii* = reoccupation of survey stations; *iii* = radiocarbon dating; *iv* = biostratigraphy calibrated with K/Ar radiometry. (After Sadler, 1981.)

these studies, on a log–log plot the sedimentation rate $R(T)$ shows an inverse linear relationship with the time span T over which it is measured, that is:

$$R(T) \propto T^A \tag{2.1.1.1}$$

where the exponent A ($A < 0$) depends on the depositional environment.

This empirical rule has clearly been demonstrated by Sadler (1981) on the basis of more than 25,000 sediment accumulation rates compiled from over 700 references (Figure 2.1). Sadler (1981) has also found that the slope of the log $R(T)$ vs. log T line (that is the exponent A) varied among depositional environments (rivers, lakes, carbonate platforms and reefs, terrigenous shelves, small basinal seas, and abyssal ocean floors; see also Schwab, 1976).

According to Sadler (1981), the decreasing accumulation rates with increasing time spans might be due to systematic measurement error, post-depositional compaction, long-term evolution of the geomorphic systems (as e.g. "speeding up of the tectonic pulse of the earth" — Salop, 1977), or episodic sedimentation. He has concluded that while post-depositional compaction certainly contributes to the relative diminution of long-term rates, it cannot be the primary factor responsible for the power-like trend (2.1.1.1) and certainly could not lead to the observed differences between depositional environments.(On the importance of adjusting to a constant porosity, that is "de-compaction", see however Behrens, 1980. Tipper, 1983, also doubts the reality of the decrease of accumulation rates with time, and claims that it is only a statistical artifact.)

Sadler (1981) also investigated the relationship between the duration and accumulation rates for eras, periods and epochs from the Palaeoprotozoic to the Quaternary (Figure 2.2) and concluded that " discontinuities exist at all scales and... sections which span longer time intervals have the opportunity to incorporate more and longer hiatuses". Curiously, though in one place of his 1981 paper

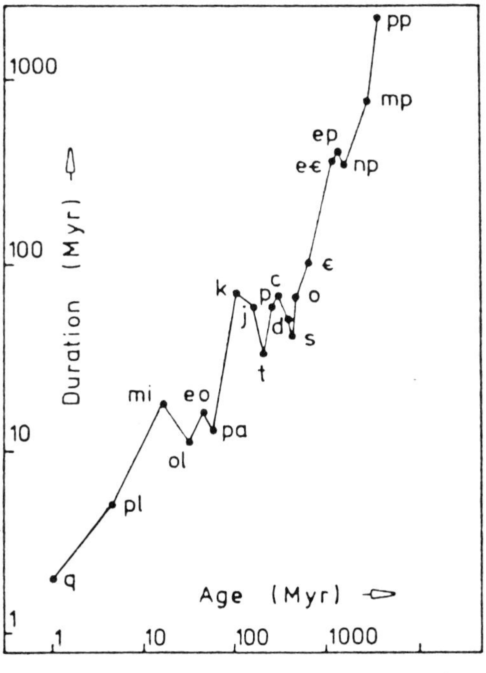

Fig. 2.2. Relationship between duration and maximum accumulation rate for eras, periods and epochs, taken from Gilluly (1949) and Salop (1977). *pp* = Palaeoprotozoic, 3500–2600 m.y.; *mp* = Mesoprotozoic, 2600–1900 m.y.; *np* = Neoprotozoic, 1900–1000 m.y.; *ep* = Epiprotozoic, 1000–650 m.y.; *e€* = Eocambrian, 650–570 m.y.; *€* = Cambrian; *o* = Ordovician; *s* = Silurian; *d* = Devonian; *c* = Carboniferous; *p* = Permian; *t* = Triassic; *j* = Jurassic; *k* = Cretaceous; *pa* = Palaeocene; *eo* = Eocene; *ol* = Oligocene; *mi* = Miocene; *pl* = Pliocene; *q* = Quaternary. (From Sadler, 1981.)

Sadler did refer to fractal analogies (viz. to the *Hurst phenomenon*, see Section 2.1.4), the first fractal model to explain the paradox of sedimentation rate has only recently been proposed, in an important research note of Plotnick (1986; discussion by Yamaji, with Plotnick's reply, in 1988).

Before discussing Plotnick's fractal model, we shall introduce a famous D-dimensional ($0 < D < 1$) fractal construction, the *Cantor set* (or Cantor "*dust*") and its higher-dimensional kins, the *Sierpinski gasket* and *-carpet* ($1 < D < 2$) and the *Menger sponge* ($2 < D < 3$).

A slight generalization of the Cantor set will yield the *Cantor bar*, and a diabolic construction, the *Devil's staircase.*

2.1.2 *From the Cantor dust to the Devil's staircase*

One time there was a picket fence / with space to gaze from hence to thence
An architect who saw this sight / approached it suddenly one night,
removed the spaces from the fence / and built of them a residence.
The picket fence stood there dumbfounded / with pickets wholly unsurrounded...
— CH. MORGENSTERN: *The Picket Fence* (translated by Max Knight, 1966)

Georg Cantor's famous construction first appeared as a footnote to his seminal "Grundlagen einen allgemeinen Mannigfaltigkeitslehre" (The Foundations of a General Theory of Sets, 1883), as an example for *perfect sets*, that is sets which are closed and dense in themselves. (For history see Meschkowski, 1967; Dauben, 1979; mathematical discussion Menger, 1967; Blumenthal and Menger, 1970; Vilenkin, 1968; Gelbaum and Olmsted, 1964; Mandelbrot, 1982; Zähle, 1984; review of applications Feder, 1988.) It is worth quoting Cantor's classical definition (in free translation): " As an example for a perfect point set, which is not everywhere dense in any interval, however short, I introduce the set of all real numbers which can be written in the form:

$$z = \frac{c_1}{3^1} + \frac{c_2}{3^2} + \cdots + \frac{c_n}{3^n} \tag{2.1.2.1}$$

where the coefficients c_i can arbitrarily assume the values 0 or 2, and the series consists of a finite or infinite number of terms". The best popular exposition of this cryptic definition is the story about tents and wet points in Vilenkin's delightful "Stories about Sets" (1968) — a tale that would have amused Edwin A. Abbott, creator of the "Flatland".

Imagine that rain is falling on the segment [0, 1] of the x axis (Figure 2.3). We go about providing shelter from the rain as follows. We divide the segment [0, 1] into three equal parts and erect a tent in the form of an equilateral triangle in the central part. It protects all the points of the central part from the rain (except the endpoints, that is the points 1/3 and 2/3). Now we divide each of the two pieces left

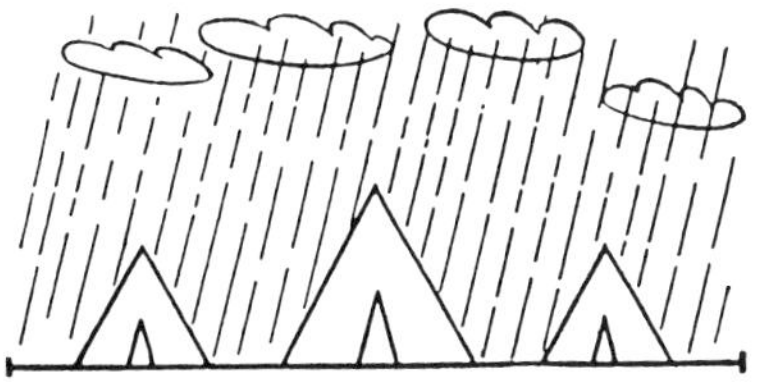

Fig. 2.3. It is raining... (from Vilenkin, 1968).

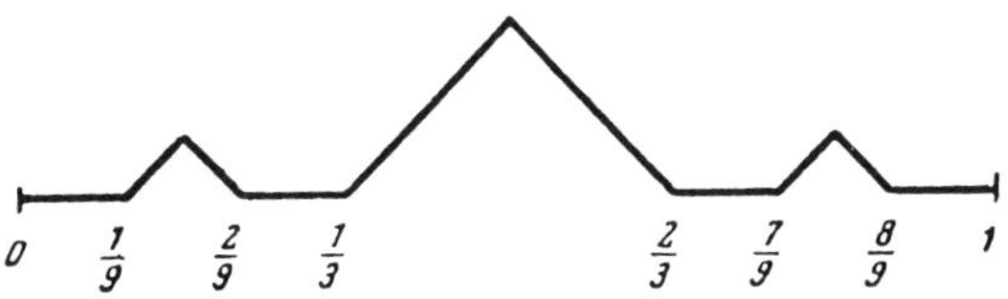

Fig. 2.4. Second step in the construction of the "wet set" (from Vilenkin, 1968).

over into three parts and protect the central parts with a tent of the same form (but only half as wide, Figure 2.4). In the next step of the procedure we erect four more tents, then eight more, etc.

Let us now determine which points remain wet after all the protective tents have been set up. We first recall that the *ternary representation* of numbers is formed the same way as the decimal representation, except that we only employ the digits 0, 1 and 2.

The first tent shelters the points lying between 1/3 and 2/3, except the endpoints, that is those points whose ternary representation starts with:

$$0.1{*}{*}{*}{*}\cdots \tag{2.1.2.2}$$

where the "$*$"'s stand for any combination of the digits 0, 1, 2. Those points still wet after the first step are those whose ternary representations have the form:

$$0.0{*}{*}{*}{*}\cdots \quad \text{or} \quad 0.2{*}{*}{*}{*}\cdots \tag{2.1.2.3}$$

It can be proved in the same way that after the two tents have been set up in the second step, only those points whose ternary representation starts with one of the following four combinations:

$$0.00{*}{*}{*}\cdots \quad \text{or} \quad 0.02{*}{*}{*}\cdots \quad \text{or} \quad 0.20{*}{*}{*}\cdots \quad \text{or} \quad 0.22{*}{*}{*}\cdots \tag{2.1.2.4}$$

will remain wet. Thus, any point in whose ternary representation a digit "1" occurs will at some stage be protected from the rain, and only those points remain wet whose ternary representation can be written without using a "1". (For example, 1/4 = 0.020202... or 3/4 = 0.202020... remain wet.)

Fig. 2.5. Construction of the triadic Cantor set. Only the first five generations are shown. $D = 0.6309$. (From Feder, 1988.)

The set of all points called "wet" in the above construction [and, more formally, defined by equation (2.1.2.1)] is called the *Cantor set* (or Cantor *dust*) and it plays a very important role in fractal theory (Mandelbrot, 1982; Feder, 1988).

As easily seen from its usual graphical representation (Figure 2.5) the Cantor set is self-similar, and it is constructed in an iterative manner. The *initiator* is the unit interval [0, 1], while the *generator* removes the open middle third (1/3, 2/3) and keeps [0, 1/3] and [2/3, 1]. By equation (1.4.1.4), we find that the fractal dimension of the Cantor set is:

$$D = -\frac{\log N}{\log r} = -\frac{\log 2}{\log(1/3)} = 0.6309 \qquad (2.1.2.5)$$

More generally, we can define the generator by dividing the unit interval [0,1] into R equal parts, and keeping N (<R) of the sub-intervals. For example, in Figure 2.6a: $R = 4, N = 2$; in Figure 2.6b: $R = 9, N = 3$.

(It is easy to check that both of the previous constructions have the same fractal dimension $D = 1/2$, in spite of the fact that the Cantor sets of Figure 2.6a and b look quite different! We shall return to this question in the Mathematical Appendix A2.1.)

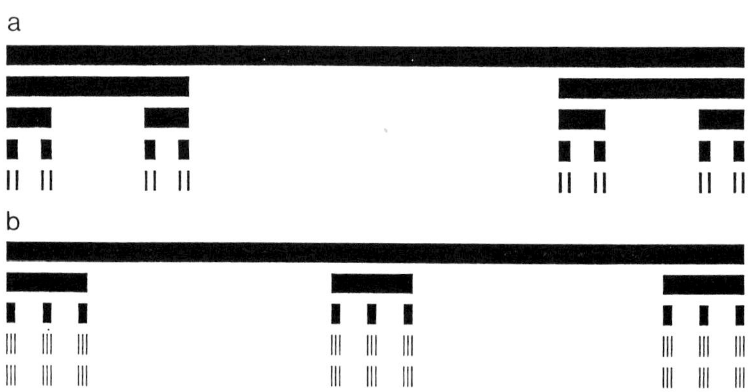

Fig. 2.6. Different Cantor sets with the same fractal dimension, $D = 0.5$. a. $R = 4, N = 2$; b. $R = 9, N = 3$. (From Feder, 1988.)

Fig. 2.7. Sierpinski gasket. The generator eliminates the central triangle. $D = 1.58$. (From Feder, 1988.)

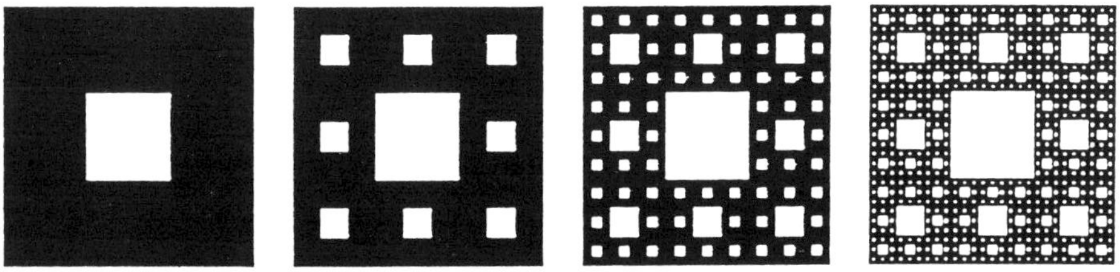

Fig. 2.8. Sierpinski carpet. The generator removes the central square. $D = 1.89$. (From Feder, 1988.)

Cantor (1884), in his inimitable style, simply remarked that "similar examples can be easily constructed for higher dimensions" ("ähnliche Beispiele lassen sich für $n > 1$ leicht bilden"). In the Euclidian plane, the $1 < D < 2$-dimensional analogues of the Cantor set are the *Sierpinski gasket* (Figure 2.7; $N = 3$, $r = 1/2$, $D = \log 3/\log 2 = 1.58$) and *Sierpinski carpet* (Figure 2.8; $N = 8$, $r = 1/3$, $D = \log 8/\log 3 = 1.89$), devised by the Polish mathematician Sierpinski (see Meschkowski, 1967; Vilenkin, 1968; Mandelbrot, 1982; Feder, 1988).

If instead of the square we take the cube, divide it to 27 equal small cubes ($r = 1/3$), discard the central cube along with its 6 neighbouring cubes ($N = 27 - 7 = 20$) and repeat this operation *ad infinitum* (Figure 2.9), we arrive at a $D = \log 20/$

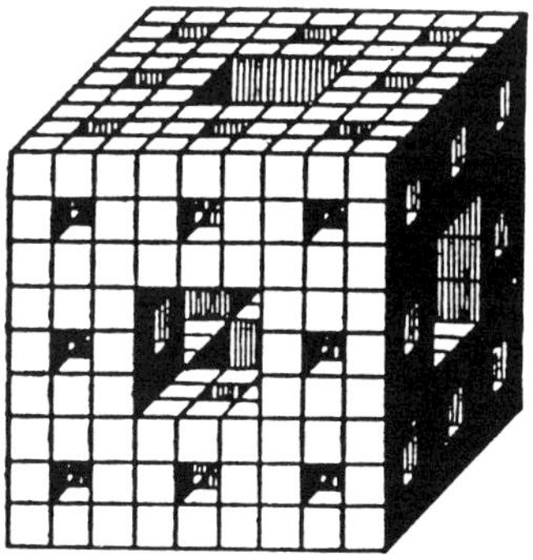

Fig. 2.9. The Menger sponge (after two operations). The generator removes the central small cube along with its 6 neighbouring cubes. $D = 2.73$. (From Vilenkin, 1968.)

log 3 = 2.7268-dimensional fractal, called *Menger's sponge* (Menger, 1967; Vilenkin, 1968; Blumenthal and Menger, 1970). What kind of a figure we would get after *all the pieces* had been discarded from the cube — a curve, a surface or a solid? — is a puzzling question, and the Reader is referred to Menger's discussion of the "universal curve" (as he calls the "sponge": Menger, 1967; Blumenthal and Menger, 1970).

We shall also need another generalization of the Cantor set, the *Cantor bar* or *Cantor's comb* (Mandelbrot, 1982; Feder, 1988; Lauwerier and Kaandorp, 1988). The *initiator* in the recursive construction of the Cantor bar is a bar of some material with unit density $\rho_0 = 1$; the bar has a length of $l_0 = 1$, that is the total mass is $m_0 = 1$.

The application of the *generator* consists of cutting the bar into two halves of equal mass $m_1 = m_2 = 1/2$ and then condensing them so that the length of each part becomes $l_1 = 1/3$ while the density of the two pieces increases to $\rho_1 = m_1/l_1 = 3/2$. In general, in the n-th step of the construction, we shall have 2^n small bars, each with the same length $l_n = 1/3^n$ and mass $m_n = 1/2^n$ (Figure 2.10). Mandelbrot (1982) picturesquely called the process "curdling", the intervals with materials "curds", and the gaps "wheys". As is easy to check, the mass of a small curd is related to its length as:

$$m_n = l_n^{\alpha} \qquad \text{with } \alpha = \frac{\log 2}{\log 3} \tag{2.1.2.6}$$

the density of the curds $\rho_n = m_n/l_n$ diverges when $l_n \to 0$ as:

$$\rho_n \propto l_n^{\alpha-1} \qquad (\alpha = \frac{\log 2}{\log 3}) \tag{2.1.2.7}$$

To find another important and interesting fractal construct, the *Devil's staircase*,

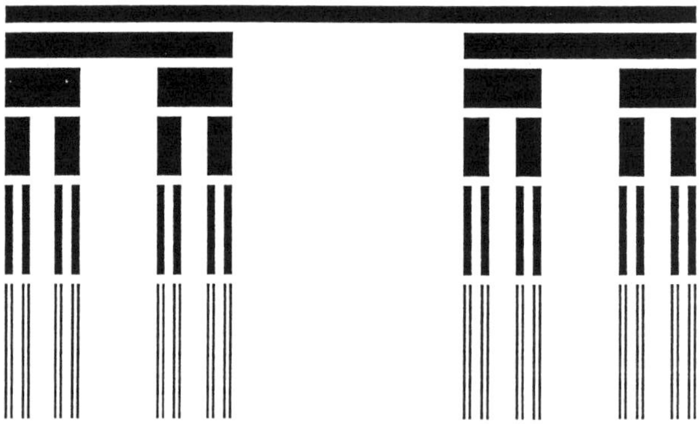

Fig. 2.10. The first five generations of the triadic Cantor bar. The height of the bars is proportional to their density. (From Feder, 1988.)

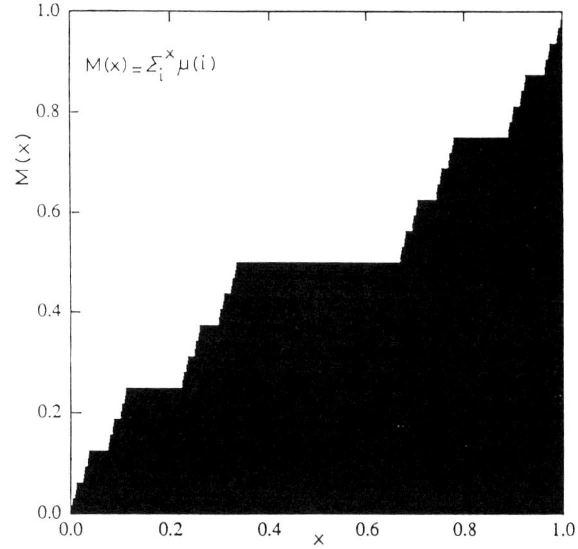

Fig. 2.11. The Devil's staircase, defined as the cumulative mass of the Cantor bar between $[0,x]$ (after Feder, 1988).

we start out from the Cantor bar of Figure 2.10 and compute the total mass contained in the segment $[0, x]$, $(x \leq 1)$:

$$M(x) = \int_0^x \rho(t)\,\mathrm{d}t \tag{2.1.2.8}$$

where the density $\rho(t)$ is zero in the gaps and, by equation (2.1.2.7), it is infinite on all points constituting the Cantor set. The resulting function $M(x)$ (Figure 2.11) is called the Devil's staircase, apparently also discovered by Cantor (Meschkowski, 1967; Vilenkin, 1968; Mandelbrot, 1982; Bak, 1986; Feder, 1988). The Devil's staircase is almost everywhere horizontal, it is self-similar (or, strictly speaking *self-affine*, see Appendix A2.1), it consists of horizontal lines over the intervals corresponding to the gaps, and of an infinite number of infinitesimal steps at the points of the Cantor set. These infinitesimal jumps however add up to $M(1) = 1$ (Vilenkin, 1968).

2.1.3 *A fractal model for stratigraphic hiatuses*

In 1986 Plotnick proposed that a fractal "Cantor bar" model can describe the statistical distribution of stratigraphic hiatuses.

In the simplest case, assume a sedimentary pile 1000 m thick, deposited over 1,000,000 years. The measured sedimentation rate for the entire pile is 1 m/1000 years (case A in the sketch below, where the bars represent sedimentation, "–" is a hiatus).

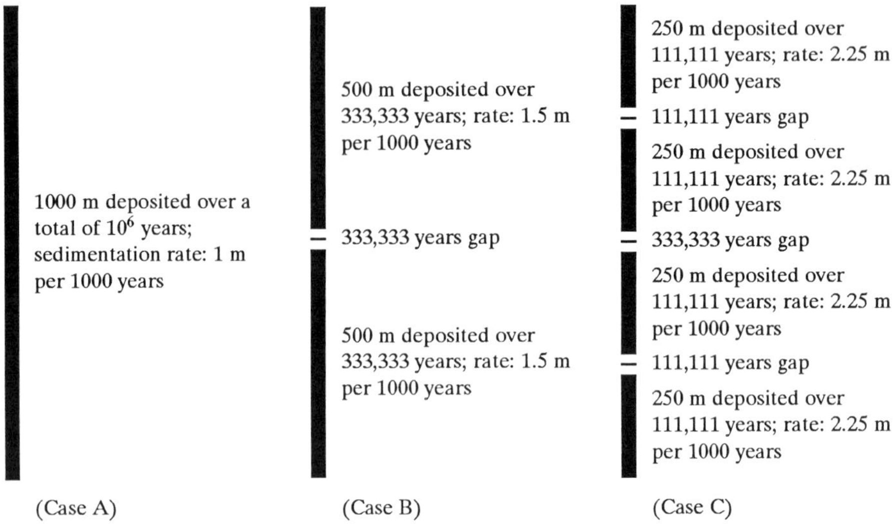

Then, assume that a hiatus of a 333,333 years exists exactly in the middle of the section (case B). It is easily seen that the measured sedimentation rate for each of the subpiles above and below the hiatus is 1.5 m/1000 years, while for the entire pile the rate is 1 m/1000 years. If we repeat the process (case C), and introduce hiatuses of 111,111 years in each of the subpiles, the sedimentation rate in the smallest units will be 2.25 m/1000 years. By iterating this process endlessly, we get subpiles representing progressively shorter periods of time, with progressively higher sedimentation rates. Obviously, this sedimentation model exactly corresponds to the "Cantor bar" discussed previously and shown in Figure 2.10. The intervals with preserved sediments are the "curds", the hiatuses the "wheys" (Mandelbrot, 1982), the length along the horizontal axis of the bar plays the role of time while the increasing heights of the bars in the subsequent generations show the increase of the sedimentation rate measured over progressively smaller spans. Note that at any particular level of resolution all periods of sedimentation are of equal length, while the hiatuses occur at all scales.

Obviously, the hiatuses should not necessarily be exactly one-third of the interval length: the relative gap size G can be any real number G $(0 < G < 1)$.

Plotnick (1986) has shown that this fractal sedimentation model satisfies the empirical rule $R(T) \propto T^A$ relating sedimentation rate and time span [cf. equation (2.1.1.1)] and proved that the coefficient A is related to the relative size of the gap G as:

$$A = -\frac{\log(1 - G)}{\log[(1 - G)/2]} \qquad (2.1.3.1)$$

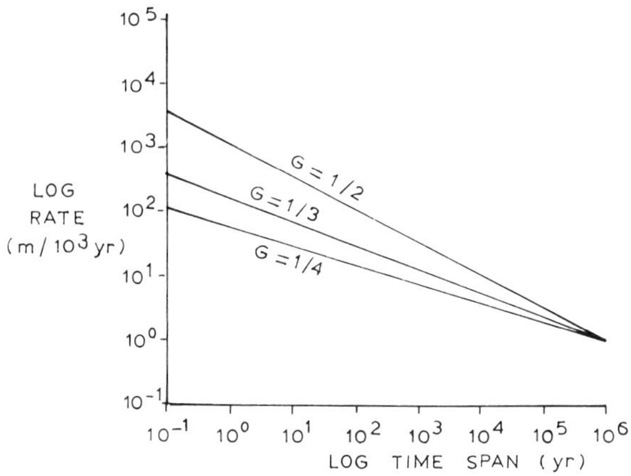

Fig. 2.12. Relationship between sedimentation rate and time span of observation for different values of the relative hiatus length G (from Plotnick, 1986).

(Figure 2.12). To prove equation (2.1.3.1), assume that in the n-th iteration the measured time interval of deposition is T_n and the thickness of sediment deposited during T_n is S_n, that is the measured sedimentation rate is $R_n = S_n/T_n$. Now, if the relative gap size is G $(0 < G < 1)$ then in the $(n+1)$-st iteration $T_{n+1} = T_n(1-G)/2$ and, as each sedimentary pile is divided to two by each iteration, $S_{n+1} = S_n/2$.

The slope of the $\log R$ vs. $\log T$ relation is:

$$A = \frac{\log R_{n+1} - \log R_n}{\log T_{n+1} - \log T_n} = \frac{\log[(S_n/2)/(T_n(1-G)/2)] - \log(S_n/T_n)}{\log(T_n(1-G)/2) - \log T_n} \tag{2.1.3.2}$$

wherefrom equation (2.1.3.1) easily follows.

From equation (2.1.3.1) we can also express the relative gap size G in terms of A, if this latter is known from the $\log R(T)$ vs. $\log T$ plot as:

$$\log(1-G) = \left[\frac{A}{A+1}\right] \log 2 \tag{2.1.3.3}$$

To find the physical meaning of equation (2.1.3.3), let us determine the fractal dimension of the Cantor bar model, for an arbitrary value of G $(0 \leq G \leq 1)$. In the generation of the Cantor bar we have kept $N = 2$ parts, deducible from the whole by a similarity of ratio $r = (1-G)/2$, that is, by equation (1.4.1.4), the fractal dimension is:

$$D = -\frac{\log N}{\log r} = -\frac{\log 2}{\log[(1-G)/2]} \tag{2.1.3.4}$$

Substituting equation (2.1.3.3) into (2.1.3.4) we find that the fractal dimension D $(0 \leq D \leq 1)$ and the slope of the $\log R(T)$ vs. $\log T$ relation, A, are related by:

$$A = D - 1 \tag{2.1.3.5}$$

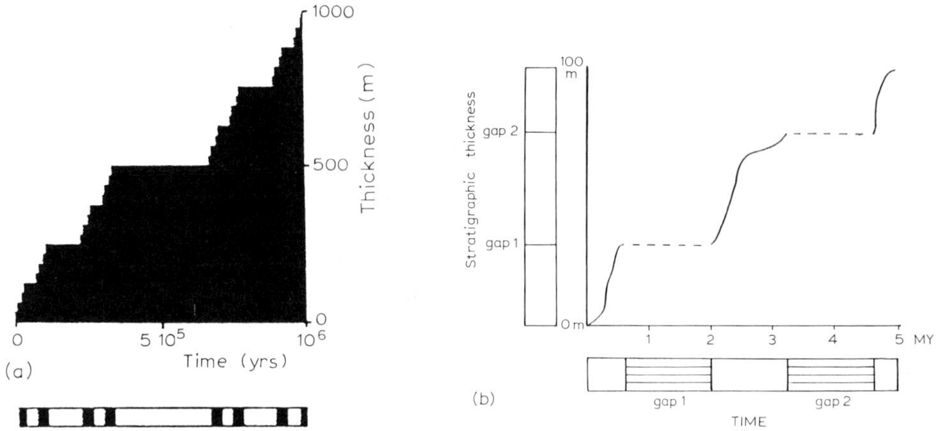

Fig. 2.13. a. Stratigraphic thickness accumulation viewed as a Devil's staircase, for $G = 1/3$; the corresponding Cantor bar is shown below the graph (from Plotnick, 1986); b. Hypothetical depositional history. The lithostratigraphic record (vertical axis) is graphed as a function of its temporal history (horizontal axis). Key: sedimentary accumulation white, non-deposition/erosion lined. (From Dingus and Sadler, 1982.)

The total thickness of preserved sediments accumulated during the time $(0, T)$ can be represented by a "Devil's staircase" (Figure 2.13a). The thickness increases in a series of short, clustered episodes of sedimentation. The plateaus of the curve represent periods of non-deposition. The general appearance of the curve is similar to the hypothetical sedimentation history in Dingus and Sadler (1982), Figure 2.13b (see also fig. 10 in Sadler and Strauss, 1990).

As discussed in Bak (1986), the "Devil's staircase" frequently arises in physical systems with non-linearly coupled oscillatory components. A few examples of relevance to the Earth Sciences are shown in Figure 2.14a–c.

The relationship between the observed long- and short-term sedimentation rates can be used to estimate the *completeness* of a given stratigraphic section (Schindel, 1982; Dingus and Sadler, 1982; Sadler and Dingus, 1982; McKinney and Schoch, 1983; Tipper, 1983).

Stratigraphical completeness is defined as *the fraction of time*, at a given level of resolution, *actually represented by rock* in the stratigraphic record. The stratigraphic completeness can be expressed as:

$$\text{completeness} = \left(\frac{t}{T}\right)^{-A} \tag{2.1.3.6}$$

where t is the time resolution, T is the time span of the whole section, and A is the slope of the line connecting the corresponding long- and short term rates on the log–log plot (Dingus and Sadler, 1982; Sadler and Dingus, 1982). By equations (2.1.3.1) and (2.1.3.5) the stratigraphical completeness can be also expressed in terms of G or D, that is the relative gap size or the fractal dimension, respectively, of the Cantor-bar sedimentation model.

If the slope A is known from empirical observations, we can determine G by

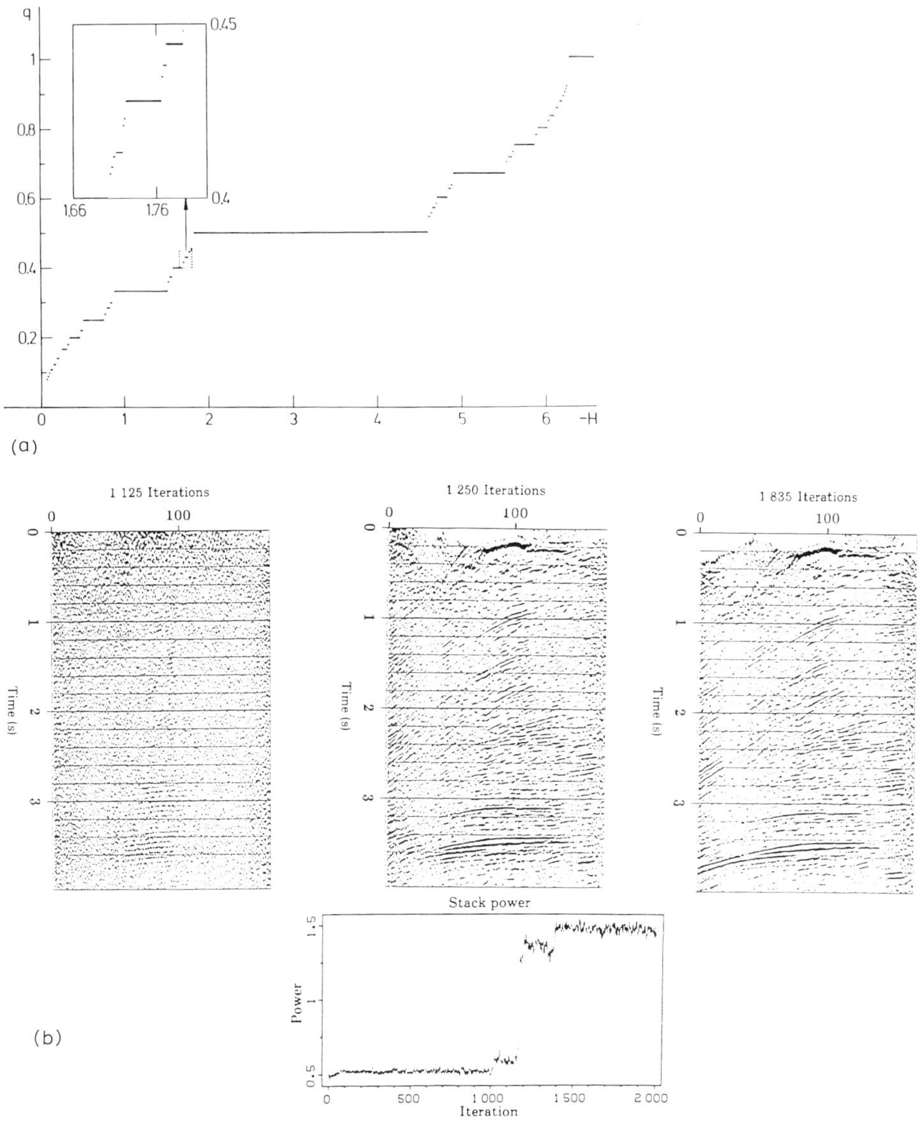

Fig. 2.14. Further examples of the Devil's staircase. a. Phase transition in a one-dimensional Ising model: the ratio q of up spins over the total number of spins, vs. the applied field H. The inset shows the area in the square magnified ten times. (From Bruinsma and Bak, 1983). b. A closely related example: the stack power measuring the overall coherency of the seismic time section increases as a Devil's staircase during successive iterations of Rothman's (1985, 1986) "simulated annealing" technique of residual statics estimation. The seismic sections correspond to the three main stages of the stack power's increase. (From Rothman, 1986).

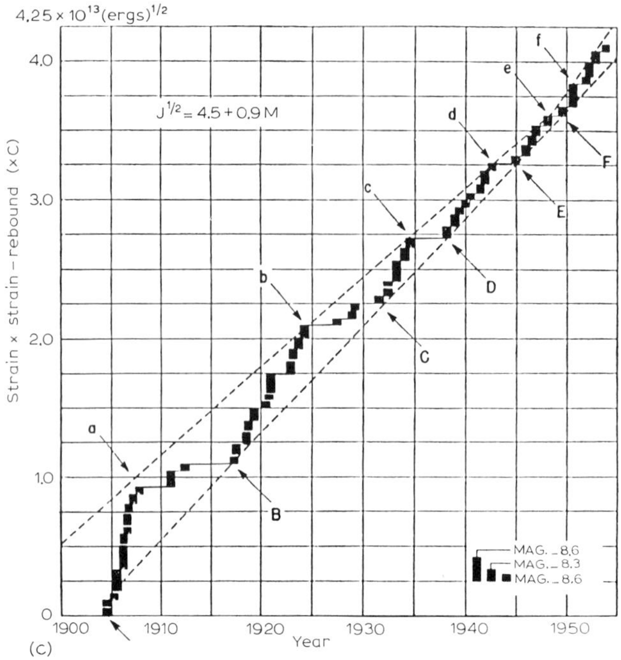

Fig. 2.14 (continued). c. Global strain release from large shallow earthquakes, 1904–1955 (from Richter, 1958).

equation (2.1.3.3) and then iteratively estimate the number and length of the gaps for any level of resolution.

If in the n-th iteration step:

– the length of a period of sedimentation is $L(n)$;

– the number of depositional units is $N(n)$;

– the total length of sedimentation periods is $T(n)$;

– the length of the shortest hiatuses is $H(n)$;

then in step $n + 1$ we have the recursive relations:

$$
\begin{aligned}
L(n+1) &= [L(n) - H(n+1)]/2 \\
N(n+1) &= 2N(n) \\
T(n+1) &= L(n+1)\,N(n+1) \\
H(n+1) &= L(n)\,G \\
\text{completeness} &= [T(n+1)]/T
\end{aligned}
\tag{2.1.3.7}
$$

with the initial conditions:

$$
L(1) = T \quad N(1) = 1 \quad T(1) = T \quad H(1) = 0 \quad (\text{completeness} = 1) \tag{2.1.3.8}
$$

TABLE 2.1

Estimates of stratigraphical completeness [a]

Iteration no.	Depositional units			Hiatus		Completeness
	Length $L(n)$ (yrs)	Number $N(n)$ (yrs)	Total length $T(n)$ (yrs)	Length	No.	
1	4 000 000	1	4 000 000	0	0	1.00
2	1 560 000	2	3 120 000	880 000	1	0.78
3	608 400	4	2 433 600	343 200	2	0.61
4	237 276	8	1 898 208	133 848	4	0.48
5	92 538	16	1 480 602	52 201	8	0.37
6	36 090	32	1 154 870	20 358	16	0.29
7	14 075	64	900 798	7 940	32	0.23
8	5 489	128	702 623	3 096	64	0.18
9	2 141	256	548 046	1 208	128	0.14
10	835	512	427 476	471	256	0.11

[a] For the Willwood Section, $G = 0.22$ (after Plotnick, 1986).

Plotnick (1986, table 1) applied this recursive scheme to estimate the distributions of the hiatuses and of the sedimentation episodes for the 4 m.y. long Willwood section [Eocene of Wyoming, described by Gingerich (1976); $A = -0.26$ at the 1,000 year level of resolution according to Sadler and Dingus (1982), that is $G = 0.22$; $D = 0.74$] and the 8.3 m.y. long DSDP 284 core [Southwest Pacific, described in Malmgren and Kennett (1981), $A = -0.05$ at the 1,000 year level of resolution according to Sadler and Dingus (1982), that is $G = 0.036$; $D = 0.95$].

For the Reader's convenience, we reproduce Plotnick's calculations for the 4 m.y. long Willwood section (Table 2.1). The exponential increase $1, 2, 4, 8, \ldots$ of the number of progressively shorter hiatuses is in conformity with the statistical analyses of deep-see sedimentary sections (Moore and Heath, 1977).

In an interesting Discussion of Plotnick's paper, Yamaji (1988) claimed that Plotnick's Cantor bar model has a transcendent, teleological stand point "... if the distribution of stratigraphic hiatuses is self-similar in the manner of a Cantor bar... the whole plan of the depositional history has to be established *a priori*".

Yamaji's statement reminds one of the anthropologist's surprise when he found that "... almost as if they had been reading the works of Claude Levi-Strauss, the Bedouins have erected a set of binary distinctions which separate the halves of the tent into the worlds of ḥaraam (Arabic for 'forbidden') and ḥalaal (Arabic for 'permissible') ... " (Randolph, 1988).

To expel the spirit of transcendence, Yamaji (1988) recommended a growth model, based on cellular automata theory (Wolfram, 1983), Figure 2.15. As Plotnick (1988) was quick to point out, the cellular automaton model would involve an even higher level of "transcendence", both in space and time.

Paraphrasing Randolph's question: Had Nature been reading Mandelbrot in or-

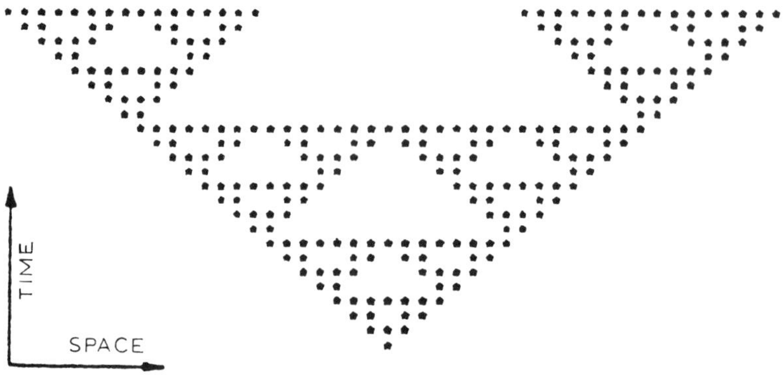

Fig. 2.15. Fractal distribution of hiatuses, both in space and time, generated by a recursive process. Sites with deposition are represented by stars. (From Yamaji, 1988.)

der to create the Cantor-bar structure of stratigraphic hiatuses, or could a hierarchically arranged set of random natural processes, acting on different time scales, *automatically* produce fractal hiatus distributions, as Plotnick (1988) has hypothetically suggested?

2.1.4 *Sadler's model of unsteady sedimentation and its fractal generalization*

> How we may steal from hence, and, for the gap
> That we shall make in time, from our hence-going
> And our return, to excuse
>
> — SHAKESPEARE: *Cymbeline*, iii. 2

In 1980 P.M. Sadler addressed the question why are the rates of accretion generally faster for shallow marine manganese nodules than for abyssal nodules? He concluded that the differences in accretion rates are due to the significantly different time intervals involved (<10,000 years for shallow marine, and 10^4–10^8 years for abyssal marine samples, Figure 2.16), rather than to environmental factors. To qualitatively explain the decrease of the measured accretion rate with the increase of the time span, Sadler (1980) proposed a "simplified growth curve" where the thickness against time function is the sum of a periodic wave function, and a straight line of positive gradient. The ascending segments of the periodic function simulate the rapid "impulse growth" (Heye, 1975), while the gradient of the straight line defines the long-term rate of accretion. Later, Sadler (1981) applied the same model for a numerical simulation of the unsteady deposition of sediments (Figure 2.17). As seen from the figure, the measured sedimentation rate decreases with increasing T and tends to the slope of the linear secular trend.

Sadler (1981) has suggested that for a more realistic simulation of the geological situations, the wave function must be allowed to assume more complex forms; namely, one should substitute the periodic part of the model by a 1-dimensional ran-

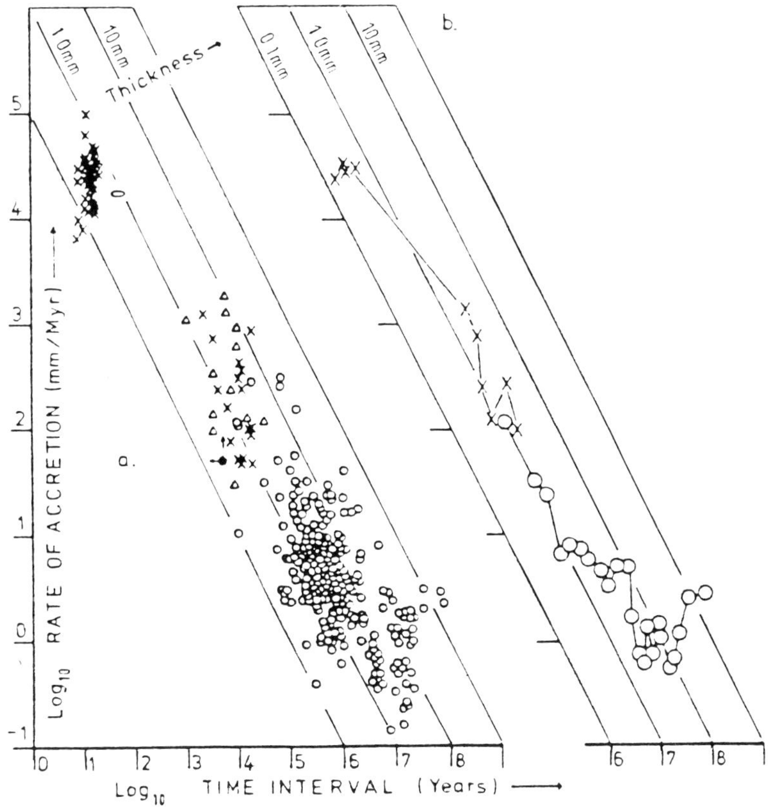

Fig. 2.16. Rates of accretion of marine manganese crusts and nodules. Key: x = shallow marine; O = abyssal marine; o = limits of Heye's (1975) "impulse growth"; △ = manganese crusts attributed to hydrothermal activity. (After Sadler, 1980.)

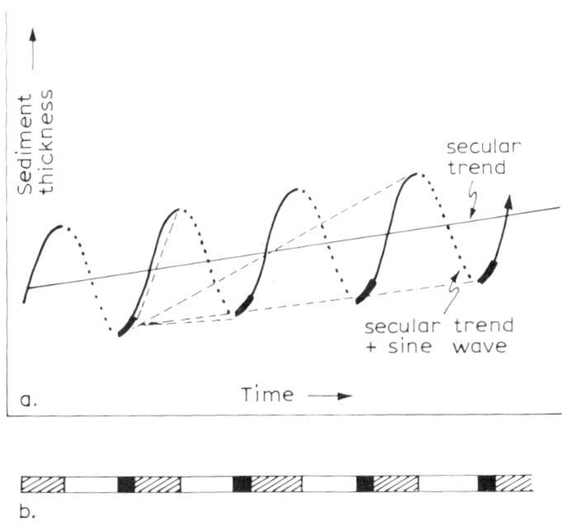

Fig. 2.17. Sadler's model of unsteady sedimentation. a. A linear secular trend added to a sine wave. The dotted portion of the curve represents erosion, the solid portion sedimentation, the thick line preserved sedimentation. b. The resulting sedimentary section along the absolute time scale: solid bar = preserved sedimentation, ruled bar = sediments that later eroded, open bar = non-depositional interval. (After Sadler, 1981.)

dom walk (Feller, 1968), or by random processes with "infinite memory", discovered by Hurst (1957) in his studies of hydrological records (see Section 5.1.1).

In order to simulate a sedimentation process:

$$x(t) = rt + s(t) \tag{2.1.4.1}$$

corresponding to Sadler's (1981) random walk model, we define $t = k\delta$ as time measured with a resolution δ, r is the long-span (asymptotic) sedimentation rate, and $s(t)$ a one-dimensional random walk:

$$s(t) = s(k\delta) = \sum_{i=1}^{k} S_i \qquad (S_i = +S \text{ or } -S \text{ with probability } 0.5) \tag{2.1.4.2}$$

The value of S must be carefully chosen to maintain that, as in any real, *observable* stratigraphical section:

$$s(k\delta) \geq 0 \qquad \text{for all } k = 1, 2, 3, \ldots \tag{2.1.4.3}$$

(Tipper, 1983).

Figure 2.18a–e shows a realization of Sadler's random walk model with $r = 1/20$, $S = 1$ and with progressively finer resolutions $\delta = 320$ (case a), $\delta = 160$ (case b), $\delta = 80$ (case c), $\delta = 40$ (case d) and $\delta = 20$ (case e). The exponential increase of

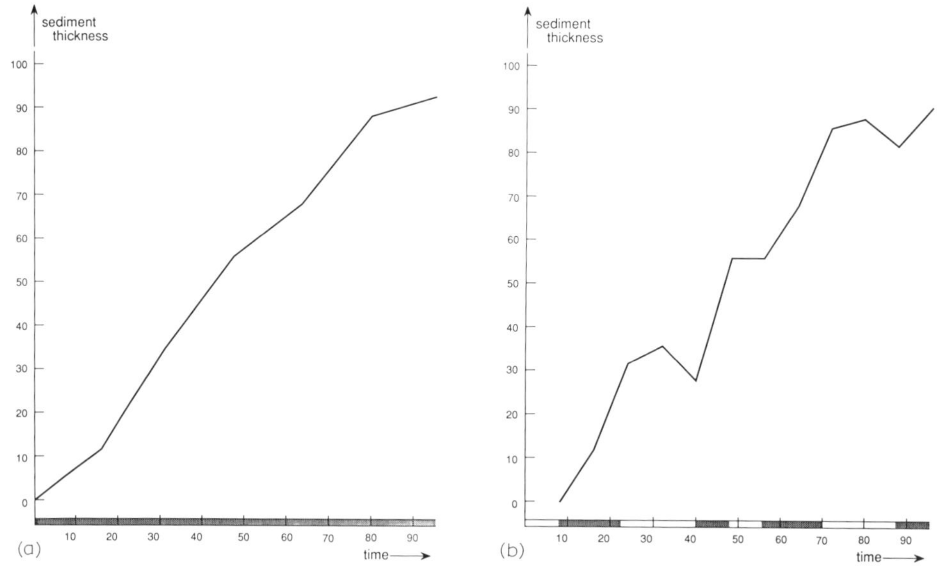

Fig. 2.18. A random generalization of Sadler's unsteady sedimentation model: $x(t) = rt + s(t)$ where $s(t)$ is a one-dimensional random walk. The solid portions of the bar along the horizontal (time) axis are the intervals of preserved sedimentation, the white portions are the hiatuses. The time resolution decreases as $\delta = 320$ (a); $\delta = 160$ (b); $\delta = 80$ (c); $\delta = 40$ (d); and $\delta = 20$ (e).

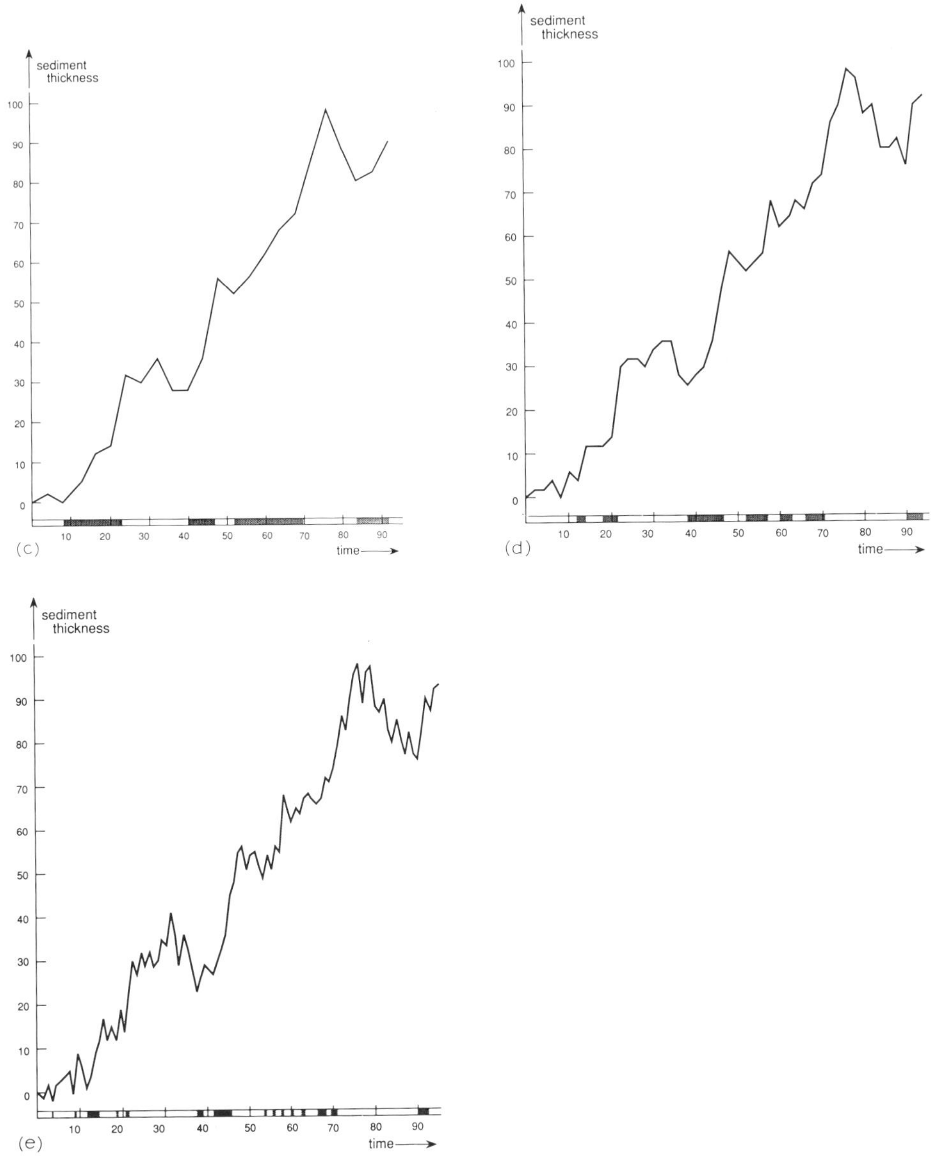

Fig. 2.18 (continued).

the number of hiatuses (0 for δ = 320, 4 for δ = 160, 5 for δ = 80, 8 for δ = 40, 15 for δ = 20) conforms with Plotnick's (1986) Cantor bar model and with published data on hiatus distribution (Moore and Heath, 1977).

The lesson from this computer simulation is that *a model consisting of a slowly varying "continuous" sedimentation and a superimposed random perturbation* (as proposed by Sadler, 1980, 1981; Strauss and Sadler, 1989; Sadler and Strauss, 1990) *can lead to a "fractal" distribution of hiatuses* — without any transcendence or teleology brought up in the Yamaji–Plotnick controversy (1988).

Physicist Readers, familiar with the Theory of Chaos, might suspect that the fractality of hiatuses and the "Devil's staircase"-like sediment accumulation curve (Figure 2.13a) would enter into the model through the non-linear interaction (phase locking) between competing frequencies (Bak, 1986). The phenomenon can be well observed on some curious swing representations of the 18th century French school of *fêtes galantes*; as in Watteau's (1684–1721) *La Balanceuse* (The Swinger), Nicholas Lancret's (1690–1743) *The Swing*, or Fragonard's (1732–1806) famous *Les Hasards Heureux de l'Escarpolette* (Wentzel, 1964, Muehsam, 1970, Posner, 1982; for the chaotic properties of the guided pendulum see Tritton, 1986a, b; 1989).

In each of these paintings we see a beautiful lady sitting on a swing, and the swing is guided by an attendant with the help of a rope. The inherent chaotic nature of the system is apparent in Fragonard's picture: one of Madame's slippers flies up into the air and caught by cupids (Figure 2.19).

In sedimentation, the archetypal pair of attendant (*driving force*) and swing (*oscillator*) would roughly correspond to the global thermal subsidence and to the sea-level changes and other short-term periodicities, respectively.

To check this possibility, we have carried out a more realistic computer simulation of the entire subsidence–deposition/erosion–compaction–isostatic compensation cycle, using the SEDPAK program, developed by Professor Christopher Kendall at the University of South Carolina (Burton et al., 1987).*

In this approach the slowly varying driving force is attributed to the joint effects of the global thermal subsidence, of another subsidence linearly increasing seaward of a given hinge point, of the isostatic accommodation to sediment loading, and of sediment compaction; while the role of the random oscillator is played by the sea-level changes. In Kendall's SEDPAK model a further non-linearity is introduced by the flip/flop-like switches from deposition to erosion and back, depending on whether the sediments are below or above sea level.

A typical sediment geometry generated with the SEDPAK package is shown in Figure 2.20, the sediment accumulation rates $R(T)$ measured over increasing time spans T are shown in Figure 2.21 for seven different locations. There is a great variation in the shape of the curves, though most of them show an approximately power-like decrease of $R(T)$ with increasing T in a limited range of T's, while $R(T)$ approaches a constant value as $T \to \infty$. On a log–log plot, the behaviour of

---

* I gratefully acknowledge the generosity of Professor Kendall in kindly sending me a copy of the SEDPAK program code for education and research purposes. The program has been installed on the University of Adelaide's VAX system by my Ph.D. student, Dr. Shaou Zhou. Dr. Zhou has also added several new routines to the SEDPAK, including one to compute the rates of sedimentation and he ran the model experiments shown in Figures 2.20 and 2.21.)

Fig. 2.19. The Swing (*Les hasards heureux de l'escarpolette*), a painting by Jean Honoré Fragonard (ca. 1767). London, Wallace Collection.

the $R(T)$ vs. T curves can be directly compared with the theoretical predictions of Strauss and Sadler derived for the Brownian sedimentation model $x(t) = rt + s(t)$ [Strauss and Sadler (1989), eq. 3.2.4 and fig. 6; Sadler and Strauss (1990), fig. 10; Strauss and Sadler's results are based on the theory of regulated Brownian motion: Harrison (1985)].

For small T's ($T < 10$ million years), most of the curves in Figure 2.21 behave as $R(T) \propto T^A$ with $A \approx -0.5$, in conformity with the results of Strauss and Sadler (1989).

Sadler (1981) also suggested that the unsteady sediment accumulation should be modelled as:

$$x(t) = r\,t + s_H(t) \tag{2.1.4.4}$$

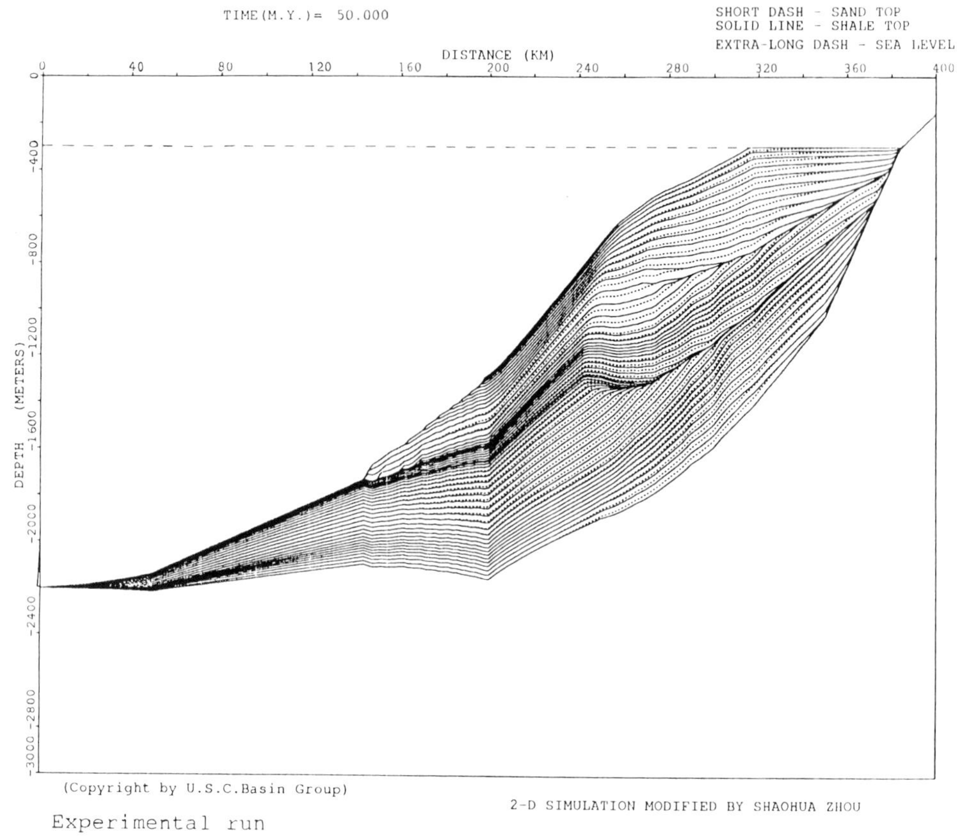

Fig. 2.20. Sedimentary geometry generated by the SEDPAK modelling.

where $s_H(t)$ is a special kind of random process, closely connected with the "Hurst phenomenon" (Hurst, 1957; Mandelbrot and Wallis, 1969; Klemes, 1974; see later in Sections 4.3 and 5.1.1).

The random function $s_H(t)$ in equation (2.1.4.4) is called a *fractional Brownian motion with Hurst exponent H* (Mandelbrot and Van Ness, 1968; Feder, 1988); it satisfies:

$$\langle s_H(t) \rangle = 0 \qquad \text{and} \qquad \left\langle [s_H(t) - s_H(t+\delta)]^2 \right\rangle = \sigma^2 \delta^{2H} \tag{2.1.4.5.}$$

The case $H = 1/2$ corresponds to the *ordinary Brownian motion*; otherwise $s_H(t)$ is called:

antipersistent if $0 < H < \dfrac{1}{2}$

persistent if $\dfrac{1}{2} < H < 1$

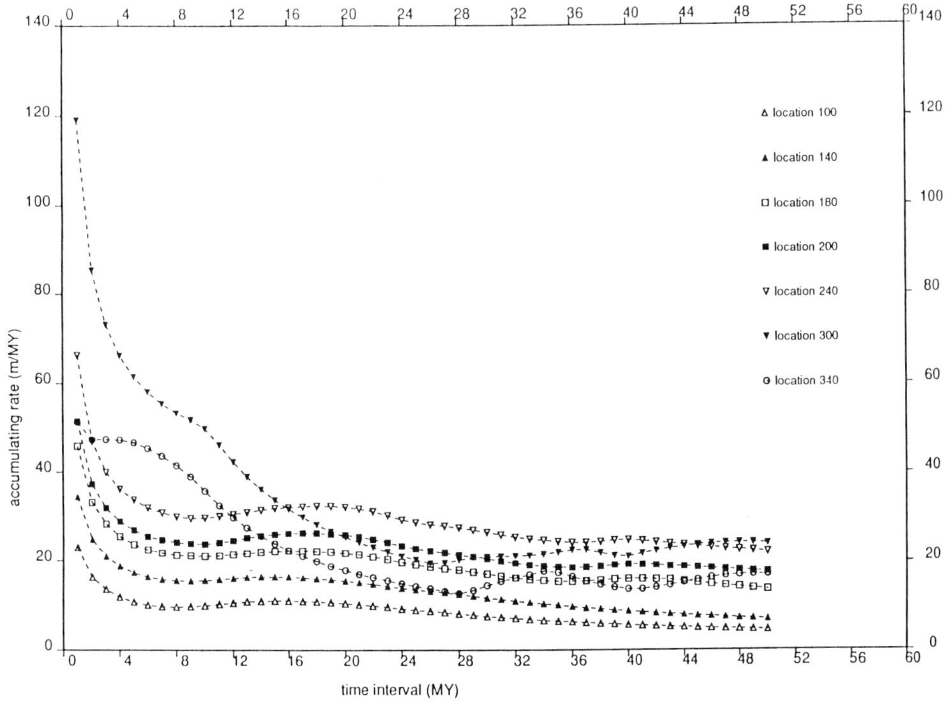

Fig. 2.21. Sediment accumulation rates vs. time-span curves for the sediment geometry of Fig. 2.20.

(see Figure 2.22). By equations (2.1.4.4)–(2.1.4.5) the sediment thickness $S(T)$ accumulated in the time interval $(0, T)$ is with a high probability within the bounds:

$$rT - T^H \leq S(T) \leq rT + T^H \tag{2.1.4.6}$$

that is:

$$\lim_{T \to \infty} R(T) = \frac{S(T)}{T} = r \tag{2.1.4.7}$$

Thus the Brownian motion models predict a convergence to the mean rate r for $T \to \infty$, both for the ordinary Brownian motion ($H = 1/2$) or the fractional Brownian motion models ($0 < H < 1/2$ or $1/2 < H < 1$).

However, as we shall indicate below, *for small values of T the measured sedimentation rate $R(T)$ of the fractional Brownian model (2.1.4.4) decreases with the time T as*:

$$R(T) = \frac{S(T)}{T} \approx T^{-H} \tag{2.1.4.8}$$

and the hiatuses form a $(1 - H)$-dimensional fractal set, where H is the Hurst exponent.

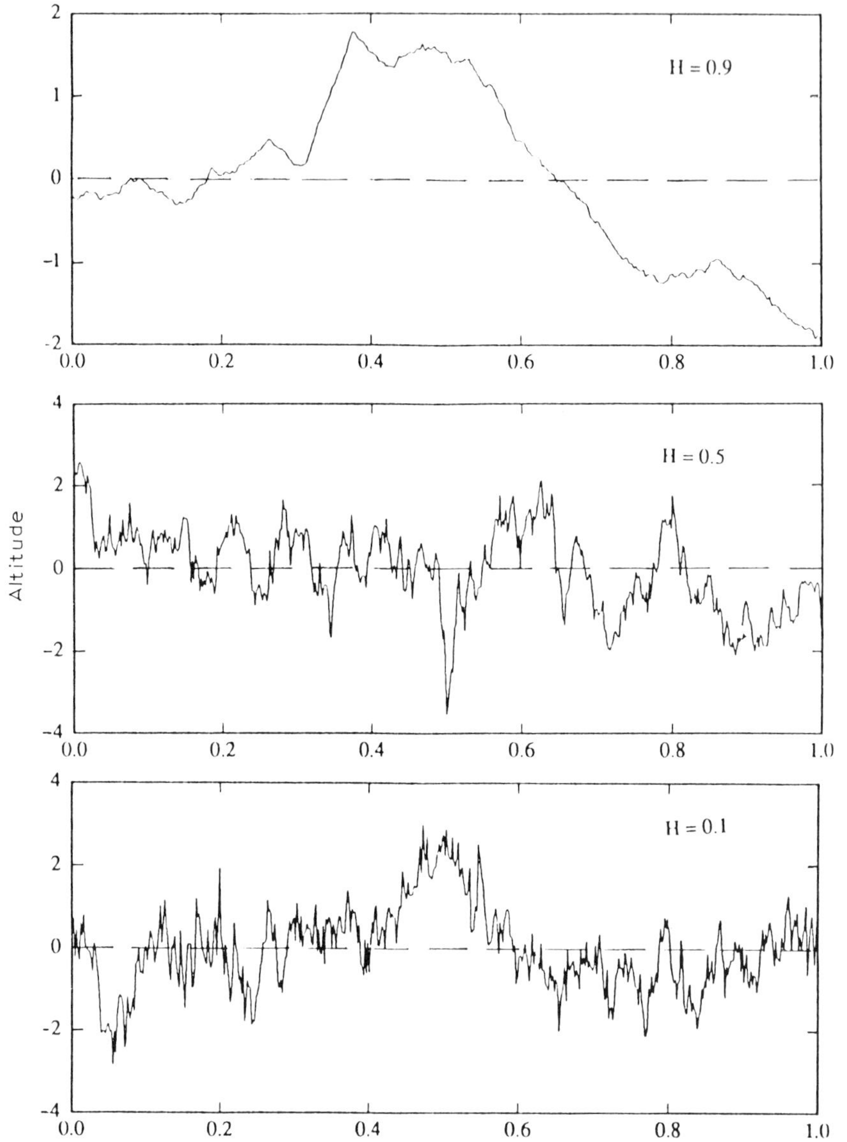

Fig. 2.22. Fractional Brownian curves for different values of the Hurst exponent H: persistent ($H = 0.9$); ordinary Brownian ($H = 0.5$); antipersistent ($H = 0.1$) (from Feder, 1988).

Consequently, the fractional Brownian motion model of sedimentation suggested by Sadler (1981) *automatically* implies the fractality of the hiatuses, in conformity with Plotnick's fractal Cantor bar model, for small time spans T.

We can only offer a plausibility argument, a rigorous proof would require the

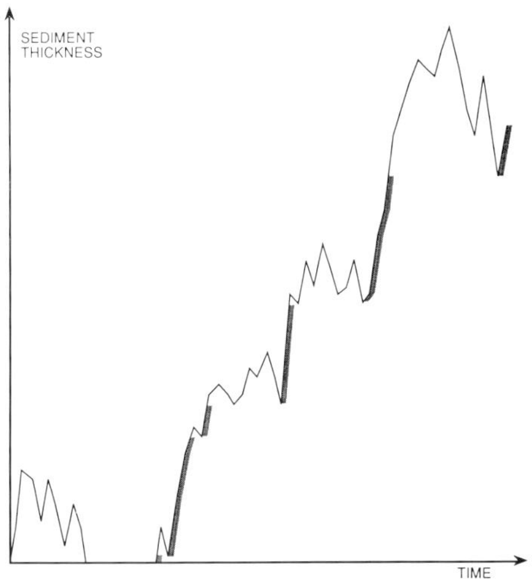

Fig. 2.23. Generalization of Sadler's (1981) model of unsteady sedimentation. The random process superimposed on the linear secular trend is a fractional Brownian motion of Hurst exponent H.

generalisation for fractional Brownian processes of some classical results due to Paul Lévy (1939, 1965) concerning ordinary Brownian motion.

Figure 2.23 shows a random variant of Sadler's periodic model: the periodic function has been substituted by a fractional Brownian process of Hurst exponent H. As in Figure 2.17, the *preserved* episodes of sedimentation are represented by the thick-line portions of the curve.

We have already seen that $S(T)/T = R(T) \to r$ if $T \to \infty$. On the other hand, there exists a T_0 (depending on the values of r, σ and H) such that for

$$t_{\mathrm{p}} - T_0 \leq t \leq t_{\mathrm{p}} \tag{2.1.4.9}$$

(where t_{p} denotes the present time) the linear term rt can be neglected and the behaviour of $x(t)$ is governed by its "unsteady" component $s_H(t)$.

We introduce an auxiliary function $m(t)$ for $t_{\mathrm{p}} - T_0 \leq t \leq t_{\mathrm{p}}$:

$$m(t) = \inf_{t \leq t' \leq t_{\mathrm{p}}} s_H(t') \tag{2.1.4.10}$$

(Figure 2.24). The function $m(t)$ is a "Devil's staircase"; it approximates the accumulation of the preserved sediment thickness between $t_{\mathrm{p}} - T_0$ and t_{p} (if the linear "drift" rt is neglected).

For ordinary Brownian motion (when $H = 1/2$, $r = 0$) Paul Lévy (1939, 1965) proved that the points where the step-function $m(t)$ increases obey the same probability distribution as the zero crossings of the process $s_{1/2}(t)$. If this holds true for *fractional* Brownian processes $s_H(t)$ (as it very likely does) we can apply Orey's theorem [Orey (1970); Voss (1985a, b); the special case $H = 1/2$ is due to Taylor (1955)]

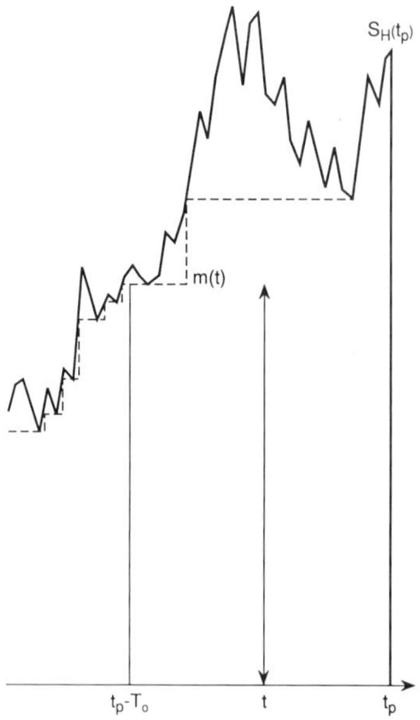

Fig. 2.24. The function $m(t)$, defined in equation (2.1.4.10).

— which states that the set of zero crossings of $s_H(T)$ is $(1 - H)$-dimensional — in order to conclude that the sediment accumulation curve $m(t)$ increases at points of a $(1 - H)$-dimensional subset $\{t_{sed}\}$ of the time instants in the interval $[t_p - T_0,\ t_p]$. In order to derive an order of magnitude estimation of the preserved sediment thickness $S(T_0)$, let us select a small resolution τ ($\tau \ll T_0$). By the definition of the fractal dimension (Appendix A2), the set $\{t_{sed}\}$ can be covered by:

$$N \propto \left(\frac{T_0}{\tau}\right)^{1-H} \tag{2.1.4.11}$$

intervals of length τ. The total thickness $S(T_0)$ of the *preserved* sediments can be estimated by equations (2.1.4.5) and (2.1.4.11) as:

$$S(T_0) \propto \left(\frac{T_0}{\tau}\right)^{1-H} \sigma\tau^H \propto T_0^{1-H} \tag{2.1.4.12}$$

that is:

$$R(T_0) = \frac{S(T_0)}{T_0} \propto T_0^{-H} \tag{2.1.4.13}$$

for smaller values of T_0, as has been stated. We have also seen that the hiatuses, which correspond to the points of increase $\{t_{sed}\}$ of $m(t)$, form a *fractal set of dimension* $1 - H$, in conformity with Plotnick's (1986) Cantor bar model. For the ordinary Brownian motion model ($H = 1/2$) equation (2.1.4.13) yields:

$$R(T_0) \approx T_0^{-1/2} \tag{2.1.4.14}$$

for small values of T_0, in agreement with the results of Strauss and Sadler (1989; Sadler and Strauss, 1990).

2.2 Fractal analysis along a line: slip lines and fractures

> I saw before me a vast multitude of small Straight Lines (which I naturally assumed to be Women) interspersed with other beings still smaller and of the nature of lustrous points — all moving to and fro in one and the same Straight Line, and, as nearly as I could judge, with the same velocity.
> A noise of confused, multitudinous chirping or twittering issued from them at intervals as long as they were moving... — E.A. ABBOTT: *Flatland. A Romance of Many Dimensions*

The stratigraphic record contains depositional units and hiatuses at all scales, any layer of a few inch thickness is made up of even thinner layers, as is well known to "everyone who has ever looked at the layering of rocks exposed at road cuts... or has seen well logs" (Rietsch, 1982). According to Plotnick's (1986) fractal model (Section 2.1) the hiatuses, separated by episodes of preserved sedimentation, form a self-similar fractal subset $H = \{t_{\text{hiat}}\}$ of the total span of geologic time T; the fractal dimension of H is $D = -\log 2/\log r$ where $r = (1 - G)/2$ and G is the relative gap size (recall Eq. 2.1.3.4). Because of self-similarity, any shorter subinterval of T, of length δ say, is a reduced replica of the whole, also consisting of hiatuses and sedimentary episodes. If we choose the time resolution τ as:

$$\tau = \delta r^n \qquad (r = \frac{[1 - G]}{2} \qquad \text{that is } r < 1;\ n \gg 1\,) \tag{2.2.1}$$

the number of hiatuses found in the interval δ will be, with the given resolution:

$$N_{\text{hiat}} = 2^n - 1 \approx 2^n \tag{2.2.2}$$

by the recursive relations (2.1.3.6) and (2.1.3.7). Eliminating n from equations (2.2.1) and (2.2.2) and using (2.1.3.4) we find that the number of hiatuses scales as:

$$N_{\text{hiat}} \propto \left(\frac{\delta}{\tau}\right)^D \tag{2.2.3}$$

Consequently, if we count the number of hiatuses for the *fixed time span* δ and *for different resolutions* τ, we would readily get the fractal dimension D as the slope of the $\log N$ vs. $\log \tau$ line:

$$\log N_{\text{hiat}} = \text{const} - D \log \tau \tag{2.2.4}$$

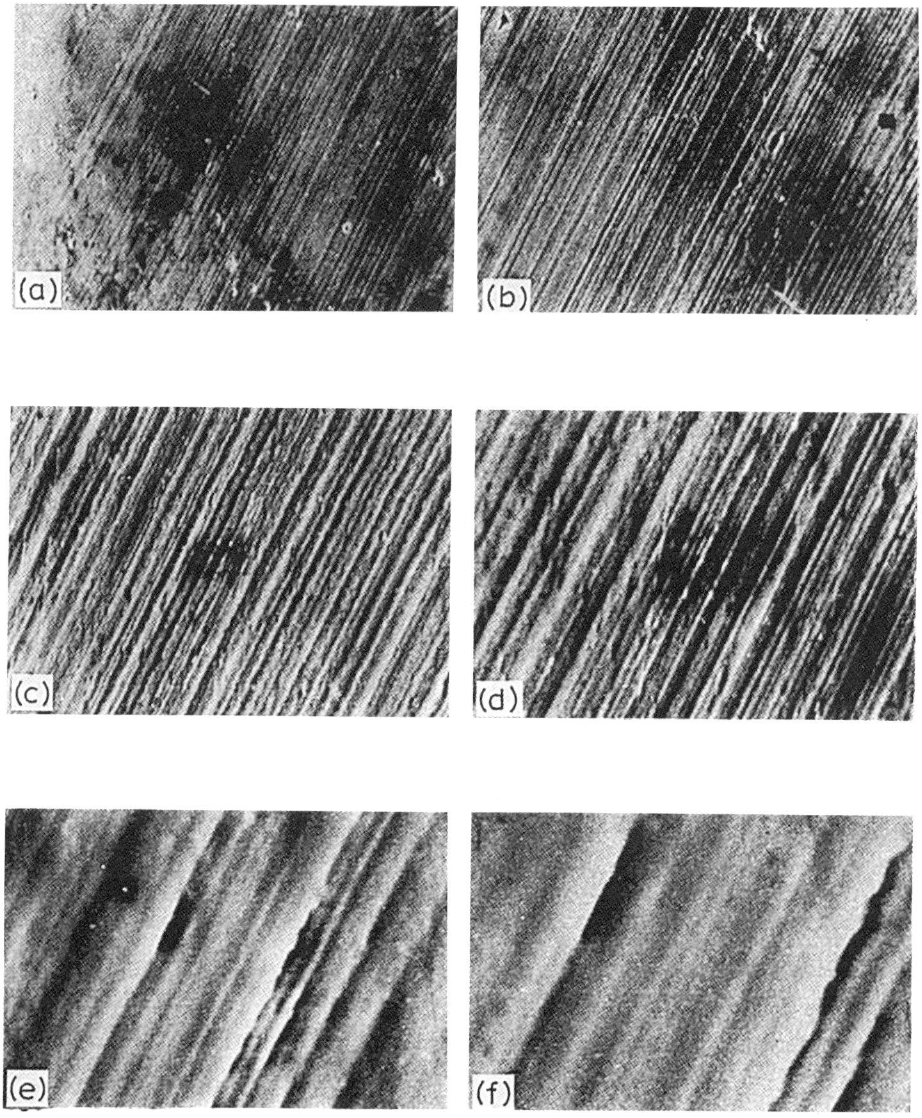

Fig. 2.25. Slip lines on the surface of Cd single crystals due to deformation by uniaxial tension. Magnifications: (a) 100; (b) 200; (c) 1,000; (d) 2,000; (e) 10,000; (f) 20,000. (From Sprušil and Hnilica, 1985.)

Equation (2.2.3) expresses a general property of the D-dimensional fractal point sets on a straight line: *any interval of length L contains*:

$$N \propto \left(\frac{L}{l}\right)^D \tag{2.2.5}$$

points of the fractal set, at the resolution l ($L/l \gg 1$).
[A good discussion of fractal point sets on a line is Cederbaum et al. (1985), with physical applications and numerical examples.]

In practical applications of equation (2.2.5) we either vary the length of the interval L or the resolution l, or both.

The Czechoslovakian physicists Sprušil and Hnilica (1985) used this technique to determine the fractal dimension of the slip line pattern of Cd single crystals. They plastically deformed a Cd single crystal in uniaxial tension to produce split lines. The crystal's surface was then shadowed with Au–Pd and observed under an angle of 45° in a scanning microscope with magnifications M_i ranging from 100 to 20,000. Figure 2.25 shows the similarity of the slip lines at different magnifications. To find the fractal dimension D of the pattern, Sprušil and Hnilica counted the number of dark lines $N_S(i)$ for the different magnifications M_i in a given fixed length L (= 4 cm) of the micrographs, perpendicularly to the slip lines.

Taking $l_i = L/M_i$ in equation (2.2.5), the fractal dimension D can be found from the plot:

$$\log N_S(i) = \text{const} - D \log M_i \tag{2.2.6}$$

($D = 0.49 \pm 0.03$, see Figure 2.26).

The appearance of the slip lines in Figure 2.25a–f is similar to the frequently

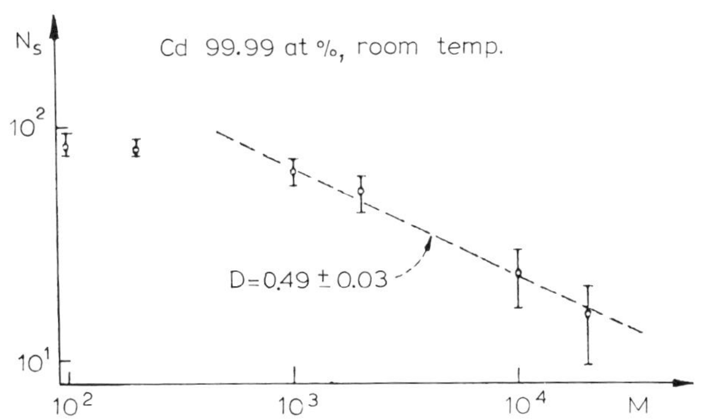

Fig. 2.26. Dependence of the number of slip lines N_S on magnification M (from Sprušil and Hnilica, 1985).

cited *strange attractor* discovered by Hénon (1976), corresponding to the recursive sequence:

$$x_{i+1} = y_i + 1 - ax_i^2 \qquad y_{i+1} = bx_i \tag{2.2.7}$$

Figure 2.27a–d shows the attractor (for $a = 1.4$, $b = 0.3$) for increasing resolutions. The fractal dimension of the Hénon attractor was determined by Russell et al. (1980) and Grassberger and Procaccia (1983a, b); for $a = 1.4$, $b = 0.3$ it is about 1.25, that is the set of its intersections with a perpendicular line is about 0.25-dimensional.

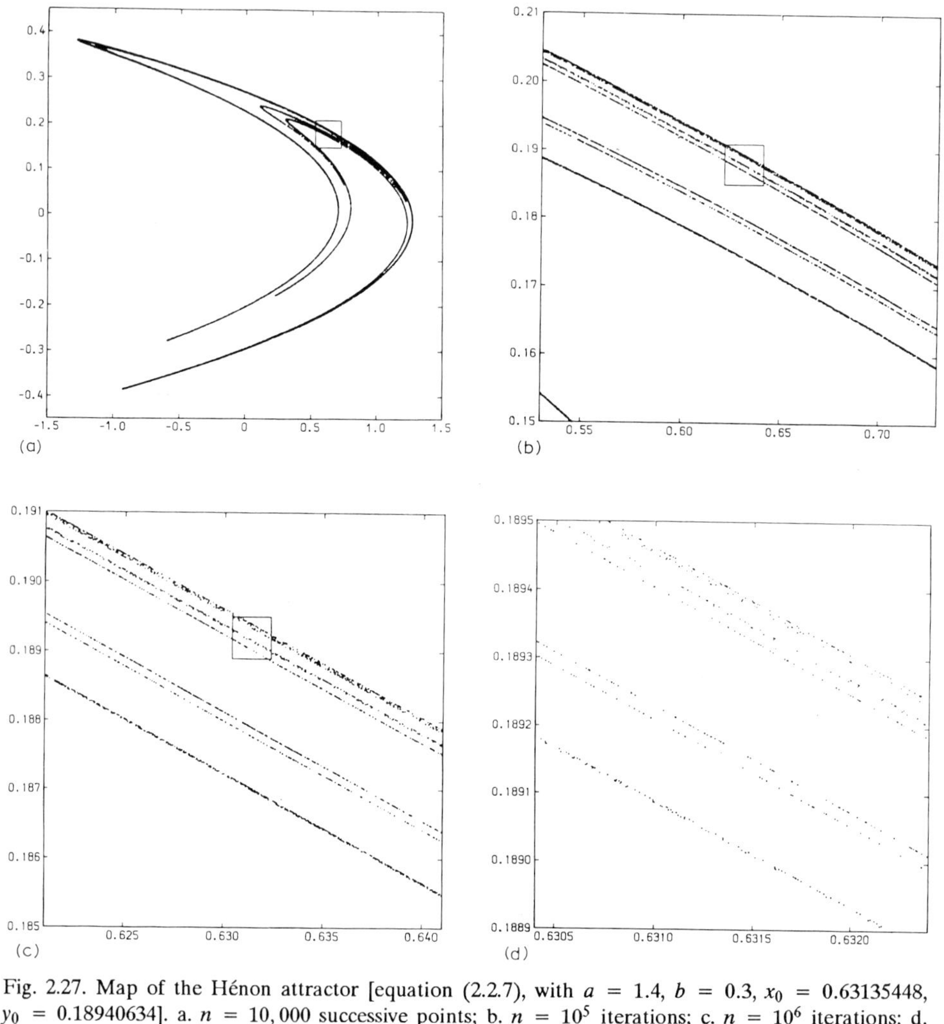

Fig. 2.27. Map of the Hénon attractor [equation (2.2.7), with $a = 1.4$, $b = 0.3$, $x_0 = 0.63135448$, $y_0 = 0.18940634$]. a. $n = 10,000$ successive points; b. $n = 10^5$ iterations; c. $n = 10^6$ iterations; d. $n = 5 \times 10^6$ iterations. (From Hénon, 1976.)

The technique used in equation (2.2.6) for the determination of the fractal dimension is a special case of *box-counting* (Appendix A2).

For the strictly self-similar Cantor dust a part of the model has the same shape as the whole, part of the part again has the same shape, and so on, as in Swift's poem:

> So, Nat'ralists observe, a Flea
> Has smaller fleas that on him prey;
> And these have smaller fleas to bite 'em
> And so proceed *ad infinitum*. — *On Poetry. A Rhapsody* (1733)

In real data sets, self-similarity is only valid between two limits, the inner and outer *cut-offs*, which correspond to the size of the smallest and largest observable features.

If the log N_i vs. log r_i plot (where N_i is the number of linear features observed with a resolution r_i in a given interval) cannot be reasonably approximated by a straight line, no unique similarity dimension can be assigned to the data. This was the case reported by J.P. Chilès, of the Bureau de Recherches Géologiques et Minières, Orléans (1988). Chilès analysed about 6,600 fracture traces recorded along the drifts of a uranium mine in a granite massif. According to their orientations the fractures could be separated to distinct sets, related to tectonic history. Chilès (1988) has only reported on the geostatistical and fractal analyses of the subvertical fractures. Figure 2.28 shows $N(r)$, the number of intersections of the subvertical fractures with the horizontal survey lines, as a function of the resolution r. The log–log plot is not a straight line, that is the fracture network *cannot be characterized by a single fractal dimension*. It is not even *bifractal* (recall Suzuki's definitions, Section 1.4.3, pp. 47–48). Rather, as shown in Figure 2.29, the fractal dimension continuously changes from 0 to 1 with increasing resolution. Consequently,

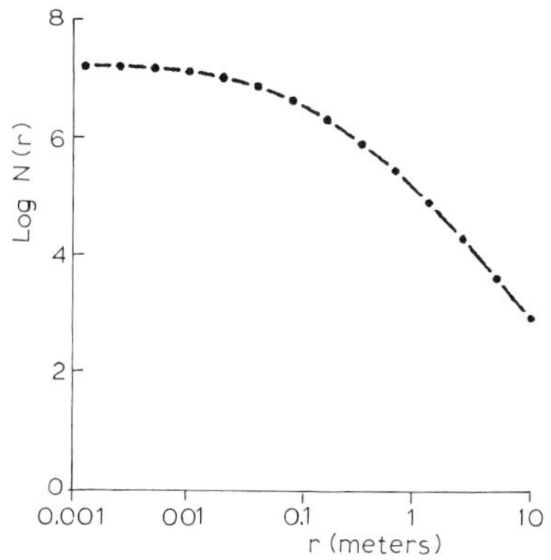

Fig. 2.28. Log–log plot of the number of intersections of the horizontal survey lines with the subvertical fractures in the drift walls of a uranium mine. $N(r)$ is number of intersections, r is resolution. (From Chilès, 1988.)

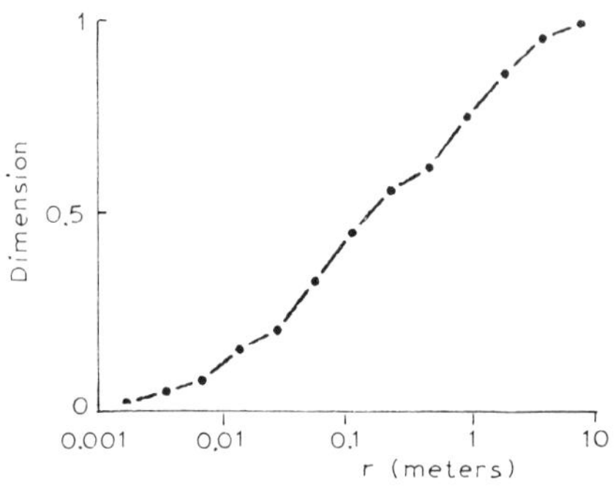

Fig. 2.29. Graph of the fractal dimension as function of the resolution, for the data of Fig. 2.28 (from Chilès, 1988).

the intersections of the subvertical faults with the horizontal survey lines form a *multifractal set* (Section 4.6) which is a union of subsets, each having its own different characteristic scale and fractal dimension.

2.3 Strange attractors, aggregates and geophysical networks

Weisst wie viele Sterne stehen
In dem blauen Himmelszelt?
Weisst wie viele Wolken gehen
Weit hinüber alle Welt?
Gott der Herr hat sie gezählet
Dass ihm auch nicht eines fehlet
Von der ganzen grossen Zahl.

Norbert Wiener begins his essay on "Newtonian and Bergsonian time" in the "Cybernetics" (1948) with this well known prayer of German children, which says in English: "Knowest thou how many stars stand in the blue tent of heaven ? Knowest thou how many clouds pass far over the world ? The Lord God hath counted them, that not one of the whole great number be lacking".

Wiener's point is that for God, in Whose sight "a thousand years are but as yesterday..." there is no difference whatever between "eternal" stars and capricious clouds. An actual enumeration (in Wiener's words "Durchmusterung') of these objects — if it were possible — would also reveal a surprising similarity: if these objects were considered point-like, and their number counted in a large sphere S of radius R around some point P, their average density:

$$\rho(P,R) = \frac{\text{number of "objects" within S}}{4R^3\pi/3} \tag{2.3.1}$$

would tend to zero, both for stars and clouds, as powers of R:

$$\rho(P,R) \propto R^{-A} \qquad \text{or simply } \rho(R) \propto R^{-A} \qquad (A > 0) \tag{2.3.2}$$

(though with different exponents A), rather than to a constant density ρ. In equation (2.3.2) it is assumed that the spatial distribution of the objects is homogeneous and isotropic. The power-like decrease of density with the increase of the linear size of the domain over which it is computed is due to the *fractality* of the spatial distribution of the point-like objects, and the exponent A in equation (2.3.2) is related to the fractal dimension D of this point set:

$$A = 3 - D \tag{2.3.3}$$

Equation (2.3.3) follows from the fact that the number $N(V)$ of points of a D-dimensional set inside a sphere of volume V scales as $N(V) \propto V^D$, that is, $= N(V)/V \propto V^{D-1} = R^{3(D-1)} = R^{-(3-D)}$.

Obviously, instead of a sphere we could have used a cube of side $2R$ ($R \to \infty$), centred around P. Equations (2.3.2) and (2.3.3) are also valid more generally: if a d_f-dimensional homogeneous isotropic fractal point set F is embedded in a d_e-dimensional Euclidian space, then the average density of F estimated over d_e-dimensional spheres or cubes of linear size $\propto R$ will decrease as:

$$\rho(R) \propto R^{-(d_e - d_f)} \qquad \text{as } R \to \infty \qquad (d_f \leq d_e) \tag{2.3.4}$$

From equation (2.3.4) we see that the fractal dimension d_f can be found from the slope of the $\log \rho(R)$ vs. $\log R$ plot.

The anomalous decrease of the estimated density of fractal point sets of the 2-dimensional plane is schematically illustrated in Figure 2.30. The phenomenon has been first observed in metallography, in the microscopic analysis of the distribution

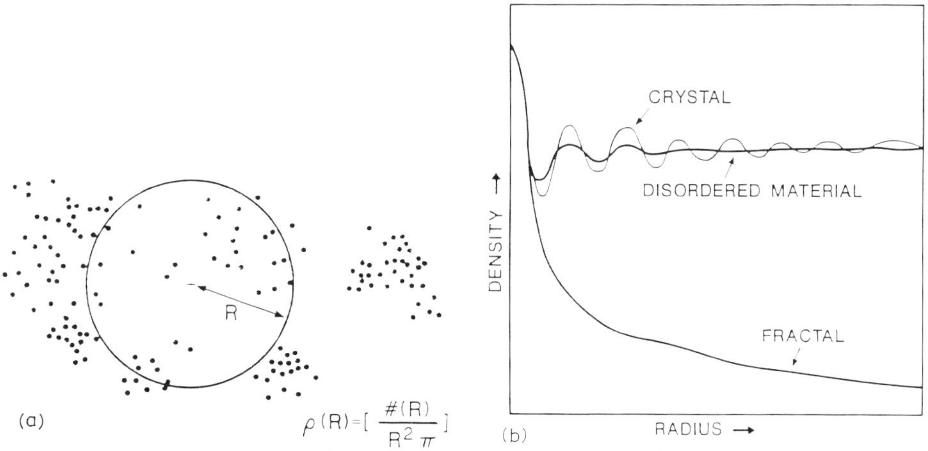

Fig. 2.30. Density of irregular planar point sets. a. Determination of the average point density. b. Average density as function of the radius. (After Sander, 1987.)

of grains and lamellae in aluminium alloys (Kostron, 1949, and Underwood, 1958, reviewed in Underwood and Banerji, 1986).

The one-dimensional variant of the method of finding the fractal dimension from the slope of the log $\rho(R)$ vs. log R curve or, equivalently, from the log $N(R)$ vs. log R curve, has already been discussed in Section 2.2 in connection with the slip lines in crystals (Sprušil and Hnilica, 1985). For planar sets the technique was first applied by Forrest and Witten (1979), to determine the fractal dimension of the electron microscope picture of an aggregate of iron particles (shown in Figure 2.31c). The method — termed "box-counting" by Forrest and Witten (1979) — and its many variants are widely used for the study of real and computer-generated fractal aggregates (for review see: Sander, 1986, 1987; Smirnov, 1986; Jullien and Botet, 1987; Meakin, 1988a, b), strange attractors (Grassberger and Procaccia, 1983a, b; reviews: Grassberger, 1986; Bai-Lin, 1984, Pt. 1, Chap. 8), geophysical measuring networks (Lovejoy et al., 1986; Korvin et al., 1990) — and raindrops caught on a blotting paper (Lovejoy and Schertzer, 1990). The practical implementation, algorithmic aspects and generalizations of box-counting will be presented in Appendix A2.

2.3.1 *Fractal characterization of geophysical measuring networks*

> The Khazars used fishing nets for sails on some of the ships of their flotilla, and these ships sailed like any other. When the Greek asked the Khazar priest how they accomplished this, a Jew standing nearby replied in their stead: 'It's simple. They catch in those nets something other than the wind.'
> — MILORAD PAVIČ: *Dictionary of the Khazars*
> (translated by Ch. Pribičevič-Zorič, 1988)

The measuring stations of a geophysical network are often spatially distributed in an inhomogeneous manner, being mainly concentrated around population centres and avoiding inaccessible areas. The distribution of *gravity stations*, for example, is a result of a hierarchy of decision processes: in Australia, for example, (1) the *reconnaissance surveys* totalling over 170,000 stations have been completed by the Bureau of Mineral Resources at stations approximately 11 km apart (in South Australia and Tasmania 7 km apart); (2) *semi-regional* surveys are usually read on a 0.5–2 km grid, according to the gravity response expected; and (3) the *detailed* gravity surveys use stations with 100–500 m spacing (Lynch and King, 1983; Fraser et al., 1976). The irregularity and sparsity of the stations imply that there are inevitable interpolation errors. Barlow (1977) estimates the interpolation error by the formula of Shazina and Grushinsky (1971): $e = 2k[\Delta x]^{1/2}$ (where $k = 0.32 \pm 0.02$; Δx is interstation distance in km) and concludes that it can be as high as 2.1 mgal for the Australian regional gravity survey, and 1.3 mgal where a more dense data coverage has been obtained.

The spatial inhomogeneity of point sets can be characterised by different statistical indices of dispersion (Ouchi and Uekawa, 1986). If the set is self-similar, that is a magnified version of any small part of it is statistically similar to the whole set, it can be characterised by its fractal dimension.

Figure 2.31a shows a typical gravity station distribution, arbitrarily chosen from a recent case history from Central India (Subba Rao et al., 1987). Figure 2.31b is an example for a dense gravity station network in and around Adelaide, South Australia (Korvin et al., 1990).

Figure 2.31c, which is surprisingly similar to the previous two, shows a punctured sea-shell. Figures 2.31d–f represent a quite different field of science: Figure 2.31d

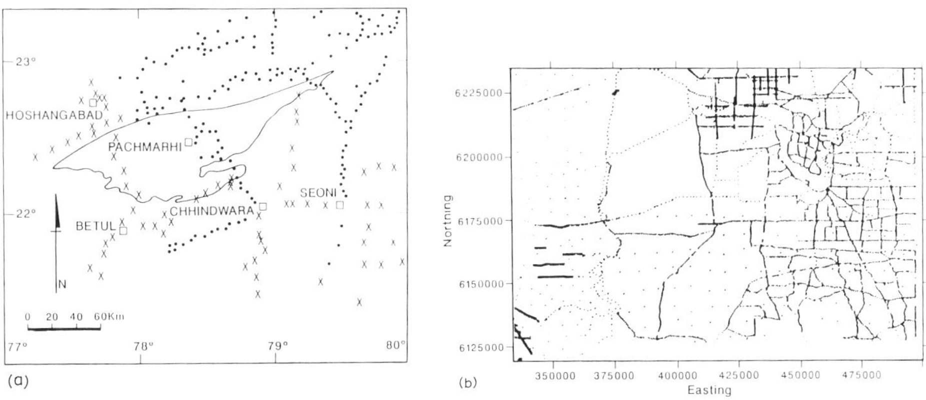

Fig. 2.31. Examples for fractal clusters. a. Gravity station distribution in the Satpura Gondwana Basin, Central India. Key: x = Survey of India stations; • = National Geophysical Research Institute stations; -- = boundary of Gondwana outcrops (after Subba Rao et al., 1987). b. A dense part of the South Australian gravity network: the Adelaide gravity stations (from Korvin et al., 1990).

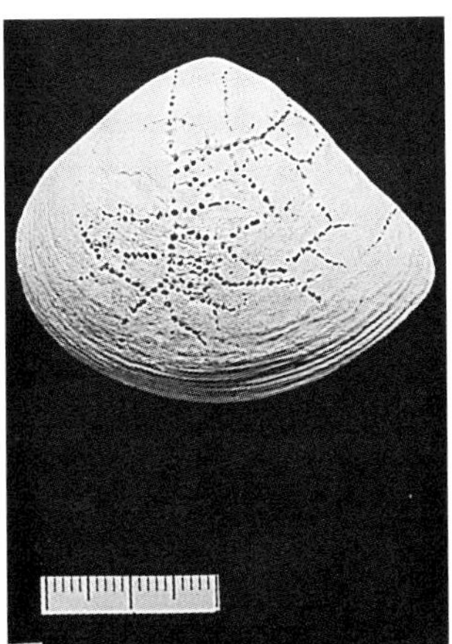

Fig. 2-31 (continued). c. Mould bio-immuration of a Bryozoa (unknown but similar to *Arachnidium sp.*) in a shell (*Mactra sp.*, found at Brighton, South Australia).

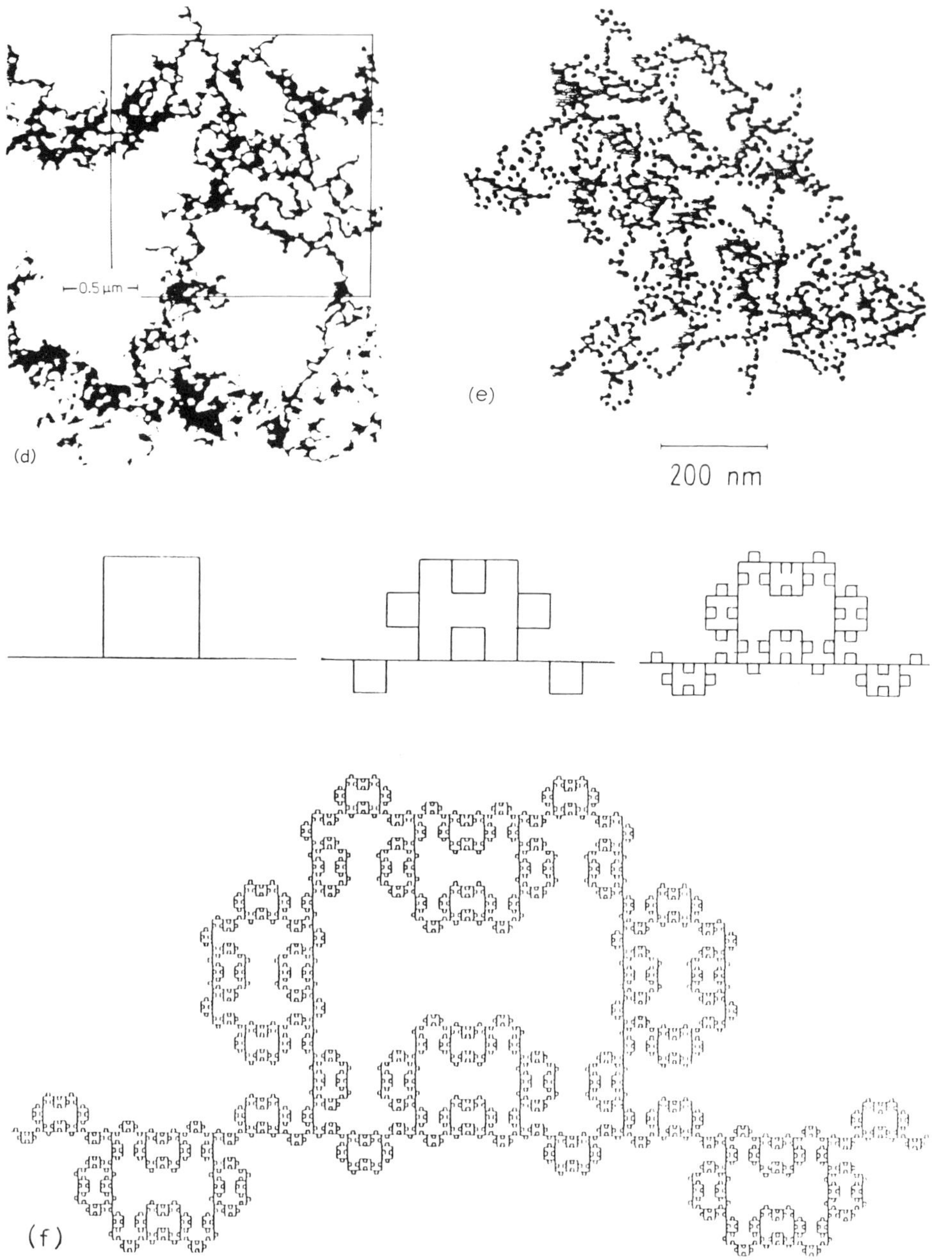

Fig. 2.31 (continued). d. Electron micrograph of an iron aggregate. The square indicates the image that was analysed (from Forrest and Witten, 1979). e. Electron micrograph of a gold colloid cluster (from Weitz and Huang, 1984). f. The Mandelbrot-Given model (1984) of the backbone of percolation clusters, and the first steps of its construction. Theoretical fractal dimension is $D = \log 6/\log 3 = 1.63$. (From Feder, 1988.)

(Forrest and Witten, 1979) is the electron microscope picture of a large smoke particle consisting of many small iron spheres of uniform size; Figure 2.31e (Weitz and Huang, 1984) is the electron micrograph of a gold colloid cluster; Figure 2.31f (Mandelbrot and Given, 1984) is a mathematical model of the "backbone" of a percolation cluster. There is a striking similarity between the general appearance of the clusters in Figures 2.31a–f. Each of them contains holes at all scales compatible with cluster size, and each is statistically self-similar.

As a matter of fact, the clusters shown in Figures 2.31d–f have been proved to be *fractals*. For the micrograph 2.31d the fractal dimension is 1.51 ± 0.05 (found by box-counting, Forrest and Witten, 1979); for the cluster in Figure 2.31e it is about 1.7–1.8 (from light scattering and small angle neutron scattering: Weitz and Huang, 1984; Weitz et al., 1985; Feder, 1988, pp. 36–40). The fractal dimension of the cluster 2.31f can be theoretically determined (Mandelbrot and Given, 1984); it is $\log 6 / \log 3 = 1.63$. Because of the similarity of the networks 2.31a–b to the clusters 2.31d–f, it is reasonable to assume that the gravity stations are also fractally distributed. The first attempt to use the fractal dimension to characterize geophysical measuring networks has been made by Lovejoy and co-workers [Lovejoy et al. (1986a), Lovejoy and Schertzer (1986); see also the interesting discussion between Hollingsworth (1986) and Lovejoy et al. (1986b)]. Lovejoy and his group proved that the World Meteorological Station Network (9,563 stations) constitutes a 1.75-dimensional fractal set, the French climatological network (3,593 stations) is 1.8-dimensional, and the Canadian meteorological network (414 stations) is only 1.5-dimensional.

They found the low dimensionality of these networks highly regrettable since "to detect phenomena, not only must a network have sufficient spatial resolution, it must also have sufficient dimensional resolution. Whenever $d_{\text{fract}} < d_{\text{Eucl}}$, sparsely distributed phenomena with dimension less than $d_{\text{Eucl}} - d_{\text{fract}}$ cannot be detected", and hinted at the difficulties in interpolating low-dimensional sparse data onto two-dimensional regular grids (Lovejoy et al., 1986a, b; Schertzer and Lovejoy, 1985, 1986).

The studies of Lovejoy and co-workers have been followed up by Korvin et al. (1990), who proved that the South Australian gravity station network (over 65,000 stations) is *only 1.4-dimensional* (Figure 2.32).

For both kinds of network the fractal dimension has been determined using the *correlation function technique* of Grassberger and Procaccia (1983a, b).

The standard method of estimating the fractal dimension of a planar point set (Mandelbrot, 1982) is to divide a large square containing the set into X^2 equal squares and to count the number $N(X)$ of those small squares containing points of the set. For fractal point sets:

$$N(X) \propto X^{d_f} \qquad (\text{as } X \to \infty) \tag{2.3.1.1}$$

where d_f ($0 < d_f < 2$) is the fractal dimension. A second method consists in taking

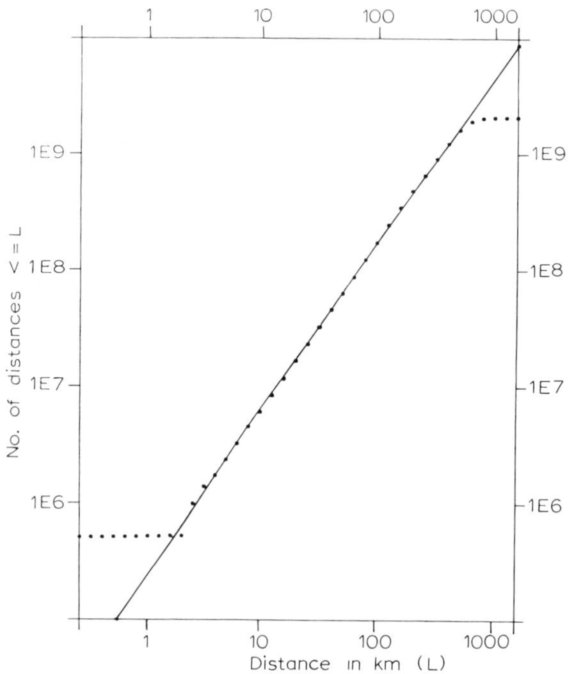

Fig. 2.32. Cumulative frequency distribution of the interstation distances for the South Australian gravity station network (from Korvin et al., 1990).

circles or squares of increasing size and measuring their contents or "mass" ("box-counting", see Forrest and Witten, 1979; Sander, 1986, 1987). For fractal point sets the number of points in a circle of radius X is:

$$N(X) \propto X^{d_{\text{box}}} \qquad (\text{as } X \to \infty) \tag{2.3.1.2}$$

where d_{box} and the d_f of equation (2.3.1.1) are not necessarily equal.

In their study of the dimension of strange attractors Grassberger and Procaccia (1983a, b) introduced the density correlation function of a finite planar point set A as:

$$C(X) = \{\text{number of pairs of points } P, Q \text{ in } A \text{ such that } |P - Q| \leq X\} \tag{2.3.1.3}$$

and they proved that for fractal sets:

$$C(X) \propto X^{d_{\text{corr}}} \qquad (\text{as } X \to \infty) \tag{2.3.1.4}$$

where the exponent d_{corr} is the same as the "box-counting" dimension d_{box}, and can be less than the "Mandelbrot" dimension d_f:

$$d_{\text{corr}} = d_{\text{box}} \leq d_f \tag{2.3.1.5}$$

(see Grassberger and Procaccia [1983a, b]; the different fractal dimensions will be compared in Appendix A2). Inequality (2.3.1.5) is quite tight in most cases if the fractal set is uniform (a counter-example is the Kaplan-Yorke attractor, analysed in Grassberger and Procaccia [1983b], where $d_{corr} = 1.42$ and $d_f = 1.43$). In most applications (for example Stauffer, 1979; Witten and Sander, 1981; Lovejoy et al., 1986a; Berry and Percival, 1986) it is tacitly assumed that $d_{corr} = d_f$.

As Lovejoy et al.(1986a) have pointed out, if a geophysical measuring network is only ($d_{net} < 2$)-dimensional (where, for simplicity, the surface of the Earth is taken two-dimensional), this imposes new detectability limits. A network of dimension d_{net} can only detect a meteorological phenomenon or geophysical anomaly with dimension d anomaly if:

$$d_{net} + d_{anomaly} > E \tag{2.3.1.6}$$

where E is the Euclidian dimension, in the present case $E = 2$. (Equation 2.3.1.6 follows from Mandelbrot's rule, which states that for two statistically independent fractal sets F' and F'' one has:

$$\dim(F' \cap F'') = \dim(F') + \dim(F'') - E \tag{2.3.1.7}$$

see Mandelbrot, 1982, p.365; Lovejoy et al., 1986a; Miyazima and Stanley, 1987; Feder, 1988, p. 206). Consequently, sparse phenomena with $d_{anomaly} < 2 - d_{net}$ will be almost certainly missed by the network.

It has recently been realised that certain geophysical fields are *multifractal*: regions of the Earth where the value of the geophysical field-parameter p exceeds a threshold T constitute a fractal set with dimension $d(T)$ decreasing with T (Schertzer and Lovejoy, 1983; Mandelbrot, 1984, 1989; Feder, 1988). In meteorology, the multifractal nature of turbulence and rainfields has been demonstrated by Lovejoy and his group. In exploration geophysics a vivid example of the Earth's multifractality is the De Wijs model of ore distribution (De Wijs, 1951, 1955; Mandelbrot, 1982, 1984, 1989; Turcotte, 1986c): every square of the surface of the earth carries some gold, but if one is only interested in gold of increasing purity, one must restrict one's search to increasingly smaller sets. Each of these sets can be approximated by a fractal whose dimension decreases with the increase of the desired purity.

There is evidence for the fractal nature of the velocity and density anomalies within the earth (faults: Aviles et al., 1987, Okubo and Aki, 1987; seismic boundaries: Barenblatt et al., 1984; density inhomogeneities: Thorarinsson and Magnusson, 1990; velocity inhomogeneities: Wu and Aki, 1985, Ojo and Mereau, 1986, Sato, 1988; hydrocarbon reservoirs: Agterberg, 1982, Hewett, 1986, 1988; etc.).

If these anomalies are *multifractal* (which has not been proved yet), the most intensive changes of the geophysical parameters would occur along a set of points of low fractal dimension $d_f \leq E$ which will eventually be missed as soon as $d_{net} + d_f < E$.

The problems caused by the low fractal dimensionality of a geophysical measuring network can be also understood from the point of view of Information Theory (Korvin et al., 1990). In the case of gravity networks, for example, the observed field is usually band-limited, that is there exists a geologically meaningful shortest wavelength $\lambda_{\min}$ such that the power spectrum $B(k_x, k_y)$ of the Bouguer anomaly field is practically zero for wavenumbers (k_x, k_y) such that:

$$k_x^2 + k_y^2 \leq \frac{1}{\lambda_{\min}^2} \tag{2.3.1.8}$$

If the region studied is a square of side X, then — by the two-dimensional form of Shannon's sampling theorem — the gravity field can only be restored from its sampled values if *at least*:

$$N = \frac{X^2}{\lambda_{\min}^2} \tag{2.3.1.9}$$

samples are taken along a regular grid (Brillouin, 1962, chapter 8).

For a fractal network of dimension $d < 2$, the number of stations within a square of side X scales as X^d rather than X^2 for increasing X, that is the network becomes more and more sparse and falls short of the Shannon condition (2.3.1.9). Sampling along this low-dimensional irregular grid will preserve some aliasing frequencies which, in their turn, will lead to the well known spurious anomalies if one attempts to interpolate onto a dense regular grid. Though all wavenumbers will be affected, the high wavenumber (short wavelength) information will be the most seriously distorted.

These considerations seem to apply to seismological stations, seismic profiles and other surveys as well. In these applications, one generally uses *arrays* of sensors (seismometers, geophones, hydrophones, etc.) and of energy sources (airguns, vibrators, etc.) rather than single receivers or sources.

Considering the negative effects associated with the fractally distributed measuring stations discussed above, it is surprising to learn that random fractal antenna arrays have advantageous, robust, low side-lobe radiation patterns. This was discovered by Kim and Jaggard (1986) who have started out from the observation that large size arrays usually are "arrays of arrays" (Schelkunoff, 1943), that is hierarchical structures.

2.4 Fractals in the plane: fractures–earthquakes–volcanoes

This new learning amazes me.
Tell me again how Sheep's Livers may be used to predict Earthquakes?

— *Monty Python and the Holy Grail*

2.4.1 *Cellular structures*

Polygonal networks of cracks and fractures can be observed in ceramic glazes, concrete, varnish and paint, mud, frozen ground, limestone and other media which undergo volume change. Such a pattern is generally the result of contraction of a layer of homogeneous material perpendicular to a cooling or shrinking surface (Lachenbruch, 1961, 1962; Plummer and Gostin, 1981; Walker, 1986). Though these polygonal networks can be observed on all scales in a wide variety of materials (Figure 2.33) and they show surprising similarities to the patterns in living tissues (Figures 2.34, 2.35), these structures are generally not fractal as they have a *characteristic length scale.*

According to Lachenbruch (1961, 1962), who studied the geometry of the ice-wedge polygons in the Arctic permafrost, cracks are generally initiated at a surface of great stress (often at or near the ground surface) and are propagated downward toward the interior of the medium where the tension diminishes and ultimately passes into compression.

The depth of the tension cracks is thus controlled by the thickness and rheology of the contracting plate (Figure 2.36). It has also been observed that the crack spacing is of the same order of magnitude or by at most an order of magnitude larger than crack depth (Lachenbruch, 1961; Neal et al., 1968), that is the contraction-crack polygons have a characteristic size, depending on the properties of the contracting plate (see Korvin's, 1989, statistical study of the fault blocks in the Gulf of Suez basement).

According to Lachenbruch (1961, 1962), within any medium subjected to volumetric decrease, one of two crack systems can be generated depending on the homogeneity and plasticity of the medium. Within inhomogeneous, plastic media an *orthogonal system* of cracking develops (i.e. cracks intersect at 90° angles, Figure 2.37). In such a system cracks form at loci of low strength or high stress concentrations but they are not propagated simultaneously. In *non-orthogonal systems*, however (Figure 2.38), very homogeneous, relatively non-plastic media are involved, with the cracks propagating laterally until a limiting velocity is reached at which time the cracks bifurcate at obtuse angles (Figure 2.34). Unlike the orthogonal system, all elements of non-orthogonal intersections are generated virtually simultaneously.

The structures shown in Figures 2.33, 2.35, 2.37 and 2.38 are common to geology, ecology, geography, histology, metallography and many other fields. They are aggregates of non-overlapping cells which, along with their boundaries, cover the whole space. Each cell of the aggregate is a (possibly curvilinear) polygon. *Edges* are common to two cells, *vertices* are common to three or more. In the practically most important case exactly three edges meet at any vertex (as in Figure 2.38b).

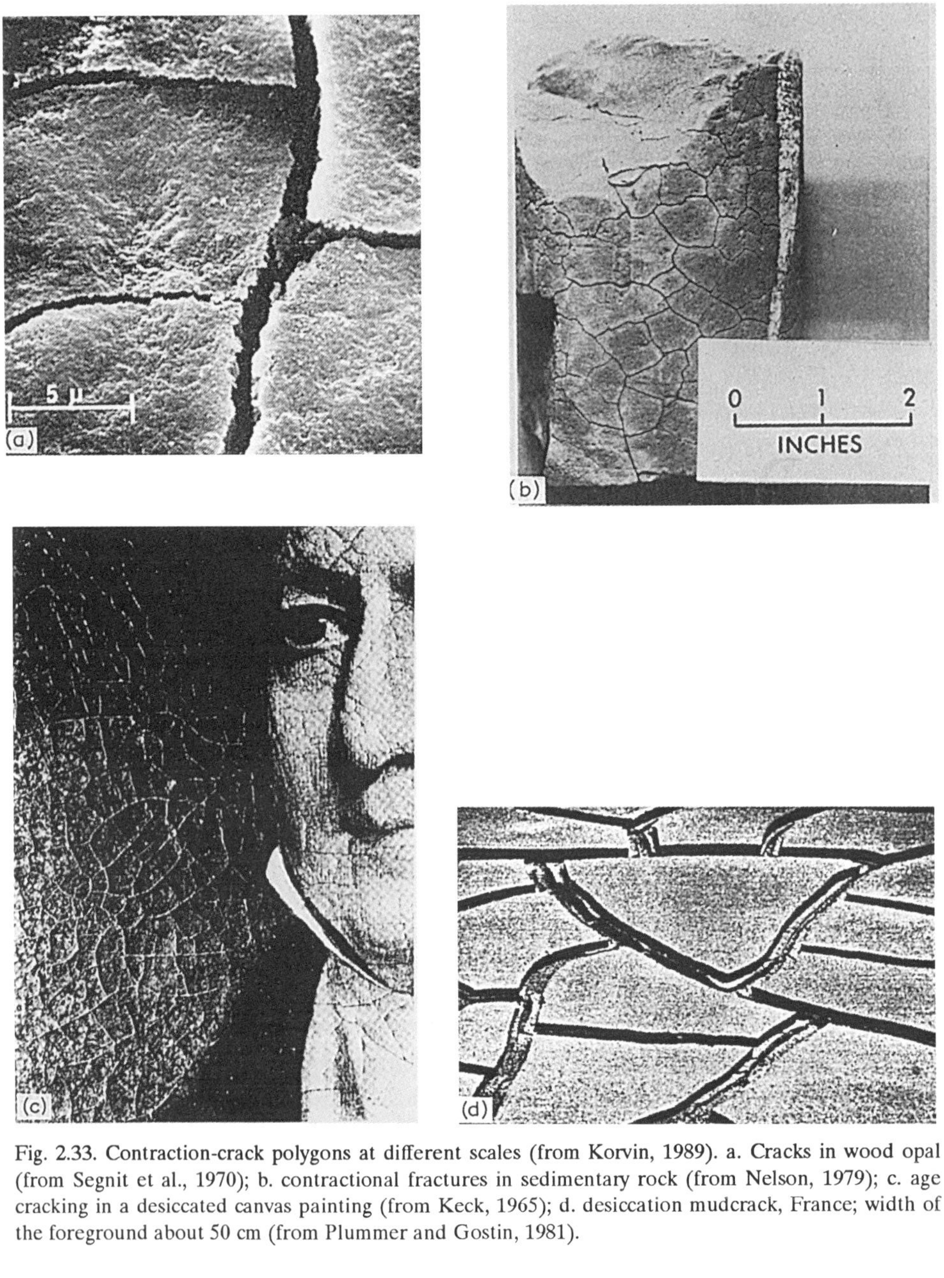

Fig. 2.33. Contraction-crack polygons at different scales (from Korvin, 1989). a. Cracks in wood opal (from Segnit et al., 1970); b. contractional fractures in sedimentary rock (from Nelson, 1979); c. age cracking in a desiccated canvas painting (from Keck, 1965); d. desiccation mudcrack, France; width of the foreground about 50 cm (from Plummer and Gostin, 1981).

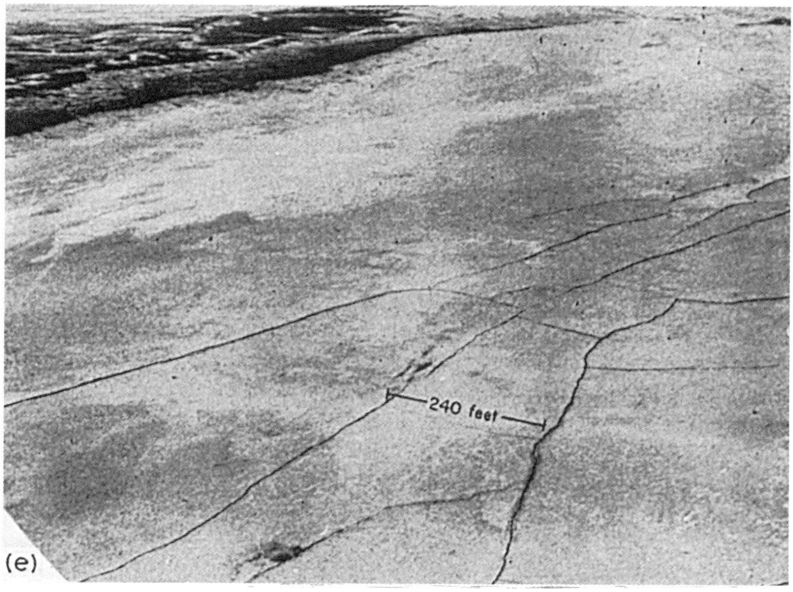

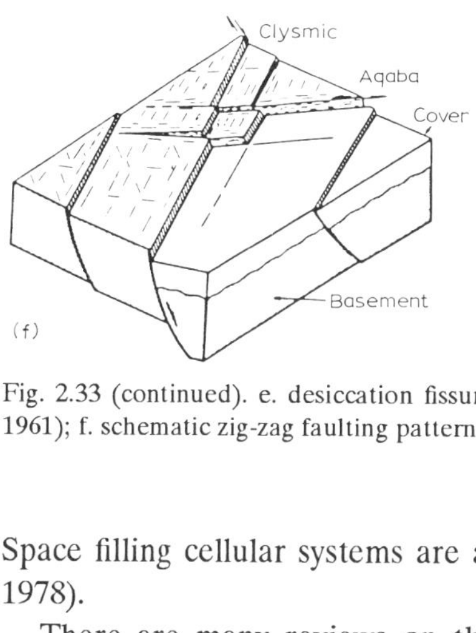

Fig. 2.33 (continued). e. desiccation fissures, Black Rock Desert, Nevada (from Willden and Mabey, 1961); f. schematic zig-zag faulting pattern of the Gulf of Suez Basement (from Jarrige et al., 1986).

Space filling cellular systems are also called *tessellations* (Miles, 1964; Lantuejoul, 1978).

There are many reviews on the properties and applications of cellular structures: D'Arcy Thompson's (1942) classic "On Growth and Form", Stevens (1974, with beautiful photographs!) and Dormer (1980) on biological applications; Smith (1952, 1954), Weaire (1983), Weaire and Rivier (1984) on metallurgical structures and topological principles; Getis and Boots (1979) on the geographical perspectives; Plummer and Gostin, 1981, Walker, 1986 on contraction-crack polygons in earth materials; while Ogawa's fascinating essay (1983) spans an enormous range from the physics of amorphous matter to the marking of giraffes, mud cracks, and comparative culture.

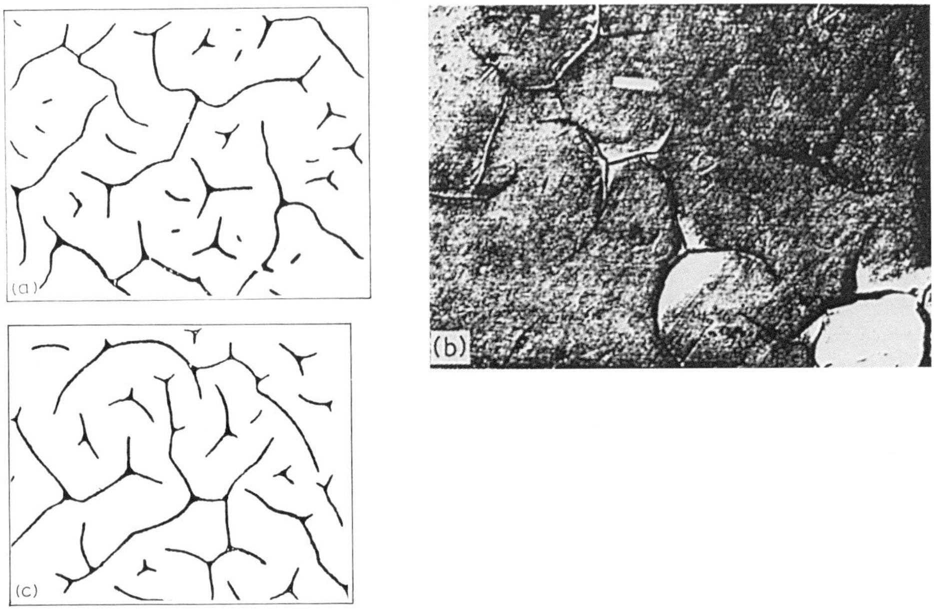

Fig. 2.34. Similar patterns in earth materials (a, b) and the human brain (c). a. Patterns formed on the surface of a contracting sphere (from Clark, 1947). b. Incomplete desiccation mudcracks showing triple junctions; scale is 5 cm long (from Plummer and Gostin, 1981). c. The sulcal pattern of the inferior part of the parietal lobe of the human brain (from Clark 1947). Clark (1947) has made use of geotectonics to explain the deformation patterns of the cerebral cortex.

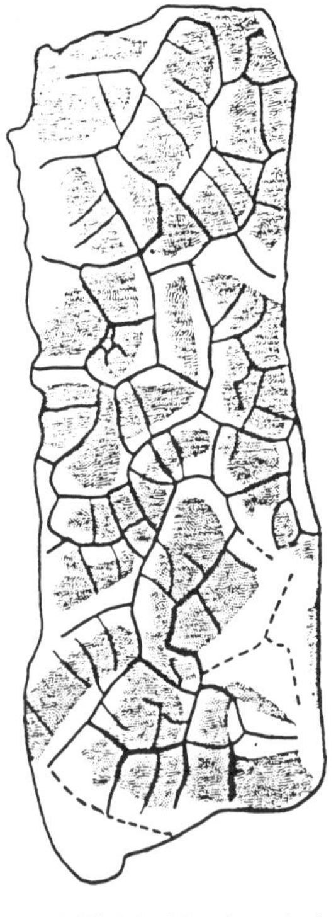

Fig. 2.35. Moulds of cracked curled up layers of thin mud in the overlying sand often resemble parts of arthropod skeletons. This polygonal mud crack system (reconstructed by Chowns, 1973) has lead to the amusing "pseudofossil" shown in the Geotimes, December 1972, cover photograph.

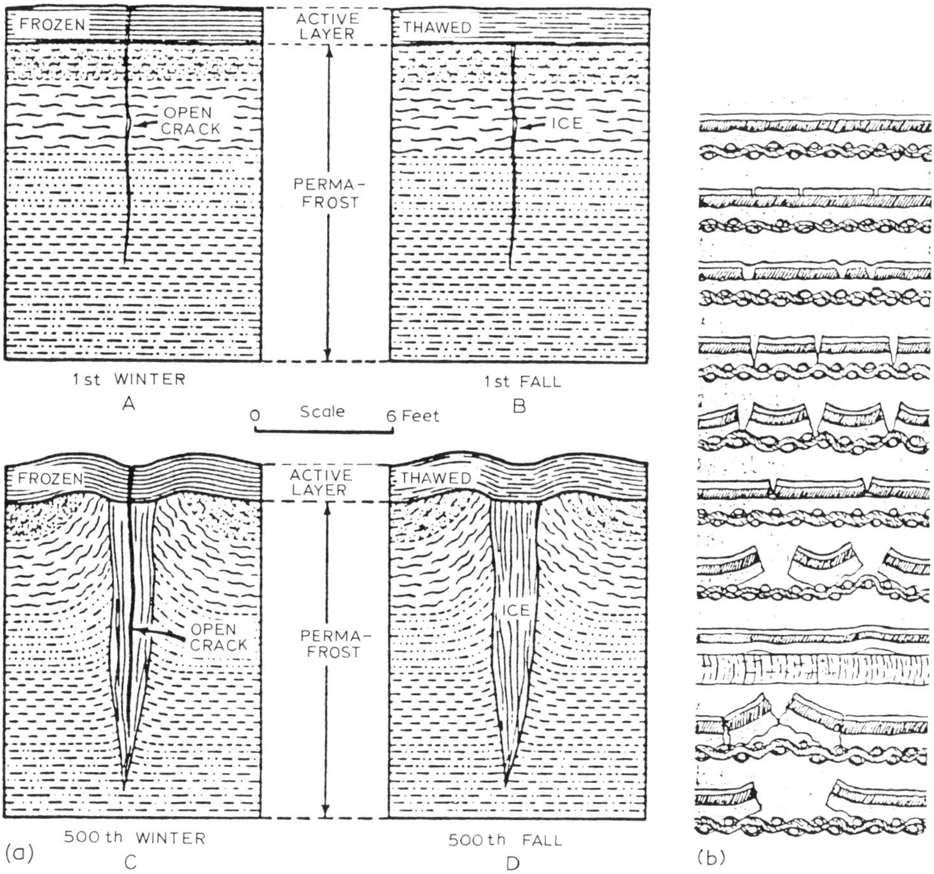

Fig. 2.36. a. Evolution of an ice-wedge according to Leffingwell's (1915) contraction theory. Width of the crack is exaggerated (from Lachenbruch, 1962). b. Some varieties of structural deterioration in paintings. The top diagram shows a new oil painting: top layer = varnish, second layer = paint films, third layer = ground, fourth layer = support (from Keck, 1965).

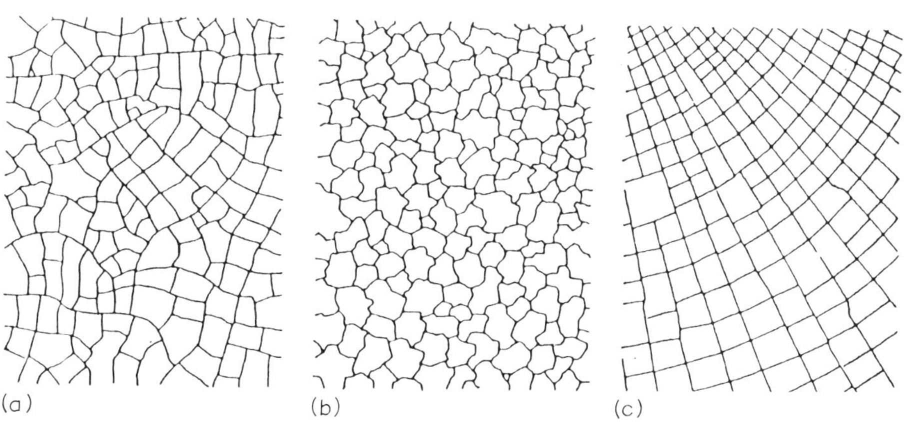

Fig. 2.37. Orthogonal crack networks: a. random and regular; b. random and irregular; c. oriented (from Walker, 1986).

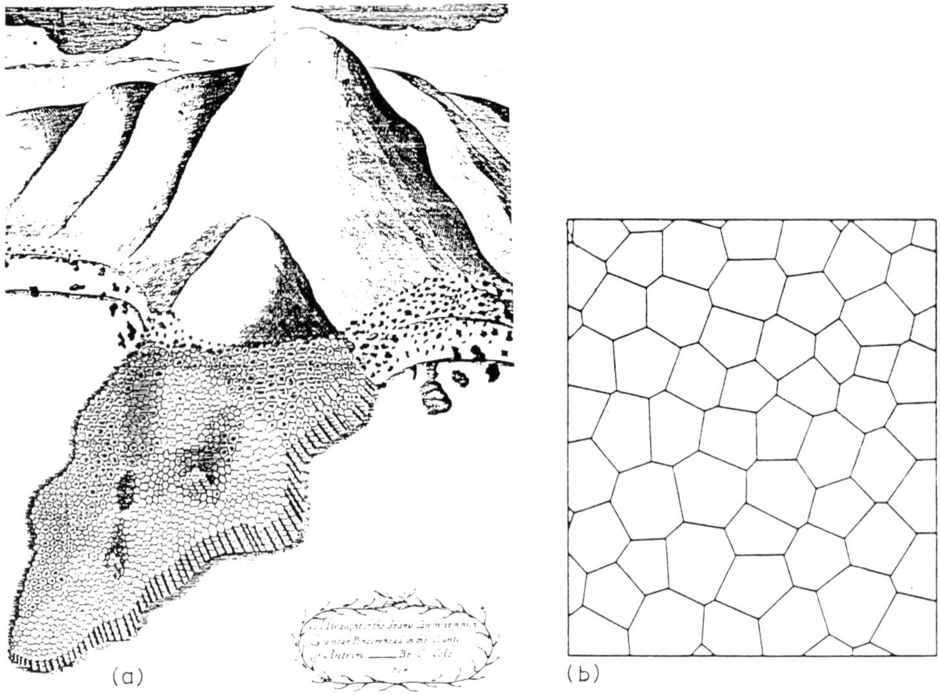

Fig. 2.38. Sketch of the Giant's Causeway, N. Ireland, from 1694 (a), and (b) a more accurate map of its detail (1879) (from Weaire and Rivier, 1984).

In any arrangement of non-overlapping space-filling cells the numbers of cells, edges and vertices satisfy Euler's equation:

$$C - E + V = \text{const} \tag{2.4.1.1}$$

where C is the number of cells, E of edges and V of vertices. The constant on the right-hand-side is a topological invariant of the surface on which the tissue is put. For example it is 2 for a sphere or a rugby ball, 0 for a torus, doughnut or tea cup, and 1 for a plane. Usually, Euler's relation is applied for planar patterns where:

$$C - E + V = 1 \tag{2.4.1.2}$$

Equation (2.4.1.2) can be checked on the network of Figure 2.39, based on a Picasso painting. [Incidentally, counting C, E and V is by no means an easy task. In an often cited study, Cox and Agnew (1974) constructed a theoretical partition of Ireland into thirty-two "ideal counties" which they miscounted as thirty-one.]

An immediate consequence of Euler's relation is that the average number of edges $\langle n \rangle$ surrounding a cell is related to the average coordination number of

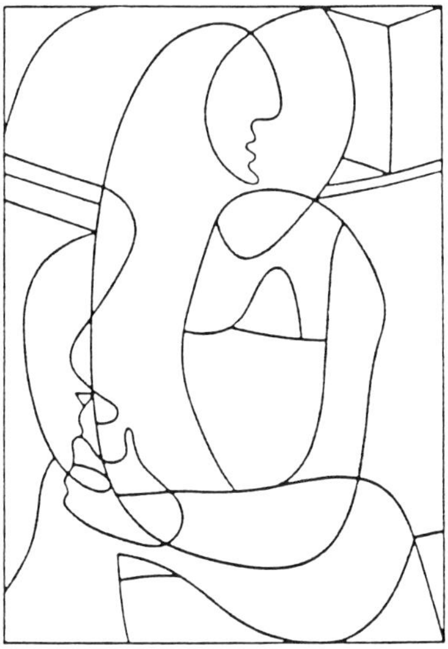

Fig. 2.39. Space filling by irregular polygons. $C - E + V = 35 - 86 + 52 = 1$. (From Smith, 1952.)

vertices $\langle Z \rangle$ as:

$$\langle n \rangle = \frac{2\langle Z \rangle}{\{\langle Z \rangle - 2\}} \tag{2.4.1.3}$$

where the *coordination number* of a given vertex is the number of edges meeting at the vertex, and — as throughout this book — angular brackets denote expectation or mean value. Indeed, denoting by C_n the number of n-sided cells one has the relations $\langle Z \rangle V = 2E$ and $\sum n\, C_n = 2E$, because on the average $\langle Z \rangle$ edges meet at a vertex and an edge links two vertices, whereas it separates two cells. Euler's relation can therefore be written as:

$$C \left[2\langle Z \rangle - (2 - \langle Z \rangle) \frac{\sum n\, C_n}{C} \right] = 1 \tag{2.4.1.4}$$

In the limit of a large system, when $C \to \infty$, equation (2.4.1.4) can only hold if the bracketed expression tends to zero. Observing that $\sum n\, C_n / C \to \langle n \rangle$ as $C \to \infty$, we get equation (2.4.1.3). In the usual constructions of random networks the average cell is either 6-sided (Voronoi polygons with $\langle Z \rangle = 3$, $\langle n \rangle = 6$; see e.g. Weaire and Rivier, 1984) or 4-sided ($\langle Z \rangle = 4$, $\langle n \rangle = 4$, for intersecting Poisson distributed random straight lines; Kendall and Moran, 1963; Miles, 1964), Figure 2.40a.

In the measured side-frequency distributions for polygons formed by tensional cracking in columnar basalts the modes of 5 and 6 are both prevalent (Figure 2.40b). For a few well known localities Beard (1959) found:

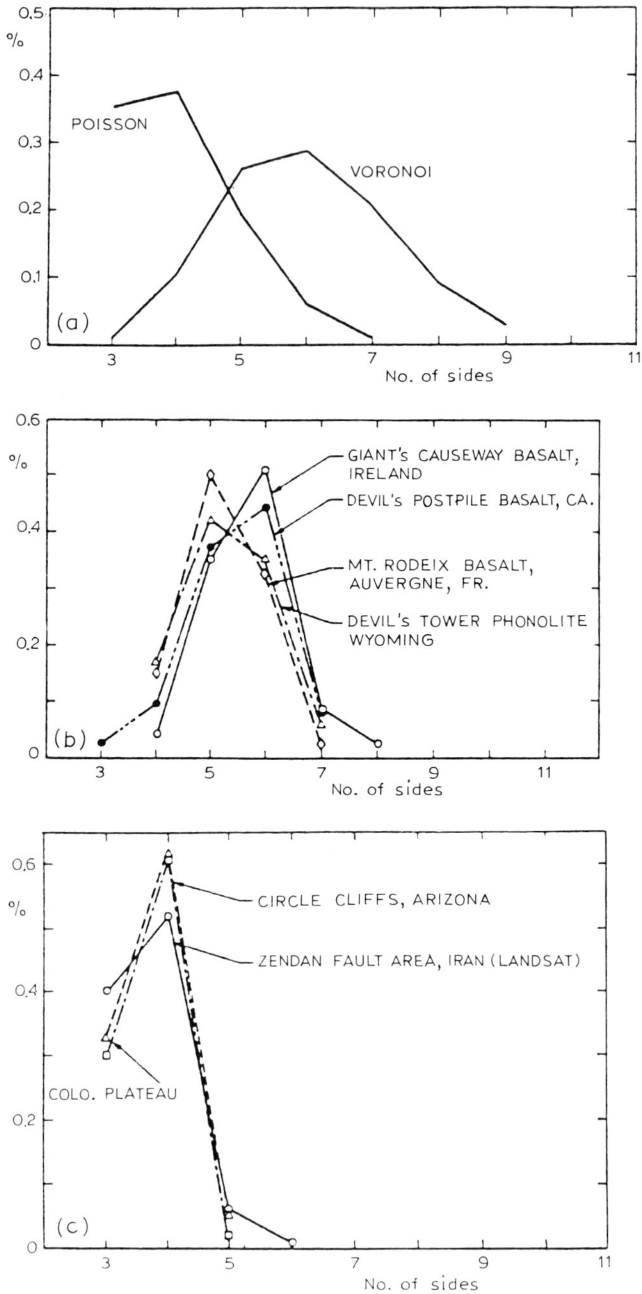

Fig. 2.40. a. Frequency distributions for the number of sides of polygons formed by the intersection of random lines, and for Voronoi polygons. b. Side-frequency distribution for polygons formed by tensional cracking in columnar basalts (compiled from Beard, 1959). c. Side- frequency distribution for intersecting faults in complexly faulted terrain (from Pieri, 1981).

$\langle n \rangle$ = 5.23 for the Mt. Rodeix basalt, France,
$\langle n \rangle$ = 5.30 Pleistocene basalt flow, Dunsmuir, California,
$\langle n \rangle$ = 5.50 Devil's Postpile basalt, California,
$\langle n \rangle$ = 5.66 Giant's Causeway, N.Ireland,

that is — in spite of the "scientific folklore" — the Giant's Causeway does not have a perfectly hexagonal pattern (an important statistical analysis of the Giant's Causeway is Weaire and O'Carroll, 1983).

Intersecting faults in complex faulted terrain generally form a network with $\langle n \rangle = 4$ (Figure 2.40.c), as in the Poisson pattern (Figure 2.41a).

By far the most commonly studied ideal structure is the Voronoi construction mentioned above, and its popularity is reflected by the number of names it has been given in its various applications: *Voronoi polygons, Wigner-Seitz cells, Dirichlet tessellations, Thiessen polygons, etc.* (Weaire and Rivier, 1984; Getis and Boots, 1979). In this construction an array of random points is first defined as "centres", then each of these centres is assigned a cell containing all points of the plane nearest to it. Specifically, one draws the perpendicular bisecting lines between any two centres, and the smallest convex polygon surrounding a given centre contains all points of the plane closest to this centre. It can be proved that the construction is unique, space filling and that for Voronoi cells constructed around Poisson-distributed centres the average number of sides is 6 (Miles, 1964). Two examples are shown in Figure 2.41b, c.

It should be mentioned that the side distribution of both the natural and artificial polygonal patterns usually has a substantial scatter, for example in the Giant's Causeway basalt 4.5% of the sides is 4-sided, 35% is 5-sided, 51% is 6-sided, 9.25% is 7-sided, and 0.25% is 8-sided (Beard, 1959; Smalley, 1966). As another example, Boots (1980) reports that out of the 215 civil parishes of the English county Somerset (as they existed at the time of the 1961 census) there were:

2	1-sided (i.e. completely surrounded),
4	2-sided,
8	3-sided,
42	4-sided,
43	5-sided,
47	6-sided,
32	7-sided,
23	8-sided,
8	9-sided,
3	10-sided,
1	11-sided,
1	13-sided,
1	14-sided

($\langle n \rangle$ = 5.8). There have been various suggested analytical forms for the side distri-

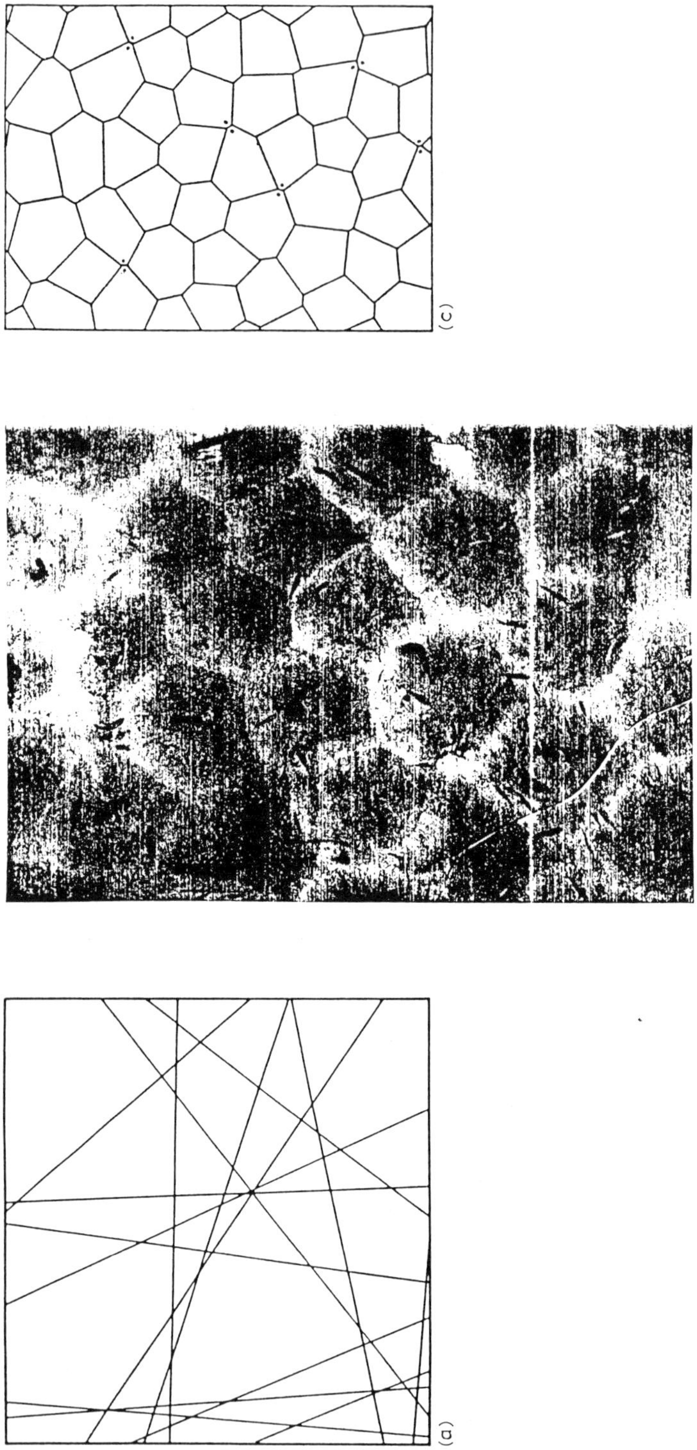

Fig. 2.41. Examples for polygonal networks. a. Poisson field of random lines (from Rothrock and Thorndike, 1984). b. Top view photograph of the territories of the males of the African mouth-breeding fish *Tilapia mossambica*. Each territory is a pit dug in the sand by the occupant. The breeding males are the black fish, 15 to 20 cm in size (from Barlow, 1974). c. Modified Voronoi construction for the simulation of crack network in basalt flows. Very short sides are indicated by dots. The side distribution compares well with the Giant's Causeway (from Smalley, 1966).

bution $f(n)$, including the Poisson distribution and lognormal distribution for grain structures, the lognormal distribution for the structure of glasses, and the gamma distribution for the cellular structures observed in soap films, epidermal plant cells and basalt columns (for references see Weaire, 1983), though none of these distributions seems to be universally applicable.

The topological and geometric properties of planar cellular structures are intertwined. The American botanist F.T. Lewis (1944) found the surprising relation that the *the mean area $\langle A(n)\rangle$ of the n-sided cells in various planar cellular patterns (cucumber epidermis, pigmented epithelium of the retina, etc.) varies linearly with n*:

$$\langle A(n)\rangle = \frac{\langle A\rangle(n-2)}{4} \tag{2.4.1.5}$$

where $\langle A\rangle$ (= total area/number of cells) is the average cell area. This relation does not follow from any elementary considerations of geometry.

Rivier and Lissowski (1982) have shown however, by very subtle maximum-entropy arguments, that if the side distribution function $\{f(n)\}$ is only subject to the geometric constraints of space filling:

$$\textstyle\sum f(n) = 1 \qquad \text{(normalisation)} \tag{2.4.1.6}$$

$$\textstyle\sum A(n)f(n) = A_{\text{total}} \qquad \text{(conservation of total area)} \tag{2.4.1.7}$$

$$\textstyle\sum (n-6)f(n) = 0 \qquad \text{(a consequence of Euler's relation (2.4.1.3) if } \langle Z\rangle = 3) \tag{2.4.1.8}$$

then the "most arbitrary" random cellular arrangement [where arbitrariness is characterised by the Shannon entropy $-\sum p(n)\log p(n)$] satisfies Lewis' law (2.4.1.5). If a mosaic does not obey Lewis' law, then the average area of its constituent cells is not regulated simply by the geometrical requirements (2.4.1.6)–(2.4.1.8) but also by specific local dynamic interactions between neighbouring cells. Sometimes the space-filling constraints act in a higher dimension: plane sections of three-dimensional cellular aggregates have different statistical properties from the two-dimensional tessellations (Smith, 1952).

In a statistical analysis of the detailed map of the Giant's Causeway made by O'Really in the 1870's, Weaire and O'Carroll (1983) found an interesting correlation between length of sides and the neighbouring angles:

$$\sigma = 90 + \frac{30L}{\langle L\rangle} \qquad \text{(angles in degrees, } \langle L\rangle \text{ is average side length)} \tag{2.4.1.9}$$

If we write out equation (2.4.1.9) for each side of an "average" n-sided polygon (Figure 2.42) and observe that $2\sum \sigma_i = (n+2)180$ we get:

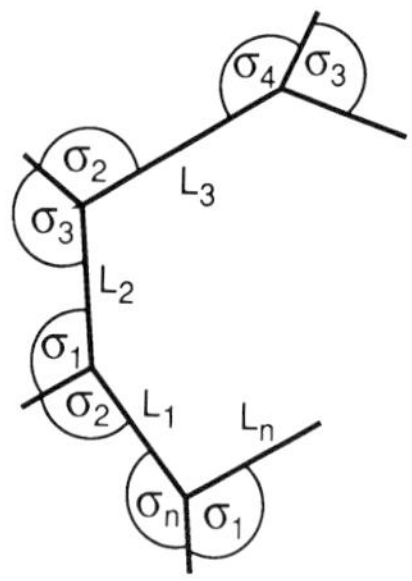

Fig. 2.42. To the derivation of equation (2.4.1.10).

$$2\sum\left[90+\frac{30L_i}{\langle L\rangle}\right]=(n+2)180$$

that is, for the Giant's Causeway basalt the average perimeters of n-sided cells are equal:

$$\langle P(n)\rangle = 6\langle L\rangle = \langle P(6)\rangle \tag{2.4.1.10}$$

independently of n. Lewis' law (2.4.1.5) and the empirical rules (2.4.1.9), (2.4.1.10) contain the length parameters $\langle A\rangle^{1/2}$, $\langle L\rangle$ or $\langle P(6)\rangle = 6\langle L\rangle$, respectively. Consequently, even though these patterns can be observed in all scales (Figure 2.33), any given realization has its characteristic scale, that is the structures are not self-similar fractals (with the useful distinction of Mandelbrot, 1981, they are *scale-bound* rather than *scaling*).

Sometimes we can observe several generations of polygonal networks in the same material, each generation having a successively smaller characteristic length scale (Figure 2.43; further examples can be found in Plummer and Gostin, 1981).

The crack pattern shown in Figure 2.43 suggests a simple way to generate a sequence of random Voronoi tessellations, each being a refinement of the previous one, which results in a self-similar fractal structure.

We shall use Voronoi polygons constructed around Poisson-distributed random points as centres, as their statistical properties (such as the mean values of the number of sides n, perimeter P and area A) are known exactly (Miles, 1964):

$$\langle n\rangle = 6 \tag{2.4.1.11a}$$

$$\langle P\rangle = \frac{4}{\rho} \tag{2.4.1.11b}$$

$$\langle A\rangle = \frac{1}{\rho^2} \tag{2.4.1.11c}$$

[The results hold for the case when the points are distributed on a unit area ac-

Fig. 2.43. Two generations of cracks in cornstarch (from Walker, 1986)

cording to a Poisson process of density ρ^2. Higher moments of the distribution have been determined by Monte Carlo simulation (Crain, 1972).]

We take a unit area and put Poisson distributed points at random everywhere, with a density ρ^2. Then we construct the *first generation* of Voronoi cells around these points as centres. There will be about ρ^2 cells, of average area $1/\rho^2$ and average perimeter $4/\rho$. Next (*second generation*) we randomly select a fraction f of the cells, that is altogether $f\rho^2$ cells, distribute in each of them ρ^2 Poisson random points with density $1/\rho^4$, and construct the corresponding Voronoi sub-cells. We assume that:

$$0 < f \leq 1 \qquad \rho^2 \gg 1 \qquad f\rho^2 \geq 1 \tag{2.4.1.12}$$

so that at least one cell of the first generation will be further divided. There will be about $f\rho^4$ cells of second generation, of average perimeter $\propto 1/\rho^2$. Continuing this procedure, the number of *n-th generation sub-cells* (within certain $[n-1]$-*th generation cells*) will be $f^{(n-1)}\rho^{2n}$, their average perimeter being $\propto 1/\rho^n$.

To prove that the construction is fractal, and to determine its fractal dimension we shall measure the total length $l(\varepsilon)$ of the whole network with a resolution:

$$\varepsilon = \frac{1}{\rho^n} \tag{2.4.1.13}$$

and study the behaviour of $l(\varepsilon)$ as $\varepsilon \to 0$. Obviously, using this resolution, only the cells and sub-cells belonging to at most the n-th generation will contribute to the total length:

total length for 1st generation cells is $\quad l_1 \propto \dfrac{\rho^2}{\rho} = \rho$

total length for 2nd generation cells is $\quad l_2 \propto \dfrac{f\rho^4}{\rho^2} = f\rho^2$

$\vdots$

total length for n-th generation cells is $\quad l_n \propto f^{(n-1)}\rho^n$

that is, the total length measured with resolution ε is:

$$l(\varepsilon) = l_1 + l_2 + \cdots + l_n \approx (f\rho)^n = \frac{f^n}{\varepsilon} = \varepsilon^{m-1} \tag{2.4.1.14}$$

where:

$$m = -\frac{\log f}{\log \rho} \tag{2.4.1.15}$$

[Observe that when calculating the total length, most of the edges have been counted twice, but this does not affect the order-of-magnitude estimation (2.4.1.14).]

Recalling that for a fractal curve of dimension d the measured length diverges as $l(\varepsilon) \propto \varepsilon^{1-d}$ [see equation (1.2.11)] we find that the fractal dimension of the network is:

$$d = 2 + \left(\frac{\log f}{\log \rho}\right) \tag{2.4.1.16}$$

The fractal dimension d is between 1 and 2 because of the inequalities (2.4.1.12). We shall get $d = 2$ in the case when $f = 1$, that is all Voronoi polygons of any given generation are further divided. (Fractals whose dimension equals the Euclidean dimension of the embedding space are called *fat fractals*.)

The above-described tessellation can be frequently observed on Moorish, Persian, Indian or Chinese ornaments. For Chinese works Jones (1867) describes the construction as follows:

"The position of the larger flowers was first fixed in the position most suited to develope the peculiar form of the vase, and the whole surface was set out by these flowers into symmetrical proportional areas; here law and order were abandoned, and the instinct and caprice of the artist came into play, in uniting all these fixed centres by a flowing line. This flowing line then dividing the different triangular spaces irregularly, masses of intermediate size, either as flowers or large leaves, were put in, springing from the continuous line; these secondary masses also balance triangularly, but in a less rigid manner than with the larger flowers: the process is continued by the introduction within the intermediate spaces of still smaller forms, buds, or stalks, till the whole is filled up, and repose is obtained by evenness of tint" (Figure 2.44).

Fig. 2.44. Ornament from a painted china bottle. (From Owen Jones, 1867, plate LXXXVII.)

Another remarkable property of cellular structures is the correlation between the neighbouring cell types. It had already been noted (in botany) by Lewis in the early 1930's that "there is a marked tendency for few-sided cells to be in contact with many-sided cells and vice-versa", but the quantitative formulation of this law is due to Aboav (1970) who examined a sample of 3,000 cells produced by sections through polycrystalline magnesium oxide and found that *the average number of sides* $\langle m(n)\rangle$ *of the cells adjoining another cells of a given number of sides n is*:

$$\langle m(n)\rangle = 5 + \frac{8}{n} \tag{2.4.1.17}$$

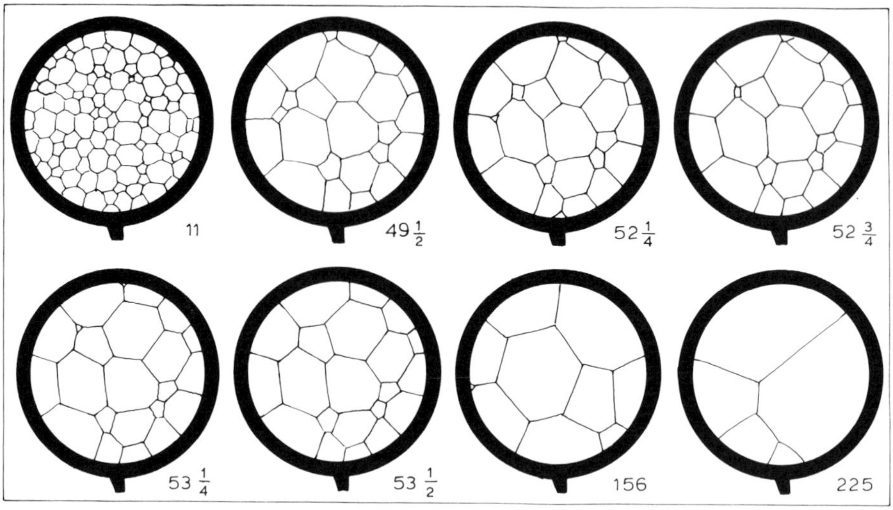

Fig. 2.45. Growth and disappearance of soap bubbles in a flat cell. Numbers indicate times in minutes after cessation of agitation. (From Smith, 1952.)

Aboav's rule has been later improved by Weaire (see Weaire, 1983, Weaire and Rivier, 1984) for structures with $\langle n \rangle = 6$ as:

$$\langle m(n) \rangle = 5 + \frac{(6 + \mu_2)}{n} \tag{2.4.1.18}$$

where:

$$\mu_2 = \sum (n - 6)^2 f(n) \tag{2.4.1.19}$$

is the second moment of the side-distribution $f(n)$ around its mean.

Aboav's law seems to be rather general, though it has never been rigorously established. It was originally found for ceramics (Aboav, 1970), but has also been verified for soap-film networks (Weaire, 1983; Weaire and Rivier, 1984) as well as in the quite different context of locational geography (for the artificial network of civil parishes in the English counties Devon, Somerset and Wiltshire: Boots, 1980).

The rule, however, is not generally valid, a notable exception being the Voronoi model (Weaire and Riviers, 1984).

When C.S.Smith (1952) presented his famous series of photographs (Figure 2.45) on the evolution of a two-dimensional soap bubble-froth in a flat cell (a readily available reference is Smith, 1954), Von Neumann (1952; a modern treatment is Bolton and Weaire, 1991) pointed out in an elegant written Discussion that the *rate of growth of the area of any soap cell is proportional to its number of sides minus six*. Thus, many-sided cells grow, few-sided cells decrease and finally disappear and six-sided cells retain their constant area. As time proceeds, the soap froth becomes

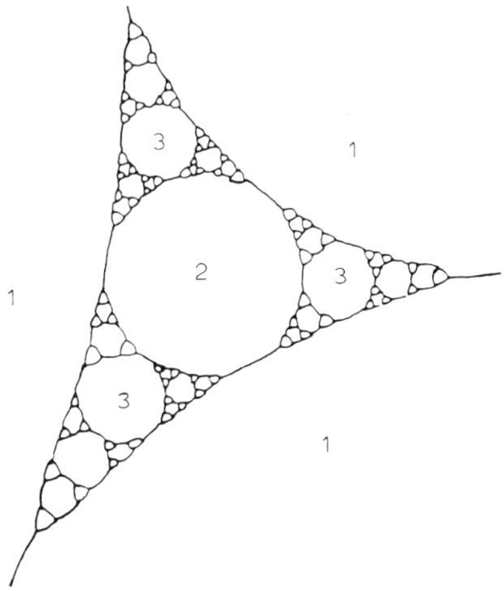

Fig. 2.46. Hypothetical fractal limiting structure for the soap cell system (from Weaire and Rivier, 1984).

progressively more disordered, the second moment of the side distribution, μ_2, increases and eventually might become infinite as $t \to \infty$.

Weaire and Kermode (1983) have conjectured that the soap froth at large times might tend to a fractal arrangement (Figure 2.46), where the more-sided a cell the larger its area (Von Neumann's growth law and Lewis' law), few-sided cells are surrounded by many-sided ones or by circles [by Aboav's law (2.4.1.18) if we let $\mu_2 = \infty$], and the side distribution falls off as a power-law: $p(n) \propto n^{-\alpha}$ with $\alpha \leq 3$.

Weaire and Rivier stated in their review (1984) that "such a structure would be... well worth pursuing experimentally, even though this would not be easy". The hypothetical pattern in Figure 2.46 is similar to the famous fractal Apollonian gasket (Figure 2.47a); an interesting realisation of this structure is the *fragmentation of sea ice* (Figure 2.47b, c) analysed by Rothrock and Thorndike (1984) and Matsushita (1985), even though it may seem a far cry from drift ice to soap froths.

If we consider larger areas of the ice (Figure 2.47d) at low resolution, we find that it could also have resulted from fragmentation, in a way akin to the previously described sequential cell divisions (compare the scaling properties of Figures 2.47d and 2.44!). Indeed, Matsushita (1985) has derived the measured *size distribution* of drift sea ice by assuming a fractal fragmentation process (see Section 3.2).

As Rothrock and Thorndike (1984) pointed out, while the floes are broken by waves, currents, abrasion by adjacent floes, thermal cracking, etc., in the freezing season some of the adjacent ones are welded together again and it is not known which process is more important.

The previous analysis suggests that the observed patterns (like that in Figure 2.47d) should be considered as the result of simultaneously or alternately acting fragmentation *and* (Von Neumann type?) welding/melting growth processes, and

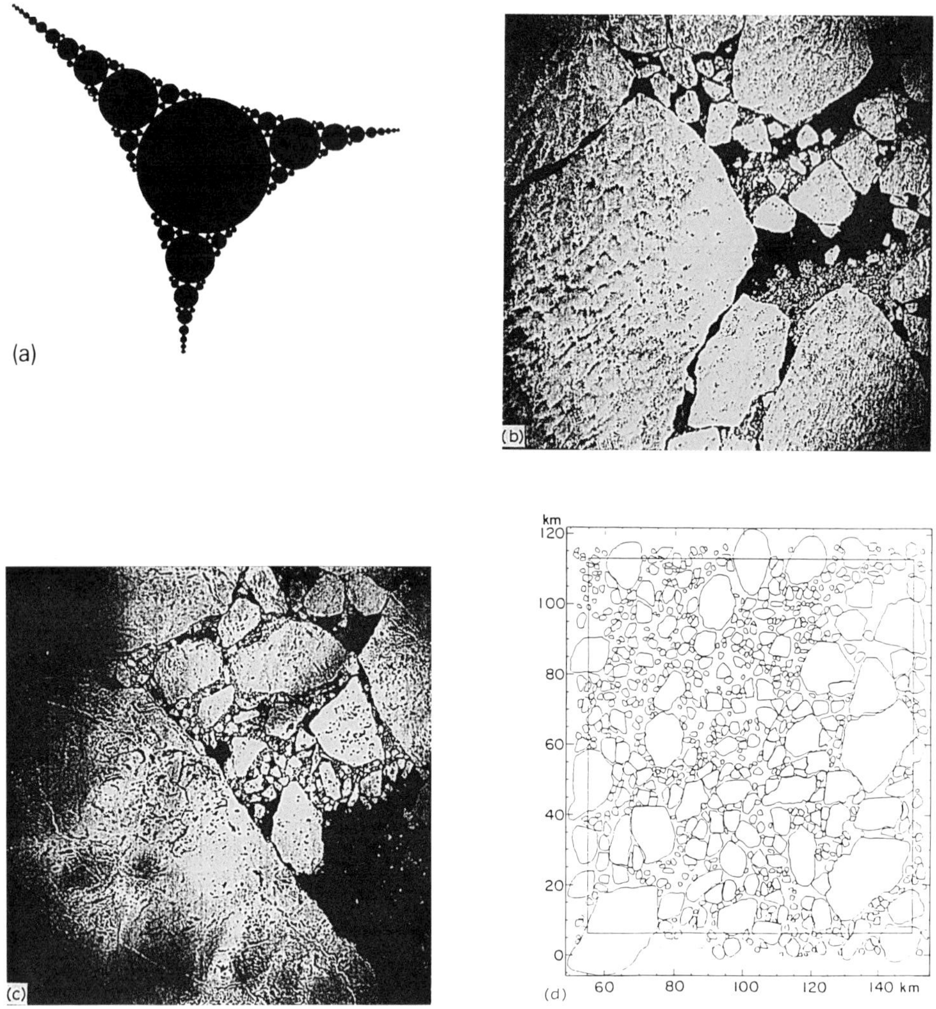

Fig. 2.47. a. Apollonian gasket. The disc sizes are distributed as power law; its construction is described in Mandelbrot (1982). b, c. Views of broken sea ice on two different scales. The frame of (b) is about 28 km wide, that of (c) 2.4 km wide. d. Digitised flow boundaries from aerial photographs of summer pack ice (from Rothrock and Thorndike, 1984).

that the geometrical (size, area, perimeter, etc.) *and* topological (number of sides, type of the adjacent cells, etc.) properties should be jointly analysed.

2.4.2 *Fracture networks, faults and earthquakes*

In a historical review of the first thirty years of Information Theory, Tribus (1979) found it worth mentioning that in the late 1960's an American geologist, J.P. Dowds, had not only used concepts of Markov chains and entropy, but — based upon such an analysis — he actually *purchased* the rights to an abandoned oil well in Oklahoma, which he christened "Rock Entropy #1". (John Dowds had been given a hard time by the oil community, as judged from his bitter remark in his important "Oil Rocks: Information Theory: Markov Chains: Entropy", 1969: "... in the report, information theory was mentioned, but I decided it best to avoid unfamiliar words such as entropy, ergodic, Markovian, etc., as these words can cause antagonisms".)

In the history of *acceptance* of fractals in Earth Sciences, it had been a similar milestone when, around 1984, U.S. Geological Survey scientist Dr. Christopher Barton could raise the money to take out a dozen fire engines to a Nevada desert to hose clean a square mile of an outcrop in order to map, and to prove the fractal nature of, the fracture network in rocks considered suitable for nuclear waste disposal (Barton and Larsen, 1985; Barton et al., 1986; for the story see Mary Fritz's "Fractals find niche in oil exploration" in the July, 1988 issue of the AAPG Explorer).

Barton first prepared a map of a few subhorizontal pavements of 200–2000 m^2 areal extent of the Miocene ash-flow tuffs at a scale of 1:50, including all fractures with trace length greater than 20 cm. After consultations with the topmost authority, Professor Mandelbrot himself, Barton used a version of the *box-counting* technique and found that the fracture network was indeed fractal and the fractal dimensions ranged from 1.49 to 1.91, the mean being 1.744. Next, he analysed the fault traces in the same tuffs at Yucca Mountain mapped at 1:12,000 scale, and found again a fractal dimension 1.773. The obvious practical significance of this result has been that in areas where the fracture and fault patterns have the same fractal dimension over a wide range of scales, the density and other spatial characteristics at the scale of interest can be predicted from the patterns mapped at another scale. [In an earlier paper, Barton and Larsen (1985) reported a fractal dimension 1.12–1.16 for fracture traces of length between 0.2 and 25 m. This much lower fractal dimension corresponds to the irregular *shape* of the individual fractures, while the higher dimension $\approx$1.7 is due to the self-similarity in their *mutual arrangement*. This *bifractality* of fracture networks has been recognised by most investigators and will be discussed later.] It is easy to find examples for self-similar networks of fractures (Figure 2.48) and lineaments (Figure 2.49).

The most popular way of estimating the fractal dimension of fracture networks is *box-counting*. In this technique the study domain is subdivided to small squares of size r ($r \to 0$) and the number of squares $N(r)$ intersected by at least one fracture is counted. In case of self-similarity $N(r)$ scales as $N(r) \propto r^{-D}$, where D is interpreted as the fractal dimension. The method was used by Barton (Barton and Larsen, 1985; Barton et al., 1986) for the fracture network of the Yucca Mountain, Nevada, volcanic tuffs (D = 1.49–1.91); by Chilès (1988) to study the fractures on drifts of a uranium mine in a granite massif (D = 1.05–2) and by Hirata (1989) for fault systems in Japan (D = 0.7–1.6), see Figure 2.50.

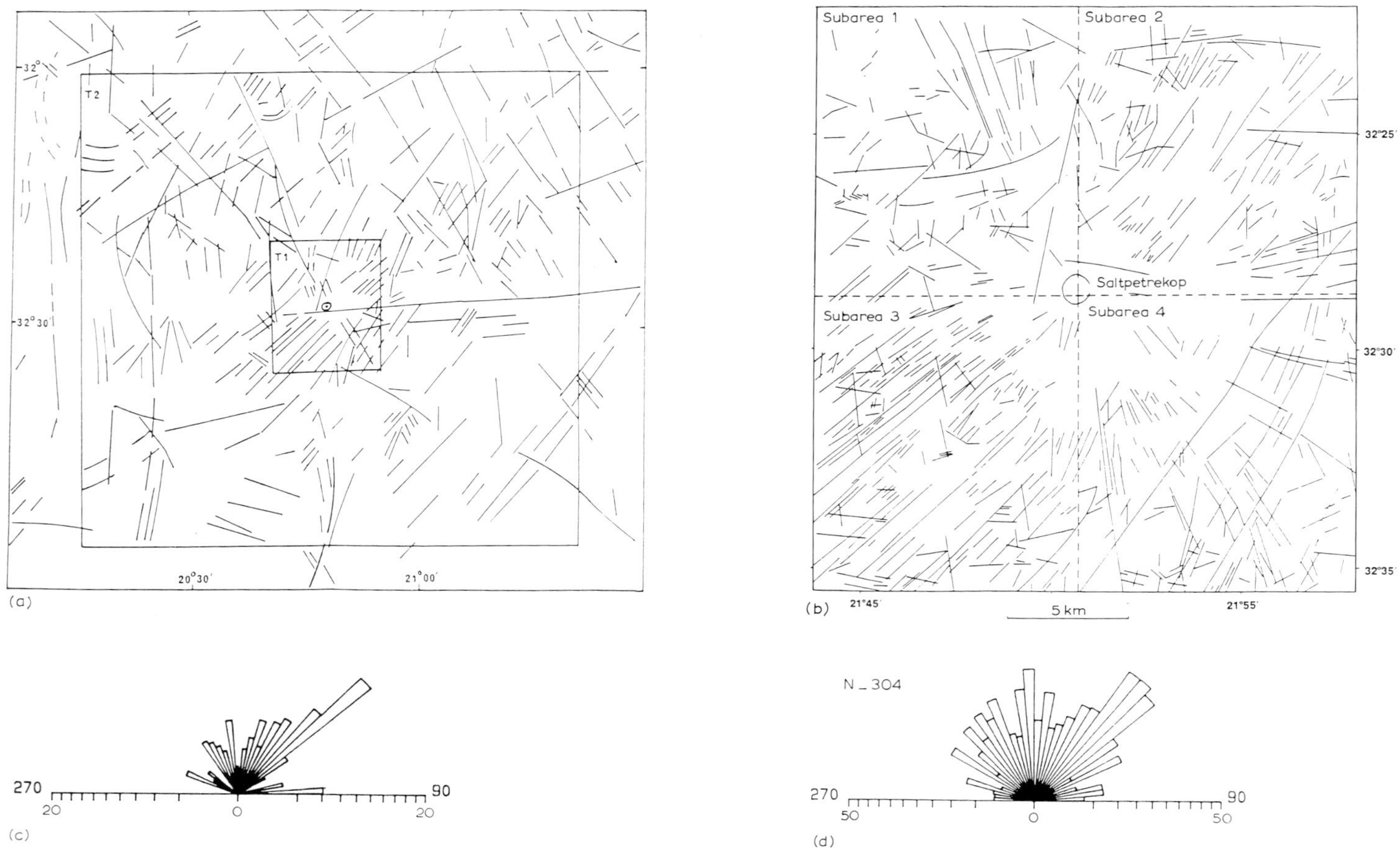

Fig. 2.48. a. Fractures traced from Landsat image around the Sutherland diatreme, Cape Province, South Africa; *T1*: air photo area. b. Fracture traces in area *T1* interpreted from air photos. c and d. Rosetta diagrams showing the similarity in the directional distribution of the fractures for the area *T1* (c) and *T2* (d). (From Newton, 1987.)

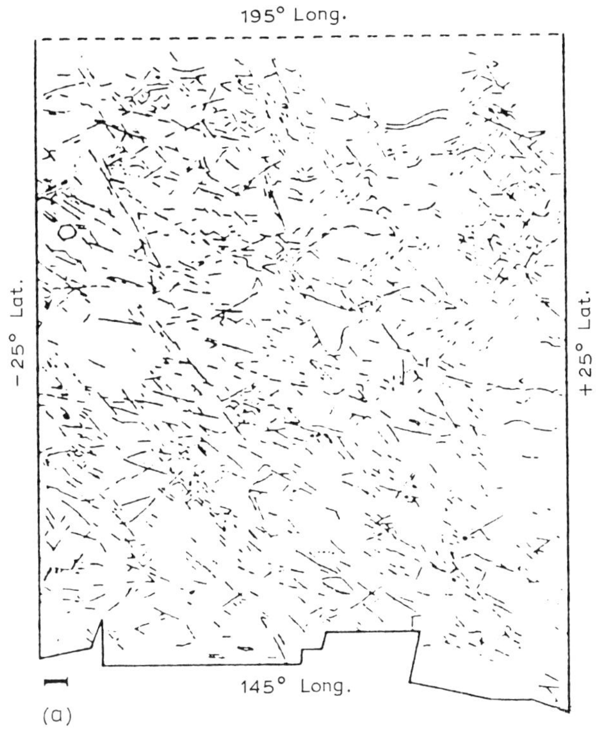

Fig. 2.49. a. Lineament map for the explored surface of the planet Mercury (area mapped is 2,000 km across). (From Thomas and Masson, 1983.)

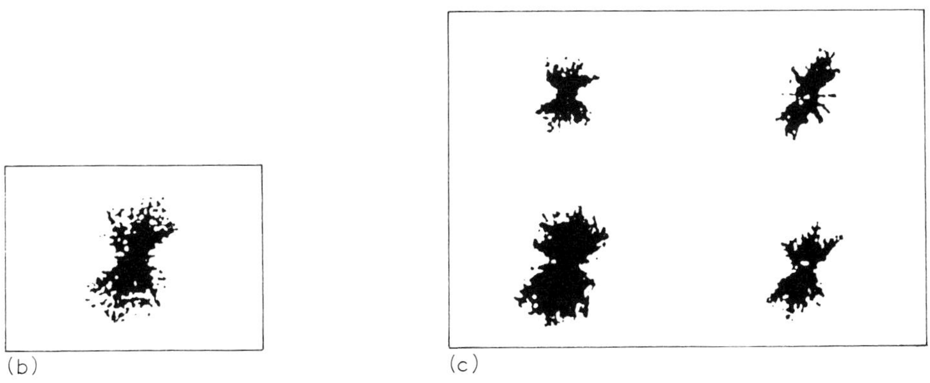

Fig. 2.49 (continued). b and c. Directional analysis of the lineament map by two-dimensional Fourier transform: Fourier transform of the entire map (b) and Fourier transform of each quarter of the map (c). (From Thomas and Masson, 1983.)

A different technique has been applied by Thomas (1987), who analysed fractures in limestones and granites in France, using a random generalisation of the Cantor construction. He observed that in the i-th generation of the Cantor set the number of those "cuts" (that is the removed white sections in Figure 2.5) which are greater than or equal to $u_i = 1/3^i$ is:

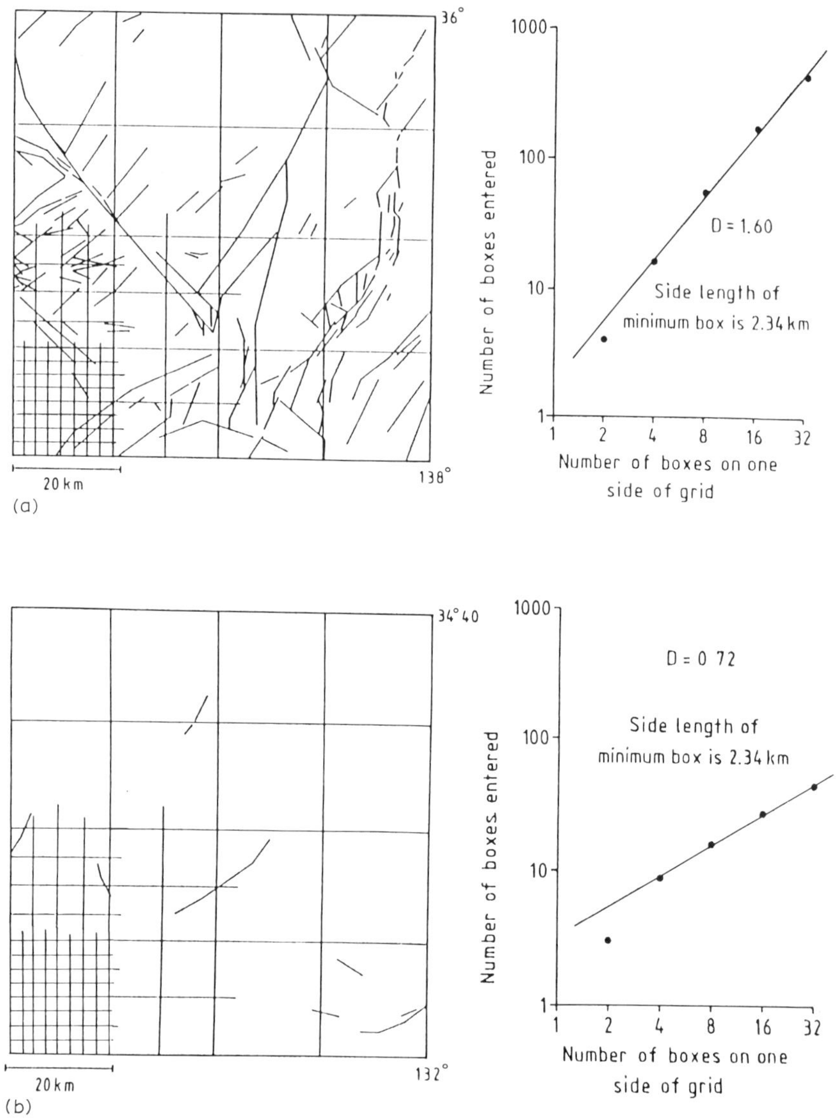

Fig. 2.50. Fractal analysis of the fault system in Japan, using box-counting. a. Fault system of the Iida region ($D = 1.6 \pm 0.10$); b. fault system of the Yamaguchi region, where the fractal dimension is the lowest in Japan ($D = 0.72 \pm 0.01$) (from Hirata, 1989).

$$N(\text{cuts} \geq u_i) \propto 2^i = u_i^{-D} \qquad (D = -\frac{\log 2}{\log 3}) \tag{2.4.2.1}$$

and, generalising this observation, he determined the fractal dimension from the length distribution of the intersections on random straight lines passing through the fracture field.

It has long been recognised (Sawada et al., 1982; Takayasu, 1985a, b, 1986) that the patterns formed by electric breakdown in an insulator (Figure 2.51) are similar to the fracture network in brittle materials. In fact, in Takayasu's computer experiments (1985b, 1986) the model of electric breakdown differs from that of brittle fracture in a single parameter (Figure 2.52).

Because of this similarity, the techniques developed by physicists for the determination of the fractal dimension of discharge patterns are directly applicable to fracture networks, as the already discussed *box-counting* and its different variants (Sawada et al., 1982; Niemeyer et al., 1984; etc.), *covering with circles* or the *perimeter–area analysis*.

Fujimori (1985) counted the minimum number $N(\varepsilon)$ of circles of radius ε which are necessary to cover an electric discharge pattern and defined the fractal dimension D as:

$$D = -\lim_{\varepsilon \to 0} \frac{\log N(\varepsilon)}{\log \varepsilon} \tag{2.4.2.2}$$

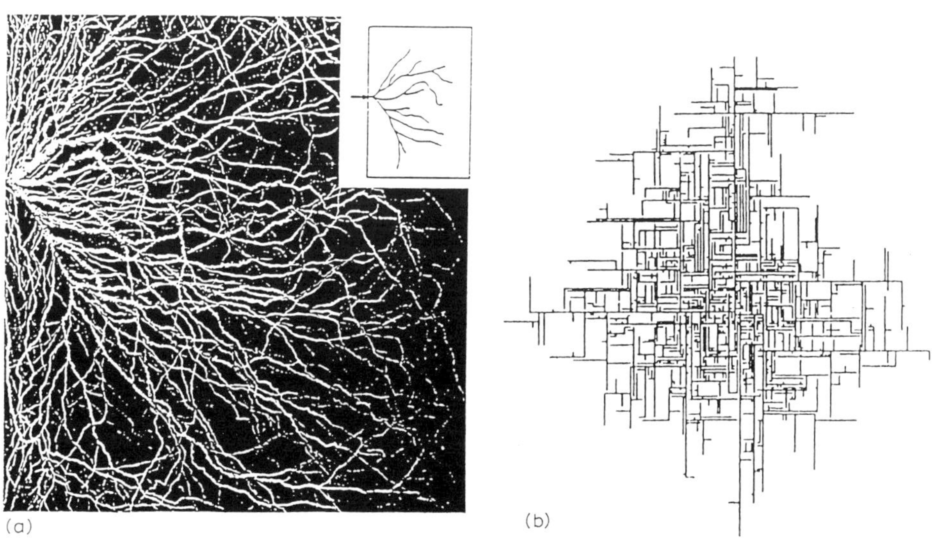

Fig. 2.51. Observed and simulated electric discharge patterns. a. Electric breakdown in an insulator. White lines are the pathways by which charges escaped to the ground. The inset shows the point where the charged insulator was initially grounded. b. Discharge pattern obtained by computer simulation. (From Sawada et al., 1982.)

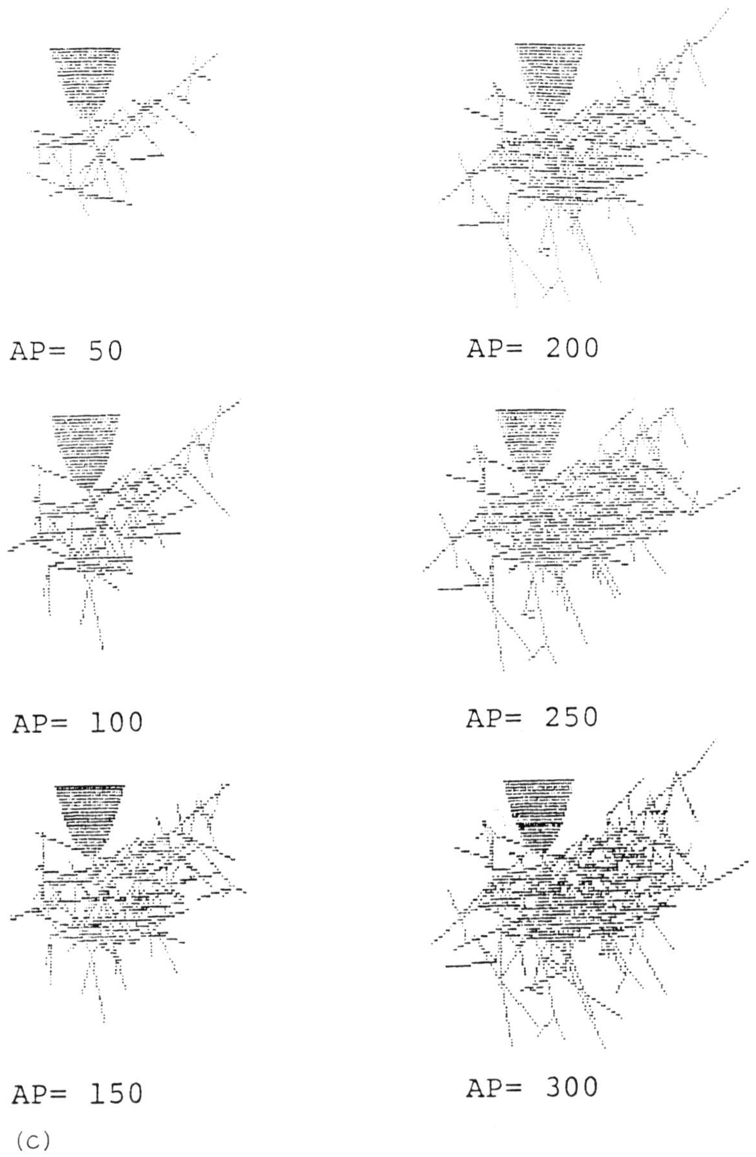

Fig. 2.51 (continued). c. Computer simulation of discharge figures developing in the gap between a positive needle electrode and a negative plane electrode. AP = number of streamers. The fractal dimension increases from $D = 1.192$ (for $AP = 50$) to $D = 1.443$ ($AP = 300$). (From Fujimori, 1985.)

A similar technique has been used by Okubo and Aki (1987) to study the fractal geometry of the San Andreas fault system (Figure 2.53).

Takayasu (1985b, 1986) used Mandelbrot's area–perimeter relation (see Sec-

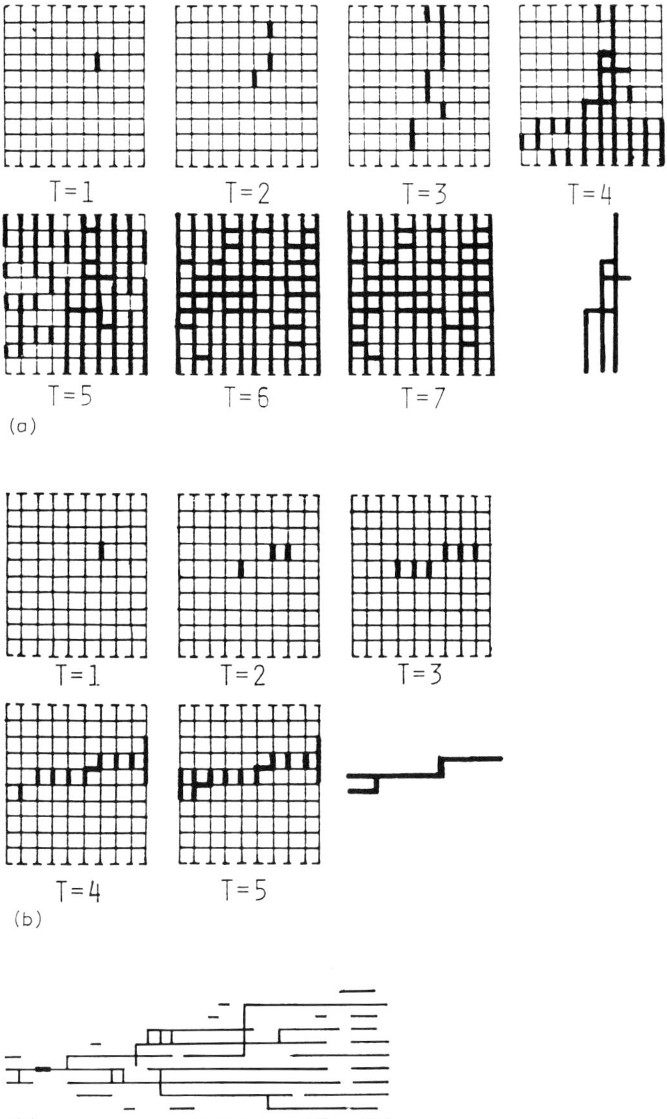

Fig. 2.52. a. Computer model of the evolution of an electric breakdown pattern on a 10 × 10 net. The last figure shows the percolation cluster at time $T = 4$. b. Computer model of the evolution of brittle fracture. The last figure shows the percolation crack. c. The final fracture pattern on a 32 × 32 net. Fractal dimension is $D = 1.65 \pm 0.05$. (a, b from Takayasu, 1986; c from Takayasu, 1985b.)

tion 1.5) in the form $(\text{area})^{1/2} \propto (\text{length})^{1/D}$ where the "length" was the longer dimension of the computer-generated elongated crack patterns (see Figure 2.52c). Korvin (1989) also applied the area–perimeter rule in a study of the fragmenta-

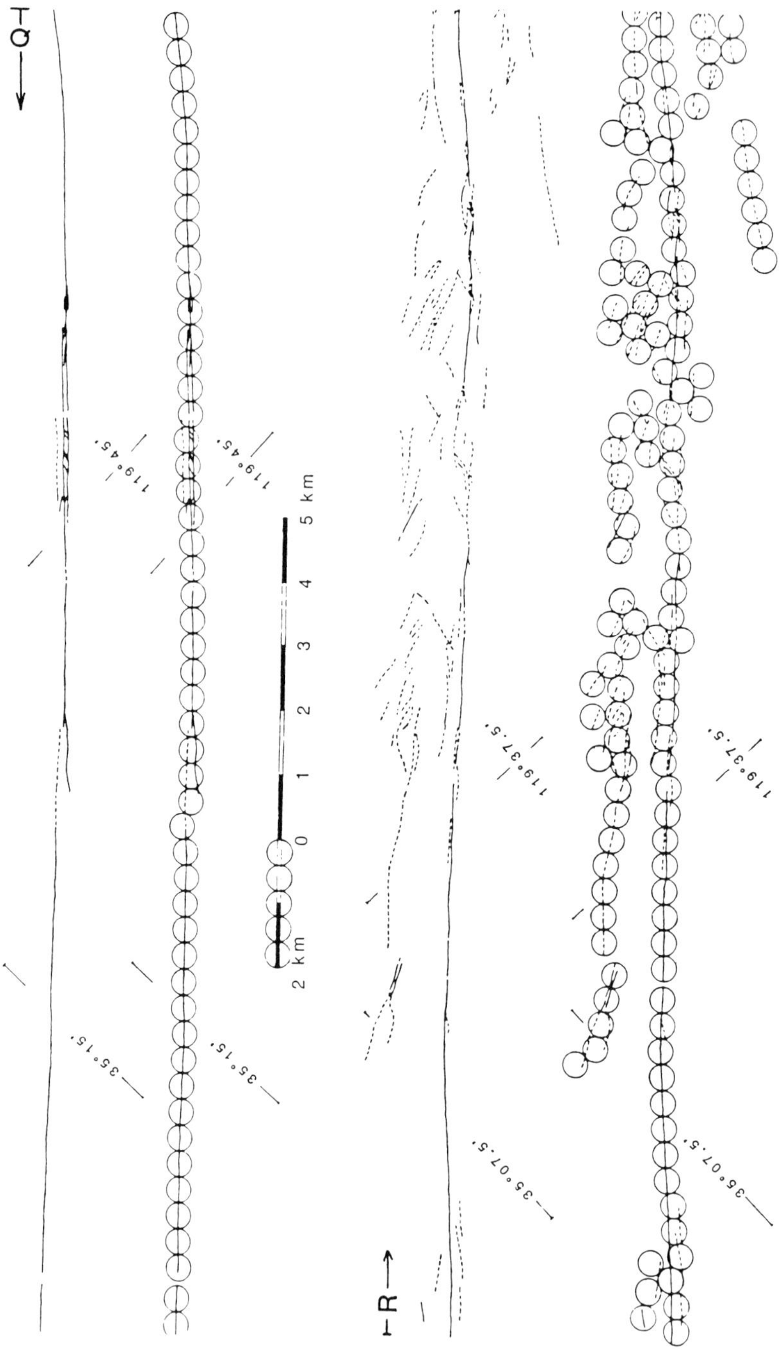

Fig. 2.53. A part of the San Andreas fault system mapped at 1 : 24,000 and its covering by circles of radius r = 200 m. The fractal dimension was determined from the slope of the log $N(r)$ vs. log r line, where $N(r)$ is the number of circles of radius r. The pattern is bifractal: D = 1.2 for radii less than 400 m and D = 1 for r > 400 m. (From Okubo and Aki, 1987.)

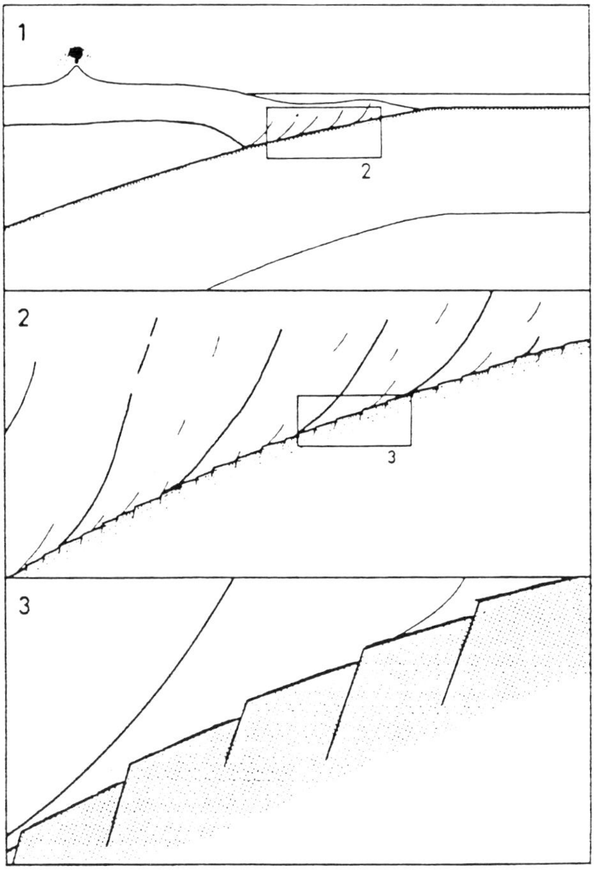

Fig. 2.54. The highest orders in the hierarchy of fault irregularity: 1. order of the width of the plate interface (≈100–200 km); 2. order of the depth of the plate interface (≈20–50 km); 3. order of the thickness of the oceanic crust (≈5–10 km) (from Fukao and Furumoto, 1985).

tion of the Gulf of Suez basement, by seeking a connection between the areas and perimeters of the fault blocks.

There have been many investigations concerning the fractal nature of *individual* faults, with a view to connecting the hierarchical features of faults (Figure 2.54) with the hierarchy of earthquakes. (The fractal models of fault propagation and rock failure will be discussed in Sections 5.2 and 3.2.1, respectively.)

The methods used have included the "traditional" ruler-and-divider technique *à la Richardson* [Aviles et al. (1987), for characteristic segments of the San Andreas fault system], covering the fault traces with circles of decreasing radii [Okubo and Aki (1987), for the San Andreas fault system] and the power spectrum analysis of the topography of faults and fractures [Brown and Scholz (1985), for a variety of natural rock surfaces such as a joint in the Palisades diabase, Palisades, N.Y. over the band 10^{-5} to 1 m; and Scholz and Aviles (1986), for a trace of the San Andreas fault ranging in wavelength from 10^3 to 10^5 m. Professor Scholz's early fascination with fractals is documented in James Gleick's bestselling "Chaos" (1987); for an illuminating discussion see Scholz (1989)].

The estimation of the fractal dimension D of the trace of a random function $x(t)$

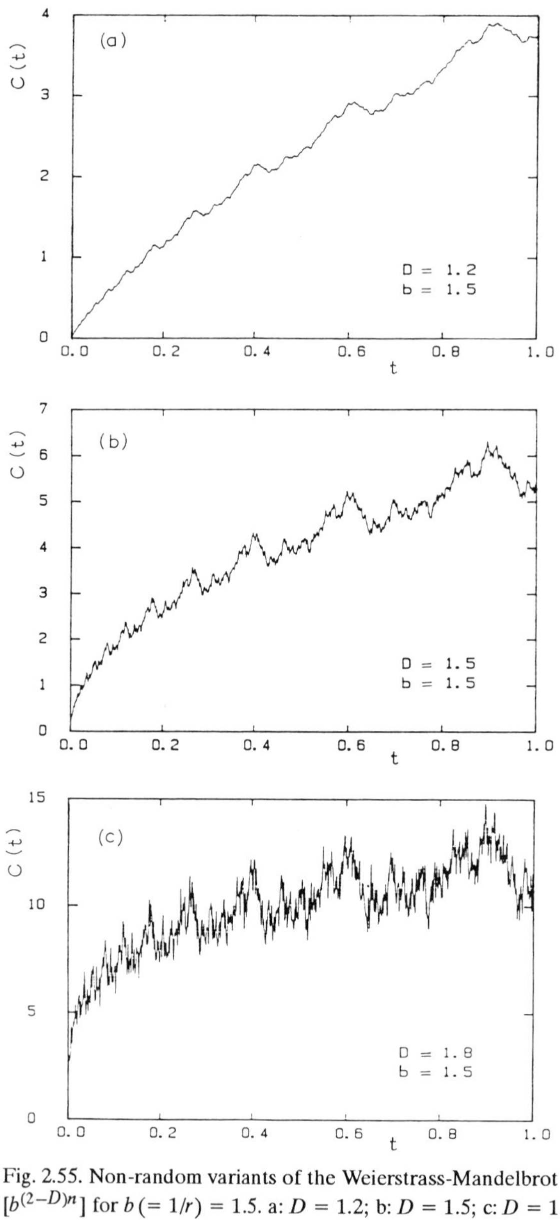

Fig. 2.55. Non-random variants of the Weierstrass-Mandelbrot fractal function: $C(t) = \sum[1 - \cos b^n t]/[b^{(2-D)n}]$ for $b\,(= 1/r) = 1.5$. a: $D = 1.2$; b: $D = 1.5$; c: $D = 1.8$. (From Feder, 1988.)

from the power spectrum $X(f)$ of $x(t)$ is based on the observation that for *self-similar* (or more exactly, *self-affine*, see Appendix A2) *random functions satisfying the scaling relation*:

$$x(rt) = r^H x(t) \qquad (r > 0) \tag{2.4.2.3}$$

where H (0 < H < 1) is a constant, the graph of the function x(t) is fractal with "box"-dimension D, and the power spectrum X(f) falls off with increasing frequencies as:

$$X(f) \propto f^{-\beta} \tag{2.4.2.4}$$

where H, D, and β are related by:

$$D = 2 - H = \frac{(5-\beta)}{2} \tag{2.4.2.5}$$

[Voss (1985a, b, 1988). The equality (2.4.2.3) must only be valid in a statistical sense, that is the functions $x(rt)$ and $r^H x(t)$ should be realizations of the same stochastic process.]

Well-known examples for random functions satisfying the above conditions are the *Weierstrass-Mandelbrot function* (Mandelbrot, 1982; Berry and Lewis, 1980; Voss, 1985b, 1988; Feder, 1988) and the *fractional Brownian motion* (Mandelbrot and Van Ness, 1968; Mandelbrot, 1982; Voss, 1985a, b, 1988; Feder, 1988; see also Section 2.1.4).

The Weierstrass-Mandelbrot fractal function is defined as:

$$x_{\mathrm{WM}}(t) = \sum_{n=-\infty}^{\infty} A_n r^{nH} \sin(2\pi r^{-n} t + \varphi_n) \tag{2.4.2.6}$$

where r and H are constants ($0 < r, H < 1$); A_n ($\langle A_n \rangle = 0$; $\langle A_n^2 \rangle = A^2$) are independent Gaussian variables with the same mean and variance; φ_n is a random phase uniformly distributed on $(0, 2\pi)$. The Weierstrass-Mandelbrot function is continuous but nowhere differentiable, it has been widely used to describe irregular surfaces (fracture surfaces: Mandelbrot et al., 1984; seafloor topography: Fox and Hayes, 1985). A few non-random variants of the Weierstrass-Mandelbrot function are shown in Figure 2.55 for different fractal dimensions.

The statistical fullfilment of the scaling relation (2.4.2.3) can be checked by direct substitution, the fractal dimension of the graph of $x_{\mathrm{WM}}(t)$ is discussed in Mauldin (1986), while the spectral density $X_{\mathrm{WM}}(f)$ can be immediately found from equation (2.4.2.6). Indeed, $x_{\mathrm{WM}}(t)$ contains only discrete frequencies $f_n = 1/r^n$; the mean square amplitude of the n-th component is $A^2 r^{2Hn} \propto 1/f_n^{2H}$ over a bandwidth $\Delta f_n = f_{n+1} - f_n \approx f_n$, that is:

$$X_{\mathrm{WM}}(f) \propto \frac{\text{mean square amplitude}}{\Delta f} \propto \frac{1}{f^{2H+1}} \tag{2.4.2.7}$$

in agreement with equations (2.4.2.4) and (2.4.2.5).

The fractional Brownian motion $x_H(t)$ has been introduced in Section 2.1.4 (equation 2.1.4.5); it is a random function for which:

$$\langle s_H(t)\rangle = 0 \qquad \langle [s_H(t') - s_H(t'')]^2\rangle \propto (t' - t'')^{2H} \tag{2.4.2.8}$$

where H $(0 \le H \le 1)$ is called the *Hurst exponent*. Because of equation (2.4.2.8), for any given positive number $r > 0$ the random functions:

$$s_H(t) \qquad \text{and} \qquad \frac{s_H(r\,t)}{r^H}$$

are statistically equivalent (compare with equation 2.4.2.3); that is they share the same power spectrum. Because of the time-scaling property of the Fourier transform (see e.g. Papoulis, 1962) according to which *if $X(f)$ is the Fourier transform of $x(t)$ then $X(f/r)/r$ $(r > 0)$ is the transform of $x(rt)$*, it is easy to prove that *for fractional Brownian functions with Hurst exponent H the power spectrum behaves as*:

$$S_H(f) \propto \frac{1}{f^\beta} \qquad \textit{with} \quad \beta = 2H + 1 \tag{2.4.2.9}$$

It can also be shown by box-counting arguments that the fractal dimension of the trace of $s_H(t)$ is $D = 2 - H$ (for a rigorous treatment see Orey, 1970), that is β, H and D are related again as:

$$D = 2 - H = \frac{(5 - \beta)}{2}$$

in conformity with equation (2.4.2.5) (Voss, 1985a, b, 1988).

According to equation (2.4.2.5) the fractal dimension D of a self-similar function $s(x)$ of the coordinate x can be determined from the slope of the log power spectral density vs. log spatial frequency plot (Figure 2.56).

Scholz and co-workers (Brown and Scholz, 1985; Scholz and Aviles, 1986) applied the power spectrum technique, and concluded that the topography of natural faults and fractures is fractal over a very wide band 10^{-5} to 10^5 m (Figure 2.57), though the fractal dimension D has been found to change with frequency:

Microscale (joint, Palisades diabase, N.Y.)	
Single crystal scale (wavelength <0.5 mm)	$D = 1$
Hand specimen scale (wavelength 0.5 mm–1 cm)	$D = 1.15 \pm 0.03$
Outcrop scale (wavelength 1–20 cm)	$D = 1.26 \pm 0.04$

Megascale (traces of the San Andreas fault)	
1:24,000 map scale (wavelength 10^3–10^5 m)	$D =$ slightly greater than 1

(The Richardson-Mandelbrot technique applied to characteristic segments of the San Andreas fault system in the 0.5–1,000 km band has also resulted in very low fractal dimensions $1 < D < 1.01$, see Aviles et al., 1987)

A discussion on the significance of the self-similarity of faulting at all scales

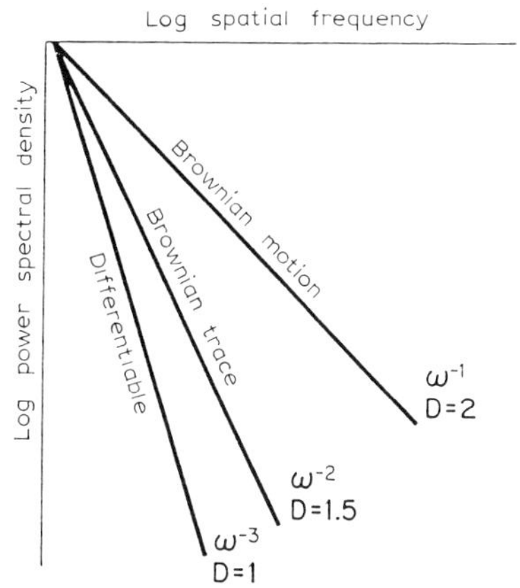

Fig. 2.56. Schematic power spectra of fractal functions of different dimensionalities (from Scholz and Aviles, 1986).

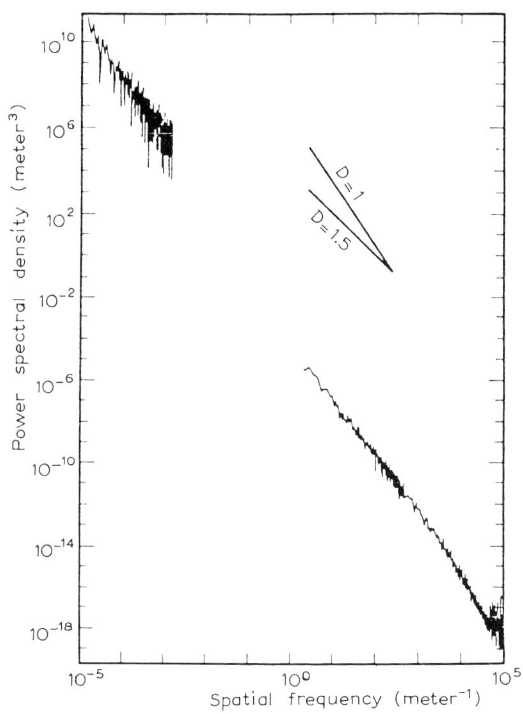

Fig. 2.57. The power spectrum of the San Andreas fault from Pt. Arena to Parkfield and of a joint in the Palisades diabase, N.Y., plotted on the same scale (from Scholz and Aviles, 1986).

(down to the grain-size distribution of fault gouges: Sammis et al., 1986; Sammis, 1988) and its connection with the hierarchy of earthquakes can be found in Scholz and Aviles (1986).

There is an apparent contradiction between the low (generally less than 1.2) fractal dimensions of the *single faults* obtained from their Richardson-Mandelbrot plots or power spectra (Aviles et al., 1987; Scholz and Aviles, 1986) and the much higher dimensions (1.3–1.7) of *fault networks* found by circle-coverings or box-counting (Okubo and Aki, 1987; Barton and Cameron, 1986; Chilès, 1988; Hirata, 1989).

In the latter techniques fault irregularity and the density of branching are *both* measured: en echelon faults are covered by the same circle if the radius is sufficiently large, as in Figure 2.53, or different branches belong to the same small square (Figure 2.50b).

Fault networks usually turn out to be *bifractal* if analysed over a wide range of lengths (Aviles et al., 1987; Okubo and Aki, 1987; Hirata, 1989; for simulated electric discharge patterns: Sawada et al., 1982) or their fractal dimension is an even more complicated function of the length scale (Scholz and Aviles, 1986; Chilès, 1988).

In a recent analysis of the active faults in Japan (that is faults that have moved during the Quaternary), Hirata (1989) has distinguished two types of fault maps: in type I fault maps there are large main faults (Figure 2.50a) percolating from side to side as cracks in the mirror of the Lady of Shallott; the type II fault maps (Figure 2.50b) contain no main faults. Hirata found that for type I faults the fractal dimension assumes an upper limit of about $D = 1.6$ at the central part of the Japan Arc (Figure 2.58) and decreases with the distance from the centre. This is due to

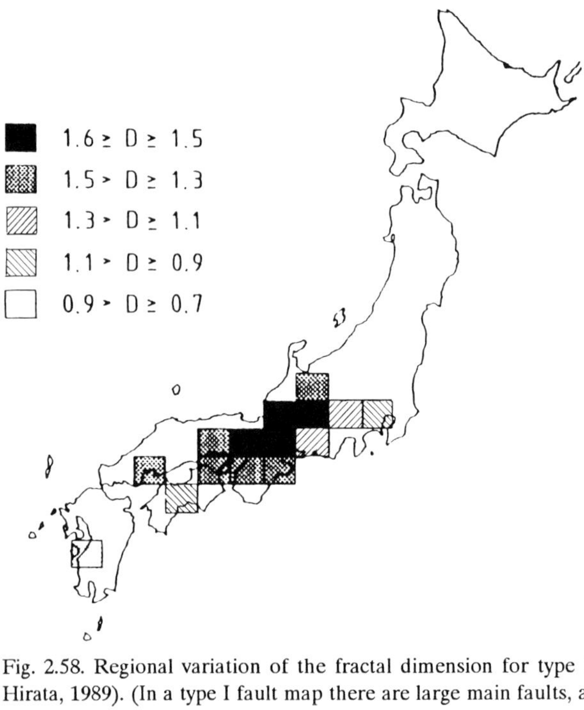

Fig. 2.58. Regional variation of the fractal dimension for type I fault maps in the Japan Arc (from Hirata, 1989). (In a type I fault map there are large main faults, as in Fig. 2.50a.)

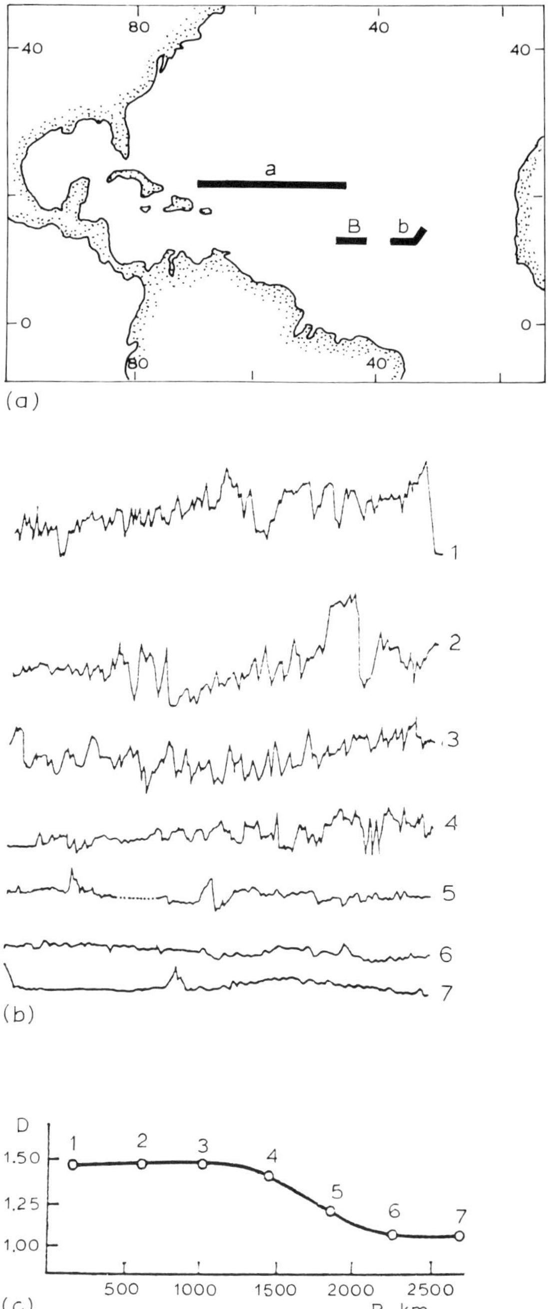

Fig. 2.59. Fractal analysis of the relief of the ocean floor around the Mid-Atlantic Oceanic ridge. a. Location of the profiles analysed; b. characteristic sections of the ocean-floor relief along profile *a*; c. their fractal dimension determined from length–resolution analysis, as a function of the distance from the axis of the ridge. (From Barenblatt et al., 1984.)

the fact that the outer regions have few branch faults, and the geometry of the fault system approaches a main fault line.

The gradual decrease of the fractal dimension when moving away from the place of the strongest tectonic activity has also been observed in another study, concerning the fractal dimension of the ocean-floor relief at various distances from the Mid-Atlantic Oceanic ridge (Barenblatt et al., 1984; Figure 2.59). Though the phenomenon seems to be of general validity, apparently it has never been studied. (In Mandelbrot's, 1975, important early paper on the Earth's relief, coastlines, and the number–area rule for islands, there is a passing reference to a referee according to whom there had been "a narrow focus on purely tectonic processes". Very likely, the

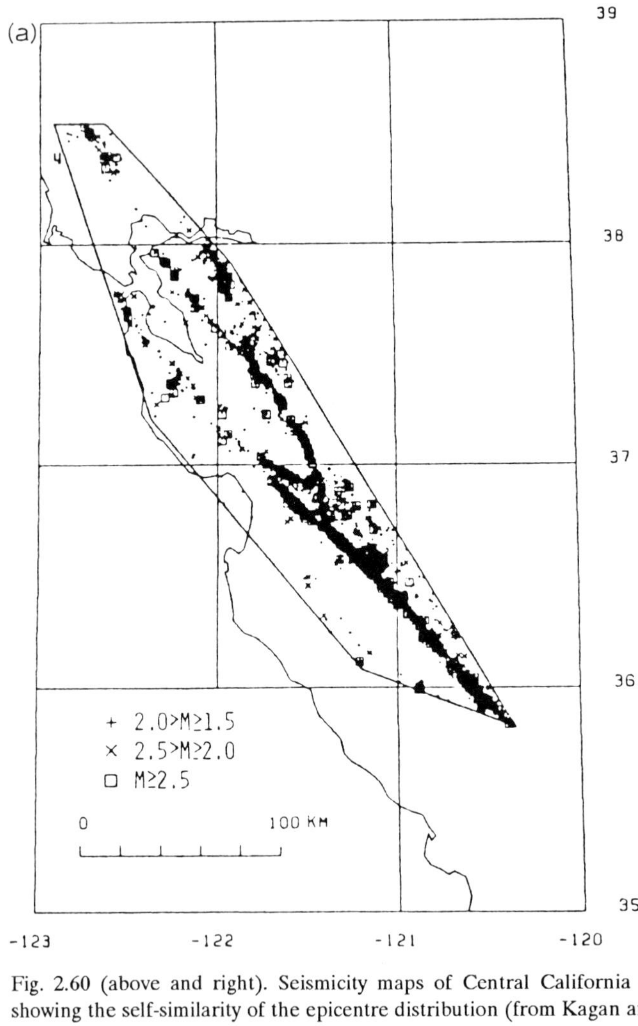

Fig. 2.60 (above and right). Seismicity maps of Central California (1971–1975) at different scales, showing the self-similarity of the epicentre distribution (from Kagan and Knopoff, 1980).

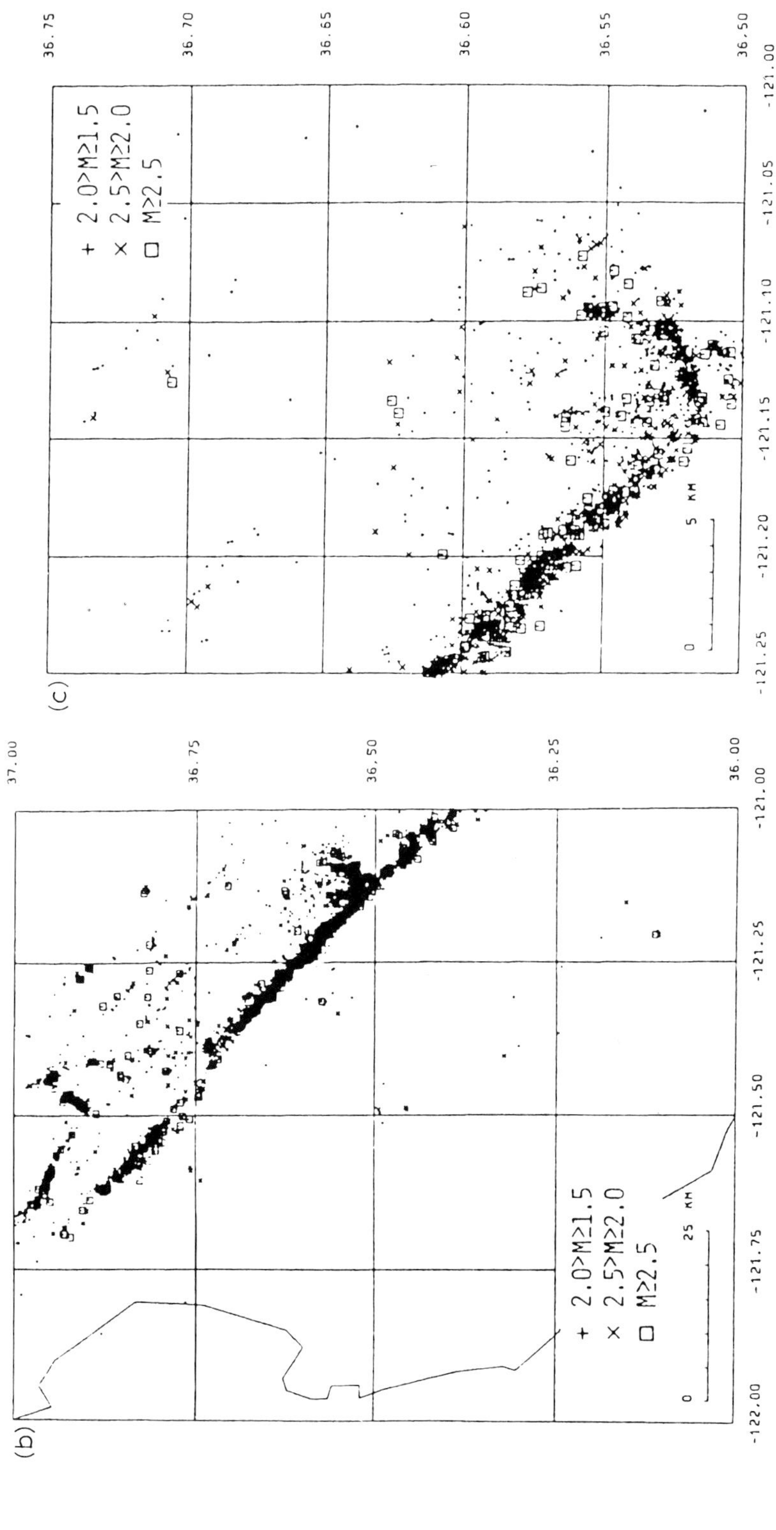

Fig. 2.60 (continued).

systematic spatial decay of the fractal dimension could only be explained in terms of processes which the anonymous referee obviously had had in mind: isostasy, erosion, sedimentation, elastic/plastic behaviour, etc.)

The upper limit of $D = 1.6$ found by Hirata for type I fault maps at the central part of the Japan Arc possibly corresponds to Takayasu's (1985b; 1986) theoretical value of $D = 1.65$, derived for the percolating crack network.

In most cases it is difficult to explain the observed fractal dimensions. Aviles et al. (1987) comment on the general lack of overall correlation between the seismic behaviour of the fault segments and their fractal dimension. For type I fault maps Hirata (1989) found that the fractal dimension roughly correlates with the density of faulting. As a similar trend can be observed on Fujimori's (1985) simulated discharge patterns (Figure 2.51) where the fractal dimension increases with the density of streamers, the possibility of sampling errors cannot be ruled out.

For type II maps, especially for those with $D < 1$ as in Figure 2.50b, the fractal dimension is controlled by the *spatial distribution of faults* and, according to Hirata (1989), it probably coincides with the fractal dimension of the clustering of the earthquake epicentres.

Earthquakes represent a five-dimensional set of events: the time of their occurrence, three spatial dimensions of the hypocentre, and the magnitude. Self-similarity can be studied in any subsets of these dimensions (for a recent review, see Smalley et al., 1987). There have been a large number of recent studies establishing the self-similarity in the spatial clustering of earthquake hypocentres and epicentres (Kagan and Knopoff, 1980, see Figure 2.60; Sadovskiy et al., 1984; Ouchi and Uekawa, 1986) and the fractality of their temporal occurrence (Smalley et al., 1987). The self-similarity in *magnitude distribution* is expressed by the celebrated Gutenberg-Richter relation:

$$\log N_T = a - bm \tag{2.4.2.10}$$

where N_T is the total number of shocks in a given region having magnitude m or greater. In 1983 Geoffrey King of the Bullard Laboratories, Cambridge, derived equation (2.4.2.10) from a few simple assumptions concerning the self-similarity of fault geometry, and has shown that the "b-value" in the Gutenberg-Richter law can be directly related to the fractal dimension of the active fault system involved in the seismic process.

The self-similarity found in the spatial, temporal and magnitude distribution of earthquakes has also been verified in microfracturing experiments on rock samples (Scholz, 1968; Nishizawa et al., 1984/85; Hirata et al., 1987; for a general discussion see Allègre et al., 1982; Scholz, 1989). In these studies cylindrical rock samples are kept under constant pressure [$\approx$200 MPa in case of Nishizawa and group's (1984/85) experiment with Yugawara andesite; 40 MPa in the study of Hirata et al.

(1987), using Oshima granite]. The rock subjected to pressure shows a characteristic time-dependent strain, known as *creep*.

Creep can usually be divided into three phases: (a) transient creep, when the strain is proportional to the logarithm of time; (b) steady creep, which is a deformation at a constant strain rate; and (c) acceleration creep when the strain change accelerates until the ultimate fracture (Nishizawa et al., 1984/85).

The *acoustic emissions* (AE) generated in the rock during creep are recorded by piezoelectric transducers, and their temporal behaviour and the spatial distribution of the AE hypocentres analysed.

Figure 2.61a shows the changes of stress and volumetric strain and the AE rate during creep as functions of time for Nishizawa and co-workers' experiment; the temporal clustering of the acoustic emissions is similar to that observed for earthquakes (Figure 2.62). Figure 2.61b shows the variation of the spatial distribution of AE hypocentres during the experiment.

The fractal nature of the temporal distribution of seismicity has been first demonstrated by Kagan and Knopoff (1978, 1981). Smalley et al. (1987) analysed the temporal clustering of earthquakes for four zones near the Efate Island in the New Hebrides Island Arc, where there had been a continuous recording of events between 1978 and 1984 (Figure 2.62). They have made use of the fact that in case of fractal clustering the fraction x of the intervals of length τ containing earthquakes is related to τ as:

$$x \propto \tau^{1-D} \tag{2.4.2.11}$$

where D is the fractal dimension ($0 < D < 1$). The value of τ went from 60 to 2^{21} s ($\approx 2 \times 10^6$ min), increased by factors of two.

For each zone shown in Figure 2.62a the fraction of intervals x that included earthquakes was plotted and the fractal dimension determined from the slope of the $\log x$ vs. $\log \tau$ plot. [The Reader might observe that the same technique had been applied by Sprušil and Hnilica (1985), to the analysis of slip lines, and by Chilès (1988), to study the intersection of subvertical fractures with the survey line, see Section 2.2.] For the four zones in Figure 2.62a, Smalley et al. (1987) obtained:

Zone A	1,330 events	$D = 0.128$;
Zone B	379 events	$D = 0.229$;
Zone C	158 events	$D = 0.126$;
Zone D	49 events	$D = 0.026$.

In spite of the similar temporal behaviour of earthquakes and acoustic emissions, the fractality of the latter apparently has not been established yet.

The spatial distribution of earthquakes also shows statistical self-similarities. Kagan and Knopoff (1980) computed the two-point correlation function (see Section 2.3.1 and Appendix A2) for four local (two Californian, one New Zealand and one

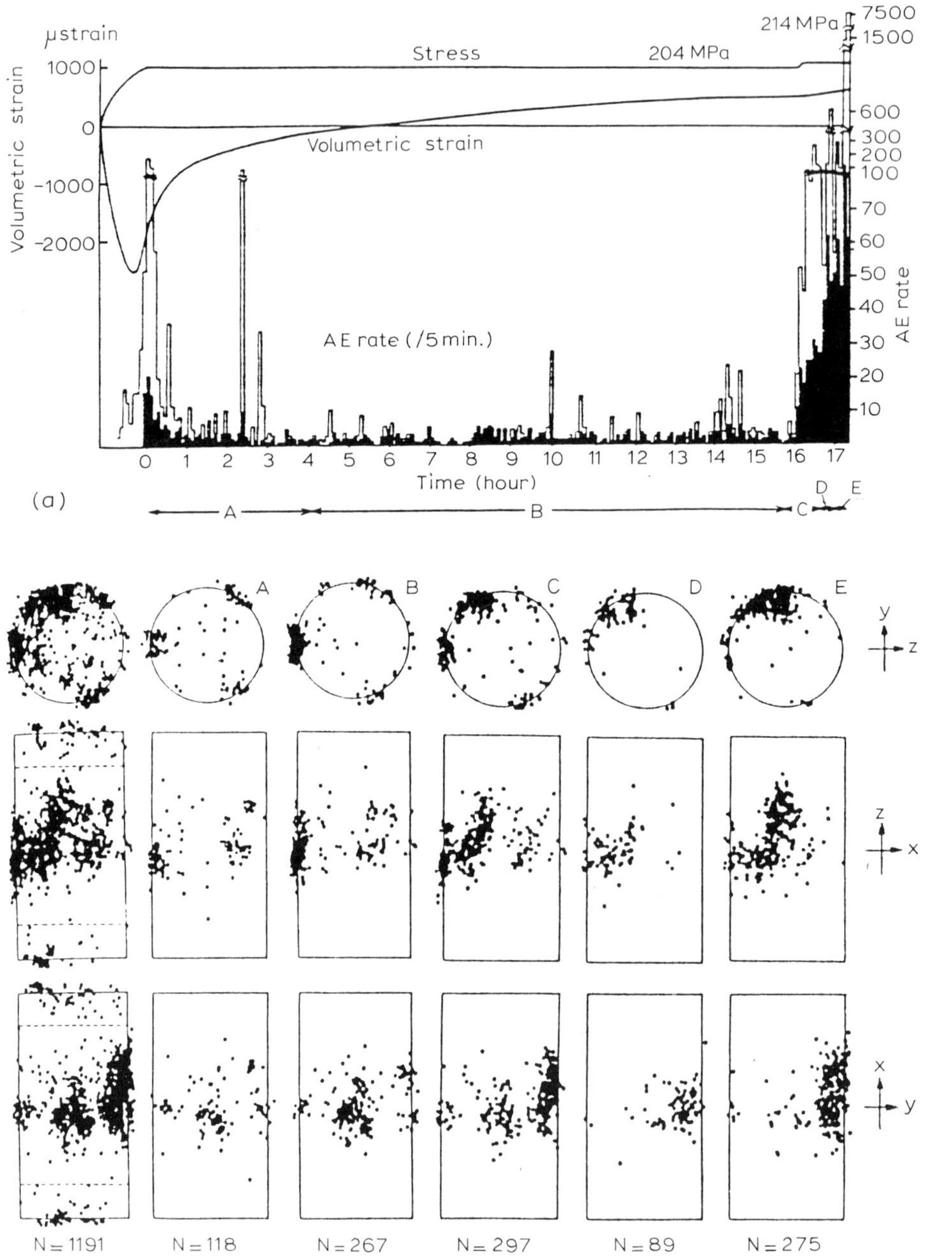

Fig. 2.61. a. Changes of stress, average volumetric strain and acoustic emission (AE) rate as a function of time for an andesite sample. Frequencies of AE for every 5 min are shown by open bars, the number of determined hypocentres by solid bars. *A*: transient creep; *B*: steady creep; *C–E*: acceleration creep leading to ultimate fracture. b. Orthographic projections of the AE hypocentres for all the 1,191 events located throughout the experiment (on the left), and for the periods *A–E*. (From Nishizawa et al., 1984/85.)

Japanese) and two world-wide earthquake catalogues, and found that the fractal dimension of the epicentres changes from 1 to 1.5 with increasing epicentral depth. Sadovskiy et al. (1984) have also verified by means of box-counting (Appendix A2)

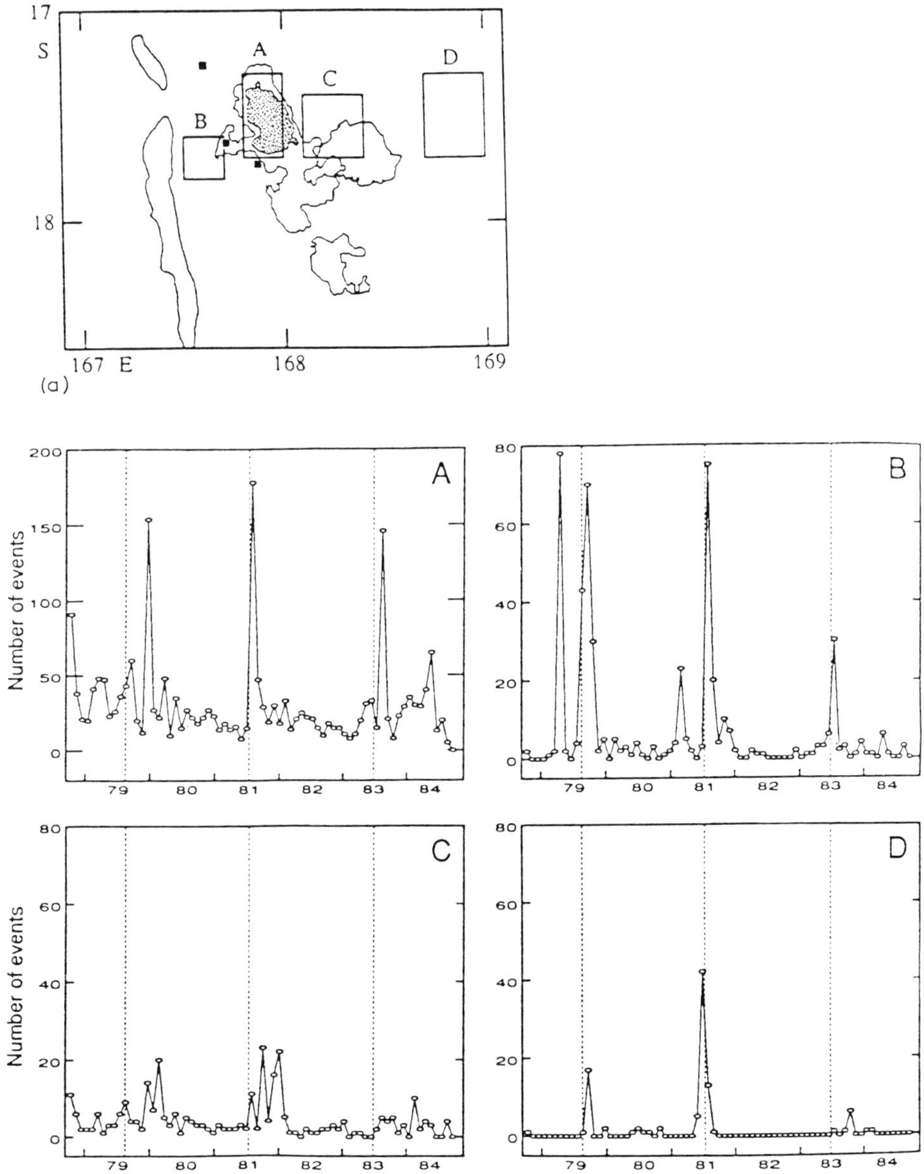

Fig. 2.62. a. Four seismic zones in the New Hebrides island arc, and b. histograms of the number of earthquakes per month. Dashed lines indicate the occurrence of major earthquakes. (From Smalley et al., 1987.)

that the earthquake epicentres are fractal both on a world-wide scale and in a local catalogue (Nurek region, Soviet Union), with the respective dimensions $D = 1.6$ and 1.4.

Hirata et al. (1987) studied the microfracturing of a sample of Oshima granite under 40 MPa confining pressure, recorded the acoustic emissions and analysed the spatial distribution of their hypocentres using the three-dimensional correlation integral:

$$C(r) = \frac{2}{n(n-1)} N(R < r) \tag{2.4.2.12}$$

where $N(R < r)$ is the number of pairs of hypocentres with a distance R smaller than r, and n was 353, 273, and 1438 for the transient creep, steady creep, and acceleration creep, respectively.

The correlation integral had the power function form $C(r) \propto r^D$, yielding the fractal dimensions:

$D = 2.75$ for the transient creep;
$D = 2.66$ for the steady creep; and
$D = 2.25$ for the acceleration creep.

Since the dimension of an "average" planar section of a D-dimensional fractal set embedded in the three-dimensional space is $D - 1$ (Mandelbrot, 1982), the fractal dimension of the AE epicentres ranges between 1.25 and 1.75, roughly corresponding with the fractal dimension of earthquake epicentres ($1 \leq D \leq 1.6$; Kagan and Knopoff, 1980; Sadovskiy et al., 1984).

The most important finding of Hirata et al.'s (1987) study has been that as the creep progressed, the degree of clustering of the AE hypocentres has gradually increased, that is their originally almost three-dimensional uniform random distribution has gradually become clustered with a much lower fractal dimension. The increased clustering of the hypocentres in the final acceleration creep stage has also been recognised in other microfracturing experiments (Mogi, 1968; Nishizawa et al., 1984/85), and a similar phenomenon has been observed before large earthquakes (Ouchi and Uekawa, 1986). Hirata et al. (1987) have proposed that *the use of the decrease in the fractal dimension of earthquake hypocentres may be valuable for predicting large earthquakes.* Though any further discussion of earthquake prediction would be outside the scope of this book, the method used by Ouchi and Uekawa (1986) to measure the degree of clustering deserves a closer scrutiny, because of its general applicability.

The distribution of earthquake epicentres can be specified by describing how N events observed over a limited area are sorted to Q non-overlapping quadrates. In case of the Polya-Eggenberger statistics (Eggenberger and Polya, 1923; Polya, 1930; Johnson and Kotz, 1969, volume I), the number of quadrates with n events is given by:

$$P(n) = \frac{1}{n!}\left[\frac{n}{Q}\right]^n (1+c)(1+2c)\dots(1+nc)\left\{1+\frac{cN}{Q}\right\}^{1/(c-n)} \qquad (2.4.2.13)$$

where the coefficient c represents the contagious properties of the events, and is called the "contagious coefficient". Special cases of equation (2.4.2.13) are the Poisson distribution:

$$P(n) = \frac{\exp(-\lambda)(\lambda)^n}{n!} \qquad (2.4.2.14)$$

for $c = 0$; and the geometrical distribution:

$$P(n) = \frac{n}{(1+\lambda)^{n+1}} \qquad (2.4.2.15)$$

for $c = 1$, where $\lambda = N/Q$. When the value of c becomes large, $P(n)$ approaches a power-law distribution, which is a characteristic of self-similar (fractal) clustering (Figure 2.63). Thus, to study the clustering of earthquakes, we divide the territory into meshes of equal area, count the number of events in each mesh and fit the distribution $P(n)$ with the Polya-Eggenberger model assuming different values of c. The approach has been used in a number of studies in Japan: Tamaki (1961) derived the geometrical distribution, Suzuki and Suzuki (1965) the power law. Finally, in an important paper Ouchi and Uekawa (1986) have shown that the contagious coefficient c depends on the size of the mesh, the same data set can be fitted with different Polya-Eggenberger distributions when using different mesh size (Figure 2.64). To circumvent this difficulty, Ouchi and Uekawa (1986) have proposed to use the *Morishita index of dispersion* (Morishita, 1959; Simpson, 1949), defined as:

$$I_\delta = \frac{\sum_{i=1}^{Q} n_i(n_i-1)}{N(N-1)} \qquad (2.4.2.16)$$

where N is the total number of sample events, Q is the number of meshes and n_i ($i = 1, 2, \dots, Q$) is the number of events in the i-th mesh.

When the spatial distribution of the earthquakes follow the Polya-Eggenberger model, the Morishita index I_δ and the coefficient c are related as:

$$I_\delta \approx c + 1 \qquad (2.4.2.17)$$

in particular, $I_\delta = 1$ for Poisson-random epicentres, $I_\delta \gg 1$ for fractal clustering. The Morishita index is easily computed and is also applicable when the epicentre distribution does not follow the Polya-Eggenberger model. The Morishita index depends on mesh size, according to Ouchi and Uekawa this dependence is charac-

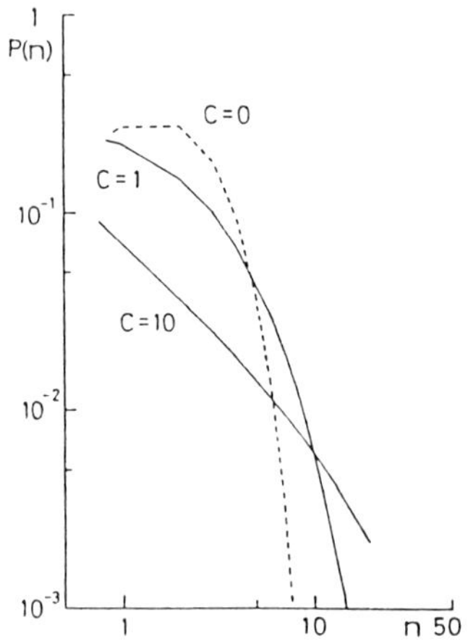

Fig. 2.63. Probability distribution of the Polya-Eggenberger model. The values of the contagious parameter c ($c = 0$, $c = 1$, $c = 10$) correspond to the Poisson, geometrical, and power type distributions, respectively. (From Ouchi and Uekawa, 1986.)

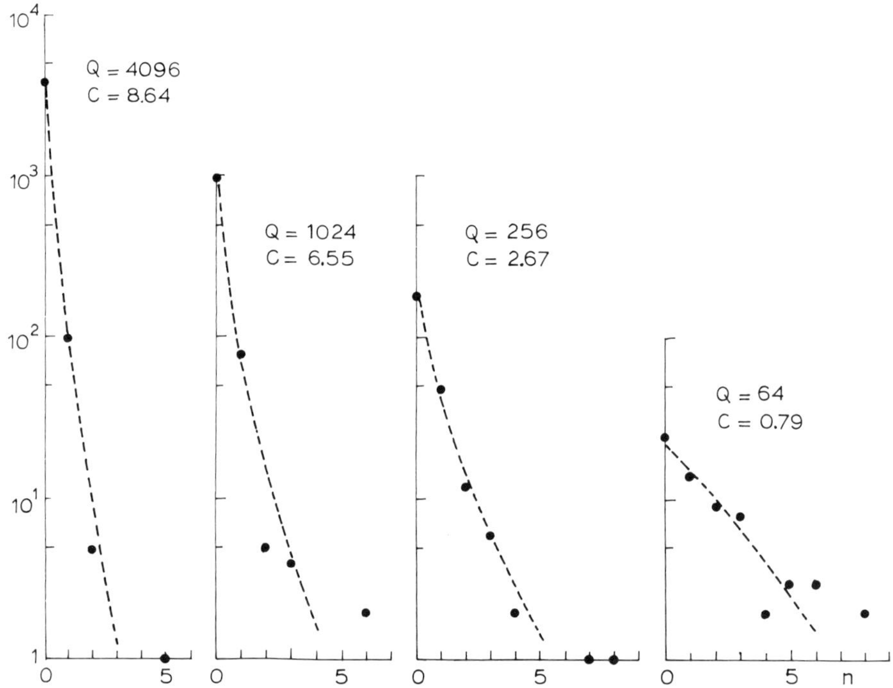

Fig. 2.64. Approximations of the epicentre distribution shown in Figure 2.66b by the Polya-Eggenberger model using meshes of different size. The study area was divided into 4096, 1024, 256, and 64 meshes, respectively. (From Ouchi and Uekawa, 1986.)

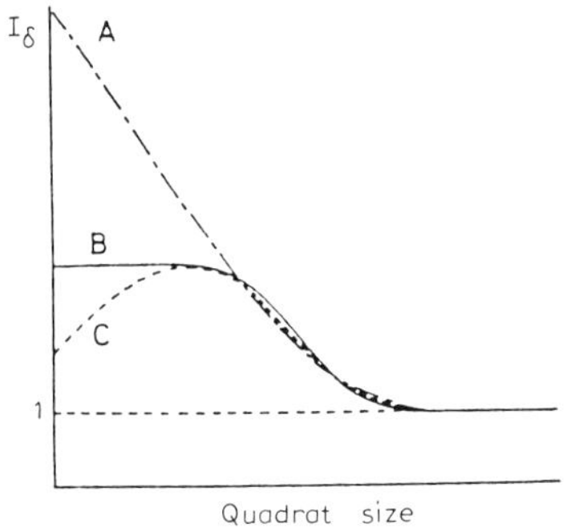

Fig. 2.65. Schematic representation of the Morishita index I_δ as a function of mesh size for three characteristic patterns of the epicentres. *A*: contagious distribution; *B*: the intra-cluster distribution random; *C*: the intra-cluster distribution uniform. (From Ouchi and Uekawa, 1986.)

teristically different for uniform and contagious distributions, respectively (Figures 2.65, 2.66). Alas, in Ouchi and Uekawa (1986) no theoretical discussion has been given of the schematic "Morishita-diagrams" shown in Figure 2.65, neither on the relation between fractal dimension and Morishita index.

The fractal dimension is only the first of a series of statistics needed to describe the geometry of random point-clusters. Higher-order statistics include the many-point correlation function (Fry and Peebles, 1978; Torquato and Stell, 1982), the clustering exponent (Thompson et al., 1987), and the elusive *lacunarity* (Allain and

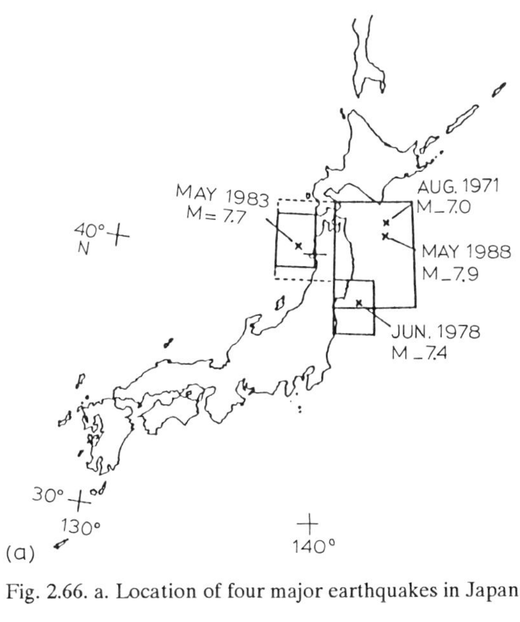

Fig. 2.66. a. Location of four major earthquakes in Japan.

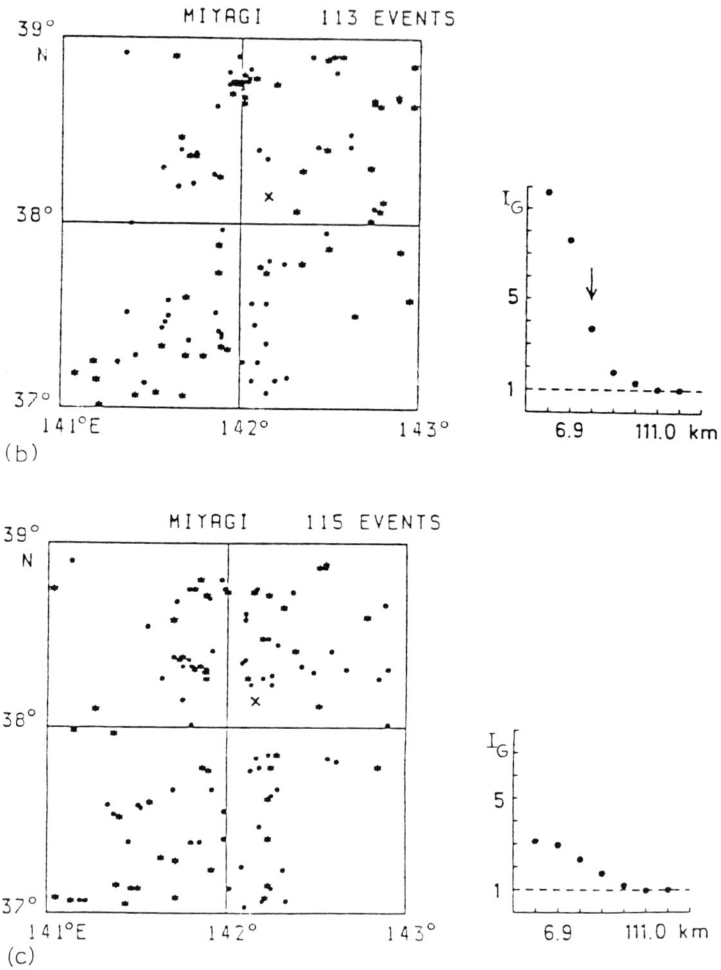

Fig. 2.66 (continued). b, c. Spatial distribution of earthquakes with amplitude $M > 3.5$ before (b) and after (c) the June 1978 Miyagi-oki earthquake, and their I_δ vs. mesh-size dependence. Crosses show the location of the main shock, the arrow indicates the characteristic cluster size. Note that, by the terminology of Figure 2.65, the Morishita diagram is of type A before the main shock, and type B afterwards. (From Ouchi and Uekawa, 1986.)

Cloitre, 1991) and *succolarity* (= nearness to percolation) introduced in the painfully short Chapter 34 of Mandelbrot (1982) devoted to textures. Apparently, these more advanced concepts have not been used yet in earthquake clustering studies.

A2 Mathematical appendix

A2.1 *Different kinds of fractal dimensions and their numerical determination*

> If you are an experimentalist, you try to measure the fractal dimension of things in nature. If you are a theorist, you try to calculate the fractal dimension of models chosen to describe experimental situations; if there is no agreement then you try another model.
>
> — H.E. STANLEY (1986)

The numerical determination of the fractal dimension of an object embedded into the d-dimensional Euclidian space is generally based on the following simple rule (Stanley, 1986): Suppose we plot on log–log paper some quantity q that can be interpreted as "number of objects" or "mass" against a characteristic length scale L.

If there is an asymptotic ($L \to \infty$) region in which this plot becomes straight, then the slope is termed the fractal dimension d_q characterising the distribution of the quantity q. Alternately, we can keep the size of the object, L, fixed, and determine q for different resolutions l. If we plot q against l, and the plot is a straight line in a range $l < l_0 \ll L$, the fractal dimension d_q will be the negative slope of this line. For example, if q is the area of a square of side L, it scales with L as L^2, if it is the number of small squares of side $l \ll L$ contained in the large square, it scales as:

$$q = \left(\frac{L}{l}\right)^2 \propto l^{-2}$$

Less trivial examples are:

– Length of a curve segment $l(A,B)$ connecting two points as function of the distance $L = \overline{AB}$ (Katz, 1985; Stanley, 1986).

– Number of steps made when walking a ruler of length l along the endpoints A and B of a statistically self-similar curve, $l \ll \overline{AB} = L$ [the "Richardson-Mandelbrot" (R-M) plot, see Section 1.4; Mandelbrot (1967, 1982), Schwarz and Exner (1980), Kennedy and Lin (1986), Underwood and Banerji (1986), Turcotte (1989). Generalisation of the R-M plot to fractal surfaces can be found in Clarke (1986), Creutzburg and Ivanov (1989)].

– The minimum number of d-dimensional spheres or cubes of characteristic size l needed to cover the object (d is dimension of the embedding space, $l \to 0$).

– The number of details found in the image of an object at a given resolution (Thompson et al., 1987; Turcotte, 1989).

– The "mass"of the object as function of its size (Termonia and Alexandrowicz, 1983; Meakin, 1988a, b; Voss, 1988) or, alternately, the "radius of gyration" of the object as function of its mass ($R_g \propto \text{mass}^{1/D}$; Meakin, 1988a, b).

– The number of branches of a "star-like" fractal object as function of its radius ($N \propto r^{(D-1)}$; Niemeyer et al., 1984).

– The information content of a fractal object as function of its size L or of the resolution l (Farmer, 1982a, b; Farmer et al., 1983; Grassberger and Procaccia, 1983a, b; Bai-Lin, 1984; Voss, 1988; Theiler, 1990).

Some closely related techniques are based on the *area–perimeter* relation (Section 1.5; Mandelbrot, 1982, Mandelbrot et al., 1984, Turcotte, 1989) or *surface–volume relation* (Turcotte, 1989) of fractal objects.

The most important concepts of fractal dimension are the Kolmogorov capacity D_c (Kolmogorov, 1958), the Hausdorff-dimension D_H (Hausdorff, 1918), the correlation dimension D_{corr} (Forrest and Witten, 1979, Grassberger and Procaccia, 1983a, b) and the information dimension D_I (Balatoni and Rényi, 1956; Farmer, 1982a, b; Farmer et al., 1983; Grassberger and Procaccia, 1983a, b). In the following definitions d will denote the dimension of the embedding space.

(a) The *capacity* of a set is given by:

$$D_c = \lim_{\varepsilon\to 0} \frac{\log N(\varepsilon)}{\log(1/\varepsilon)} \tag{A2.1.1}$$

where, if the set is a bounded subset of the d-dimensional Euclidian space, $N(\varepsilon)$ is the minimum number of d-dimensional cubes of side ε needed to cover the set.

(b) Capacity can be viewed as a simplified version of the Hausdorff dimension D_H , where the covering is done with d-dimensional cubes of variable edge length less than or equal to ε. Theoretically, $D_c \geq D_H$ (Farmer et al., 1983, a simple example where strict inequality holds is the set of numbers $\{1, 1/2, 1/3, 1/4, \ldots\}$, where the Hausdorff dimension is zero while equation (A2.1.1) yields $D_c = 1/2$); these two dimensions are however equal in most cases occurring in practice so that the Hausdorff dimension is sometimes simply defined by equation (A2.1.1) (Grassberger and Procaccia, 1983a, b).

(c) Historically, one of the first approaches to determine the dimension of fractal aggregates (Forrest and Witten, 1979) was based on the *box-counting* principle, according to which the number of points belonging to the fractal object within a d-dimensional "box" (cube or sphere) of size L centred around a randomly selected point X scales, at a given level of resolution $l \ll L$, as:

$$N_X(L,l) \propto \left(\frac{L}{l}\right)^{D_{box}} \tag{A2.1.2}$$

If we take the mean value of $N_X(L)$, averaged over all possible points X of the fractal, we get the *correlation integral*:

$$C(L) = \lim_{N\to\infty} \frac{1}{N^2}\{\text{number of pairs } (X_i,X_j)_1^N \text{ whose distance } \overline{X_iX_j} < L\} \tag{A2.1.3}$$

For fractals, $C(L)$ scales as:

$$C(L) \propto L^{D_{corr}} \tag{A2.1.4}$$

and the *correlation dimension* can be determined from its double logarithmic plot (Forrest and Witten, 1979; Grassberger and Procaccia, 1983a, b; Takens, 1984; Thompson et al., 1987; Henderson and Wells, 1988; Theiler, 1990).

(d) The *information dimension* measures how many bits of information are needed to specify a point of the fractal with a given accuracy. In case of a d-dimensional cube of unit size, there are 2^{dk} subcubes of side $r = 2^{-k}$ that is we need $S(r) = dk$ bits to specify the location of any points inside the d-cube with an accuracy r, wherefrom:

$$d = \frac{-S(r)}{\log_2 r} \qquad \text{(A2.1.5)}$$

Let us generalise this observation. If a fractal is partitioned into boxes B_i of size r, then the average information needed to specify the particular box containing the point is given by Shannon's formula:

$$S(r) = -\sum_i p_i \log_2 p_i \qquad \text{(A2.1.6)}$$

where p_i is the probability that a randomly selected point is in the i-th box. By equations (A2.1.5) and (A2.1.6) the information dimension is defined as:

$$D_\mathrm{I} = \lim_{r \to 0} \frac{\sum_i p_i \log_2 p_i}{\log_2 r} \qquad \text{(A2.1.7)}$$

(Farmer et al., 1983; Grassberger and Procaccia, 1983a, b; Bai-Lin, 1984; Voss, 1988; Theiler, 1990).

The different fractal dimensions D_H, D_corr, D_I are generally not equal, but satisfy the inequality:

$$D_\mathrm{corr} \leq D_\mathrm{I} \leq D_\mathrm{H} \qquad \text{(A2.1.8)}$$

where equality occurs if and only if the fractal is strictly self-similar and homogeneous. [Inequality (A2.1.8) was first observed experimentally by Forrest and Witten (1979); mathematical proofs can be found in Grassberger and Procaccia (1983a, b), Hentschel and Procaccia (1983), Henderson and Wells (1988).] Hentschel and Procaccia (1983) introduced an infinite number of fractal dimensions D_q ($q > 0$) describing the scaling of the multipoint correlation functions, where $D_0 = D$, $D_1 = D_\mathrm{I}$, $D_2 = D_\mathrm{corr}$, and they generalised inequality (A2.1.8) by showing that if $q' > q$ then $D_q < D_{q'}$ and that all the D_q's are equal for an homogeneous fractal (see also Voss, 1985a, 1988). Since inequality (A2.1.8) is quite tight in most cases, in most applications (e.g. Stauffer, 1979; Witten and Sander, 1981; Lovejoy et al., 1986a; Berry and Percival, 1986) it is tacitly assumed that $D_\mathrm{corr} = D_\mathrm{H}$ mainly because the numerical determination of D_corr is much easier than that of D_H. Mandelbrot (1984) has warned that "when a point set of interest to physics is fully specified, there is

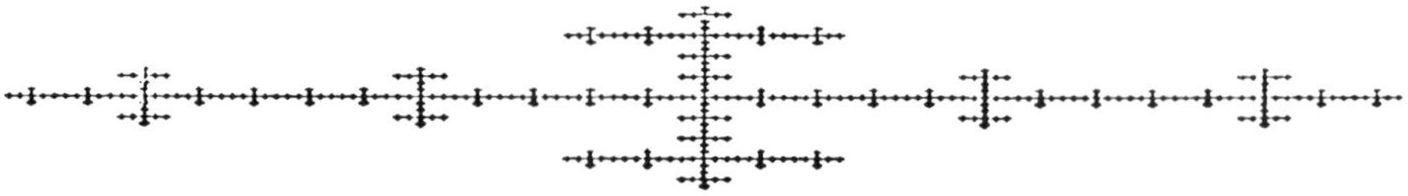

Fig. A2.1. Construction of a self-affine fractal aggregate (from Jullien and Botet, 1987).

a single well-defined fractal dimensionality D", and he paraphrased Occam's Razor ("Entia non sunt multiplicanda praeter necessitatem", a maxim attributed to William Ockham or Occam, 1285–1349) to advise that "dimensions that multiply without necessity *must* not be encouraged".

The previous definitions of fractal dimension have assumed that the object in question is (statistically) *self-similar.* A more general concept of the structure of fractal objects allows for different scaling behaviour in different "directions" (or, more precisely, along different coordinates). Figure A2.1 (from Jullien and Botet, 1987) shows the construction of a typical *self-affine* fractal. In each iteration one adds the same pattern in the $\pm x$-directions, and simultaneously adds two patterns along the directions $\pm y$. In order to get an enlargement identical with an original part of the object, one must enlarge by 3 in the x-direction and simultaneously by 7 in the y-direction; the total mass will then increase by a factor of 7. The object is not self-similar but *self-affine*; it is characterised by two fractal dimensions $D_x = \log 7/\log 3 = 1.77$ and $D_y = \log 5/\log 7 = 1.21$. The mass contained in a rectangle of sides x and y scales as:

$$m(x,y) \propto Ax^{D_x} + By^{D_y} \tag{A2.1.9}$$

where A and B are constants. For a square or circle of size r:

$$m(x,y) \propto Cr^{D_x}\left\{1 + Er^{-(D_x - D_y)}\right\} \tag{A2.1.10}$$

For $r \gg 1$ the mass asymptotically scales as r^{D_x}, that is according to the larger of the two exponents, while exponent D_y only appears in the correction term.

Rigorous definitions of *self-similarity* and *self-affinity* can be found in Mandelbrot (1982, chapter 39), Voss (1985b), Feder (1988, chapter 10). A further much-discussed example for self-affine fractals is the Brownian process $B(t)$ (Voss, 1985a, b; Meakin, 1988a, b) which describes the position of a random walker along a line starting at position $B(0) = 0$ at time $t = 0$. The statistical properties of $B(t)$ and $b^{-1/2}B(bt)$ are indistinguishable for any $b > 0$ (i.e. $B(t)$ is invariant under a change in the horizontal [time] scale by a factor of b and a simultaneous change in the vertical scale by a factor of $b^{1/2}$). Since t and $B(t)$ scale differently, the Brownian process is self-affine. Most of the observed natural surfaces (mountains, electro-deposits, fractures) are self-affine, because of the special role of gravity affecting the vertical coordinate (Meakin, 1988b; Lovejoy et al., 1987; Turcotte, 1989). For self-affine fractals the fractal dimension is not uniquely defined (Mandelbrot, 1985, 1986; Feder, 1988); the different techniques result in different dimensions.

Turcotte (1989) reports that the R-M plot for specified topographic contours in several mountain belts resulted in a fractal dimension $D \approx 1.2$, while the fractal dimension obtained from the power spectral density of the Earth's topography is $D \approx 1.5$. Turcotte attributes this discrepancy to the self-affinity of the surface of the earth: the vertical coordinate is not similar to the horizontal ones as gravity results in low angle slopes.

Numerical algorithms of the different methods of finding the fractal dimension can be found in a number of research papers and reviews, but their discussion would be beyond the scope of this book.

The practical aspects of the automatic determination of the perimeter of an irregular closed curve for a series of step sizes are discussed in Schwarz and Exner (1980), Kennedy and Lin (1986). They observe that the fractal dimension estimated from the R-M plot is generally not invariant to rotations and reflections of the closed curves, but also depends on the starting point location and on the direction of circumambulation (clockwise or counterclockwise). Also, spurious values arise if the step-length is of the order of the curve's irregularities. Kennedy and Lin (1986) present a FORTRAN subroutine (FRACT) claimed to be free of these problems.

Underwood and Banerji (1986) review the application of R-M plots to describe the fracture profiles of steel (Figure A2.2). They find that the "profile roughness parameter" R_L (defined as the apparent profile length divided by its projected length) is non-linear on the double-logarithmic R-M plot, that is, it cannot be approximated by the form:

$$\log R_L(\eta) = \log C - (D - 1)\log \eta \tag{A2.1.11}$$

and recommend to treat these plots as *reversed sigmoid curves* (known from phase transformation studies, cf. Burke, 1965) described by the equation:

$$\log\log\left\{\frac{(R_L)_0 - 1}{(R_L)_\eta - 1}\right\} = \log\left(\frac{\alpha}{2.3}\right) \pm \beta \log \eta \tag{A2.1.12}$$

where α and β are constants, β being related to the fractal dimension D_{IP} at the inflexion point through the empirical equation:

$$D_{IP} = 1.1286\,\beta^{0.02754} \tag{A2.1.13}$$

(for details see Underwood and Banerji, 1986).

The Richardson-Mandelbrot technique can be generalised to find the fractal dimension of topographic surfaces (Clarke, 1986) or grey-level image surfaces (Creutzburg and Ivanov, 1989). Here we only describe the algorithm of Creutzburg and Ivanov, as it is claimed to be by a factor of two faster than Clarke's method. The technique uses a three-dimensional equivalent of the "walking dividers" method in

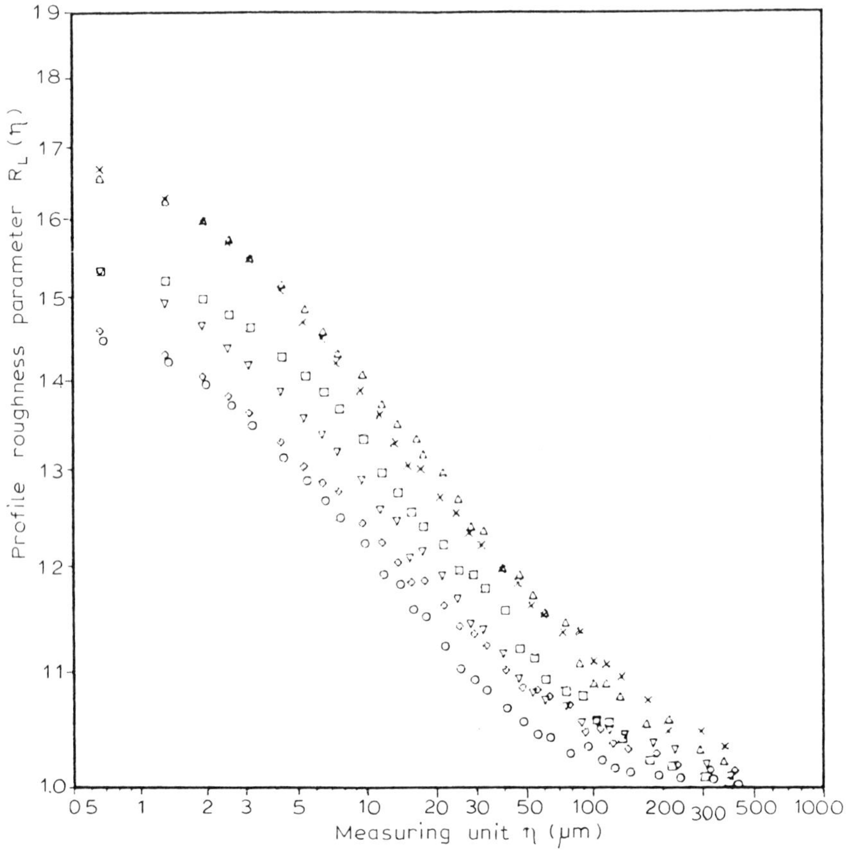

Fig. A2.2. Richardson-Mandelbrot plot of fracture profiles of a series of AISI 4340 steel specimens tempered at various temperatures: 200°C; 300°C; 400°C; 500°C; 600°C; 700°C (from Underwood and Banerji, 1986).

two dimensions. Let us consider a grid of a given spacing ε, let A, B, C, D be the four corner points of a square of the grid; let E be the centre of the square, and let us denote by a, b, c, d and e the corresponding amplitudes (elevations, grey-levels, etc.), where e is the interpolated value $e = (a + b + c + d)/4$.

Thus we can divide the square $ABCD$ into four triangles ABE, BCE, CDE, DAE. The triangles with their corner values define triangular prisms and we can easily calculate the surface area of the tops of these prisms by elementary vector algebra. If a triangle is spanned by two vectors $\boldsymbol{a}$ and $\boldsymbol{b}$ with Cartesian coordinates (x_a, y_a, z_a) and (x_b, y_b, z_b), respectively, then the area of the triangle is given by:

$$S = \frac{1}{2}\left\{(y_a z_b - y_b z_a)^2 + (x_a z_b - x_b z_a)^2 + (x_a y_b - x_b y_a)^2\right\}^{1/2} \qquad \text{(A2.1.14)}$$

The total surface area of the top of a grid cell is obtained by adding the top surface areas of the triangular prisms:

$$S_{ABCD} = S_{ABC} + S_{BCE} + S_{CDE} + S_{DAE} \tag{A2.1.15}$$

and the total top surface area of the whole topographic surface or grey-level image will be the sum of the contributions from the individual grid cells. Repeating this operation for decreasing grid spacing ε, the fractal dimension D of the surface $(2 \leq D \leq 3)$ can be determined from the log $S_{\text{total}}(\varepsilon)$ vs. log ε generalised R-M plot.

The most popular way of estimating the fractal dimension of point sets is based on *box-counting* or its generalised version, the *correlation integral*, defined in equations (A2.1.2) and (A2.1.3) (see the references above). It has been frequently noted that the box-counting algorithm is impractical for the determination of fractal dimensions greater than three because of the enormously large number of points needed (Farmer, 1982; Bai-Lin, 1984; Greenside et al., 1982; Theiler, 1990). A statistical discussion of the convergence and error of the estimation of the fractal dimension as dependent on the number of points used was given by Takens (1984), a frequently cited estimate is that $n \sim 10^d$ points are needed if the fractal point set is embedded into the d-dimensional Euclidian space (Henderson and Wells, 1988), that is around 10,000 points if $3 < D < 4$.

An important generalisation of box-counting is *functional box-counting*, introduced by Lovejoy et al. (1987). They recognised that for most physical fields encountered in meteorology the spatial points:

$$S(T) = \{X : f(X) > T\}$$

for which the measured values $f(X)$ of some physical quantity (temperature, radar reflectivity, etc.) exceed a given threshold T, constitute a fractal set of dimension $D(T)$ generally dependent on T. The points of the set $S(T)$ within a cube of size L of the embedding Euclidian space scales as:

$$N(L,l) = \left(\frac{L}{l}\right)^{-D(T)}$$

and the fractal dimension $D(T)$ can be determined by box-counting.

By varying T, we obtain the (decreasing) function $D(T)$. In many important cases the dependence of the dimension $D(T)$ on T can be described by simple analytical expressions (Gabriel et al., 1988; Schertzer et al., 1991; Lavallée et al., 1992).

Further details on *multifractal analysis* will be found in Section 4.6 (see also Mandelbrot, 1982, 1984, 1989; Feder, 1988; Sreenivasan, 1991).

The correlation integral (A2.1.3) for N points can be expressed as:

$$C(N,r) = \frac{\text{number of distances less than } r}{\text{number of distances altogether}} \propto r^{D_{\text{corr}}} \tag{A2.1.16}$$

for some finite range $r_{\min} < r < r_{\max}$. It is readily seen that the values of the correlation integral range from $2/N^2$ to 1 while the results of box counting only range from 1 to N. Different algorithms to compute $C(N,r)$ and some of the pitfalls due to finite sampling, edge effects and noise are reviewed in Theiler (1990).

The most time-consuming part of the computation of equations (A2.1.3) or (A2.1.16) is not the computation of the $N(N-1)/2$ distances but the determination of their cumulative sum. Grassberger and Procaccia (1983a) proposed an efficient algorithm for this task, claimed to accelerate computations by an order of magnitude. [Grassberger and Procaccia report that typical runs with 20,000 points took between 15 and 30 minutes on an IBM 370/165 using the fast algorithm. A more demanding computation, for 65,000 geographical locations defined by spherical co-ordinates, required more than 100 hour computer time on a VAX/VMS 11/780, with straightforward application of equation (A2.1.16); see Korvin et al. (1990).]

It should be noted that, in accordance with the rules of time series analysis (Jenkins and Watts, 1968), the estimation of the autocorrelation function is not reliable for distances r comparable with the size of the whole region. A practical rule of thumb is that the autocorrelation must not be computed for lags greater than about one third of the image size (Forrest and Witten, 1979; Korvin et al., 1990).

Autocorrelation functions are especially advantageous to describe the statistical geometry of the pore space of sedimentary rocks (Fara and Scheidegger, 1961; Armstrong and Korvin, 1981; Armstrong, 1984; Berryman, 1985; Berryman and Blair, 1986; Thompson et al., 1987). Autocorrelation measurements are conveniently made on polished thin sections, using image processing techniques. If we introduce a function:

$$f(P) = \begin{cases} 1 & \text{if the point } P \text{ is in the rock matrix} \\ 0 & \text{if the point } P \text{ is in the pore space} \end{cases} \tag{A2.1.17}$$

then the correlation function:

$$C(R) = \langle f(P)f(Q) \rangle \qquad (\overline{PQ} = R) \tag{A2.1.18}$$

will fall off for large values of R as R^{D-2} for fractal pore spaces of dimension D (Thompson et al., 1987). If the point P is located in pore space, then $C(R)$ expresses the probability of finding pore space at a distance R (Thompson et al., 1987; see Figures A2.3. and A2.4). The computation of (A2.1.18) for a digitised micrograph can be carried out using FFT (Fast Fourier Transform); optical determination of the correlation function (Allain and Cloitre, 1986; Lee and Moon, 1986) is a viable alternative.

As previously noted, for non-uniform fractals the different dimensions (Hausdorff-, correlation-, etc.) are not equal [cf. inequality (A2.1.8)].

Hentschel and Procaccia (1983) introduced an infinite number of fractal dimensions D_n based on the scaling of the n-point correlation function; Stanley (1986) compiled a "Rosetta Stone" of the 12 different fractal dimensions used to describe

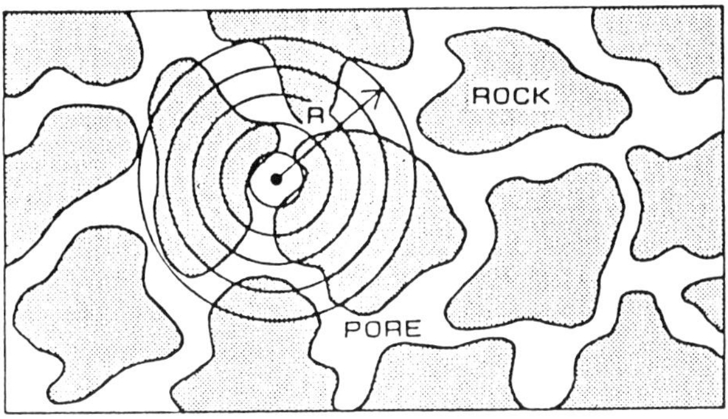

Fig. A2.3. Schematic of the autocorrelation function measurement on thin sections (from Thompson et al., 1987).

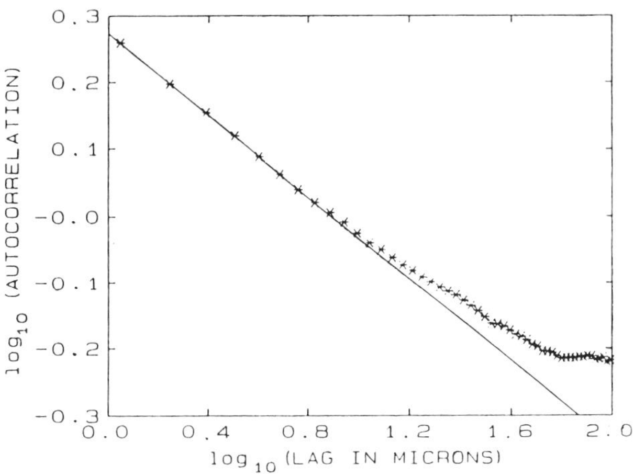

Fig. A2.4. Autocorrelation function computed from the scanning electron micrograph of a low-permeability highly altered sandstone sample (Mesa Verde formation, Utah). The slope is -0.27, that is $D = 1.73$. (From Thompson et al., 1987.)

percolation clusters. For self-affine fractals we also need a set of fractal dimensions for their characterisation (as e.g. the *elliptical dimensions* used to describe stratified fields; Lovejoy et al., 1987).

But even in uniform, self-similar fractals the (unique) fractal dimension is not sufficient to characterize the sets. [Other properties that have been proposed include *lacunarity*, see Mandelbrot (1982), Feder (1988), Voss (1985a, 1988), Allain and Cloitre (1991); and the *clustering exponent*, Thompson et al. (1987).] As Chorin (1982) noted: "Dimension is but one property of a set, and thought is required in each special case".

References

Abbott, E.A. (1932). Flatland. A Romance of Many Dimensions. 4th, Revised Edition, Basil Blackwell, Oxford.

Aboav, D.A. (1970). The arrangement of grains in a polycrystal. Metallography, 3: 383–390.

Ager, D.V. (1981). The Nature of the Stratigraphic Record. 2nd Edition Halsted Press, New York.

Agterberg, F.P. (1982). Recent developments in geomathematics. Geo-Processing, 2: 1–32.

Allain, C. and Cloitre, M. (1986). Optical diffraction on fractals. Phys. Rev. B, 33: 3566–3569.

Allain, C. and Cloitre, M. (1991). Characterizing the lacunarity of random and deterministic fractal sets. Phys. Rev. A., 44: 3552–3558.

Allègre, C.J., Le Mouel, J.L. and Provost, A. (1982). Scaling rules in rock fracture and possible implications for earthquake prediction. Nature, 297: 47–49.

Armstrong, B.H. and Korvin, G. (1981). Discussion. Geophysics, 46: 1314–1315.

Armstrong, B.H. (1984). Models for thermoelastic attenuation of waves in heterogeneous solids. Geophysics, 49: 1032–1040.

Aviles, C.A., Scholz, C.H. and Boatwright, J. (1987). Fractal analysis applied to characteristic segments of the San Andreas Fault. J. Geophys. Res., 92, B1: 331–344.

Avnir, D., Farin, D. and Pfeifer, P. (1985). Surface geometric irregularity of particulate materials: The fractal approach. J. Colloid Interface Sci., 103, 112–123.

Bai-lin, H. (Editor) (1984). Chaos. World Scientific, Singapore.

Bak, P. (1986). The Devil's staircase. Physics Today, Dec.1986, pp. 38–45.

Balatoni, J. and Rényi, A. (1956). On the notion of entropy. *In*: P. Turán (Editor) , Selected Papers of A. Rényi. Vol. I. Academic Press, Budapest, pp. 558–587.

Barenblatt, G.I., Zhivago, A.V., Neprochnov, Yu.P. and Ostrovsky, A.A. (1984). Fractal dimension: A quantitative characteristic of the ocean bottom geomorphological evolution processes. Okeonologiya, 24: 924–928 (in Russian).

Barlow, B.C. (1977). Data limitations on model complexity; 2-D gravity modelling with desk-top calculators. Bull. Austr. Soc. Explor. Geophys., 8: 139–143.

Barlow, G.W. (1974). Hexagonal territories. Anim. Behav., 22; 876–878.

Barton, Ch.C., Cameron, B.G. and Montgomery, J.R. (1986). Fractal scaling of fracture and fault maps at Yucca Mountain, Southern Nevada. Eos, 67, No. 44: 871 (Abstract No. U31A-08).

Barton, C.C. and Larsen, E. (1985). Fractal geometry of two-dimensional fracture networks at Yucca Mountain, Southwestern Nevada. *In:* O. Stephanson (Editor), Proc. Int. Symp. on Fundamentals of Rock Joints, pp. 77–84.

Beard, C.N. (1959). Quantitative study of columnar jointing. Bull. Geol. Soc. Am., 70: 379–381.

Behrens, E.W. (1980). On sedimentation rates and porosity. Marine Geol. 35: M11–M16.

Berry, M.V. and Lewis, Z.V. (1980). On the Weierstrass-Mandelbrot fractal function. Proc. R. Soc. Lond., A, 370: 459–484.

Berry, M.V. and Percival, I.C. (1986). Optics of fractal clusters such as smoke. Optica Acta, 33: 577–591.

Berryman, J.G. (1985). Measurement of spatial correlation functions using image processing technique. J. Appl. Phys., 57: 2374–2384.

Berryman, J.G, and Blair, S.C. (1986). Use of digital image analysis to estimate fluid permeability of porous materials: Application of two-point correlation functions. J. Appl. Phys., 60: 1930–1938.

Blumenthal, L.M. and Menger, K. (1970). Studies in Geometry. Freeman and Co., San Francisco.

Bolton, F and Weaire, D. (1991). The effects of Plateau borders in the two-dimensional soap froth. I. Decoration lemma and diffusion theorem. Phil. Mag. B, 63: 796–809.

Boots, B.N. (1980). Packing polygons: Some empirical evidence. Can. Geogr., 24: 406–411.

Brillouin, L. (1962). Science and Information Theory. 2nd Edition, Acad. Press, New York.

Brown, S.R. and Scholz, Ch.H. (1985). Broad bandwidth study of the topography of natural rock surfaces. J. Geophys. Res., 90: 12575–12582.

Bruinsma, R. and Bak, P. (1983). Self-similarity and fractal dimension of the devil's staircase in the one-dimensional Ising model. Phys. Rev., B, 27: 5824–5825.

Burke, J. (1965). The Kinetics of Phase Transformations in Metals. Pergamon, Oxford.

Burton, R., Kendall, Ch.G.St.C. and Lerche, I. (1987). Out of our depth: on the impossibility of fathoming eustasy from the stratigraphic record. Earth-Sci. Rev., 24: 237–277.

Cantor, G. (1883–84). Grundlagen einen allgemeinen Mannigfaltigkeitslehre. I: Math. Ann., 21: 545–586; II: Math. Ann., 23: 453–488.

Cederbaum, L.S., Haller, E. and Pfeifer, P. (1985). Fractal dimension function for energy levels. Phys. Rev., A, 31: 1869–1871.

Chilès, J.P. (1988). Fractal and geostatistical methods for modeling of a fracture network. Math. Geol., 20: 631–654.

Chorin, A.J. (1982). Numerical estimates of Hausdorff dimension. J. Comp. Phys., 46: 390–396.

Chowns, T.M. (1973). Fossils vs mud rolls. Geotimes, 18, No.3: 11.

Clark, W.E. Le Gros (1947). Deformation patterns in the cerebral cortex. *In*: W.E. Le Gros Clark and P.B. Medawar (Editors), Essays on Growth and Form. Clarendon Press, Oxford.

Clarke, K.C. (1986). Computation of the fractal dimension of topographic surfaces using the triangular prism surface area method. Comp. Geosci., 12: 713–722.

Cox, K.R. and Agnew, J.A. (1974). Optimal and non-optimal territorial partitions: a possible approach towards conflict. Papers of the Peace Sci. Soc. (Int.), 23: 123–138.

Crain, I.K. (1972). Monte-Carlo simulation of the random Voronoi polygons: Preliminary results. Search, 3: 220.

Creutzburg, R, and Ivanov, E. (1989). Fast algorithm for computing fractal dimensions of image segments. *In*: V. Cantoni, R. Creutzburg, S. Levialdi and G. Wolf (Editors), Recent Issues in Pattern Analysis and Recognition. Lecture Notes in Comp. Sci., 399. Springer, New York–Berlin–Heidelberg–London–Paris–Tokyo–Hong Kong, p. 42–51.

Dauben, J.W. (1979). Georg Cantor. His Mathematics and Philosophy of the Infinite. Harvard Univ. Press, Cambridge, Mass.

De Wijs, H.J. (1951 and 1955). Statistics of ore distribution. Pt. 1. Frequency distribution of assay values. Geol. Mijnbouw, 13: 365–375; Pt. 2. Theory of binomial distribution applied to sampling and engineering problems. Geol. Mijnbouw, 17: 12–24.

Dingus, L. and Sadler, P.M. (1982). The effect of stratigraphic completeness on estimates of evolutionary rates. Syst. Zool., 31: 400–412.

Dormer, K.J. (1980). Fundamental Tissue Geometry for Biologists. Cambridge Univ. Press.

Dowds, J.P. (1969). Oil rocks: Information Theory: Markov chains: Entropy. Q. Col. Sch. Mines, 64: 275–293.

Eggenberger, F. and Polya, G. (1923). Über die Statistik verketteter Vorgänge. Z. Angew. Math. Mech., 1: 179–289.

Fara, H.D. and Scheidegger, A.E. (1961). Statistical geometry of porous media. J. Geophys. Res., 66: 3279–3284.

Farmer, J.D. (1982a). Information dimension and the probabilistic structure of chaos. Z. Naturforsch., 37a: 1304–1325.

Farmer, J.D. (1982b). Dimension, fractal measures, and chaotic dynamics. *In*: H. Haken (Editor), Evolution of Order and Chaos. Springer, Berlin, pp. 228–248.

Farmer, J.D., Ott, E. and Yorke, J.A. (1983). The dimension of chaotic attractors. Physica, 7D: 153–180.

Feder, J. (1988). Fractals. Plenum Press, New York–London.

Feller, W. (1968). An Introduction to Probability Theory and its Applications. 3rd Edition, Wiley, New York.

Forrest, S.R. and Witten, T.A., Jr. (1979). Long-range correlation in smoke-particle aggregates. J. Phys., A, 12: L109–L117.

Fox, Ch.G. and Hayes, D.E. (1985). Quantitative methods for analyzing the roughness of the seafloor. Rev. Geophys., 23: 1–48.

Fraser, A.R., Moss, F.J. and Turpie, A. (1976). Reconnaissance gravity survey of Australia. Geophysics, 41: 1337–1345.

Fritz, M. (1988). Fractals find niche in oil exploration! AAPG Expl. 9, No. 7: 4–5.

Froehling, H., Crutchfield, J.P., Farmer, D., Packard, N.H. and Shaw, R. (1981). On determining the dimension of chaotic flows. Physica, 3D: 605–617.

Fry, J.N. and Peebles, P.J.E. (1978). Statistical analysis of catalogs of extragalactic objects. IX. The four-point galaxy correlation function. Astrophys. J., 221: 19–33.

Fujimori, S. (1985). Electric discharge and fractals. Jpn J. Appl. Phys., 24: 1198–1203.

Fukao, Y. and Furumoto, M. (1985). Hierarchy in earthquake size distribution. Phys. Earth Planet. Inter., 37: 149–168.

Gabriel, P., Lovejoy, S., Schertzer, D. and Austin, G.L. (1988). Multifractal analysis of resolution dependence in satellite imagery. Geophys. Res. Lett., 15: 1373–1376.

Gelbaum, B.R. and Olmsted, J.M.H. (1964). Counterexamples in Analysis. Holden-Day, Inc., San Francisco–London–Amsterdam.

Getis, A. and Boots, B. (1979). Models of Spatial Processes. Cambridge Univ. Press.

Gilluly, J. (1949). Distribution of mountain building in geologic time. Geol. Soc. Am. Bull., 60: 561–590.

Gingerich, P.D. (1976). Paleontology and phylogeny: patterns of evolution at the species level in Tertiary mammals. Am. J. Sci., 276: 1–28.

Gleick, J. (1987). Chaos. Making a New Science. Cardinal, London.

Grassberger, P. (1986). Estimating the fractal dimensions and entropies of strange attractors. *In*: A.V. Holden (Editor), Chaos. Princeton Univ. Press, Princeton, NJ.

Grassberger, P. and Procaccia, I. (1983a). Measuring the strangeness of strange attractors. Physica, 9D: 189–208.

Grassberger, P. and Procaccia, I. (1983b). Characterization of strange attractors. Phys. Rev. Lett., 50: 346–349.

Greenside, H.S., Wolf, A., Swift, J. and Pignataro, T. (1982). Impracticality of a box-counting algorithm for calculating the dimensionality of strange attractors. Phys. Rev., A, 25: 3453–3456.

Harrison, J.M. (1985). Brownian Motion and Stochastic Flow Systems. Wiley, New York.

Hausdorff, F. (1918). Dimension und Ausseres Mass. Math. Ann., pp. 157–179.

Henderson, H.W. and Wells, R. (1988). Obtaining attractor dimensions from meteorological time series. Adv. Geophys., 30: 205–237.

Hénon, M. (1976). A two-dimensional mapping with a strange attractor. Commun. Math. Phys., 50: 69–77.

Hentschel, H.G.E. and Procaccia, I. (1983). The infinite number of generalized dimensions of fractals and strange attractors. Physica, 8D: 435–444.

Heye, D. (1975). Wachstumsverhätnisse von Manganknollen. Geol. Jahrb., E5: 3–122.

Hewett, T.A. (1986). Fractal distributions of reservoir heterogeneity and their influence on fluid transport. SPE Paper No. 15386.

Hewett, T.A. (1988). Conditional simulation of reservoir heterogeneity with fractals. SPE Paper No. 18326.

Hirata, T. (1989). Fractal dimension of fault systems in Japan: Fractal structure in rock fracture geometry at various scales. PAGEOPH, 131: 157–170.

Hirata, T., Satoh, T. and Ito, K. (1987). Fractal structure of spatial distribution of microfracturing in rock. Geophys. J. R. Astr.Soc., 90: 369–374.

Hollingsworth, A. (1986). Storm hunting with fractals. Nature, 319: 11–12.

Hurst, H.E. (1957). A suggested statistical model of some time series which occur in Nature. Nature, 180: 494.

Jarrige, J.J., Ott D'Estevou, P., Burollet, P.F., Thiriet, J.-P., Icart, J.C., Richert, J.P., Sehans, P., Montenat, C. and Prat, P. (1986). Inherited discontinuities and Neogene structure: the Gulf of Suez and the northwestern edge of the Red Sea. Phil. Trans. R. Soc. Lond., A, 317: 129–139.

Jenkins, G.W. and Watts, D.G. (1968). Spectral Analysis and its Applications. Holden-Day, San Francisco.

Johnson, N.L. and Kotz, S. (1969). Distributions in Statistics. Vol.I. Discrete Distrbutions. Houghton Mifflin Co., Boston.

Jones, O. (1867). The Grammar of Chinese Ornaments. S. and T.Gilbert, London. (Reprint Edition by Studio Editions, London, 1987).

Jullien, R. and Botet, R. (1987). Aggregation and Fractal Aggregates. World Scientific Publ. Co., Singapore.

Kagan, Y.Y. and Knopoff, L. (1978). Statistical study of the occurrence of shallow earthquakes. Geophys. J.R. Astron. Soc., 55: 67–86.

Kagan, Y.Y. and Knopoff, L. (1980). Spatial distribution of earthquakes: the two-point correlation function. Geophys. J. R. Astr. Soc., 62: 303–320.

Kagan, Y.Y. and Knopoff, L. (1981). Stochastic synthesis of earthquake catalogs. J. Geophys. Res., 86: 2853–2862.

Katz, M.J. (1985). How straight do axons grow? J. Neurosci., 5: 589–595.

Keck, C.K. (1965). A Handbook on the Care of Paintings. American Association for State and Local History.

Kendall, M.G. and Moran, P.A.P. (1963). Geometrical Probability. Griffin and Co., London.

Kennedy, S.K. and Lin, W.-H. (1986). FRACT — a FORTRAN subroutine to calculate the variables necessary to determine the fractal dimension of closed forms. Comp. Geosci., 12: 705–712.

Kim, Y. and Jaggard, D.L. (1986). The fractal random array. Proc. IEEE, 74: 1278–1280.

King, G. (1983). The accomodation of large strains in the upper lithosphere of the Earth and other solids by self-similar fault systems: the geometrical origin of b-value. PAGEOPH, 121: 761–815.

Klemes, V. (1974). The Hurst phenomenon: A puzzle? Water Resour. Res., 10: 675–688.

Kolmogorov, A.N. (1958). A new invariant for transitive dynamical systems. Dokl. Akad. Nauk SSSR, 119: 861–864.

Korvin, G. (1989). Fractured but not fractal: Fragmentation of the Gulf of Suez Basement. PAGEOPH, 131: 289–305.

Korvin, G., Boyd, D.M. and O'Dowd, R. (1990). Fractal characterization of the South Australian gravity station network. Geophys. J. Int., 100: 535–539.

Lachenbruch, A.H. (1961). Depth and spacing of tension cracks. J. Geophys. Res., 66: 4273–4292.

Lachenbruch, A.H. (1962). Mechanics of thermal contraction cracks and ice-wedge polygons in permafrost. Geol. Soc. Am., Spec. Pap. No. 70.

Lantuejoul, C. (1978). Computation of the histograms of the number of edges and neighbours in a tessellation. *In*: R.E. Miles and J. Serra (Editors), Geometrical Probability and Biological Structures. Lecture Notes in Biomaths, Springer, Berlin–Heidelberg–New York, pp. 323–329.

Lauwerier, H.A. and Kaandorp, J.A. (1988). Fractals (Mathematics, Programming and Applications). *In*: M.M. de Ruiter (Editor), Advances in Computer Graphics III. Springer, Berlin–Heidelberg–New York–London–Paris–Tokyo, pp. 177–205.

Lavallée, D., Lovejoy, S., Schertzer, D. and Ladoy, P. (1992). Nonlinear variability of landscape topography: Multifractal analysis and simulation. *In*: L. de Cola and N. Lam (Editors), Fractals in Geography. Prentice Hall, N.Y. (in press).

Lee, C.-K. and Moon, F.C. (1986). An optical technique for measuring fractal dimensions of planar Poincare maps. Phys. Letts., 114A: 222–226.

Leffingwell, E. de K. (1915). Ground-ice wedges; the dominant form of ground-ice on the north coast of Alaska. J. Geol., 23: 635–654.

Lévy, P. (1939). Sur certaines processus stochastiques homogènes. Comp. Math., 7: 283–339.

Lévy, P. (1965). Processus Stochastiques et Mouvement Brownien. Gauthier-Villars, Paris.

Lewis, F.T. (1944). The geometry of growth and cell division in columnar parenchyma. Am. J. Botany, 31: 619–629.

Lovejoy, S. and Schertzer, D. (1986). Scale invariance, symmetries, fractals, and stochastic simulations of atmospheric phenomena. Bull. Am. Meteorol. Soc., 67: 21–32.

Lovejoy, S. and Schertzer, D. (1990). Fractals, raindrops and resolution dependence of rain measurements. J. Appl. Meteorol., 29: 1167–1170.

Lovejoy, S., Schertzer, D. and Ladoy, P. (1986a). Fractal characterization of inhomogeneous geophysical measuring networks. Nature, 319: 43–44.

Lovejoy, S., Schertzer, D. and Ladoy, P. (1986b). Outlook brighter on weather forecasts. Nature, 320: 401.

Lovejoy, S., Schertzer, D. and Tsonis, A.A. (1987). Functional box-counting and multiple elliptical dimensions in rain. Science, 235: 1036–1038.

Lynch, A.M. and King, A.R. (1983). A review of parameters affecting the accuracy and resolution of gravity surveys. Bull. Austr. Soc. Explor. Geophys., 14: 131–142.

Malmgren, B.A. and Kennett, J.P. (1981). Phyletic gradualism in a late Cenozoic planktonic foraminiferal lineage: DSDP site 284, southwest Pacific. Paleobiology, 7: 230–240.

Mandelbrot, B. (1967). How long is the coast of Britain? Statistical self-similarity and fractional dimension. Science, 156: 636–638.

Mandelbrot, B.B. (1975). Stochastic models for the Earth's relief, the shape and the fractal dimension of coastlines, and the number–area rule for islands. Proc. Nat. Acad. Sci. USA, 72: 3825–3828.

Mandelbrot, B.B. (1981). Scalebound or scaling shapes: a useful distinction in the visual arts and in the natural sciences. Leonardo, 14: 45–47.

Mandelbrot, B.B. (1982). The Fractal Geometry of Nature. Freeman, San Francisco.

Mandelbrot, B.B. (1984). Fractals in physics: Squig clusters, diffusions, fractal measures, and the unicity of fractal dimensionality. J. Stat. Phys., 34: 895–930.

Mandelbrot, B.B. (1985). Self-affine fractals and fractal dimension. Phys. Scr., 32: 257–260.

Mandelbrot, B.B. (1986). Self-affine fractal sets. *In*: L. Pietronero and E. Tosatti (Editors), Fractals in Physics. North-Holland, Amsterdam, pp. 3–28.

Mandelbrot, B.B. (1989). Multifractal measures, especially for the geophysicist. PAGEOPH, 131: 5–42.

Mandelbrot, B.B. and Given, J.A. (1984). Physical properties of a new fractal model of percolation clusters. Phys. Rev. Lett., 52: 1853–1856.

Mandelbrot, B.B. and Van Ness, J.W. (1968). Fractional Brownian motions, fractional noises and applications. SIAM Rev., 10: 422–437.

Mandelbrot, B.B. and Wallis, J.R. (1969). Some long-run properties of geophysical records. Water Resour. Res., 5: 321–340.

Mandelbrot, B.B., Passoja, D.E. and Paullay, A.J. (1984). Fractal character of fracture surfaces of metals. Nature, 308: 721–722.

Matsushita, M. (1985). Fractal viewpoint of fracture and accretion. J. Phys. Soc. Jpn., 54: 857–860.

Mauldin, R.D. (1986). On the Hausdorff dimension of graphs and random recursive objects. *In*: G. Mayer-Kress (Editor), Dimensions and Entropies in Chaotic Systems. Springer, Berlin–Heidelberg, pp. 28–33.

McKinney, M.L. and Schoch, R.M. (1983). A composite terrestrial Paleocene section with completeness estimates, based upon magnetostratigraphy. Am. J. Sci., 283: 801–814.

Meakin, P. (1988a). The Growth of Fractal Aggregates and Their Fractal Measures. *In*: C. Domb and J.L. Lebowitz (Editors), Phase Transitions and Critical Phenomena Vol. 12. Academic Press, London–New York, pp. 335–489.

Meakin, P. (1988b). Fractal aggregates. Adv. Colloid Interf. Sci., 28: 249–331.

Menger, K. (1967). Kurventheorie. Second, reprint edition. Chelsea Publ.Co., Bronx, NY.

Meschkowski, H. (1967). Probleme des Unendlichen. Werk u. Leben Georg Cantors. Friedr. Vieweg and Sohn, Braunschweig.

Miles, R.E. (1964). Random polygons determined by random lines in a plane. Pts.I and II. Proc. N.A.S. 52: 901–907; 1157–1160.

Miyazima, S. and Stanley, H.E. (1987). Intersection of two fractal objects: Useful method of estimating the fractal dimension. Phys. Rev., B, 35: 8898–8900.

Mogi, K. (1968). Source locations of elastic shocks in the fracturing process in rocks. Pt.I. Bull.Earthq. Res. Inst. Tokyo Univ., 46: 1103–1125.

Moore, T.C., Jr. and Heath, G.R. (1977). Survival of deep-sea sedimentary sections. Earth Planet. Sci. Lett., 37: 71–80.

Morishita, M. (1959). Measuring of the dispersion of individuals and analysis of the distribution patterns. Mem. Fac. Sci. Kyushu Univ., Ser. E, 2: 215–235.

Muehsam, G. (1970). French Painters and Paintings from the Fourteenth Century to Post-Impressionism. Unger Publ. Co., New York.

Neal, J.T., Langer, A.M. and Kerr, P.F. (1968). Giant desiccation polygons of great basin playas. Bull. Geol. Soc. Am., 79: 69–70.

Nelson, R.A. (1979). Natural fracture systems: Description and classification. Bull. AAPG, 63: 69–70.

Newton, A.R. (1987). The fracture pattern around the Sutherland diatreme, Cape Province, from remote sensing. S. Afr. J. Geol., 90: 99–106.

Niemeyer, L., Pietronero, L. and Wiesmann, H.J. (1984). Fractal dimension of dielectric breakdown. Phys. Rev. Lett., 52: 1033–1036.

Nishizawa, O., Onai, K. and Kusunose, K. (1984/85). Hypocenter distribution and focal mechanism of AE events during two stress stage creep in Yugawara andesite. PAGEOPH, 122: 36–52.

Ogawa, T. (1983). Problems in a digital description of a configuration of atoms and some other geometrical topics in physics. *In*: F. Yonezawa and T. Ninomiya (Editors), Topological Disorder in Condensed Matter. Springer, Berlin–Heidelberg–New York–Tokyo, pp. 60–77.

Ojo, S.B. and Mereau, R.F. (1986). The effect of random velocity functions on the travel times and amplitudes of seismic waves. Geophys. J. R. Astr. Soc., 84: 607–618.

Okubo, P.G. and Aki, K. (1987). Fractal geometry in the San Andreas fault system. J. Geophys. Res. 92, B1: 345–355.

Orey, S. (1970). Gaussian sample functions and the Hausdorff dimension of level crossings. Z. Wahrsch. Verw. Geb., 15: 249–256.

Ouchi, T. and Uekawa, T. (1986). Statistical analysis of the spatial distribution of earthquakes — variation of the spatial distribution of earthquakes before and after large earthquakes. Phys. Earth Planet. Int., 44: 211–225.

Papoulis, A. (1962). The Fourier Integral and its Applications. McGraw-Hill Book Co., New York.

Pavič, M. (1988). Dictionary of the Khazars. (Translated by Ch. Pribičevič-Zorič). Hamish Hamilton, London.

Pieri, D.C. (1981). Lineament and polygon patterns on Europa. Nature, 289: 17–21.

Plotnick, R.E. (1986). A fractal model for the distribution of stratigraphic hiatuses. J. Geol. 94: 885–890 (Discussion: (1988), 96: 101–102).

Plummer, P.S. and Gostin, V.A. (1981). Shrinkage cracks: Desiccation or synaeresis? J. Sediment. Petrol., 51: 1147–1156.

Polya, G. (1930). Sur quelques points de la théorie des probabilitées. Ann. Inst. Henri Poincaré, 1: 117–161.

Posner, D. (1982). The swinging women of Watteau and Fragonard. Art Bull., 64: 75–88.

Randolph, R.A. (1988). The Beni Maklaab over the horizon: Males and females, dogs and beduin. *In*: R.A. Randolph, D.M. Schneider and M.N. Diaz (Editors), Dialectics and Gender. Anthropological Approaches. Westview Press, Boulder and London.

Richter, Ch.F. (1958). Elementary Seismology. Freeman and Co., San Francisco.

Rietsch, E. (1982). Deconvolution in the presence of noise and a higher than Nyquist signal rate. Geoexploration, 20: 61–73.

Rivier, N. and Lissowski, A. (1982). On the correlation between sizes and shapes of cells in epithelial mosaics. J. Phys., A, 15: L143–L148.

Rothman, D.H. (1985). Nonlinear inversion, statistical mechanics, and residual statics estimation. Geophysics, 50: 2784–2796.

Rothman, D.H. (1986). Automatic estimation of large residual static corrections. Geophysics, 51: 332–346.

Rothrock, D.A. and Thorndike, A.S. (1984). Measuring the sea ice floe size distribution. J. Geophys. Res., 89, C4: 6477–6486.

Russell, D.A., Hanson, J.D. and Ott, E. (1980). Dimension of strange attractors. Phys. Rev. Lett., 45: 431–434.

Sadler, P.M. (1980). The significance of time scale for the rate of accretion of marine manganese nodules and crusts. Mar. Geol., 35: M27–M32.

Sadler, P.M. (1981). Sediment accumulation rates and the completeness of stratigraphic sections. J. Geol., 89: 569–584.

Sadler, P.M. and Dingus, L.W. (1982). Expected completeness of sedimentary sections: estimating a time-scale dependent, limiting factor in the resolution of the fossil record. Proc. 3rd N. Am. Paleontol. Conv., 2: 461–464.

Sadler, P.M. and Strauss, D.J. (1990). Estimation of the completeness of stratigraphical sections using empirical data and theoretical models. J. Geol. Soc. Lond., 147: 471–485.

Sadovskiy, M.A., Golubeva, T.V., Pisarenko, V.F. and Shnirman, M.G. (1984). Characteristic dimensions of rock and hierarchical properties of seismicity. Izvestiya, Earth Phys., 20: 87–96.

Salop, L.J. (1977). Precambrian of the Northern Hemisphere. Elsevier, N.Y.

Sammis, C.G. (1988). The micromechanics of fractal fault-gauge generation. Geol. Soc. Am. Meet., Denver, Oct. 31–Nov. 3, 1988, Abstract No. 27, p. 313.

Sammis, C.G., Osborne, R., Anderson, J.L., Banerdt, M. and White, P. (1986). Self-similar cataclisis in the formation of fault gouge. PAGEOPH, 124: 53–78.

Sander, L.M. (1986). Fractal growth processes. Nature, 322: 789–793.

Sander, L.M. (1987). Fractal growth. Sci. Am., 256, 1: 82–88.

Sato, H. (1988). Fractal interpretation of the linear relation between logarithms of maximum amplitude and hypocentral distance. Geophys. Res. Lett., 15: 373–375.

Sawada, Y., Ohta, S., Yamazaki, M. and Honjo, H. (1982). Self-similarity and a phase-transition-like behavior of a random growing structure governed by a nonequilibrium parameter. Phys. Rev., A, 26: 3557–3563.

Sawada, Y., Dougherty, A. and Golub, J.P. (1986). Dendritic and fractal patterns in electrolytic metal deposits. Phys. Rev. Lett., 56: 1260–1263.

Sazhina, N. and Grushinsky, N. (1971). Gravity Prospecting. Mir Publishers, Moscow.

Schelkunoff, S.A. (1943). A mathematical theory of linear arrays. Bell Syst. Technol. J., 22: 80–107.

Schertzer, D. and Lovejoy, S. (1983). The dimension and intermittency of atmospheric dynamics. *In*: L.J.S. Bradbury, F. Durst, B.E. Launder, F.W. Schmidt and J.H. Whitelaw (Editors), Turbulent Shear Flows 4. Springer, Berlin–Heidelberg–New York–Tokyo, pp. 7–33.

Schertzer, D. and Lovejoy, S. (1985). Generalised scale invariance in turbulent phenomena. Phys. Chem. Hydrodyn., 6: 623–635.

Schertzer, D. and Lovejoy, S. (1986). Generalised scale invariance and anisotropic inhomogeneous fractals in turbulence. *In*: L. Pietronero and E. Tosatti (Editors), Fractals in Physics. Elsevier, Amsterdam–New York, pp. 457–460.

Schertzer, D., Lovejoy, S., Lavallée, D. and Schmitt, F. (1991). Universal hard multifractal turbulence: Theory and observations. *In*: R.Z. Sagdaev et al. (Editors), Nonlinear Dynamics of Structures. World Scientific, Singapore, pp. 213–235.

Schindel, D.E. (1980). Microstratigraphic sampling and the limits of paleontological resolution. Paleobiology, 6: 408–426.

Schindel, D.E. (1982). Resolution analysis: a new approach to gaps in the fossil record. Paleobiology, 8: 340–353.

Scholz, C.H. (1968). Microfractures, aftershocks, and seismicity. Bull. Seismol. Soc. Am., 58: 1117–1130.

Scholz, C.H. (1989). Global perspectives of chaos. Nature, 338, 459–460.

Scholz, C.H. and Aviles, C.A. (1986). The fractal geometry of faults and faulting. *In*: S. Das, J. Boatwright and C.H. Scholz (Editors), Earthquake Source Mechanics. Geophys. Monograph. 37, Maurice Ewing Vol. 6, Am. Geophys. Union, Washington, D.C., pp. 147–155.

Schwab, F.L. (1976). Modern and ancient sedimentary basins: Comparative accumulation rates. Geology, 4: 723–727.

Schwarz, H. and Exner, H.E. (1980). The implementation of the concept of fractal dimension on a semi-automatic image analyser. Powder Technol., 27: 207–213.

Segnit, E.R., Anderson, C.A. and Jones, J.B. (1970). A scanning microscope study of the morphology of opal. Search, 1: 349–351.

Simpson, E.H. (1949). Measurement of diversity. Nature, 163: 688.

Smalley, I.J. (1966). Contraction crack networks in basalt flows. Geol. Mag., 103: 110–114.

Smalley, R.F., Jr., Chatelain, J.-L., Turcotte, D.L. and Prevot, P. (1987). A fractal approach to the clustering of earthquakes: Application to the seismicity of the New Hebrides. Bull. Seism. Soc. Am., 77: 1368–1381.

Smirnov, B.M. (1986). Fractal clusters. Sov. Phys. Usp., 29: 481–505.

Smith, C.S. (1952). Grain shapes and other metallurgical applications of topology. *In*: Metal Interfaces. Am. Soc. Metals, Cleveland, Ohio, pp. 65–109 (Discussion by John von Neumann on pp. 108–110).

Smith, C.S. (1954). The shape of things. Sci. Am., 190: 58–64.

Sprušil, B. and Hnilica, F. (1985). Fractal character of slip lines of Cd single crystals. Czech. J. Phys., 35: 897–900.

Sreenivasan, K.R. (1991). Fractals and multifractals in fluid turbulence. Ann. Rev. Fluid Mech., 23: 539–600.

Stanley, H.E. (1986). Form: An introduction to self-similarity and fractal behavior. *In*: H.E. Stanley and N. Ostrowsky (Editors), On Growth and Form. Fractal and Non-Fractal Patterns in Physics. Martinus Nijhoff Publ., Dordrecht–Boston–Lancester, pp. 21–53.

Stauffer, D. (1979). Scaling theory of percolation clusters. Phys. Rep., 54: 1–74.

Stevens, P.S. (1974). Patterns in Nature. An Atlantic Monthly Press Book. Little, Brown and Co., Boston–Toronto.

Strauss, D. and Sadler, P.M. (1989). Stochastic models for the completeness of stratigraphic sections. Math. Geol., 21: 37–59.

Subba Rao, D.V., Bapu Reddy, C. and Krishna Brahman, N. (1987). Gravity modelling of Satpura Gondwana Basin (Central India). J. Assoc. Explor. Geophys., 8: 169–178.

Suzuki, M. (1984). Finite-size scaling for transient similarity and fractals. Prog. Theor. Phys., 71: 1397–1400.

Suzuki, Z. and Suzuki, K. (1965). On space distribution functions of earthquakes. Sci. Rep. Tohoku Univ., 5, No 17: 9–23.

Takayasu, H. (1985a). Simulation of electric breakdown and resulting variant of percolation fractals. Phys. Rev. Lett., 54: 1099–1101.

Takayasu, H. (1985b). A deterministic model of fracture. Prog. Theor. Phys., 74: 1343–1345.

Takayasu, H. (1986). Pattern formation of dendritic fractals in fracture and electric breakdown. *In*: L. Pietronero and E. Tosatti (Editors), Fractals in Physics. Elsevier, Amsterdam–New York, pp. 181–184.

Takens, F. (1984). On the numerical determination of the dimension of an attractor. *In*: B.L.J. Braaksma, H.W. Broer and F. Takens (Editors), Dynamical Systems and Bifurcations. Springer, Berlin–Heidelberg–New York–Tokyo, pp. 99–106.

Tamaki, I. (1961). Seismicity in relation to the crust structure, with special reference to the Japanese area. Mem. Osaka Inst. Technol., A7: 1–92.

Taylor, S.J. (1955). The alpha-dimensional measure of the graph and set of zeroes of a Brownian path. Proc. Cambridge Phil. Soc., 51, Pt II: 265–274.

Termonia, Y. and Alexandrowicz, Z. (1983). Fractal dimension of strange attractors from radius versus size of arbitrary clusters. Phys. Rev. Lett., 51: 1265–1268.

Theiler, J. (1990). Estimating the fractal dimension of chaotic time series. Lincoln Lab. J., 3: 63–85.

Thomas, A. (1987). Structure fractale de l'architecture de champs de fractures en milieu rocheux. C.R. Acad. Sci. Paris, Ser. II, 304: 181–186.

Thomas, P.G. and Masson, P. (1983). Tectonic evolution of Mercury; comparison with the Moon. Ann. Geophys., 1: 53–58.

Thompson, A.H., Katz, A.J. and Krohn, C.E. (1987). The microgeometry and transport properties of sedimentary rock. Adv. Phys., 36: 625–694.

Thompson, D'Arcy W. (1942). On Growth and Form. 2nd Edition, Cambridge Univ. Press.

Thorarinsson, F. and Magnusson, S.G. (1990). Bouguer density determination by fractal analysis. Geophysics, 55: 932–935.

Tipper, J.C. (1983). Rates of sedimentation, and stratigraphic completeness. Nature, 302: 696–698.

Torquato, S. and Stell, G. (1982). Microstructure of two-phase random media. I. The n-point probability functions. J. Chem. Phys., 77: 2071–2077.

Tribus, M. (1979). Thirty years of Information Theory. *In*: R.D. Levine and M. Tribus (Editors), Maximum Entropy Formalism. MIT Press, Cambridge, Mass.

Tritton, D. (1986a). Chaos in the swing of a pendulum. New Scientist, 24 July 1986, pp. 37–40 (Erratum: 7 Aug. 1986, p. 57).

Tritton, D.J. (1986b). Ordered and chaotic motion of a forced spherical pendulum. Eur. J. Phys., 7: 162–169.

Tritton, D.J. (1988). Physical Fluid Dynamics. 2nd Edition, Oxford Univ. Press, pp. 400–402.

Tritton, D.J. (1989). Deterministic chaos, geomagnetic reversals, and the spectral pendulum. *In*: F.J. Lowes et al. (Editors), Geomagnetism and Palaeomagnetism. Kluwer Acad. Publ., pp. 215–226.

Turcotte, D.L. (1986a). Fractals and fragmentation. J. Geophys. Res., 91, B2: 1921–1926.

Turcotte, D.L. (1986b). A fractal model for crustal deformation. Tectonophysics, 132: 261–269.

Turcotte, D.L. (1986c). A fractal approach to the relationship between ore grade and tonnage. Econ. Geol., 81: 1528–1532.

Turcotte, D.L. (1989). Fractals in geology and geophysics. PAGEOPH, 131: 171–196.

Underwood, E.E. and Banerji, K. (1986). Fractals in fractography. Mater. Sci. Eng., 80: 1–14.

Vilenkin, N.Ya. (1968). Stories About Sets. Academic Press, New York–London.

Von Neumann, J. (1952). Discussion. Metal Interfaces. Am. Soc. Metals, Cleveland, Ohio, pp. 108–110.

Voss, R.F. (1985a). Random fractals: Characterization and measurement. *In*: R. Pynn and A. Skjeltorp (Editors), Scaling Phenomena in Disordered Systems. Plenum Press, New York–London.

Voss, R.F. (1985b). Random fractal forgeries. *In*: R.A. Earnshaw (Editor), Fundamental Algorithms for Computer Graphics. NATO ASI Series, Vol. F17. Springer, Berlin–Heidelberg, pp. 805–835.

Voss, R.F. (1988). Fractals in nature: From characterization to description. *In*: H.-O. Peitgen and D. Saupe (Editors), The Science of Fractal Images. Springer, Berlin–Heidelberg, pp. 21–70.

Walker, J. (1986). The amateur scientist. Sci. Am., 255, No. 4: 178–183.

Weaire, D. (1983). Random two-dimensional cellular structures. *In*: F. Yonezawa and T. Ninomiya (Editors), Topological Disorder in Condensed Matter. Springer, Berlin–Heidelberg, pp. 51–59.

Weaire, D. and Kermode, J.P. (1983). The evolution of the structure of two-dimensional soap froth. Phil. Mag., B, 47: L29–L31.

Weaire, D. and O'Carroll, C. (1983). A new model for the Giant's Causeway. Nature, 302: 240–241.

Weaire, D. and Rivier, N. (1984). Soap, cells and statistics — Random patterns in two dimensions. Contemp. Phys., 25: 59–99.

Weitz, D.A. and Huang, J.S. (1984). Self-similar structures and the kinetics of aggregation of gold colloids. *In*: F. Family and D.P. Landau (Editors), Aggregation Gelation. North-Holland, Amsterdam.

Weitz, D.A., Lin, M.Y., Huang, J.S., Witten, T.A., Sinha, S.K., Gethner, J.S. and Ball, R.C. (1985). Scaling in colloid aggregation. *In*: R. Pynn and A. Skjeltorp (Editors), Scaling Phenomena in Disordered Systems. Plenum Press, New York–London, pp. 171–188.

Wentzel, H. (1964). Jean-Honoré Fragonards "Schaukel". Bemerkungen zur Ikonographie der Schaukel in der Bildenden Kunst. Wallraf-Richartz-Jahrbuch, 26: 187–218.

Wiener, N. (1948). Cybernetics. John Wiley and Sons, New York.

Willden, R. and Mabey, D.R. (1961). Giant desiccation fissures on the Black Rock and Smoke Creek deserts, Nevada. Science, 133: 1359–1360.

Witten, T.A., Jr. and Sander, L.M. (1981). Diffusion-Limited Aggregation, a kinetic critical phenomenon. Phys. Rev. Lett., 47: 1400–1403.

Wolfram, S. (1983). Statistical mechanics of cellular automata. Rev. Mod. Phys., 55: 601–644.

Wu, R.S. and Aki, K. (1985). The fractal nature of the inhomogeneities in the lithosphere evidenced from seismic wave scattering. PAGEOPH, 123: 805–818.

Yamaji, A. (1988). A fractal model for the distribution of stratigraphic hiatuses: A discussion. J. Geol., 96: 101–103.

Zähle, U. (1984). Random fractals generated by random cutoffs. Math. Nachr., 116: 27–52.

3 Korčak's law and fragmentation theory

'Tis all in peeces, all cohaerence gone. — JOHN DONNE: *An Anatomie of the World*

(In Julie's case)... the overall unity of her being had broken up into several 'partial assemblies' or 'partial systems' (quasi-autonomous 'complexes', 'inner objects'), each of which had its own little stereotyped 'personality' (molar splitting). In addition, any actual sequence of behaviour was fragmented in a much more minute manner (molecular splitting).

— R.D. LAING: *The Divided Self. An Existential Study in Sanity and Madness*

3.1 The size–frequency relation for islands, lakes and caves

The logical counterpart of the Richardson-Mandelbrot paradox concerning the length of Britain's coastline (Section 1.4.1; Richardson, 1961; Mandelbrot, 1967) is the question "How many islands does Britain have?". As Mandelbrot (1982) pointed out, the number of islands is both very large and very ill-determined, and tends to infinity with increasing resolution. Korčak (1940) analysed the size distribution of geographical objects (areas of islands and lakes, length of rivers, etc.) and found that the areas of islands follow a hyperbolic (or "power-law", or "Pareto") distribution:

$$Pr(A > a) = ka^{-b} \tag{3.1.1}$$

where k and b are constants. While Korčak claimed that $b = 1/2$ for islands, Fréchet (1941) regressed Korčak's data and obtained $b = 0.48$. A more careful examination of the data for the whole Earth (Mandelbrot, 1975a) resulted in $b = 0.65$, with considerable variation for the different regions (0.5 for Africa, 0.75 for Indonesia and North America).

Mandelbrot (1982) offered two explanations for the power-law (3.1.1).

Firstly, generalising the construction of fractal Von Koch "coastlines" (Section 1.4.1, Figure 1.27), he defined fractal "continents" and "archipelagos" as shown in the examples of Figure 3.1a, b. The generator consists (in these simple cases) of a connected broken line of N_c links called coastline generator, and a closed loop of N_i links called island generator. The fractal dimension of any individual island will be:

$$D_c = \frac{\log N_c}{\log (1/r)} \qquad (r \text{ is the scale factor}) \tag{3.1.2a}$$

while the joint coastline of all islands lumped together is a disconnected fractal

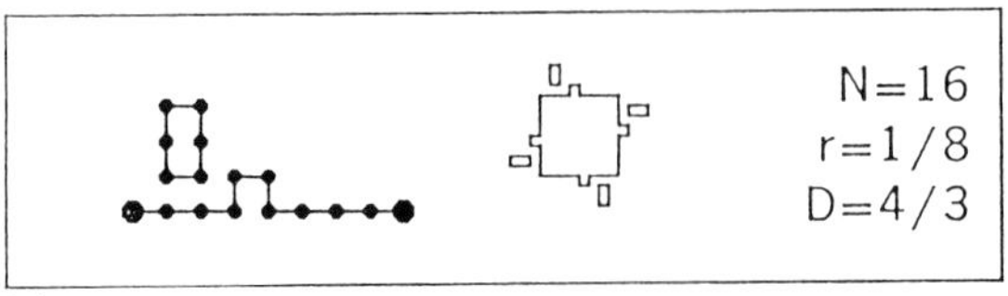

(a)

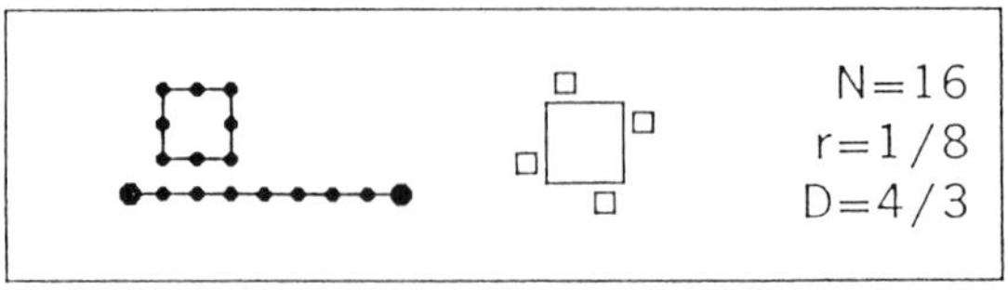

(b)

Fig. 3.1. a, b. Generation of fractal Von Koch continents and islands. Observe that in (b) the islands' coastlines are one-dimensional rectifiable curves. (From Mandelbrot, 1982.)

curve of dimension:

$$D = \frac{\log N}{\log(1/r)} \qquad (N = N_c + N_i) \tag{3.1.2b}$$

where:

$$1 \leq D \leq D_c \tag{3.1.2c}$$

It is easy to prove that Korčak's law (3.1.1) holds for this construction, with $b = D/2$. Indeed, the first step of the construction creates one island of linear size l and area $a = l^2$; the second step creates N islands of size $rl, \ldots$, the m-th step N^m islands of size $r^m l$. That is, the number of islands greater than a given size $x = r^m l$ will be:

$$Nr(X > x) \propto x^{-D} \tag{3.1.3a}$$

or, in terms of areas:

$$Nr(A > a) \propto a^{-D/2} \tag{3.1.3b}$$

where $D = \log N/\log(1/r)$; which is equivalent with Korčak's law. Observe that in Figure 3.1b the coastline of the individual islands are smooth straight lines; $D_c = 1$ by equation (3.1.2a). The dimension D (which happens to be the same, 4/3, for both Figures 3.1a and b) is not a measure of irregularity of the islands' shapes, but only of their *fragmentation*.

Another way of heuristically deriving Korčak's law (Mandelbrot, 1975a) is to model the Earth's surface with a random fBm (fractional Brownian motion) obeying for any two points $P' = (x',y')$; $P''(x'',y'')$:

$$\langle B_H(P)\rangle = 0$$

$$\langle\{B_H(P') - B_H(P'')\}^2\rangle \propto \overline{P'P''}^{2H} \qquad (0 < H < 1) \tag{3.1.4}$$

and to define islands and continents by the inequality:

$$B_H(x,y) \geq 0 \tag{3.1.5}$$

While a box-counting argument easily yields Korčak's law in the form:

$$Nr(A > a) \propto a^{-(2-H)/2} \tag{3.1.6}$$

the fractal dimension D_c of the coast of the Brown islands themselves is not known.

According to equation (3.1.6), the Earth-wide estimate $b = 0.65$ (Mandelbrot, 1975a) corresponds to $D \approx 1.3$ which would arise if the surface of the Earth were a 2.3-dimensional self-affine fractal, or a fBm of Hurst exponent $H = 2 - 1.3 = 0.7$ [Goodchild (1988) also observed, in another context, that fBm surfaces with $H = 0.7$ have the greatest similarity to real terrains]. The relation (3.1.6) between the fractal dimension $3 - H$ of a fBm surface and the exponent $b = (2 - H)/2$ describing the size distribution of the islands very likely remains true for any fractal surface of dimension D ($2 \leq D \leq 3$), in the form:

$$Nr(A > a) \propto a^{-(D-1)/2} \tag{3.1.7}$$

and can be used to estimate D from the size distribution of the islands. Alternately, one can analyse the *shape* of the islands, determine their fractal dimension D_c from area–perimeter analysis or a Richardson-Mandelbrot plot and estimate the dimension of the surface as $D' \approx D_c + 1$ [See Mandelbrot et al. (1984) and other works on the "slit island analysis" of fracture surfaces cited in Section 1.5.3]. Because of inequality (3.1.2c), the estimated D' will be generally less than the real one.

Islands that "never were" are well known from fiction and mythology. Figure 3.2a shows some of them visited by Gulliver [Laputa, Balnibarbi, Luggnag, Glubbdubdrid, and a very distorted Japan; a fractal computer vision of Laputa, due to Voss (1988), is displayed on the back cover of Peitgen and Saupe (1988)]. Islands are also encountered under the microscope. Figure 3.2b is a schematic representation of a thin section of reservoir rock where the entrapped discrete oil ganglia form islands during enhanced oil recovery by chemical flooding.

Sherwood (1986) used stochastic simulation on a 361×361 size grid in a study of the radial motion of water injected into a porous reservoir to flush out the more viscous oil, and found that the number of "islands" formed by s entrapped oil particles scales according to Korčak's law as:

$$n_s \propto s^{-b}$$

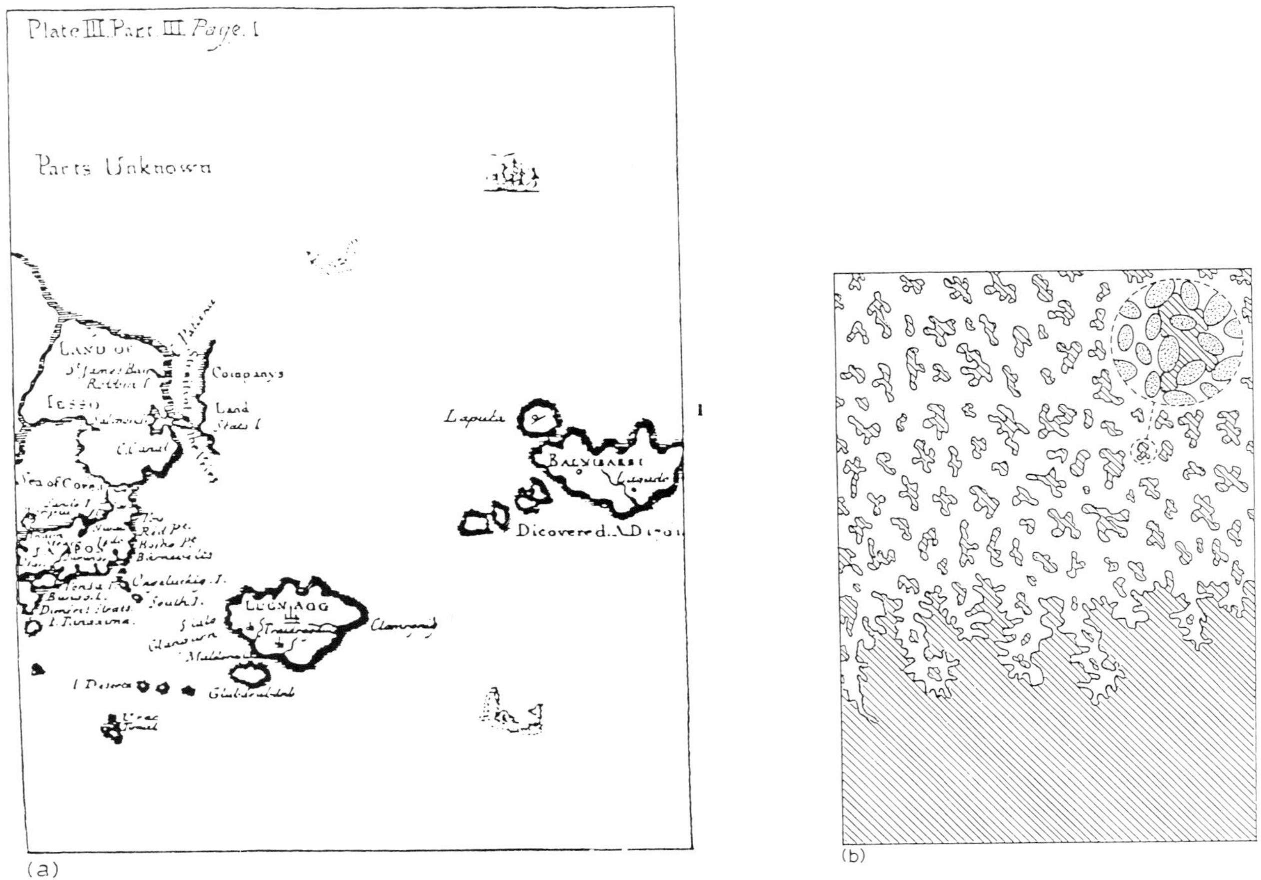

Fig. 3.2. a. Some of the islands described in "Gulliver's Travels". b. Schematic representation of the formation of oil ganglia during immiscible oil displacement by waterflooding. Hatched areas = oil; white areas = water; dotted areas = grains (in the inset). (From Payatakes, 1982.)

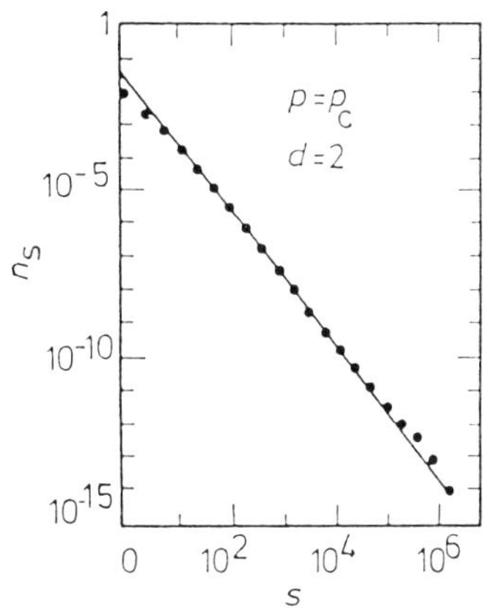

Fig. 3.3. Variation of n_s , the average number (per lattice site) of clusters with s occupied sites each, with size s at the two-dimensional site percolation threshold $p_c = 1/2$ (from Hoshen et al., 1979).

where $b = 2.07 \pm 0.07$. As, on the average, two thirds of the oil remains entrapped in the reservoirs after secondary flooding (Payatakes, 1982), the understanding of its spatial and size distribution is of the utmost practical importance.

It should be noted that Sherwood's (1986) simulation of radial fluid invasion is based on the same technique as Niemeyer et al.'s (1984) dielectric breakdown model, mentioned in Section 2.4.2. Also, the exponent $b = 2.07$ found by Sherwood (1986) is close to that figuring in the power-law cluster-size distribution in site percolation *at* the percolation threshold p_c (Figure 3.3):

$$n_s \propto s^{-\tau} \tag{3.1.8}$$

where n_s is the number of s-element clusters per lattice site and $\tau \approx 2.05$ for two-dimensional lattices, 2.135 for three-dimensional lattices. [See e.g. Hoshen et al. (1979), Stauffer (1986), where generalizations of equation (3.1.8) for *at and above* the percolation threshold are also discussed. Feder (1988, chapter 7.5) proves that the cluster-size distribution (3.1.8) is a consequence of the self-similarity of clustering at the percolation transition.]

There is empirical evidence that in any region the number of lakes $N(A) = Nr(a > A)$ with area greater than A is also given by the Korčak-Mandelbrot law:

$$F(A) \propto kA^{-D/2} \tag{3.1.9}$$

Kent and Wong (1982) checked the fulfilment of equation (3.1.9) for more than 2,500 Canadian Shield lakes in the Muskoka and Haliburton regions of Ontario. They rewrote equation (3.1.9) as:

$$\frac{D}{2} = \log_{10}\left[\frac{F(A) - F(10A)}{F(10A) - F(100A)}\right] \tag{3.1.10}$$

TABLE 3.1

Areal distribution of some Canadian Shield lakes

Lake group	No. of lakes in size class 10–99 ha	No. of lakes in size class 1–9 ha	Estimations of D from (3.1.10)
Kawartha, Muskoka	75	426	1.51
Algonquin	111	724	1.63
North Bay, Sudbury	150	1022	1.66

Source: Kent and Wong (1982).

and found that for three groups of lakes with areas between 1–9 ha and 10–99 ha the fractal dimensions were 1.51, 1.63, and 1.66, respectively (Table 3.1). They also determined D from shoreline length–resolution analysis (R-M plots) and lake area versus shoreline length analysis. The three methods (R-M plot, A-P cross-plot, Korčak's law) have resulted in different estimates of the fractal dimension, as a matter of fact, the Richardson-Mandelbrot plot revealed that the shoreline of the lakes is *bifractal* (see Figure 1.37 in Section 1.4.3). [Kent and Wong's (1982) paper contains the most charming "Acknowledgements" in the theory of fractals. The authors, then postgraduates at the University of Toronto, thank a friend for having lent them a digitizer, and another "for his loan of Mandelbrot's book".]

Goodchild (1988) tested Korčak's law for computer-generated lakes on fractal fBm surfaces. Realizations of fBm were generated for values of H between 0.3 and 0.7 on square arrays up to 256×256 size.

To understand the motivation for these studies, let us first observe (Mandelbrot, 1982) that the power-law distribution of lake areas does *not* follow from the similar distribution of islands. Lakes on a fBm surface cannot be simply defined by:

$$B_H(x,y) < 0 \tag{3.1.11}$$

as mirror images of islands. The areas satisfying (3.1.11) are not always lakes: some of them contain water only to a level below that of the ocean (Dead Sea, Caspian Sea) or are dry (Death Valley, Quattara Depression). Also, there are lakes whose bottom is above sea level.

In Goodchild's (1988) simulation, first the pits of a fBm surface were flooded to form lakes, then these initial lakes were allowed to flood and expand through outlets. (One result for a 50×50 array of the $H = 0.6$ simulation was illustrated in Figure 1.55 and discussed in Section 1.5.3.)

Goodchild found that the lake areas on fBm surfaces fit the Korčak model well. On the other hand, estimates of the exponent b were inconsistent with the theoretical expectation that $b = D/2$, and $D = 2 - H$ (Table 3.2).

As Mandelbrot (1982) noted, the "distribution of lake areas remains mysteri-

TABLE 3.2

The Korčak law fitted to lake-area distributions on computer-generated fBm surfaces

H	Sample size	b	$D_{est} = 2b$	$D_{theor} = 1 - H$
0.3	481	1.02	2.04	1.7
0.4	485	1.18	2.36	1.6
0.5	358	1.11	2.22	1.5
0.7	35	1.37	2.74	1.3
0.7	120	1.09	2.18	1.3
0.7	109	1.31	2.62	1.3

Source: Goodchild (1988), for a 100 × 100 realization.

ous. Perhaps its being hyperbolic is merely due to the robustness of the hyperbolic distribution under diverse forms of torture". As shown in Mandelbrot (1963), the product of a hyperbolic random variable and a largely arbitrary random multiplier is itself hyperbolic. If, as in Goodchild's (1988) model, lake areas had been hyperbolically distributed in their "original" state, that is on the primeval fBm surface of the Earth hypothesized by Mandelbrot (1975a), then they have very likely retained this distribution through geological time during the subsequent geomorphological "tortures".

In a series of important studies, Curl (1960, 1966, 1986) has established that the distribution of lengths of all caves in a region also follows Korčak's rule:

$$N(l) = N(l_0) \left\{ \frac{l}{l_0} \right\}^{-b} \tag{3.1.12}$$

where l_0 is some reference length, $N(l)$ is the number of caves longer than l. Curl's results for ten geographical regions are shown in Figure 3.4. Except for the group of caves marked as "1883" and "Clare" the exponent b lies between 1.2 and 1.6.

There are two technical difficulties in verifying the law (3.1.12):

(1) In the light of Richardson's paradox the "length" of a cave is rather meaningless without specifying the size of the "explorer" or the measuring rod (Curl, 1986; Laverty, 1987) — as a matter of fact $N(l)$, $N(l_0)$ can also change with changing resolution as caves can be connected, fragmented, created or lost.

(2) Korčak's law (3.1.12) can only be expected to hold if we include "unobservable" caves, that is caves without entrances.

As for the first difficulty, Curl (1986) defined the *standard explorer* as a sphere, the diameter of that sphere, μ, is called the *modulus* of the cave. The *cave space* is the envelope of volume that can be explored by a spherical explorer of modulus μ (Figure 3.5). The cave length, $L(\mu)$, with respect to a given modulus, is defined as the "minimum traversable horizontal plan length" including all passages and separate levels (Curl, 1966).

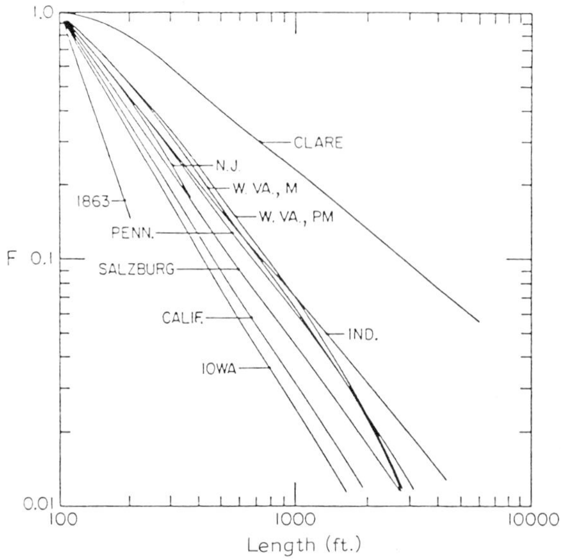

Fig. 3.4. Smoothed cumulative length distribution of all caves longer than 100 ft in several regions: California, U.S.A.; Indiana, U.S.A.; Iowa, U.S.A.; New Jersey, U.S.A.; Pennsylvania, U.S.A.; West Virginia, U.S.A., Mississippian; West Virginia, U.S.A., Pre- Mississippian; "Group 1863", SW of Vienna, Austria; Salzburg, Austria; County Clare, Ireland (from Curl, 1986).

The distribution (3.1.12) is conditional on the modulus, and should read as:

$$N(l\mu) = N(l_0\mu)\left(\frac{l}{l_0}\right)^{-b} \tag{3.1.13}$$

where possibly b also depends on μ. For example, the distribution for the cave lengths in Pennsylvania ($l_0 = 15.24$ m; $\mu = 0.5$ m) is given by:

$$N(0.5l) = 2526\left(\frac{l}{15.24}\right)^{-1.4} \qquad (l > 0.5\ \text{m}) \tag{3.1.14}$$

(Curl, 1986). For the caves of Figure 3.4 the explorer is human and the modulus is the size of a human.

In order to estimate the number of entranceless caves, Curl (1966) assumed a stochastic "immigration–emigration" process of cave development: if l is the length of the cave, the rate at which the cave gains an entrance is proportional to its length, rl; the rate at which the cave is losing an entrance is proportional to the number of entrances the cave has at the moment, sn. When stochastic equilibrium has been reached, the joint probability density distribution of caves with n entrances and of length l will become:

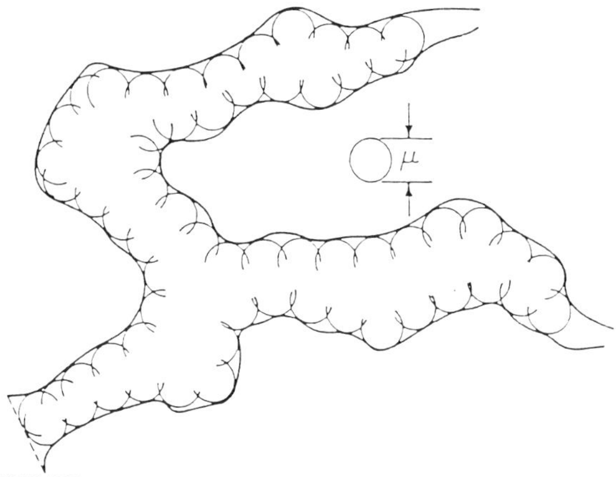

Fig. 3.5. Cave space defined by a specified modulus μ (after Curl, 1986).

$$p(n,l) = \frac{(\lambda l)^n e^{-\lambda l} p(l)}{n!} \qquad (n = 0, 1, 2, \ldots) \tag{3.1.15}$$

where $\lambda = r/s$ is the *karst constant* (varying from 0.0001 to 0.01/ft) and $p(l)$ is the distribution of cave lengths. The distribution (3.1.15) is Poissonian for any given l, and the number $p(0, l)$ of the (usually unobservable) caves without entrances can be estimated by extrapolation.

Curl (1966) analysed 820 caves in ten regions using this technique; his results (Figure 3.4) refer to *all* caves in a given region, including those without entrance.

Laverty (1987) applied resolution–length analysis over the range of resolutions 1–100 m to the line skeleton of cave passages, for two caves in Sarawak, Borneo, and two in Spain, in the alpine karst of the Picos de Europa (Figure 3.6). His range of results ($D = 1$ to 1.5) gives rise to interesting speculations, because it includes the fractal dimensions of Brownian motion (3/2) and of the self-avoiding random walk (4/3 in 2-D; 5/3 in 3-D).

Curl (1986) offers three possible causes for the observed power-law size distribution of caves:

(a) It might be due to fractal fragmentation (see Section 3.2).

(b) It can be a consequence of a growth process (Curl, 1960), where caves grow at a rate proportional to their size, the proportionality constant being arbitrarily but independently distributed. If the size distribution *ever had a hyperbolic form, it will always have that form.*

(c) A third interpretation of the hyperbolic distribution arises by observing that all moment integrals are undefined for $0 < l$. That is, the distribution *has no characteristic length scale.*

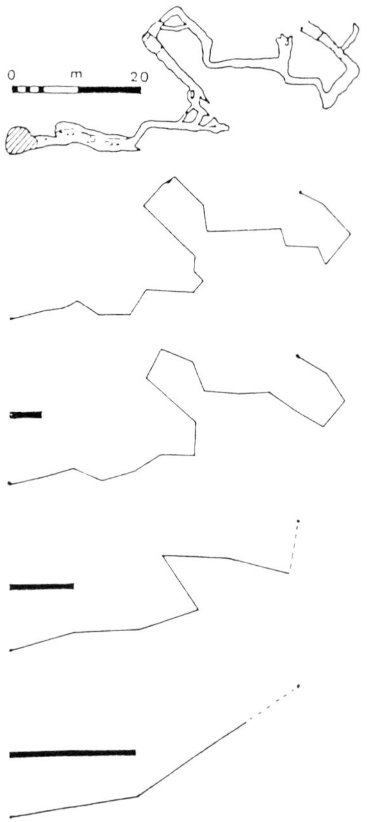

Fig. 3.6. Length versus resolution analysis for the skeleton line of the passage of a cave in Penyok Mts., Borneo (from Laverty, 1987).

3.2 Fragmentation: from broken sea ice to the distribution of galaxies

> ...it is utterly hopeless to expect a theoretical basis for distribution functions of random variables such as strength properties of materials, or of machine parts or particle sizes, the 'particles' being fly ash, Cyrtoideae, or even adult males, born in the British Isles.
>
> — W. WEIBULL (1951)

3.2.1 *The fractal theory of fragmentation*

Fragmentation can be observed in all materials and on all scales: rocks fragmented by joints and weathering (Allègre et al., 1982), ejecta produced by impacts (Fujiwara et al., 1977), drift sea ice (Rothrock and Thorndike, 1984; Matsushita, 1985), asteroids and meteorites (Hartmann, 1969) and even the mass distribution of galaxies (Brown et al., 1983; Vicsek and Szalay, 1987) are well-known examples (Figure 3.7).

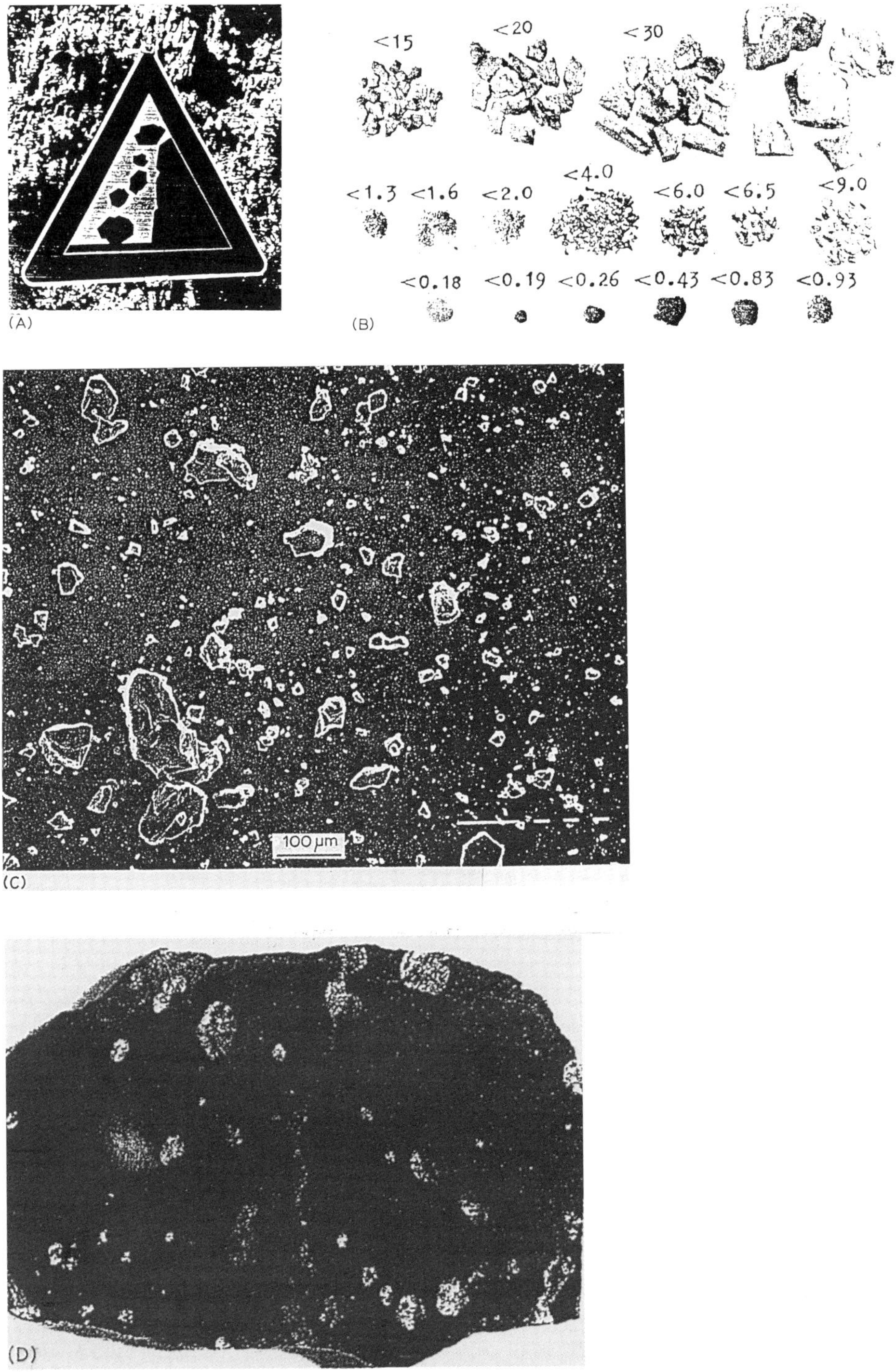

Fig. 3.7 Fragmentation at all scales. A. German traffic sign (from Bevlin, 1984). B. Destruction of a basaltic body by high velocity impact (from Fujiwara et al., 1977). C. Scanning electron micrograph of shock fragmented quartz (from Grady and Kipp, 1987). D. Felsic clusters in the Italian basaltic magma (from Lucido, 1985).

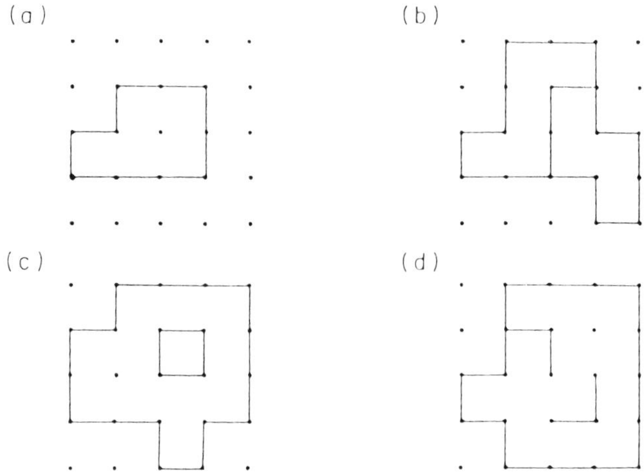

(E)

(F)

Fig. 3.7 (continued). E. "Fragmentation" in two-dimensional bond percolation: (a) one fragment, (b) two fragments, (c) one fragment inside another, (d) internal "cracks" in a fragment (from Englman et al., 1984). F. Computer model of the galaxy distribution (from Vicsek and Szalay, 1987).

The cumulative frequency distribution of fragment size is generally approximated by one of the following functions:

– by an exponential distribution (Brown et al., 1983):

$$N(r) \propto \exp\left[-\left(\frac{r}{r_0}\right)^{\nu}\right] \tag{3.2.1.1}$$

where $N(r)$ is number of fragments greater than r, r_0 and ν are constants;

– by a logarithmic normal distribution (Epstein, 1947; Johnson and Kotz, 1970; Shimiziu and Crow, 1988; Rendu, 1988; Crow, 1988):

$$N(r) \propto \int_r^{\infty} \frac{1}{\sigma(x - r_1)} \exp\left[-\frac{1}{2\sigma^2}\left\{\log\frac{x - r_1}{b}\right\}^2\right] \mathrm{d}x \tag{3.2.1.2}$$

(σ, r_1 and b are constants); or

– by the power-law (or "Pareto", or "hyperbolic") distribution (Pareto, 1897; Mandelbrot, 1963, 1982; Bader, 1970; Johnson and Kotz, 1970):

$$N(r) \propto \left(\frac{r}{r_0}\right)^{-D} \tag{3.2.1.3}$$

In equations (3.2.1.1)–(3.2.1.3) the "size" of the fragments is not uniquely defined except perhaps for the most simple geometrical objects such as a sphere or a cube. For irregular particles the size commonly depends on the method of measurement. Allen (1968) lists eight differing definitions for the size of sedimentary particles. Rothrock and Thorndike (1984) in their study of sea ice floe distribution use several measures of size (area, perimeter, diameter) and prove that these quantities are highly correlated and lead to the same type of probability distribution. In experimental studies of small particles the most commonly used measure of size is the "sieve diameter": for particles passing through the sieve with opening diameter d their size is regarded as less than d (see, e.g. Epstein, 1947; Fujiwara et al., 1977). In certain cases equations (3.2.1.1)–(3.2.1.3) are expressed in terms of total weight or mass of particles larger than a given mass m. For example, Fujiwara et al. (1977) describe the destruction of basaltic bodies by high velocity impacts by the distributions:

$$M(<s) \propto s^k \tag{3.2.1.4}$$

or:

$$N(>m) \propto m^{-k} \tag{3.2.1.5}$$

where $M(<s)$ is the total mass of fragments of size less than s and $N(>m)$ is the number of fragments with mass larger than m.

An alternative representation of the size–frequency relation is the Weibull distribution (Weibull, 1951; Johnson and Kotz, 1970):

$$\frac{M(r)}{M_{\mathrm{T}}} = 1 - \exp\left[\left(\frac{-r}{\sigma}\right)^{\alpha}\right] \tag{3.2.1.6}$$

where $M(r)$ is the cumulative mass of fragments with a radius less than r, M_{T} is the total mass, and σ is a characteristic size related to the mean size. For $r/\sigma \ll 1$, equation (3.2.1.6) approximately becomes:

$$\frac{M(r)}{M_T} \propto \left(\frac{r}{\sigma}\right)^{\alpha} \tag{3.2.1.7}$$

which is equivalent with a Korčak-type "fractal" relation:

$$N(r) \propto r^{-D} \qquad \text{if we put} \quad D = 3 - \alpha \tag{3.2.1.8}$$

[for the derivation of equation (3.2.1.8) see e.g. Turcotte (1986)].

Table 3.3 (after Hartmann 1969; Turcotte 1986, 1989) lists more than twenty objects whose fragment-size distribution behaves according to the Korčak law. I added

TABLE 3.3

Fractal size distribution of various fragmented objects

Object	D [from $N(>r) \propto r^{-D}$]	Reference
Projectile fragmentation of gabbro,		
with steel	1.44	Lange et al. (1984)
with lead	1.71	Lange et al. (1984)
Projectile fragmentation of basalt	2.56	Fujiwara et al. (1977)
Artificially crushed quartz	1.89	Hartmann (1969)
Renormalization group model of rock fragmentation	2.84	Allègre et al. (1982)
Renormalization group model of rock fragmentation	1.97	Turcotte (1986)
Disaggregated gneiss	2.13	Hartmann (1969)
Disaggregated granite	2.22	Hartmann (1969)
Granite fragmented in		
0.2 kt chemical explosion	2.42	Schoutens (1979)
61 kt nuclear explosion	2.50	Schoutens (1979)
Asteroids (theoretical)	2.48	Hellyer (1971)
Asteroids (theoretical)	2.51	Dohnanyi (1969)
Asteroids (measured)	3.05	Donnison and Sugden (1984)
Asteroids (simulated)	2.5	Capaccioni et al. (1986)
Asteroids (simulated),		
30% fragmented	2.76	Bianchi et al. (1984)
90% fragmented	3.06	Bianchi et al. (1984)
Galaxy distribution (comp. simulation)	1.9	Vicsek and Szalay (1987)
Interstallar grains	2.50	Mathis (1979)
Meteorites	3.00	Hawkins (1960)
Meteorites	1.86	McCrosky (1968)
Broken coal	2.50	Bennett (1936)
Sandy clays	2.61	Hartmann (1969)
Terrace sands and gravels	2.82	Hartmann (1969)
Ash and pumice	3.54	Hartmann (1969)
Glacial till	2.88	Hartmann (1969)
Broken sea ice	2.16	Matsushita (1985)
Broken sea ice	1.7–2.5	Rothrock and Thorndike (1984)
Natural fault gouge	2.60 ± 0.11	Sammis (1988)

Sources: Hartmann (1969); Turcotte (1986, 1989), modified.

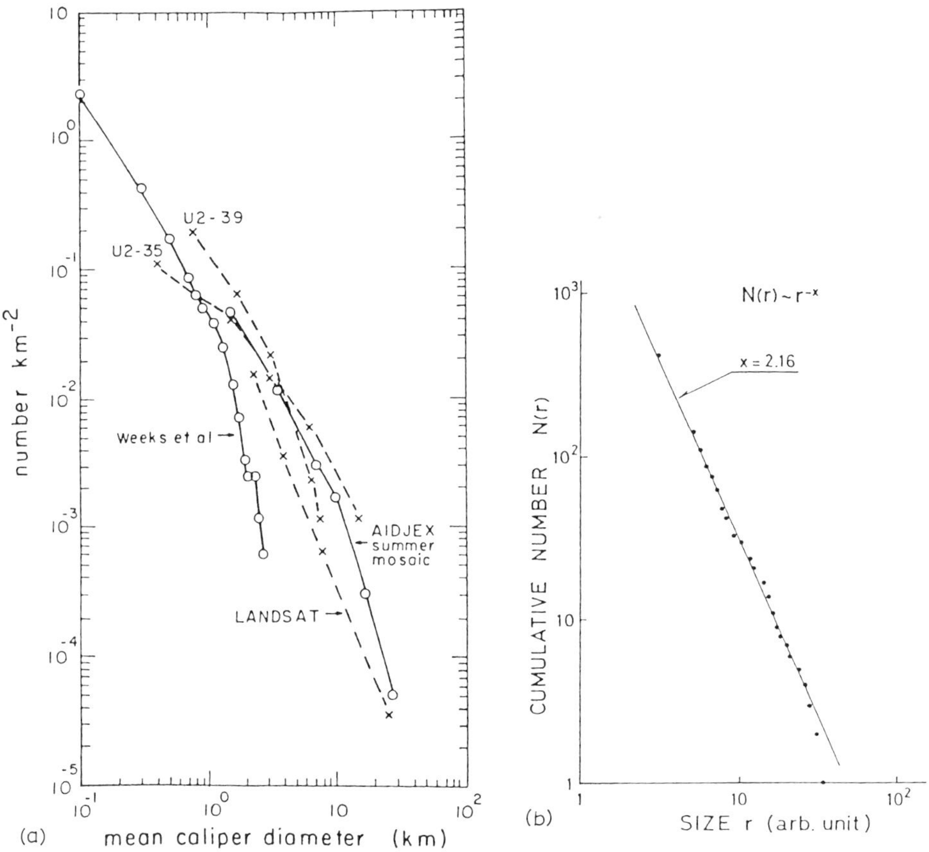

Fig. 3.8. Cumulative size distribution of the arctic drift ice. a. Size distribution of the mean caliper diameter (from Rothrock and Thorndike, 1984). b. Size distribution, Okhotsk Sea, size being defined as the mean of the length of the longest axis of a piece of drift ice and the length measured perpendicular to that axis (from Matsushita, 1985).

some recent data on the simulation of asteroid impacts (Capaccioni et al., 1986; Bianchi et al., 1984); on the fragmentation of sea ice (Rothrock and Thorndike, 1984; Matsushita, 1985; see Figure 3.8); the size distribution of natural fault gouge (Sammis, 1988) and the size distribution of the galaxies according to the stochastic cellular-automaton model of Vicsek and Szalay (1987).

Observe that the fractal dimensions in Table 3.3 all fall between 1.4 and 3.6. (A closer study of the original papers shows that in some cases the approximation of the size distributions by a simple power law is questionable, as in the situation shown in Figure 3.8a, or the distribution is *bifractal* : one obtains different fractal dimensions for the small- and large particle sizes, respectively, see Figure 3.9.)

It has recently been recognised that the power-law size distribution observed in the fragmentation of earth materials is a consequence of the scale invariance of

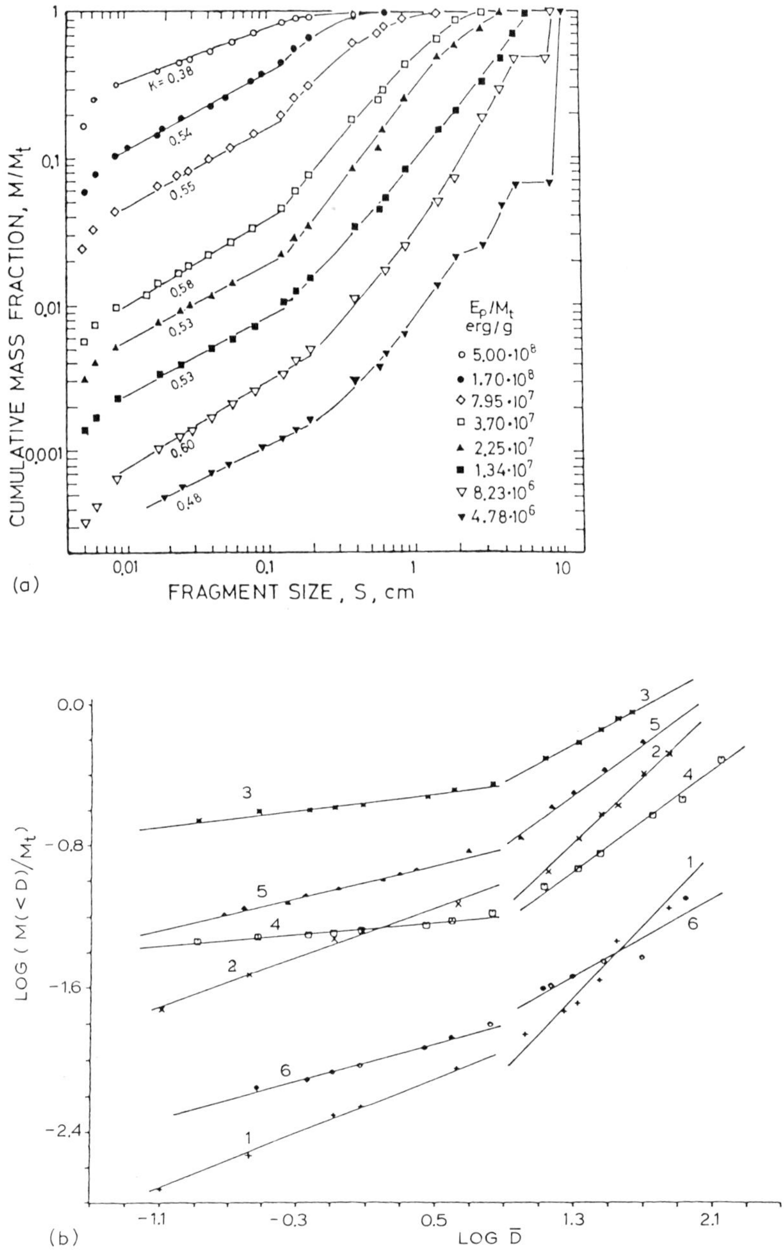

Fig. 3.9. Bifractal size distributions: a. For basaltic targets, destroyed by high-velocity impacts. E_p/M_t is imparted energy per unit mass of target (from Fujiwara et al., 1977). b. For six fragmentation experiments carried out with concrete and basalt targets. $M(<\overline{D})$ is cumulative mass of the pieces with mean diameter $<\overline{D}$; M_t is target mass (from Capaccioni et al., 1986).

Fig. 3.10. Fracture cascade in two dimensions. The hatched squares are arranged regularly for convenience. (From Matsushita, 1985.)

the fragmentation mechanism, that is the pre-existing zones or planes of weakness where breakage occurs, can be found on all scales (Matsushita, 1985; Turcotte, 1986). In these fractal fragmentation theories it is assumed that the flaws have a hierarchical structure where a fracture at the macroscopic scale is a consequence of the accumulation of fractures at lesser scales. The formalization of this principle has lead to the powerful *renormalization group methods* of predicting rock failure (Section 3.2.2; Allègre et al. 1982; Madden, 1983; Turcotte et al., 1985; Turcotte, 1986; Smalley et al. 1985; Solla, 1986a, b).

It is easy to construct a simple fractal fragmentation process which is analogous to the Cantor-Sierpinski-Menger constructions and leads to the "Korčak-type" size distribution $N(>r) \propto r^{-D}$ (Matsushita 1985; Turcotte 1986, 1989).

In the simplest case in two-dimensional space (Figure 3.10) we take a square with a side of unit length. At the first step we divide the square into four equal subsquares of side 1/2, and choose randomly one of them to hatch. At the second step we take the remaining three (non-hatched) subsquares, divide each into four equal sub-subsquares of side $(1/2)^2$, and choose randomly inside each of the non-hatched subsquares one sub-subsquares to hatch, etc. By repeating the procedure *ad infinitum*, we obtain a collection of an infinite number of hatched squares of various sizes (in Figure 3.10 they are arranged in a regular manner for sake of simplicity).

As easily seen, at the n-th step there are 3^{n-1} newly hatched squares of side 2^n. For $n \gg 1$ the cumulative number $N(r)$ of hatched squares greater than $r = 2^n$ is:

$$N(r) = 1 + 3 + 3^2 + \cdots + 3^{n-1} \approx 3^n = r^{-D} \qquad (3.2.1.9a)$$

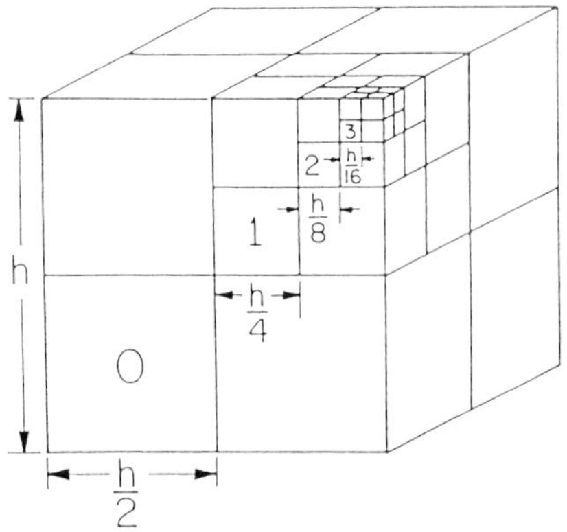

Fig. 3.11. Idealized model for fractal fragmentation in three dimensions (from Turcotte, 1989).

with:

$$D = \frac{\log 3}{\log 2} = 1.585 \tag{3.2.1.9b}$$

Matsushita (1985) used the planar fragmentation model to explain the power-law size distribution of the broken sea-ice (Figure 3.8b).

The model can be easily generalised for the fragmentation of solid bodies. We take a three-dimensional cubic cell with a side of length h.

At the first step we divide it into b^3 equal-sized 1st-order cells of side h/b, and choose randomly i 1st-order cells to hatch, where b and i are arbitrary integers satisfying $1 < b$, $1 < i < b^3$ (in Figure 3.11, $b = 2$). At the second step we take the remaining $b^3 - i$ first-order cells, divide each of them to b^3 2nd-order cells and choose randomly i out of each b^3 2nd-order cells to hatch. We shall now have $i(b^3 - i)$ newly hatched 2nd-order cells. At the n-th step ($n \gg 1$) the cumulative number of hatched cells larger than $r = 1/b^n$ is:

$$\begin{aligned} N(r) &= i\{1 + (b^3 - i) + (b^3 - i)^2 + \cdots + (b^3 - i)^{n-1}\} \\ &\approx (b^3 - i)^n = r^{-D} \end{aligned} \tag{3.2.1.10a}$$

where:

$$D = \frac{\log (b^3 - i)}{\log b} \tag{3.2.1.10b}$$

For the fractal fragmentation of d-dimensional hypercubes we can similarly prove that:

$$N(r) \propto r^{-D} \quad \text{with } D = \frac{\log (b^d - i)}{\log b} \quad \text{if } 1 < b; \quad 1 < i < b^d \tag{3.2.1.11}$$

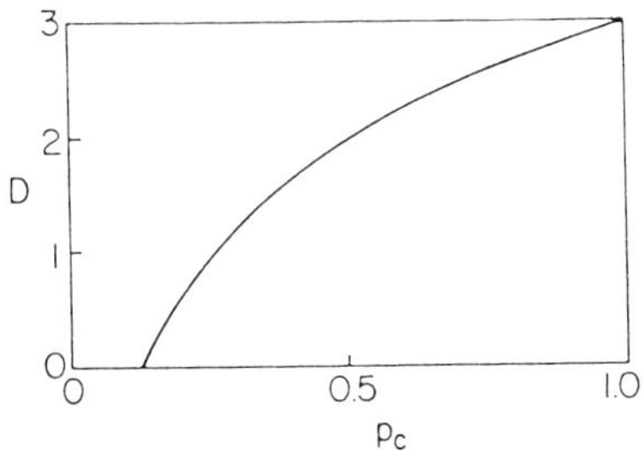

Fig. 3.12. Dependence of the fractal dimension D on the probability of fragmentation p_c (from Turcotte, 1986).

The simple planar case described above [equations (3.2.1.9a) and (3.2.1.9b)], corresponds to $d = 2, b = 2, i = 1$; equation (3.2.1.11) correctly gives the dimensions of the Cantor bar ($d = 1$), the Sierpinski carpet ($d = 2$), and the Menger sponge ($d = 3$).

In the above examples the self-similarity of distributions (3.2.1.9)–(3.2.1.11) has followed from the fact that at any order n of the construction always the same fraction $(b^d - i)/b^d$ of the n-th order cells of size h/b^n will be further fragmented, independently of n. Denoting this fraction by p_c, that is letting:

$$p_c = \frac{b^d - i}{b^d} \tag{3.2.1.12}$$

the fractal dimension D in equation (3.2.1.11) becomes:

$$D = \frac{\log(b^d p_c)}{\log b} \tag{3.2.1.13}$$

Because of the inequalities (3.2.1.11) we have:

$$\frac{1}{b^d} < p_c < 1 \qquad (0 < D < d) \tag{3.2.1.14}$$

In particular, for three-dimensional fragmentation:

$$D = \frac{\log 8p_c}{\log 2} \qquad (\frac{1}{8} < p_c < 1; \quad 0 < D < 3) \tag{3.2.1.15}$$

(Turcotte, 1986), the dependence of D on p_c is shown in Figure 3.12.

It should be noted that p_c in the above derivation has been defined as a fraction of cells fragmented in each step of the process rather than a probability describing the event that a given n-th order cell will fragment into $n - 1$ smaller pieces with probability p_c and remain intact ("hatched") with probability $1 - p_c$. If this were the case, there would be situations where all n-th order cells further fragment or all become hatched with non-zero probability, which obviously cannot occur in the fractal fragmentation model. According to an important theorem of Epstein (1947)

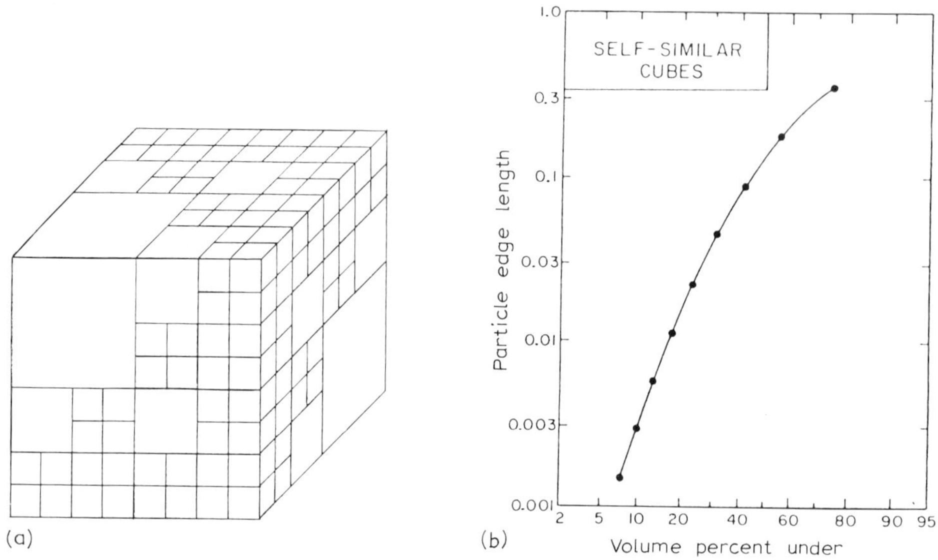

Fig. 3.13. Mimicking the lognormal distribution by a power law. a. Self-similar fragmentation cascade, $D = \log 6/\log 2 = 2.58$. b. The power-law size distribution corresponding to (a), plotted on lognormal probability grid. (From Sammis et al., 1986.)

this probabilistic variant of the fragmentation process would lead to a lognormal particle-size distribution instead of the power-law one. However, the difference between the two models is subtle, and in many practical situations the lognormal and the power-law size distributions cannot be distinguished [see Figure 3.13, from Sammis et al. (1986); Montroll and Shlesinger (1983) contains a general discussion on how lognormal distributions can mimic power-law ones].

The fractal size distribution found in many examples of fragmentation (see Table 3.3) and over a wide range of scales suggests that the fragmentation process is a scale-invariant catastrophic phenomenon with a well-defined *phase-transition* between the "sound" and the "broken" states of the rocks. (For example, if the stress on a rock is increased slowly, a critical stress will be reached at which sudden failure occurs; see Section 2.4.2.) Turcotte (1986) recognised that the quantity p_c, determining the fractal dimension D of the fragmentation through equations (3.2.1.12)–(3.2.1.13), can be considered as a "critical parameter" in the sense of the Theory of Phase Transitions, and — following earlier work of Allègre et al. (1982) — used a renormalization group (RNG) model for its determination.

3.2.2 *The renormalization group (RNG) model of rock fragmentation*

According to K.G.Wilson, who won the 1982 Nobel Prize in physics for the application of the RNG to critical phenomena of statistical mechanics and quantum physics, "the renormalization group is a method for dealing with some of the most difficult problems of physics" (Wilson and Kogut, 1974). The standard technical introduction is Pfeuty and Toulouse (1977), or the review papers Fisher (1974) and Wilson and Kogut (1974). There are a few popular expositions: Wilson (1979), Bruce and Wallace (1989), those who read Russian would enjoy Bogolyubov and Shirkov's "Renormgroup — it is simple" (1984), — the title refers to Pierre Bezoukhov's utterance "How simple and clear ... How come I didn't know it before?" An application of RNG to two-dimensional bond-percolation has been discussed in Section 1.3.2 (see also Young and Stinchcombe, 1975; Harris et al. 1975, Reynolds et al., 1977; Feder, 1988); for the renormalization of probability distributions and random processes see Shlesinger and Hughes (1981), Procaccia and Schuster (1983), West et al. (1986).

In geophysics, RNG techniques have been used by Madden (1983) to predict the onset of electric conductivity and the onset of failure in a rock with random distribution of microcracks; by Allègre et al. (1982) to study the coalescence of faults; by Turcotte (1986) for catastrophic rock fragmentation. Smalley et al. (1985), Turcotte et al. (1985), Solla (1986a, b) applied RNG to describe the failure of asperities under stress ("stick/slip" transition); their approach ("collapse of loaded fractal trees") is of general validity, "applicable to a variety of problems including cardiac arrest, fracture of bronchial systems, failure of electrical distribution systems and the instabilities resulting in earthquakes" (Turcotte et al., 1985).

As Turcotte (1989) noted, fractals inevitably arise as solutions to problems obtained by using the RNG approach (a formal treatment of the connection between RNG and fractals is Suzuki, 1983). Indeed, the RNG approach is based on the concepts of *hierarchy* and *scale invariance*, which are the main ideas of fractal theory.

In the RNG model in order to study the critical behaviour of a large system at phase transition, we first consider small subsystems and find their properties. Each subsystem is then treated as an element of a larger-scale subsystem. The properties of this larger-scale subsystem are determined — in terms of its elements — and the subsystem is used as element of an even larger-scale subsystem, and this *hierarchical* process is repeated again.

Suppose that each of the n-th order subsystems can be in one of the physical states $S_1, S_2, \ldots, S_K$ with the respective probabilities:

$$p_1^{(n)}, p_2^{(n)}, \ldots, p_K^{(n)} \qquad \left(p_1^{(n)} + p_2^{(n)} + \ldots + p_K^{(n)} = 1\right)$$

independently of each other. Similarly, any $(n+1)$-th order subsystem can belong to the states S_i with probability:

$$p_i^{(n+1)} \qquad (i = 1, 2, \ldots, K; \quad \sum_{i=1}^{K} p_i^{(n+1)} = 1)$$

The distributions $\{p_i^{(n+1)}\}$ and $\{p_i^{(n)}\}$ are related by the equations:

$$\begin{aligned} p_1^{(n+1)} &= F_1\left[p_1^{(n)}, p_2^{(n)}, \ldots, p_K^{(n)}\right] \\ p_2^{(n+1)} &= F_2\left[p_1^{(n)}, p_2^{(n)}, \ldots, p_K^{(n)}\right] \\ &\ldots\ldots\ldots\ldots \\ p_K^{(n+1)} &= F_K\left[p_1^{(n)}, p_2^{(n)}, \ldots, p_K^{(n)}\right] \end{aligned} \tag{3.2.2.1}$$

where the functions F_i depend on the physical meaning of state S_i and on the geometry of "coarse-graining" (that is how the n-th order subsystems are combined to form an $[n + 1]$-st order subsystem).

Because of the assumed *scale invariance*, at phase transition the probabilities do not change during coarse-graining, that is, the system will be in the states $\{S_i\}_{i=1}^K$ with the respective critical probabilities $\{p_{c,1}, p_{c,2}, \ldots, p_{c,K}\}$, where $\{p_{c,i}\}_{i=1}^K$ satisfy the non-linear system of equations:

$$\begin{aligned} p_{c,1} &= F_1(p_{c,1}, p_{c,2}, \ldots, p_{c,K}) \\ &\ldots\ldots\ldots\ldots \\ p_{c,K} &= F_K(p_{c,1}, p_{c,2}, \ldots, p_{c,K}) \end{aligned} \tag{3.2.2.2}$$

In the most simple cases there are only two possible states [for example permeable/impermeable or conductive/non-conductive in Madden (1983); fragile/sound in Allègre et al. (1982), Turcotte (1986)], with the respective probabilities $p^{(n)}$ and $q^{(n)}$, $p^{(n)} + q^{(n)} = 1$ in the n-th step of coarse-graining. The system of equations (3.2.2.1) reduces to the recursive relation:

$$p^{(n+1)} = F[p^{(n)}] \tag{3.2.2.3}$$

and the critical probability p_c satisfies the equation:

$$p_c = F[p_c] \tag{3.2.2.4}$$

The fragmentation theory of Allègre et al. (1982) and Turcotte (1986) is based on the laboratory observation (Section 2.4.2) that the failure of the rock is a consequence of fractures, which are preceded by the concentration of a swarm of microfissures which are themselves the result of an accumulation of microcracks.

The rock is divided to small elementary cubical cells ("order 1 cells", or "1-cells" for short) and each cell is considered to be either *fragile* or *sound* with

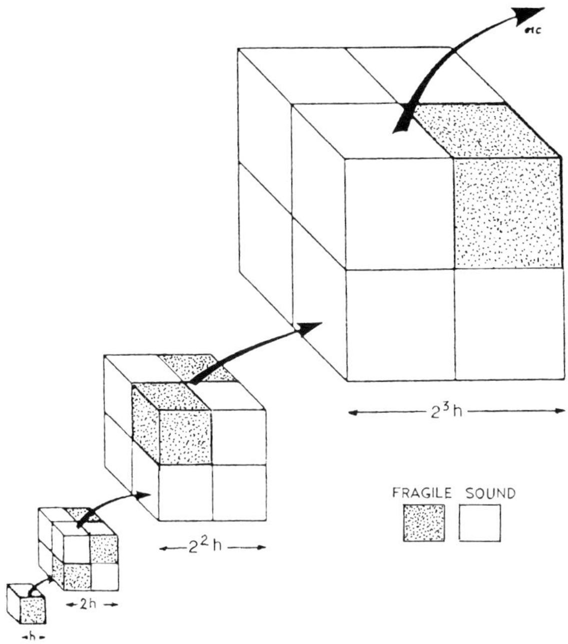

Fig. 3.14. Geometry of the RNG model (from Allègre et al., 1982).

the respective probabilities $p^{(1)}$ and $1 - p^{(1)}$. (Physically, the value of $p^{(1)}$ might depend on microcrack density, on the applied stress, or some other macroscopic parameter.) Next we unite 2^3 1-cells to form a 2-cell, $\ldots 2^3$ n-cells to form an $(n + 1)$-cell (Figure 3.14). The fragility of an $(n + 1)$-cell is defined in terms of the fragilities of its n-th order subcells. First, observe that from the $2^8 = 256$ possible combinations of n-cells there are only 22 topologically different, labelled 0, 1, 2a, $\ldots$, 8 in Figure 3.15a. In the figure solid dots indicate fragile elements, the first number below a configuration indicates the number of fragile elements, letters denote the topologically different configurations with the same number of fragile elements, and the number in parentheses gives the multiplicity of that configuration.

The probability of a configuration with f fragile elements and of multiplicity m is:

$$m[p^{(n)}]^f[1 - p^{(n)}]^{8-f} \tag{3.2.2.5}$$

Allègre et al. (1982) defines a cell "fragile" if it contains no continuous "pillar" linking any two opposite faces. Otherwise, the cell is "sound" (Figure 3.15b). The configurations 4f, 5c, 6b, 6c, 7 and 8 (underlined by a solid line in Figure 3.15a) are fragile according to this definition; that is, by summing up the corresponding

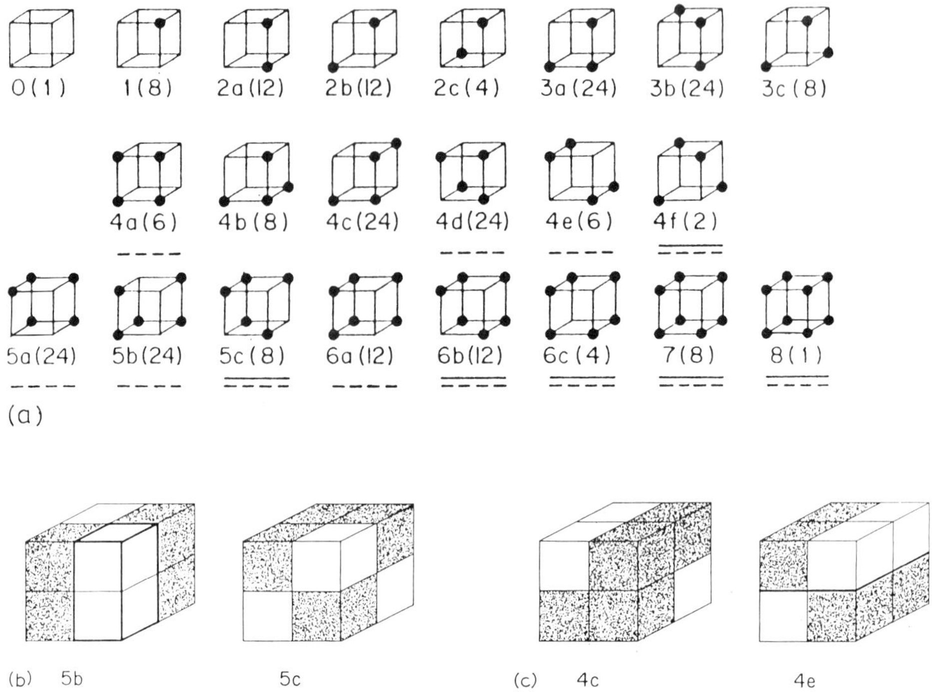

Fig. 3.15. Illustration of the RNG technique of Allègre et al. (1982) and Turcotte (1986). a. The 22 topologically different configurations. Fragile elements are indicated by the solid dots. The meaning of the letters is explained in the text. b. Illustration of the "pillar of strength" condition of Allègre et al. (1982). Fragile elements are shaded. Configuration *5b* is "sound" according to this condition, while *5c* is "fragile". Configurations which are fragile according to this convention, are underlined by solid line in (a). c. Illustration of Turcotte's "plane of weakness" condition. Fragile elements are shaded. Configuration *4c* is "sound" according to this definition, while *4e* is "fragile". Configurations which are fragile according to this convention are underlined by dashed line in (a). (From Allègre et al., 1982; Turcotte, 1986.)

probabilities (3.2.2.5) we obtain:

$$
\begin{aligned}
p^{(n+1)} &= \left[p^{(n)}\right]^8 + 8\left[p^{(n)}\right]^7\left(1-\left[p^{(n)}\right]\right) + 16\left[p^{(n)}\right]^6\left(1-\left[p^{(n)}\right]\right)^2 + \\
&\quad + 8\left[p^{(n)}\right]^5\left(1-\left[p^{(n)}\right]\right)^3 + 2\left[p^{(n)}\right]^4\left(1-\left[p^{(n)}\right]^4\right) \\
&= \left[p^{(n)}\right]^4\left(3\left[p^{(n)}\right]^4 - 8\left[p^{(n)}\right]^3 + 4\left[p^{(n)}\right]^2 + 2\right)
\end{aligned}
\tag{3.2.2.6}
$$

that is, by equation (3.2.2.4) the critical probability p_c satisfies the algebraic equation:

$$p = p^4(3p^4 - 8p^3 + 4p^2 + 2) \tag{3.2.2.7}$$

The non-trivial solution of equation (3.2.2.7) is $p_c = 0.896$ and, substituting this value to equation (3.2.1.15), one finds that the fractal dimension is $D = 2.84$ for this model of fragmentation.

An alternative approach (Turcotte, 1986) is based on a different definition of the fragility of cells. Consider an $(n + 1)$-th order cell, assume again that each of its eight n-cells can be fragile with probability $p^{(n)}$. The sides of a fragile n-cell are considered "planes of weakness", and the $(n + 1)$-cell is defined as fragile if the sides of its fragile elements form an internal plane through the cell (Figure 3.15c). The configurations indicated by dashed underlining in Figure 3.15a (that is 4a, 4d, 4e, 4f, 5a, 5b, 5c, 6a, 6b, 6c, 7, 8) are fragile, and a similar reasoning as above leads to the equation:

$$p = 3p^8 - 32p^7 + 88p^6 - 96p^5 + 38p^4 \qquad (3.2.2.8)$$

whose non-trivial root yields the critical probability $p_c = 0.49$ and the fractal dimension $D = \log 8p_c / \log 2 = 1.97$. Both fractal dimensions 2.84 and 1.97 lie within the range of experimental D values in Table 3.3. As the "plane of weakness" model corresponds to more fragile materials, Turcotte (1986) suggests that more fragile materials may be associated with a smaller fractal dimension.

3.2.3 *Maximum-entropy people and fractal people*

> But [Aquinas and Bonaventura]...were both right; if I may say so, it is a privilege of people who contradict each other in their cosmos to be both right.... It was in any case a time of wild disturbance, and, as is common in such times, those who were trying to put things right were most vigorously accused of putting things wrong. — G.K. CHESTERTON: *St. Thomas Aquinas*

In spite of the elegant fractal models of catastrophic fragmentation discussed in the previous sections, their practical usefulness is still controversial. According to a recent authoritative review on Dynamic Rock Fragmentation (Grady and Kipp, 1987), "the currently popular subject of fractals has also been suggested as an underlying basis for material fragmentation. To the authors' knowledge, no significant connection has been demonstrated, and it is suspected that the only connection to dynamic fracture may just be an unfortunate similarity in names".

Without trying to settle this controversy, we only mention a few difficulties which arise in connection with the power-law (Pareto) particle–size distribution and the fractal fragmentation models.

THE GENERAL APPLICABILITY OF THE POWER-LAW DISTRIBUTION

The power-law distribution or Pareto-distribution is named after an Italian-Swiss professor of economics, Vilfredo Pareto (1897), who described the distribution of income over a population as:

$$N = Ax^{-\alpha} \qquad (3.2.3.1)$$

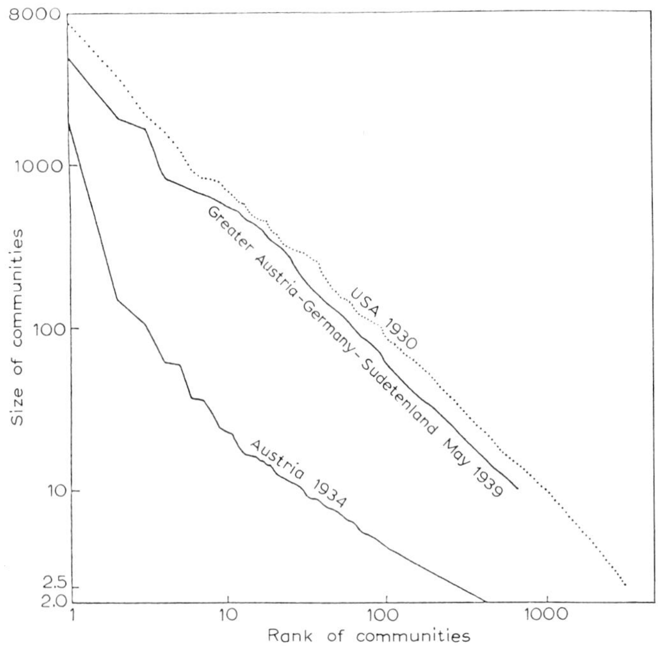

Fig. 3.16. Pareto plots of the communities ranked in decreasing order of population size for the "Greater Austria–Germany–Sudetenland" (May, 1939), compared with a similar plot for Austria before the Anschluss (1934), and for the U.S.A. (1930) (from Zipf, 1949).

where N is the number of persons having an income greater than or equal to x. Pareto felt that this law was inevitable and universal, regardless of taxation and of political and social conditions.

One of Pareto's followers, G.K. Zipf (1902–1950), for twenty years a Lecturer at Harvard, who considered himself a "statistical human ecologist", applied Pareto's law to all possible fields of human endeavours (Zipf, 1949), and he even tried to justify the "Anschluss" of Austria into Germany because it improved the fit to the Pareto law (Figure 3.16). "Refutations" of Pareto's law have been given over the past 100 years by many of the leading economists and statisticians, the most important for the present discussion is Macaulay's (1922) observation (quoted in Mandelbrot, 1963; Brock, 1971) that "the approximate linearity of the tail of a frequency distribution charted on a double logarithmic scale signifies relatively little, because it is such a common characteristic of frequency distributions of many and various types".

Indeed, there are many examples where Weibull distributions, or lognormal distributions mimic a power-law distribution over certain ranges (see e.g. Turcotte, 1986; Montroll and Shlesinger, 1983; or Fig. 3.13 from Sammis et al., 1986). The general theory of the *asymptotically Pareto distributions* satisfying:

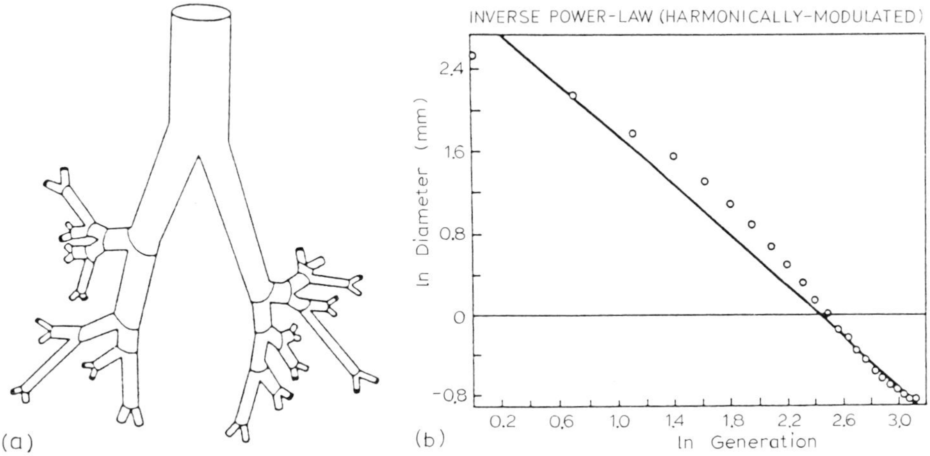

Fig. 3.17. Fractal analysis of the tracheobronchial tree. a. The first four generations for a human lung cast. b. Log–log plot of the mean bronchial diameter vs. generation number. Note the "waviness" of the data around the power-law regression line. (From Goldberger and West, 1987.)

$$P(x > r) \propto \left(\frac{r}{r_0}\right)^{-\alpha} \qquad \text{as } r \to \infty \tag{3.2.3.2}$$

is mainly due to Mandelbrot (1960), a useful summary is Johnson and Kotz (1970, vol. ii, chapter 19).

Brock (1971) also quotes Macaulay's (1922) remarks in his critical review on the size distribution of atmospheric aerosols. He also states an important rule of thumb of fractal studies: *the linearity on the double logarithmic plot should extend over at least two or three orders of magnitude of particle size.* [Brock (1971) explains the observed power-low size distribution of the atmospheric aerosols as being the result of a stochastic growth process — actually, his model is similar to Curl's (1960) "immigration–emigration" model of cave development, mentioned in Section 3.1.]

Figure 3.17 (from Goldberger and West, 1987) is an example from physiology to illustrate yet another difficulty with power laws. The average dimensions (diameter, length, and volume) of the airways in the mammalian bronchial tree have been found to follow power-laws in generation number, however the power-like decay is modulated by a harmonic variation (Figure 3.17b). According to West et al. (1986; see also Shlesinger and Hughes, 1981; Montroll and Shlesinger, 1983) the periodic modulation of the power-law is a general property of branching systems like that in Figure 3.17a, and can be derived by a careful analysis of the RNG relations between successive generations of the average variables.

Smith and Jordan (1987) analysed the size distribution of seamounts in the eastern and southern Pacific. The number of seamounts per unit area with summit

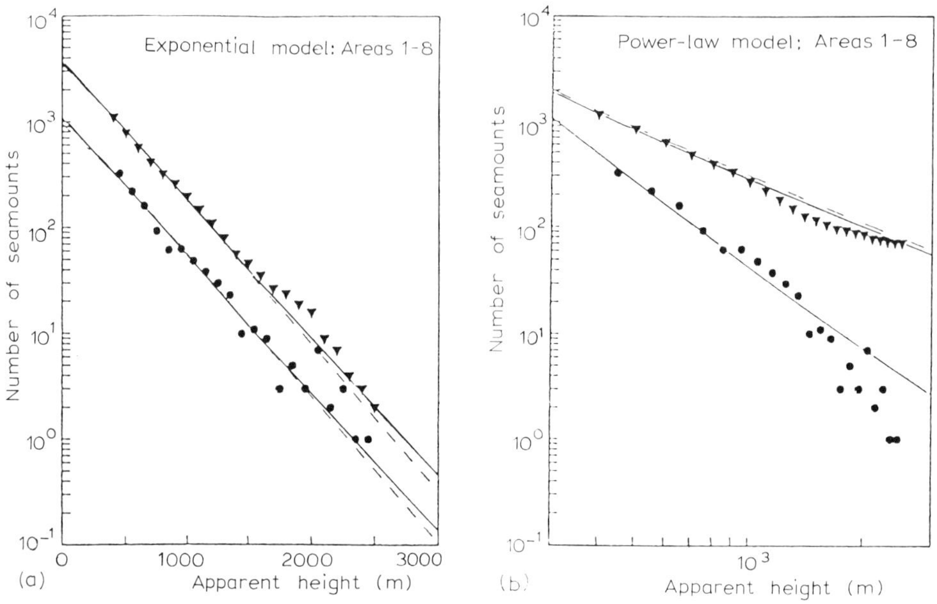

Fig. 3.18. Maximum-likelihood fits to the apparent-height data for a total of 1118 Pacific seamounts. Circles are actual seamount counts, triangles are the cumulative counts. (a) Exponential model, (b) power-law model. Compare the "waviness" of the plots with that in Figure 3.17b. (From Smith and Jordan, 1987.)

height $h \geq H$ was found to be more closely approximated by an exponential distribution:

$$n(H) \propto \exp(-\beta H)$$

than by a power-law:

$$n(h) \propto (H/H_1)^{-\gamma}$$

(Figure 3.18); actually, the power-law model was rejected by a statistical test. The close similarity of Figures 3.17b and 3.18b, however, suggests that a *periodically modulated power function* is also a valid possibility.

NON-CATASTROPHIC FRAGMENTATION

In Sections 3.2.1 and 3.2.2 the power-law for the fragment sizes was derived by assuming a self-similar distribution of the microcracks, cracks, fractures, and other inhomogeneities at the critical transition to rock failure (Allègre et al., 1982; Madden, 1983; Matsushita, 1985; Turcotte, 1986, 1989). As in these models a fragment is pictured as a piece of material completely surrounded by lines or planes of weakness, the geometry of fragmentation can be conveniently studied on two- or three-dimensional percolation models (Englman et al., 1984; Campi, 1989).

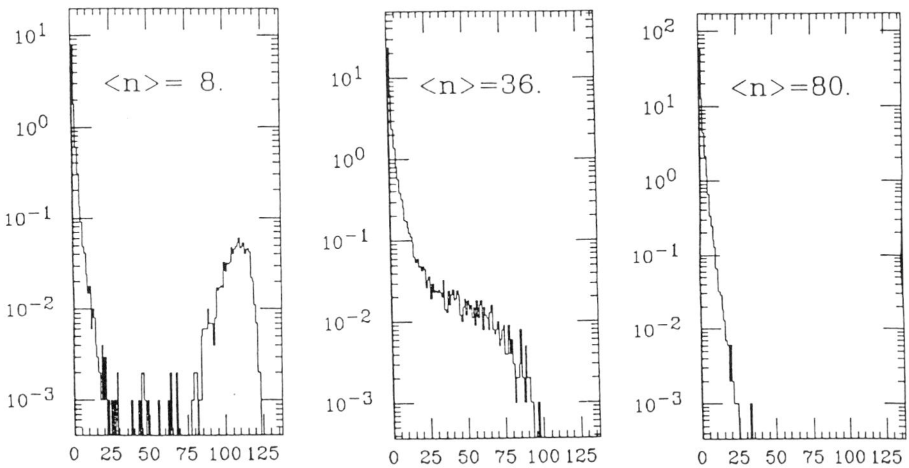

Fig. 3.19. Cluster size distribution for bond percolation for a cubic lattice with 5^3 bonds at different concentrations p of the "fragile" bonds ($p_c = 0.75$): $p = 0.55 < p_c$ (left); $p = 0.70 \approx p_c$ (middle); $p = 0.85 > p_c$ (right); $\langle n \rangle$ is mean cluster number (after Campi, 1988).

Campi (1989) studied bond percolation on a cubic lattice of 5^3 sites and analysed the size distribution $N(s,p)$ of the cluster "fragments" (Figure 3.19), where s is number of points in the cluster, p is the probability that a bond is inactive, that is "fragile", using the terminology of the previous sections. He found that while the size distribution was hyperbolic at the percolation threshold $p = p_c = 0.75$ (Figure 3.19, middle), the shape of the distributions were more complicated for $p < p_c$ or $p > p_c$.

When $p < p_c$, most points belonged to one connected cluster, and this large cluster coexisted with exponentially distributed small clusters so that the distribution was U-shaped (Figure 3.19, left). In the opposite, when $p_c < p \approx 1$, most points were isolated or belonged to exponentially distributed small clusters (Figure 3.19, right).

Englman et al. (1984) considered fragmentation in a two-dimensional percolation model (Figure 3.7E) and established that *above the percolation threshold* the cumulative fraction of fragments of area A is given by the Mott distribution:

$$f_{\text{cum}}^{\text{Mott}}(A) \propto \exp\left[-\left(\frac{2A}{\overline{A}}\right)^{1/2}\right] \tag{3.2.3.3}$$

[For linear fragment sizes, and fragment volumes, respectively, the Mott distribution is defined as:

$$f \propto \exp\left[-\left(\frac{L}{\overline{L}}\right)\right] \quad \text{and} \quad f \propto \exp\left[-\left(\frac{3V}{\overline{V}}\right)^{1/3}\right]$$

N.F. Mott found this distribution in his wartime research, in 1943, in which he established that the difference between the fragment distributions of the Allied and German shells was due to the different carbon contents of the metal from which they were made. More about the Mott distribution can be found in Grady and Kipp (1985).]

Both Englman et al. (1984) and Campi (1989) claim that effective fragmentation in rocks occurs at higher crack densities than the percolation threshold, that is, according to these authors the power-law ("fractal") fragmentation is an exception rather than a rule.

While Korčak's (1940) statement (cited in Curl, 1986), that "only two fundamental types of statistical distribution exist" (viz. the normal and the power-law) might very well be true in an asymptotic sense, in the earth sciences we encounter a plethora of further size distributions: exponential, Weibull, Mott, Rayleigh, lognormal distributions..., not to mention the more esoteric ones used for fine particle description. Figure 3.20 shows several examples:

(a) The size distribution of manganese nodules has been variably reported as *exponential* (Heath, 1979; Figure 3.20a) or *Rayleigh* (Magnuson et al., 1981; Figure 3.20b).

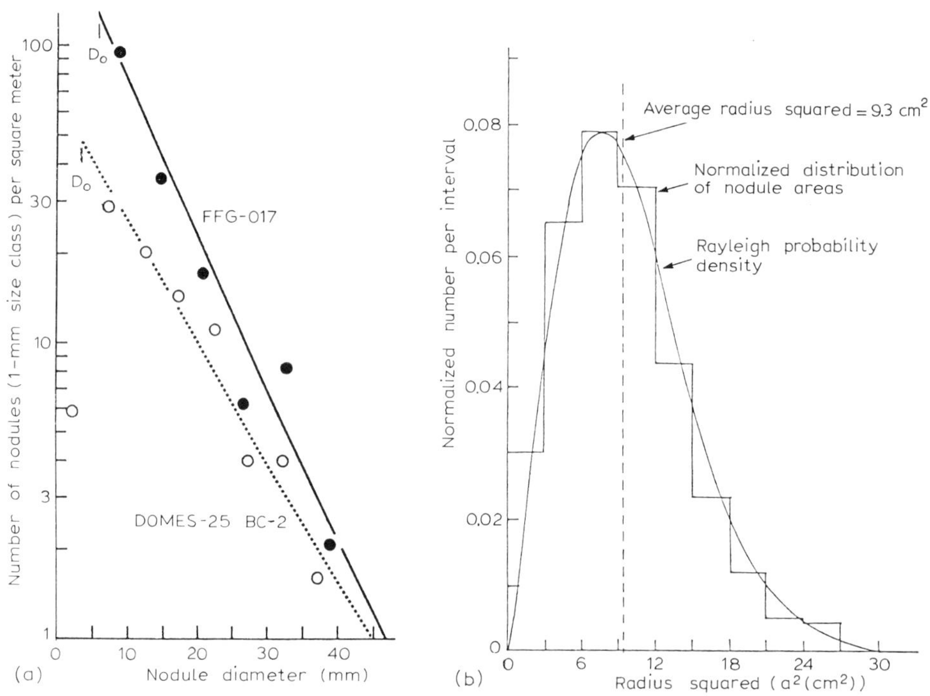

Fig. 3.20. Non-Paretian size distributions. a, b. Manganese nodules are sometimes found exponentially distributed (a, from Heath, 1979) or Rayleigh-distributed (b, from Magnuson et al., 1981). Both data sets refer to Pacific nodules.

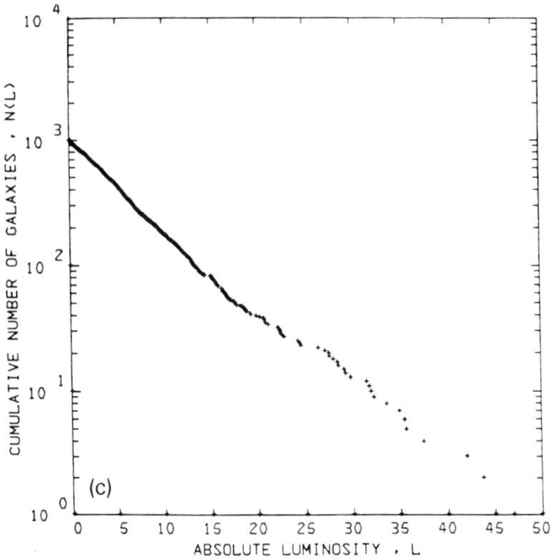

Fig. 3.20 (continued). c. Cumulative number of the nearby galaxies $N(L)$ having an absolute luminosity greater than L. Luminosity is expressed in units of solar luminosity $\times\ 10^{10}$. Galactic mass is assumed to be proportional to luminosity. (From Brown et al.,1983.)

(b) The cumulative number distribution function $N(L)$ for about a thousand nearby galaxies with absolute luminosity greater than L is exponential (Figure 3.20c). By assuming that the luminosity of a galaxy is proportional to its mass, Brown et al. (1983) concluded that the galaxies' size distribution is also *exponential.* Note that the *spatial distribution* of galaxies is known to be fractal with dimension $D \approx 1.2$ [Groth and Peebles (1977), Peebles (1980), Szalay and Schramm (1985), Lucchin (1986), Pietronero and Kupers (1986), Pietronero (1987), Vicsek and Szalay (1987), Balian and Schaeffer (1988). The most elegant model for the continuous hierarchy of galactical clustering is Mandelbrot (1975b). Lucido (1985) pursues the analogies between magma clustering and galaxy clustering]. A recent computer simulation, based on a cellular-automaton-type stochastic process (Vicsek and Szalay, 1987; see Figure 3.7F; a popular exposition is Maddox, 1987) has not only reproduced the correlation-dimension $D \approx 1.2$ but also predicted a power-law cluster-size distribution:

$$n(s) \propto s^{-1.9}$$

instead of the exponential one derived by Brown et al. (1983).

(c) A similar controversy has arisen in the statistics of cloud populations: Lopez (1977, Figure 3.20d) found that the cumulative frequency distribution of the height, horizontal size and duration of cumulus cloud populations are *lognormal,* while — according to the work of Lovejoy and his group, reviewed in Section 1.5.2 — the shape of the clouds is *fractal.*

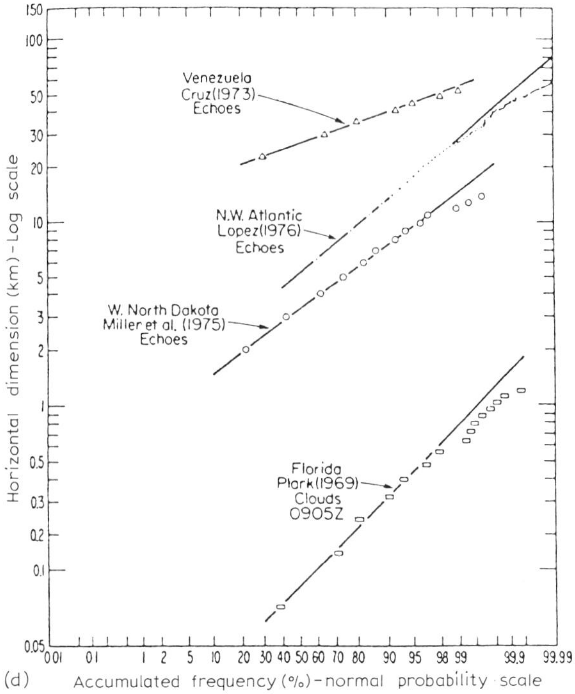

Fig. 3.20 (continued). d. Cumulative frequency distribution of cloud and radar echo horizontal dimensions. The straight lines are the best fitting lognormal distributions (detail, from Lopez, 1977).

There are cases where different size distributions can be observed in the same geological process. Korvin (1989) analysed the statistical geometry of the network of faults on the structural map of the Palaeozoic Basement of the Gulf of Suez (Figure 3.21). The "Gulf of Suez" trend sides (N10–60° W) of the fault blocks were *exponentially* distributed, the "cross trend" block sides (N20–70° E) *lognormally* distributed.

With the single exception of the power-law distribution, all other size distributions depend on some characteristic length as a parameter, that is they cannot correspond to *scale-invariant processes*. For example, contraction-crack polygons (Section 2.4.1) whose sides are exponentially or lognormally distributed, generally have a characteristic length related to the elastic properties and thickness of the contracting layer (Lachenbruch, 1961; Neal et al., 1968). Vogt (1974) also observed that the average volcano spacing is "... essentially equal to the thickness of the lithospheric plate".

According to Montroll and Shlesinger (1983), "in the world of the investigations of complex phenomena that require statistical modeling and interpretation, several competing styles have been emerging, each with its own champions", and they list the *maximum entropy formalism people* who use a "probabilistic interpretation of

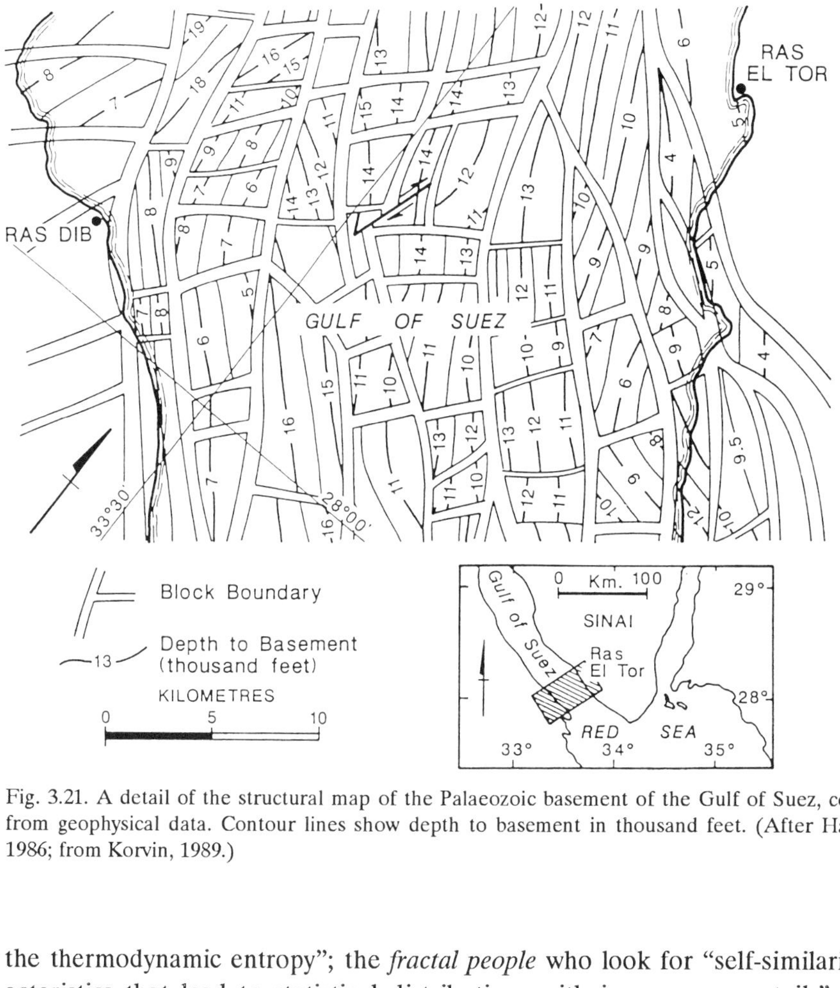

Fig. 3.21. A detail of the structural map of the Palaeozoic basement of the Gulf of Suez, constructed from geophysical data. Contour lines show depth to basement in thousand feet. (After Hammouda, 1986; from Korvin, 1989.)

the thermodynamic entropy"; the *fractal people* who look for "self-similarity characteristics that lead to statistical distributions with inverse power tails" and the *renormalization crowd* using scaling ideas in the hope "of transcending details of individual models".

In Sections 3.2.1. and 3.2.2. we have already presented examples for the style of the "fractal people" and the "renormalization group people".

The mathematics of the maximum entropy method (MEM) can be found in Jaynes (1957). Application of the technique to fragmentation (Griffith, 1943; Rivier and Lissowski, 1982; Grady and Kipp, 1985; Englman et al., 1987, 1988a, b) assumes that the actual fragment-size distribution will be the "most likely" probability distribution, subject to a set of physical constraints. Formally, one seeks the distribution function $p(x)$ that maximizes the *entropy*:

$$S = -\int p(x) \log p(x)\, dx = \max \tag{3.2.3.4}$$

subject to the normalization constraint:

$$\int p(x)\,\mathrm{d}x = 1 \tag{3.2.3.5}$$

and the physical constraints:

$$\int F_i(x)p(x)\,\mathrm{d}x = c_i \qquad (i = 2, 3, \ldots, L) \tag{3.2.3.6}$$

As well known, the solution of this problem is:

$$p(x) = \exp\left[-\{1 + \lambda_1 + \lambda_2 F_2(x) + \ldots \lambda_L F_L(x)\}\right] \tag{3.2.3.7}$$

where the Lagrange parameters λ_i can be determined in terms of the auxiliary functions $F_i(x)$ from equations (3.2.3.5)–(3.2.3.6).

An early example of this technique is Griffith (1943), who showed that if the total energy consumed in breaking is spent to the creation of new surfaces then the maximum entropy formalism leads to an exponential size distribution of the resulting particles. Grady and Kipp (1985) considered the two-dimensional fragmentation process, subject only to area preservation, and proved that the exponential distribution $f(A) \propto \exp[-\text{const}A]$ is the "most likely" solution, rather than the Mott distribution $\exp[-\text{const}A^{1/2}]$, where A is the area of the fragments. Rivier and Lissowski (1982) also dealt with planar fragmentation. They prescribed area preservation and the fulfilment of Euler's topological relation, and proved that the maximum-entropy fragmentation satisfies Lewis' law, that is the areas and numbers of sides of the pieces are linearly related (see Section 2.4.1).

Montroll and Shlesinger's essay (1983) explains why is it unlikely to obtain the empirically observed power-law or lognormal size distributions as solutions to a traditionally posed maximum entropy problem. They prove that the distribution $p(x) \propto x^{-\nu}$, for example, can only arise as a solution to equations (3.2.3.4)–(3.2.3.6) if one in an *a priori* manner prescribes a constraint (3.2.3.6) with an unconventional auxiliary function F_2 behaving as $\nu \log x$ for large x.

The first serious attempt to reconcile maximum entropy and fractals in fragmentation theory is due to Englman et al. (1987, 1988a, b), who used MEM to show that the fragment distribution is approximately power-law for not too large fragments. In an analogy with the Boltzmann theory of molecules they associated with each fragment of linear size a an energy:

$$e(a) = \left(\frac{2\pi}{3}\right)\rho u^2 a^3 + \left(\frac{2\pi}{5}\right)\rho\varepsilon^2 a^5 + \left(\frac{4\pi}{6K}\right)\sigma^2 a^3 + 4\pi\gamma a^2\left(1 - \frac{a}{R}\right) \tag{3.2.3.8}$$

where the fragments are assumed to be of roughly spherical shape. The energy consists of the following terms: the kinetic energy of the fragment of density ρ moving with speed u; the internal kinetic energy about the centre of mass, expressed in terms of ε^2, the volumetric dilation rate squared; the stress energy stored in the fragments (σ being the stress and K the bulk modulus); and the surface formation energy (γ is the surface energy density and R the radius of the body before fragmentation).

If the probability of having n_i fragments of size a is written as $p(n_i, a)$, we have:

$$\sum_i p(n_i, a) = 1 \tag{3.2.3.9}$$

The expected number of such fragments and their total weights are, respectively:

$$N_a = \sum_i n_i p(n_i, a) \tag{3.2.3.10}$$

$$w(a) = \left(\frac{4\pi}{3}\right) a^3 \rho N_a \tag{3.2.3.11}$$

Assuming energy conservation and that a fixed weight (or mass) of material is fragmented, in the sense of the MEM we have to maximise the entropy:

$$S = -\sum_a \sum_i p(n_i, a) \log\left[p(n_i, a)\right] \tag{3.2.3.12}$$

subject to the constraints:

$$\sum_a w(a) = W \quad \text{and} \quad \sum_a N_a e(a) = E \tag{3.2.3.13}$$

The solution is:

$$p(n_i, a) = \frac{\exp\left[-n_i x_a\right]}{\sum_j \exp\left[-n_j x_a\right]} \tag{3.2.3.14}$$

where:

$$x_a = \lambda e(a) + \mu \left(\frac{4\pi}{3}\right) \rho a^3 \tag{3.2.3.15}$$

(λ and μ are the Lagrange multipliers). From (3.2.3.10) and (3.2.3.14):

$$N_a = \left[\exp\left(x_a^{-1}\right) - 1\right] \tag{3.2.3.16}$$

that is, for small-sized fragments with $x_a \ll 1$:

$$N_a \propto x_a^{-1} \tag{3.2.3.17}$$

(for large particles the number-distribution is exponential). As, by equations (3.2.3.8) and (3.2.3.15), x_a is a polynomial in the fragment size a, with powers ranging from 2 to 5, in a size range where a single power θ dominates in the polynomial expression

(3.2.3.15), the number of fragments $N(a)$ greater than a is given by the power law:

$$N(a) = \int_a^\infty N_a \, \mathrm{d}a \propto a^{-[\theta-1]} = a^{-D} \tag{3.2.3.18}$$

where the fractal dimension D falls between 1 and 4, conforming to the experimental values in Table 3.3. If there are more than one dominating powers in expression (3.2.3.15), the plot of $N(a)$ will show one or more "knees", as in case of the bifractal distributions in Figure 3.9.

Which size distribution is then right, the exponential, the lognormal or the power-law? Very likely *all of them*; paraphrasing Chesterton "it is a privilege of two distributions which contradict each other... to be both right". In multiple-fragmentation processes, as e.g. in the cracking of cornstarch in Figure 2.43, in the n-th generation *selected* pieces of the $(n-1)$-th generation are broken into *exponentially* distributed parts (on strength of the maximum entropy principle of Grady and Kipp, 1985), for any sufficiently large value of n the size distribution of the n-th generation fragments will be *lognormal* (by Epstein's theorem, 1947), while the distribution for *all* fragments at the end of the process ($n \to \infty$) will be power-like (Matsushita 1985; Turcotte 1986, 1989). What we actually observe very much depends on the way of sampling. Englman et al. (1987) mention that the size distribution of all the debris around a detonation was found power-law like, though at each individual heap it was exponential, non-fractal.

References

Allègre, C.J., Le Mouel, J.L. and Provost, A. (1982). Scaling rules in rock fracture and possible implications for earthquake prediction. Nature, 297: 47–49.

Allen, T. (1968). Particle Size Measurement. Chapman and Hall, London.

Bader, H. (1970). The hyperbolic distribution of particle sizes. J. Geophys. Res., 75: 2822–2830.

Balian, R. and Schaeffer, R. (1988). Les dimensions fractales de la distribution des galaxies dans l'univers. C.R. Acad. Sci. Paris, Ser. II, 306: 689–693.

Bennett, J.G. (1936). Broken coal. J. Inst. Fuel, 10: 22–39.

Bevlin, M.E. (1984). Design Through Discovery. 4th Edition Holt Rinehart Winston, New York.

Bianchi, R., Capaccioni, F., Cerroni, P., Coradini, M., Flamini, E., Martelli, G. and Smith, P.N. (1984). Experimental simulation of asteroidal fragmentation by macroscopic hypervelocity impacts against free falling bodies. Astron. Astrophys., 139: 1–6.

Bogolyubov, N.N. and Shirkov, D.V. (1984). Renormgroup — It is simple! Pryroda, 1984, No. 8, pp. 3–13 (in Russian).

Brock, J.R. (1971). On size distributions of atmospheric aerosols. Atmos. Environ., 5: 833–841.

Brown, W.K., Karpp, R.R. and Grady, D.E. (1983). Fragmentation of the universe. Astrophys. Space Sci., 94: 401–412.

Bruce, A. and Wallace, D. (1989). Critical point phenomena: universal physics at large length scales. *In*: P. Davies (Editor), The New Physics. Cambridge Univ. Press, Cambridge.

Campi, X. (1989). Geometric and kinetic models of fragmentation. J. Phys., 50C: 183–188.

Capaccioni, F., Cerroni, P., Coradini, M., Di Martino, M., Farinella, P., Flamini, E., Martelli, G., Paolicchi, P., Smith, P.N., Woodward, A. and Zappala, V. (1986). Asteroidal catastrophic collisions simulated by hypervelocity impact experiments. Icarus, 66: 487–514.

Chesterton, G.K. (1943). St. Thomas Aquinas. Hodder and Stoughton, London.

Crow, E.L. (1988). Applications in atmospheric sciences. *In*: E.L. Crow and K. Shimiziu (Editors), Lognormal Distributions. Theory and Applications.. Marcel Dekker Inc., New York and Basel, pp. 331–356.

Curl, R.L. (1960). Stochastic models of cavern development. Bull. Nat. Speleol. Soc., 22: 66–76.

Curl, R.L. (1966). Caves as measure of karst. J. Geol., 74: 798–830.

Curl, R.L. (1986). Fractal dimensions and geometries of caves. Math. Geol., 18: 765–783.

Dohnanyi, J.S. (1969). Collisional model of asteroids and their debris. J. Geophys. Res., 74: 2531–2554.

Donnison, J.R. and Sugden, R.A. (1984). The distribution of asteroid diameters. Monthly Not. R. Astron. Soc., 210: 673–682.

Englman, R., Jaeger, Z. and Levi, A. (1984). Percolation theoretical treatment of two-dimensional fragmentation in solids. Phil. Mag., B, 50: 307–315.

Englman, R., Rivier, N. and Jaeger, Z. (1987). Fragment-size distribution in disintegration by maximum-entropy formalism. Phil. Mag., B, 56: 751–769.

Englman, R., Rivier, N. and Jaeger, Z. (1988a). Size distribution in sudden breakage by the use of entropy maximization. J. Appl. Physics, 63: 4766–4768.

Englman, R., Rivier, N. and Jaeger, Z. (1988b). The problem of priors in the maximum entropy derivation of fragment-size distributions. Nucl. Phys., B (Proc. Suppl.), 5A: 345–349.

Epstein, B. (1947). The mathematical description of certain breakage mechanisms leading to the logarithmico-normal distributions. J. Franklin Inst., 244: 471–477.

Feder, J. (1988). Fractals. Plenum Press, New York and London.

Fisher, M.E. (1974). The renormalization group in the theory of critical behavior. Rev. Modern Phys., 46: 597–616.

Fréchet, M. (1941). Sur la loi de repartition de certaines grandeurs géographiques. J. Soc. Stat. Paris, 82: 114–122.

Fujiwara, A., Kamimoto, G. and Tsukamoto, A. (1977). Destruction of basaltic bodies by high-velocity impact. Icarus, 31: 277–288.

Goldberger, A.L. and West, B.J. (1987). Fractals in physiology and medicine. Yale J. Biol. Med., 60: 421–435.

Goodchild, M.F. (1988). Lakes on fractal surfaces: a null hypothesis for lake-rich landscapes. Math. Geol., 20: 615–630.

Grady, D.E. and Kipp, M.E. (1985). Geometric statistics and dynamic fragmentation. J. Appl. Physics, 58: 1210–1222.

Grady, D.E. and Kipp, M.E. (1987). Dynamic rock fragmentation. *In*: B.A. Atkinson (Editor), Fracture Mechanics of Rock. Academic Press, London, pp. 429–475.

Griffith, L. (1943). A theory of the size distribution of particles in a comminuted system. Can. J. Res., 21A: 57–64.

Groth, F.J. and Peebles, P.J.E. (1977). Statistical analysis of catalogs of extragalactic objects. VII. Two- and three-point correlation functions for the high-resolution Shane-Wirtanen catalog of galaxies. Astrophys. J., 217: 385–405.

Hammouda, H.M. (1986). Study and Interpretation of Basement Structural Configuration in the Southern Part of Gulf of Suez using Aeromagnetic and Gravity Data. Ph.D. Thesis, Cairo University.

Harris, A.B., Lubensky, T.C., Holcomb, W.K. and Dasgupta, C. (1975). Renormalization-group approach to percolation problems. Phys. Rev. Lett., 35: 327–330.

Hartmann, W.K. (1969). Terrestrial, lunar, and interplanetary rock fragmentation. Icarus, 10: 201–213.

Hawkins, G.S. (1960). Asteroidal fragments. Astrophys. J., 65: 318–322.

Heath, G.R. (1979). Burial rates, growth rates, and size-distribution of deep-sea manganese nodules. Science, 205: 903–904.

Hellyer, B. (1971). The fragmentation of asteroids, II. Monthly Not. R. Astron. Soc., 154: 279–291.

Hoshen, J., Stauffer, D., Bishop, G.H., Harrison, R.J. and Quinn, G.D. (1979). Monte Carlo experiments on cluster size distribution in percolation. J. Phys. A, Math. Gen., 12: 1285–1307.

Jaynes, E.T. (1957). Information theory and statistical mechanics. Phys. Rev., 106: 620–630.

Johnson, N.L. and Kotz, S. (1970). Distributions in Statistics. Vol. II. Continuous Univariate Distributions — 1. Houghton Mifflin Co., Boston.

Kent, C. and Wong, J. (1982). An index of littoral complexity and its measurement. Can. J. Fish. Aquat. Sci., 39: 847–853.

Korčak, J. (1940). Deux types fondamentaux de distribution statistique. Bull. Inst. Int. Stat., 30: 295–299.

Korvin, G. (1989). Fractured but not fractal: fragmentation of the Gulf of Suez basement. PAGEOPH, 131: 289–305.

Lachenbruch, A. (1961). Depth and spacing of tension cracks. J.Geophys. Res., 66: 4273–4292.

Laing, R.D. (1965). The Divided Self. An Existential Study in Sanity and Madness. Penguin, Baltimore.

Lange, M.A., Ahrens, T.J. and Boslough, M.B. (1984). Impact cratering and spall failure of gabbro. Icarus, 58: 383–395.

Laverty, M. (1987). Fractals in karst. Earth Surf. Proc. Landforms, 12: 475–480.

Lopez, R.E. (1977). The lognormal distribution and cumulus cloud populations. Mon. Weather Rev., 105: 865–872.

Lucchin, F. (1986). Clustering in the Universe. *In*: L. Pietronero and E. Tosatti (Editors), Fractals in Physics. Elsevier, Amsterdam, pp. 313–318.

Lucido, G. (1985). Magma-clustering and galaxy-clustering: a comparison. Phys. Earth Planet. Int., 39: 134–140.

Macaulay, F. (1922). Pareto's laws and the general problem of mathematically describing the frequency distribution of income. *In*: Income in the United States, its Amount and Distribution, 1909–1919. Nat. Bur. Econ. Res., New York, Chap. XXIII.

Madden, T.R. (1983). Microcrack connectivity in rocks: a renormalization group approach to the critical phenomena of conduction and failure in crystalline rocks. J. Geophys. Res., 88B: 585–592.

Maddox, J. (1987). The Universe as a fractal structure. Nature, 329: 195.

Magnuson, A.H., Sundkvist, K., Ma, Y. and Smith, K. (1981). Acoustic sounding for manganese nodules. Offshore Tech. Conf. Pap. No. OTC 4133, pp. 147–154.

Mandelbrot, B.B. (1960). The Pareto-Lévy law and the distribution of income. Int. Econ. Rev., 1: 79–106.

Mandelbrot, B.B. (1963). New methods in statistical economics. J. Polit. Econ., 71: 421–440.

Mandelbrot, B.B. (1967). How long is the coast of Britain? Statistical self-similarity and fractional dimension. Science, 155: 636–638.

Mandelbrot, B.B. (1975a). Stochastic models for the Earth's relief, the shape and the fractal dimension of the coastlines, and the number–area rule for islands. Proc. Nat. Ac. Sci. USA, 72: 3825–3828.

Mandelbrot, B.B. (1975b). Sur un modèle décomposable d'univers hiérarchisé: déduction des correlation galactiques sur la sphère céleste. C.R., Paris, 280A: 1551–1554.

Mandelbrot, B.B. (1982). The Fractal Geometry of Nature. Freeman, N.Y.

Mandelbrot, B.B., Passoja, D.E. and Paullay, A.J. (1984). Fractal character of fracture surfaces of metals. Nature, 308: 721–722.

Mathis, J.S. (1979). The size distribution of interstellar particles, II. Polarization. Astrophys. J., 232: 747–753.

Matsushita, M. (1985). Fractal viewpoint of fracture and accretion. J. Phys. Soc. Jpn., 54: 857–860.

McCrosky, R.E. (1968). Distributions of large meteoric bodies. Smiths Astrophys. Obs., Spec. Rep. No. 280, Washington, D.C.

Montroll, E.W. and Shlesinger, M.F. (1983). Maximum entropy formalism, fractals, and $1/f$ noise: A tale of tails. J. Stat. Phys., 32: 209–230.

Mott, N.F. and Linfoot, N.H. (1943). A Theory of Fragmentation. Extra-Mural Research No.F72/80, Ministry of Supply (England).

Neal, J.T., Langer, A.M. and Kerr, P.F. (1968). Giant desiccation polygons of Great Basin playas. Geol. Soc. Am. Bull., 79: 69–70.

Niemeyer, L., Pietronero, L. and Wiesmann, H.J. (1984). Fractal dimension of dielectric breakdown. Phys. Rev. Lett., 52: 1033–1036.

Pareto, V. (1897). Cours d'Économie Politique. Rouge and Co., Lausanne and Paris.

Payatakes, A.C. (1982). Dynamics of oil ganglia during immiscible displacement in water-wet porous media. Ann. Rev. Fluid Mech., 14: 365–393.

Peebles, P.J.E. (1980). The Large-Scale Structure of the Universe. Princeton Univ. Press, Princeton, N.J.

Peitgen, H.-O. and Saupe, D., Editors (1988). The Science of Fractal Images. Springer, Berlin and Heidelberg.

Pfeuty, P. and Toulouse, G. (1977). Introduction to the Renormalization Group and to Critical Phenomena. John Wiley, New York.

Pietronero, L. (1987). The fractal structure of the universe: correlations of galaxies and clusters and the average mass density. Physica, 144A: 257–284.

Pietronero, L. and Kupers, R. (1986). Stochastic approach to large scale clustering of matter in the universe. *In*: L. Pietronero and E. Tosatti (Editors), Fractals in Physics, Elsevier, Amsterdam, pp. 319–324.

Procaccia, I. and Schuster, H. (1983). Functional renormalization-group theory of universal $1/f$ noise in dynamical systems. Phys. Rev. A, 28: 1210–1212.

Rendu, J.-M.M. (1988). Applications in Geology. *In*: E.L. Crow and K. Shimiziu (Editors), Lognormal Distributions. Theory and Applications, Marcel Dekker, Inc., New York and Basel, pp. 357–365.

Reynolds, P.J., Klein, W. and Stanley, H.E. (1977). A real-space renormalization group for site and bond percolation. J. Phys. C, 10: L167–L172.

Richardson, L.F. (1961). The problem of contiguity. An appendix to the "Statistics of Deadly Quarrels". Gen. Syst., 6: 139–187.

Rivier, N. and Lissowski, A. (1982). On the correlation between sizes and shapes of cells in epithelial mosaics. J. Phys. A, L143–L148.

Rothrock, D.A. and Thorndike, A.S. (1984). Measuring the sea ice floe size distribution. J. Geophys. Res., C89: 6477–6486.

Sammis, C.G. (1988). The micromechanics of fractal fault-gouge generation. Geol. Soc. Am. 1988 Annu. Meet., Abstract No. 27313.

Sammis, C.G., Osborne, R.H., Anderson, J.L., Banerdt, M. and White, P. (1986). Self-similar cataclisis in the formation of fault gouge. PAGEOPH, 124: 53–78.

Schoutens, J.E. (1979). Empirical analysis of nuclear and high-explosive cratering and ejecta. Nuclear Geophysics Sourcebook, Vol. 55, Pt. 2, Sect. 4, Rep. DNA OIH-4-2. Def. Nucl. Agency, Bethesda, MD.

Sherwood, J.D. (1986). Island size distribution in stochastic simulations of the Saffman-Taylor instability. J. Phys. A, Math. Gen., 19: L195–L200.

Shimiziu, K. and Crow, E.L. (1988). History, Genesis, Properties. *In*: E.L. Crow and K. Shimiziu (Editors), Lognormal Distributions. Theory and Applications. Marcel Dekker, Inc., New York and Basel, pp. 1–25.

Shlesinger, M.F. and Hughes, B.D. (1981). Analogs of renormalization group transformations in random processes. Physica, A109: 597–608.

Smalley, R.F., Jr., Turcotte, D.L. and Solla, S, A. (1985). A renormalization group approach to the stick-slip behavior of faults. J. Geophys. Res., B90: 1894–1900.

Smith, D.K. and Jordan, T.H. (1987). The size distribution of pacific seamounts. Geophys. Res. Lett., 14: 1119–1122.

Solla, S.A. (1986a). Crack propagation and onset of failure. *In*: H.E. Stanley and N. Ostrowsky (Editors), On Growth and Form. Fractal and Non-Fractal Patterns in Physics. Nijhoff, Dordrecht, pp. 260–262.

Solla, S.A. (1986b). Collapse of loaded fractal trees. *In*: L. Pietronero and E. Tosatti (Editors), Fractals in Physics. Elsevier, Amsterdam, pp. 185–188.

Stauffer, D. (1986). Percolation and cluster size distribution. *In*: H.E. Stanley and N. Ostrowsky (Editors), On Growth and Form. Fractal and Non-Fractal Patterns in Physics. Martinus Nijhoff, Dordrecht, pp. 79–100.

Suzuki, M. (1983). Phase transition and fractals. Prog. Theor. Phys., 69: 65–76.

Szalay, A.S. and Schramm, D.N. (1985). Are galaxies more strongly correlated than clusters? Nature, 314: 718–719.

Turcotte, D.L. (1986). Fractals and fragmentation. J. Geophys. Res., B91: 1921–1926.

Turcotte, D.L. (1989). Fractals in geology and geophysics. PAGEOPH, 131: 171–196.

Turcotte, D.L., Smalley, R.F., Jr. and Solla, S.A. (1985). Collapse of loaded fractal trees. Nature, 313; 671–672.

Vicsek, T. and Szalay, A.S. (1987). Fractal distribution of galaxies modeled by a cellular-automaton-type stochastic process. Phys. Rev. Lett., 58: 2818–2821.

Vogt, P.R. (1974). Volcano-spacing, fractures, and thickness of the lithosphere. Earth Planet. Sci.Lett., 21: 235–252.

Voss, R.F. (1988). Fractals in nature: from characterization to simulation. *In*: H.-O. Peitgen and D. Saupe (Editors), The Science of Fractal Images. Springer, Berlin and Heidelberg, pp. 21–70.

Weibull, W. (1951). A statistical distribution function of wide applicability. J. Appl. Mech., 18: 293–297.

West, B.J., Bhargava, V. and Goldberger, A.L. (1986). Beyond the principle of similitude: renormalization in the bronchial tree. J. Appl. Physiol., 60: 1089–1097.

Wilson, K.G. (1979). Problems in physics with many scales of length. Sci. Am., August 1979, pp. 140–157.

Wilson, K.G. and Kogut, J. (1974). The renormalization group and the ε-expansion. Phys. Reps., 12: 75–200.

Young, A.P. and Stinchcombe, R.B. (1975). A renormalization group theory for percolation problems. J. Phys. C, 8: L535–L540.

Zipf, G.K. (1949). Human Behavior and the Principle of Least Effort. An Introduction to Human Ecology. Addison-Wesley, Cambridge, Mass.

4 Fractal surfaces

The prospect of the landscape of Southern England prompts two questions. Why is it so smooth and contrariwise why is it so irregular?
— CULLING and DATKO (1987)

4.1 Fractal surfaces everywhere

There was a fractal man who walked a fractal mile
Who found a fractal sixpence upon a fractal style,
He bought a fractal cat that caught a fractal mouse
And they all lived together in a fractal little house.
— Nursery rhyme, paraphrased by B.H. KAYE (1989)

It is impossible to get research grants in Australia for fractal projects. A few years ago, when my third grant application was turned down, I found consolation in using photocopies of the letter of rejection to check whether "paper crushes fractally" (as reported by Gomes, 1987a, b). Gomes and his 89 students carried out this experiment in his freshman course Experimental Physics at the Universidade Federal de Pernambuco, Brasil; using rectangular sheets of paper of different thickness and size (the largest was 98 × 65 cm). On the basis of 4361 measurements they found that the radii of the crumpled paper balls, R, and their total surface, S, were related as:

$$S \propto R^{2.51 \pm 0.19} \tag{4.1.1}$$

A similar experiment with crumpled sheets of foil (Kantor et al., 1986) has also resulted in $S \propto R^{2.5}$, in agreement with the theoretical exponent in the scaling relation:

$$S \propto R^{(d+2)/2} \tag{4.1.2}$$

holding for *self-avoiding surfaces* embedded into the d-dimensional space, if $d = 3$ (Kantor et al., 1986; Gomes, 1987a). [Paper folding and crumpling are closely related to chaos theory and fractals — see Figure 4.1a, b and the review of Mendes France (1984).]

Equations (4.1.1) and (4.1.2) immediately suggest a number of new "fractal" paradoxes. Assuming the scaling law $S \propto R^{2.5}$ for simplicity, we find that if the outer surface of the crumpled paper ball is spherical, that is if its area is proportional to R^2, this becomes much less for a large R than the interior surface. This reminds one of the remark of a hero of the Czech writer Jaroslev Hašek, who avowed that

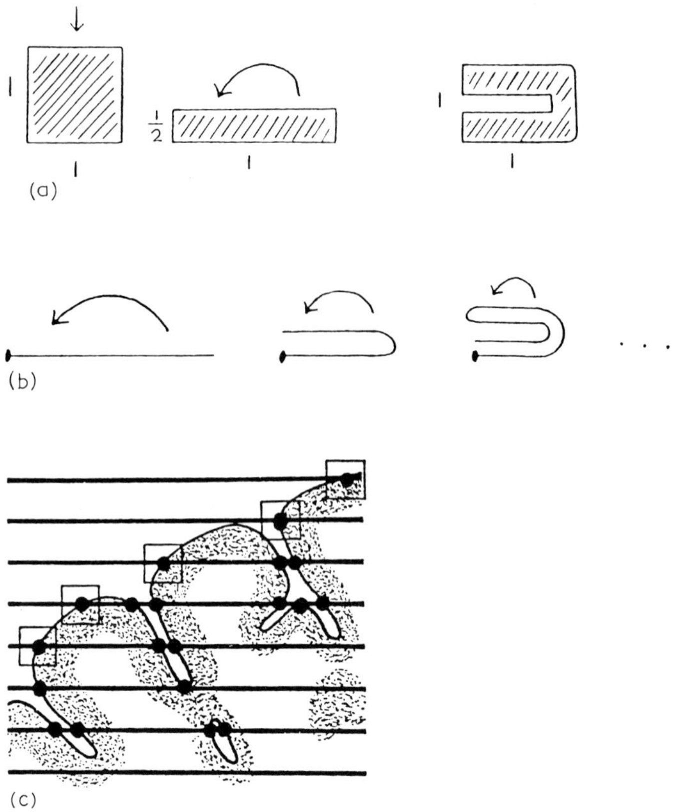

Fig. 4.1. Fractal folds. a. The famous baker transform (or Poincaré transform) of chaos theory; b. paper folding; c. mammalian brain-folds (a, b from Mendes France, 1984; c from Elias and Schwartz, 1969).

"inside the terrestrial globe there's another globe, a good deal bigger than the one outside" (see also the Verne illustration, Figure 4.14c); or of the enormous internal surfaces encountered in anatomy where e.g. the total area of membrane available for gas exchange in the human lung is about 70 square metres — some 40 times the surface area of the body (Comroe, 1966; Mandelbrot, 1982).

Another paradox arises if we recall that for a regular body of volume V and surface S conventional dimensional consideration suggests $V^{1/3} \propto S^{1/2}$. For the crumpled paper balls however, we have from equations (4.1.1) and (4.1.2):

$$S^{1/2.5} \propto R \propto V^{1/3}$$

or, more generally:

$$S^{1/D} \propto V^{1/3} \qquad (4.1.3)$$

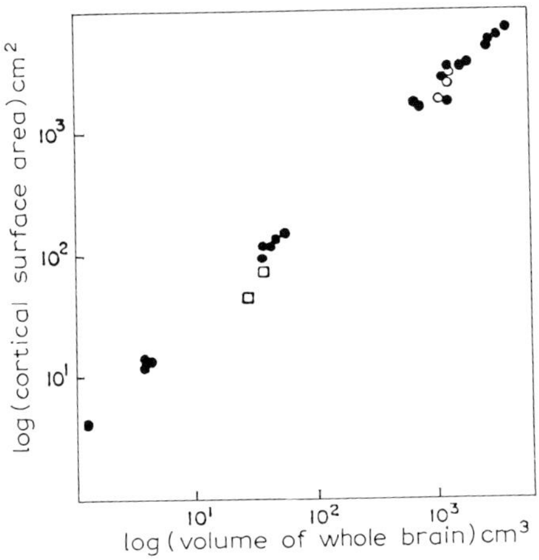

Fig. 4.2. Relation between volume and cerebrocortical surface area of mammalian brains. Upper group of points: whales and dolphins (solid circles) and man (open circles); middle group: carnivores (solid circles) and kangaroos (squares); lower group: opossums, lowest point: mouse opossum (from Elias and Schwartz, 1969).

where D ($2 \leq D \leq 3$) is the fractal dimension of the surface. Biological surfaces are often similar to crushed paper: for the surface areas of mammalian brain-folds, for example, Elias and Schwartz (1969) found $\log S \propto 0.93 \log V$, which is a special case of equation (4.1.3) with $D = 2.79$ (Figures 4.1c and 4.2).

As is well known, many physiological parameters are related to the size of the organism. The so-called "allometric" relations express relationships between body size or weight and metabolic quantities such as overall metabolic rate (Figure 4.3), and are formulated as power laws:

$$y = aM^b \tag{4.1.4a}$$

or, for the specific quantity:

$$\frac{y}{M} = aM^{b-1} \tag{4.1.4b}$$

where y is some metabolic quantity; M is body mass; a, b are constants. As Sernetz et al. (1985) noted, the very name of these empirical laws ("allometric" coming from the Greek $\alpha\lambda\lambda\tilde{\omega}\mu\epsilon\tau\rho o\nu$ = "strange scale") expresses the frustration of the earlier investigators when the traditional inference that "the metabolic rate is proportional to the available surface area, consequently the specific turnover rates of similar organisms should be proportional to $M^{2/3}$" was not satisfied and "strange" scaling laws (like the $y \propto M^{0.75}$ rule shown in Figure 4.3) were found. As Sernetz et al. (1985) point out, the "strange" exponent, $b = 0.75$ is due to the fact that the surfaces of the organisms included in Figure 4.3 are 2.25-dimensional fractals.

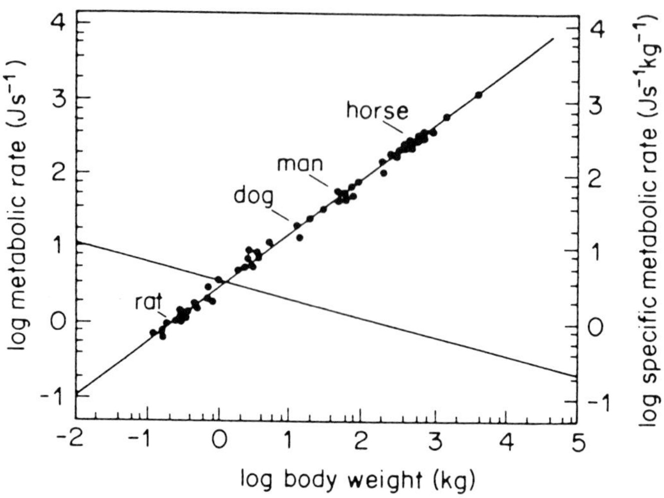

Fig. 4.3. Metabolic rate (left scale) and mass-specific metabolic rate (right scale) as function of body weight (from Sernetz et al., 1985).

Equation (4.1.3) can be generalised to an aggregate of fractal particles, such as sedimentary rocks, crushed stones, soil, powders, etc. (Avnir, 1986; Avnir et al., 1984, 1985; Ben Ohoud et al., 1988; Pfeifer et al., 1983, 1984; Van Damme et al., 1988), and this leads to a powerful method of determining the fractal dimension of the internal surface of such materials (Avnir et al., 1985). The following derivation is based on Van Damme et al. (1988) and Ben Ohoud et al. (1988).

Consider a set of particles of variable size r, of surface fractal dimension D_s, and assume that the particles are self-similar for $r_0 < r < r_1$, where r_0 and r_1 are the *inner and outer self-similarity cut-offs*, respectively. In this range the surface area scales as:

$$s = \alpha r^{D_s} \tag{4.1.5}$$

but it is constant within the range of self-similarity.

On the other hand, in the same range of self-similarity the number of particles of size r in a macroscopic volume V is:

$$N(r) = \frac{V}{\gamma r^3} \tag{4.1.6}$$

where γ is a packing factor. Since the total surface area of the sample is $S = N(s)s$, by equations (4.1.5) and (4.1.6) the *surface area per unit volume*, $S_{\text{p.u.v.}}$, is:

$$S_{\text{p.u.v.}} = \frac{S}{V} = \left(\frac{\alpha}{\gamma}\right) r^{D_s-3} \propto r^{D_s-3} \tag{4.1.7}$$

In the derivation it has been assumed that the volume of a single particle is proportional to r^3. If the mass of the particles also behaves as a fractal, that is, if:

$$s(r) \propto r^{D_s} \qquad m(r) \propto r^{D_m} \tag{4.1.8}$$

where D_m is the *mass fractal exponent* and r is some characteristic size of the particles, the surface area, S, and the mass, M, of a macroscopic sample containing $N(r)$ particles will become:

$$S(r) \propto N(r) r^{D_s} \qquad M(r) \propto N(r) r^{D_s} \tag{4.1.9}$$

and the *surface area per unit weight* of that sample will scale as:

$$S_{\text{p.u.w.}} \propto r^{D_s - D_m} \tag{4.1.10}$$

The mass fractal exponent D_m is 3 for spheroids, 2 for plates of constant thickness, and 1 for needle-like particles of constant diameter (Figure 4.4).

The *surface area per unit apparent volume*, $S_{\text{p.u.v.}}$ can be similarly determined. The total apparent volume of a macroscopic sample is:

$$V(r) \propto N(r) r^{D_c} \tag{4.1.11}$$

where D_c is a fractal exponent related to the dimension of arrangement of the particles with respect to each other at length scales larger than the size of the individual particles. The exponents D_m (characterising a single particle) and D_c are generally not equal; for example, $D_c = 3$ both for an ordered or homogeneously disordered arrangements of spheroids; $D_c = 2$ for an ordered stack of platelets and 3 for randomly oriented platelets; $D_c = 1$ for aligned rods and 2 for randomly oriented rods (Ben Ohoud et al., 1988). From equations (4.1.9)–(4.1.11) the *apparent density* and the specific (per volume) surface area can be expressed as:

$$\rho_a \propto r^{D_m - D_c} \tag{4.1.12}$$

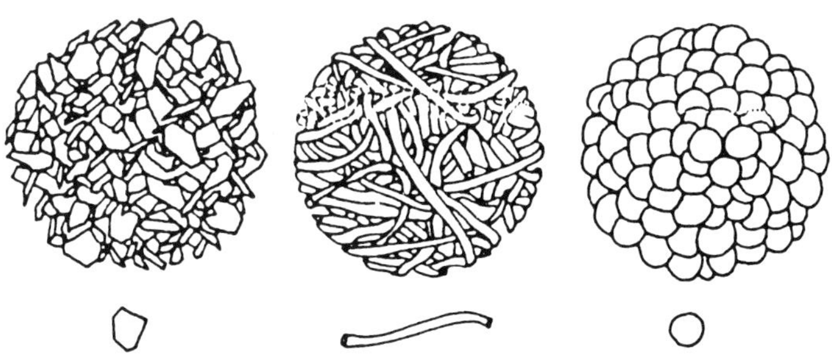

Fig. 4.4. Homogeneous aggregates of platelets ($D_m = 2$), fibres ($D_m = 1$) and spheres ($D_m = 3$), respectively. D_m is the mass fractal dimension figuring in equations (4.1.9) and (4.1.10). Several clay minerals, such as illite or xonotlite have a structure similar to the first two types, some silicas are similar to the third one. (From Van Damme et al., 1988.)

TABLE 4.1

Particle size dependence of the surface area per unit apparent volume, the surface area per unit weight, and the apparent density

Particle	Packing	D_m	D_s	D_c	$S_{p.u.v.}$	$S_{p.u.w.}$	ρa
Spheroids	Ordered or random			3	$\propto r^{D_s-3}$	$\propto r^{D_s-D_m}$	$\propto r^{D_m-3}$
Plates	Random	2	2	3	$\propto r^{-1}$	Constant	r^{-1}
Plates	Stack	2	2	2	Constant	Constant	Constant
Rods	Random	1	1	2	$\propto r^{-1}$	Constant	$\propto r^{-1}$
Rods	Aligned	1	1	1	Constant	Constant	Constant

Source: Ben Ohoud et al. (1988).

$$S_{p.u.v} \propto r^{D_s-D_c} \tag{4.1.13}$$

Observe that the apparent density of the aggregate is only independent of particle size if $D_m = D_c$. Some important cases of the above equations are summarised in Table 4.1.

Avnir and his co-workers (Avnir, 1986; Avnir et al., 1984, 1985; Pfeifer et al., 1983, 1984) have used equation (4.1.13) to determine the surface fractal dimension of solid materials. The surface areas are measured by chemical adsorption. If we vary the size of the object, R, assume that $D_m = 3$ and that the features of the surface irregularities are larger than the size of the probe molecule (usually N_2), then the number of moles n per one gram adsorbent will scale as:

$$n \propto R^{D_s-3} \tag{4.1.14}$$

The surface fractal dimension can be determined from the slope of the log n vs. log R plot (Figure 4.5a). Some geologically relevant surface-fractal dimensions obtained by this method are (Avnir et al., 1985): periclase — 1.95; quartz — 2.02 to 2.21; coal dust — 2.33 and 2.53; crushed carbonate rocks — 2.16 to 2.97; soils — 2.19 to 2.99; crushed rock from a nuclear test site — 2.7 to 3. Sometimes the log n vs. log R plot shows *bifractality* (Figure 4.5b) or is even more complicated, indicating that the simple approximation (4.1.14) is not applicable over the whole range of particle size.

What is the physical meaning of a 2.5-dimensional paper ball, or other surface? The number of inhabitants in a simple city where most of the buildings are of the same height (like the wonderful Lebanese city, Figure 4.6) is proportional to the square of the linear size of the city, $N \propto L^2$: the city is two-dimensional. In big capital cities with houses of widely distributed heights (Jensen, 1966) a person's address should be generally specified by two geographical coordinates and the apartment number referring to the storey in which he is living: the population scales as L^D $(2 < D < 3)$, and the city is D-dimensional (the example is due to Zel'dovich and Sokolov, 1985).

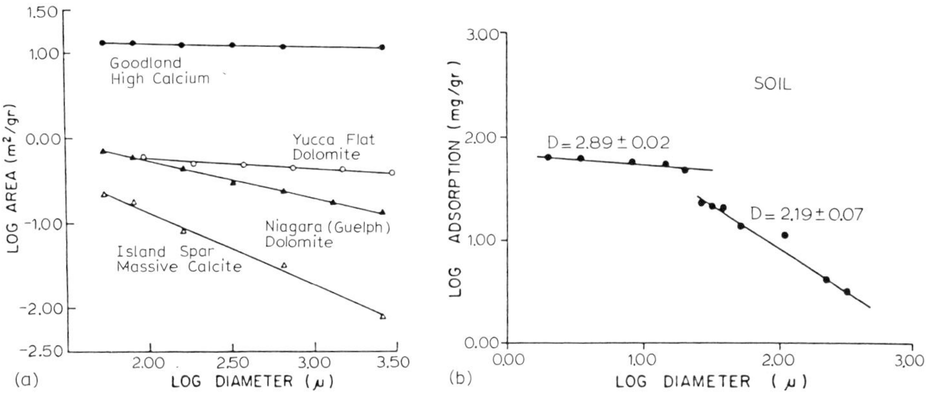

Fig. 4.5. a. Surface area (m^2/g) measured by chemical adsorption as a function of particle diameter (μm) for crushed carbonate rocks. b. Two ranges of self-similarity in a soil sample. Surface area is expressed in mg of adsorbed malachite green/g. (From Avnir et al., 1985.)

A much more complicated case arises if the dwellers of the fractal city are of widely different size, as e.g. arthropods living on a fractal vegetation (Morse et al., 1985). According to the study of Morse et al. (1985) vegetation has a fractal surface with D_s ranging from 2.28 for Virginia creeper (*Parthenocissus tricuspidata* Sieb. and Zucc.) to 2.79 for cotoneaster (*Cotoneaster horizontalis* Decaisne), that is the usable space for smaller animals of size l increases as:

$$A(l) \propto l^{2-D_s} \tag{4.1.15}$$

if $l \to 0$. [Equation (4.1.15) is the surface analogue of equation (1.4.1.3) figuring in the solution of the "coast of Britain" problem (Mandelbrot, 1967); its proof can be found e.g. in Ackermann and Spindler (1988).]

Let us denote the number of arthropods of size l living on a given D_s-dimensional vegetation by $N(l)$; let $d(l)$ denote the population density of these animals, $d(l) = N(l)/A(l)$ and $R(l)$ their metabolic rate.

As the population densities are inversely proportional to the metabolic rates: $d(l) \propto 1/R(l)$ (Peters, 1983) and the metabolic rate scales as the 0.75-th power of body weight, $R(l) \propto (l^3)^{0.75} = l^{2.25}$ (Sernetz et al., 1985; Figure 4.3), we have from equation (4.1.15) that $N(l)$ scales as:

$$N(l) \propto l^{-(D_s+0.25)} \tag{4.1.16}$$

Morse et al. (1985) have experimentally found $N(l) \propto l^{-2.75}$ which corresponds to $D_s \approx 2.5$, in fair agreement with the independent determination of D_s from the resolution analysis of photographs of different magnifications (Figure 4.7).

A consequence of the fractal nature of the surfaces occurring in chemistry, biology or geology is that the measured value of the surface areas depends on the

Fig. 4.6. Hasbaya, Lebanon, a two-dimensional city (from Ragette, 1980).

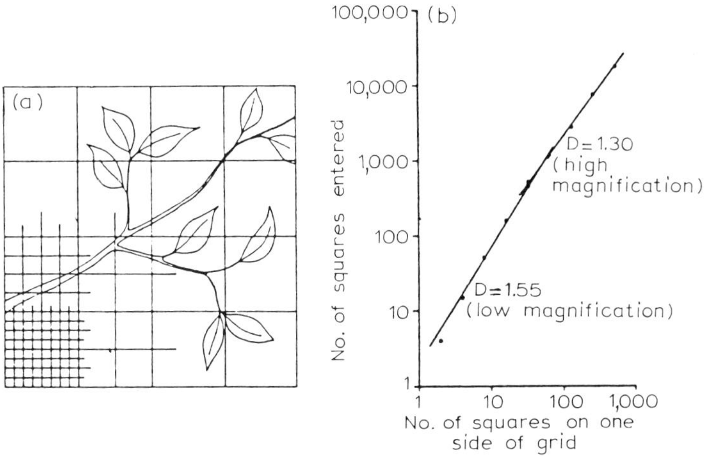

Fig. 4.7. a. Resolution analysis of photographs of plants at different magnifications. The logarithm of the number of squares entered by the outline of the plant is plotted against the logarithm of the number of divisions on one side of the grid. For fractal plant surfaces of dimension D_s a straight line results with slope $D = D_s - 1$. b. The log–log plot for the Virginia creeper, photographed without leaves in early spring. (From Morse et al., 1985.)

yardstick used. "What is the surface area of 1 g of silica gel?", ask Pfeifer et al. (1983), "This is the surface analogue of Mandelbrot's celebrated question as to the length of Britain's coast".

Paumgartner et al. (1981) observed a discrepancy in the published estimates of the surface area of the endoplasmic reticulum and mitochondrial membranes in rat liver cells; proved that it can be attributed to the fractality of these systems; and also presented a technique for measuring the fractal dimension of these surfaces using the power-law scaling of the lengths of their contour lines determined in the same plane section for various resolutions.

Professor Avnir of The Hebrew University of Jerusalem and his group (Avnir, 1986; Avnir and Pfeifer, 1983; Avnir et al., 1983, 1984; Farin et al., 1985; Pfeifer et al., 1983, 1984) have shown that one can apply the classical Richardson-Mandelbrot method to determine the surface fractal dimension of solids, by using the adsorbed molecules as yardsticks. [The same technique has later been applied by other groups: Lewis and Rees (1985), Van Damme and Fripiat (1985). Bunde et al. (1987) and De Gennes (1984) are theoretical studies. A general introduction to adsorption phenomena is Gregg and Sing (1967).]

Suppose we can cover a surface with N balls (and no less) of radius r (Figure 4.8). For smooth surfaces, $N \propto r^{-2}$, for fractal surfaces, however, N increases for decreasing r as:

$$N \propto r^{-D_s} \tag{4.1.17a}$$

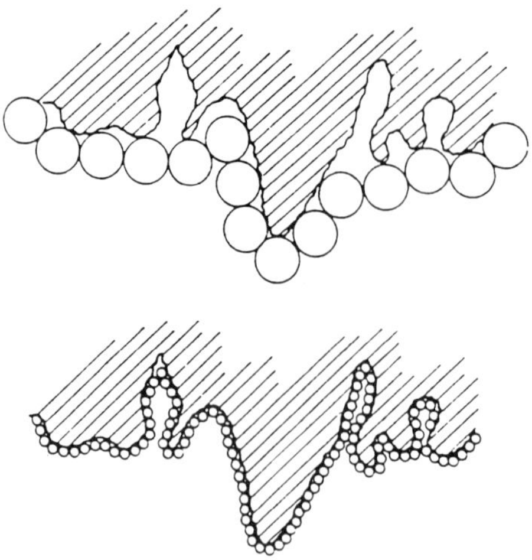

Fig. 4.8. Covering an irregular surface with a monolayer of spheres of radius r. For a fractal surface of dimension D_s the number of spheres grows as $N \propto r^{-D_s}$. (From Farin et al., 1985.)

where D_s ($2 \leq D_s \leq 3$) is the fractal dimension of the surface. In the adsorption method N is identified as the number of molecules forming a monolayer on the adsorbent's surface. If the covering molecules are not spherical, equation (4.1.17) can be expressed in terms of the effective cross-section, σ, of the probe molecules as:

$$N \propto \sigma^{-D_s/2} \tag{4.1.17b}$$

or:

$$A = A(\sigma) = N\sigma \approx \sigma^{(2-D_s)/2} \tag{4.1.17c}$$

where $A(\sigma)$ is the total surface area.

In most applications, σ ranges between 10–100 Å^2. A typical example is presented in Figure 4.9. The experimentally found fractal dimension is 2.07 ± 0.1, which is close to the theoretically expected $D = 2$ for an ideally layered graphite. [Van Damme and Fripiat (1985) used the same method and also obtained a low surface fractal dimension $D_s = 2$, or slightly below 2, for smectites, which are clay minerals with a mica-like sheet structure.] On the other hand, some of the surfaces studied by Avnir and co-workers have proved to be almost space filling with $D_s \approx 3$ ($D = 2.94 \pm 0.04$ for porous silicagel, $D_s = 2.79 \pm 0.03$ for porous alumina, D_s is up to 3 for charcoal).

In many applications, the surface fractal dimension of an object is determined from the fractal analysis of one or several of its planar sections. If the object has a D-dimensional boundary profile, its surface dimension is $D = D_s + 1$. A good review of the method is Farin et al. (1985). Three principal techniques are used:

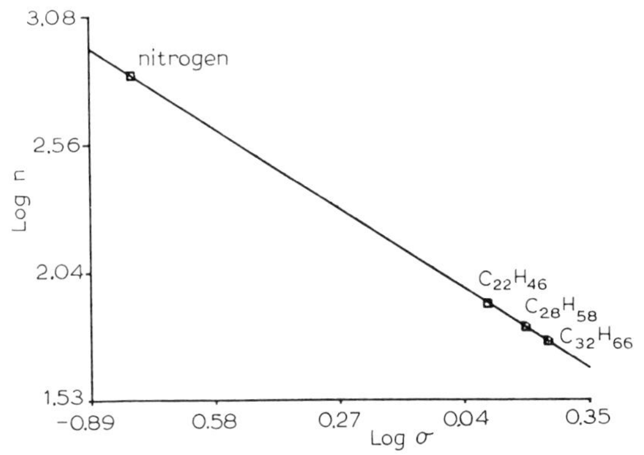

Fig. 4.9. Determination of the surface fractal dimension of graphite with long-chain n-alkanes and nitrogen. n (μmol/g) is monolayer amount of adsorbate; σ is adsorbate cross-section (nm^2). (From Avnir et al., 1983.)

(1) length measurement as a function of resolution;
(2) perimeter/area analysis at fixed resolution;
(3) Length measurement as a function of magnification, using a fixed yardstick.

As each of these techniques were already discussed in Chapter 1, we only list here a few examples of their application.

– Technique (1) can be used to estimate D_s of an irregular body from its planar boundary line. Figure 4.10a shows this for a quartz particle (Farin et al., 1985). Turcotte (1989) applied the technique for specified topographic contours in different mountain belts; Kaye (1989, p. 386) analysed the outline of the ground plan of a unique mountain form, a "Bornhardt" (see Twidale, 1982), etc. The method can also be applied to vertical profiles of surfaces (sea bottom relief: Barenblatt et al., 1984;

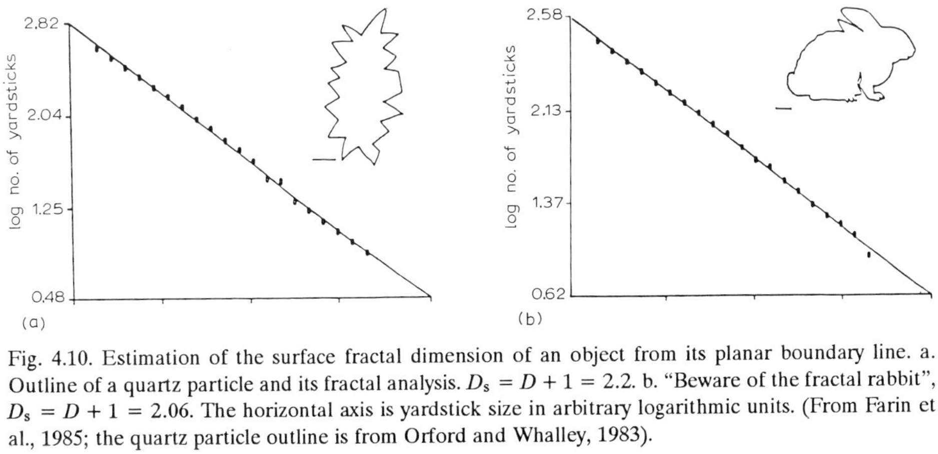

Fig. 4.10. Estimation of the surface fractal dimension of an object from its planar boundary line. a. Outline of a quartz particle and its fractal analysis. $D_s = D + 1 = 2.2$. b. "Beware of the fractal rabbit", $D_s = D + 1 = 2.06$. The horizontal axis is yardstick size in arbitrary logarithmic units. (From Farin et al., 1985; the quartz particle outline is from Orford and Whalley, 1983).

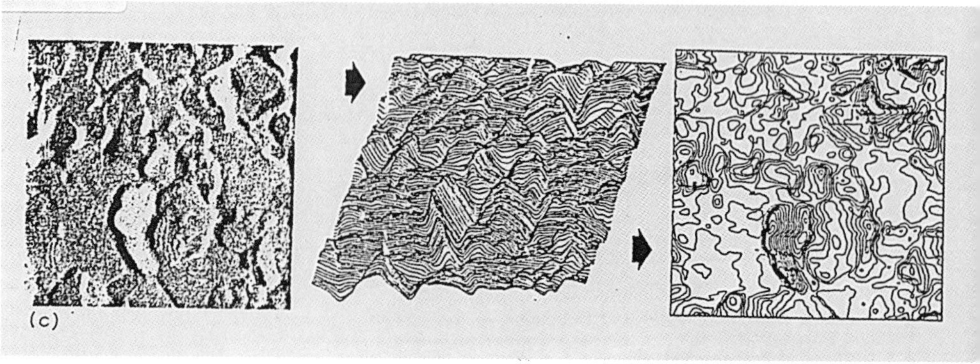

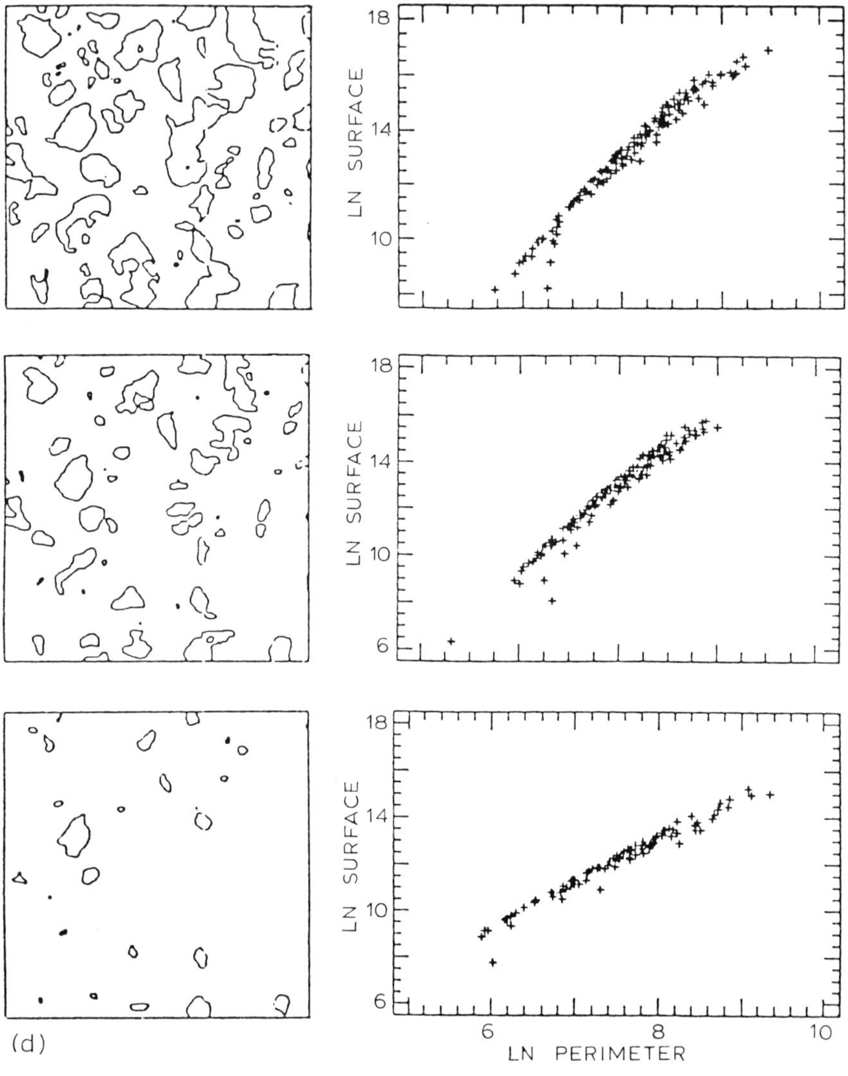

(d)

Fig. 4.10 (continued). c. Electron photomicrograph of a magnesium thin film top surface, the restored relief, and the isoheight contours, 60 Å apart from each other. d. Perimeter/area analysis of a limited part of the surface. Height levels are, from top to bottom, 45, 90, and 180 Å (D_s = 2.11, 2.13, 2.17, respectively). (From Varnier et al., 1988.)

fracture surfaces: Pande et al., 1987a, b; mountain profiles: Kaye, 1989, p. 385). As seen in Figure 4.10b, the obtained dimension is not always physically meaningful.

– The perimeter/area analysis, first proposed by Mandelbrot et al. (1984) was discussed in detail in Section 1.5.3 ("slit island" analysis of fracture surfaces). An interesting application was reported by Varnier et al. (1988), who determined the fractal dimension of thin-film top surfaces from the P/A analysis of the isoheight contours (Figure 4.10c, d). The technique seems to be applicable to the fractal analysis of geomagnetic and gravity anomaly maps.

– The third technique had been used in the previously mentioned biological examples, that is, to find D_s for the vegetation (Morse et al., 1985) or for cell membranes (Paumgartner et al., 1981); and for the estimation of the fractal dimension of the pore wall of sandstones from thin sections (Katz and Thompson, 1985; see Section 4.5, Figure 4.41).

4.2 Simple geometrical models of fractal surfaces

The easiest way to construct a fractal surface is to slide a fractal curve, say a triadic Von Koch curve, along a direction perpendicular to the plane of the curve (Figure 4.11). In order to cover a strip of width l of this surface with small rectangular tiles of area $\varepsilon \times \varepsilon$ we shall need:

$$N(\varepsilon) \propto \frac{l}{\varepsilon}\frac{1}{\varepsilon^{D}} \propto \frac{1}{\varepsilon^{D+1}} \tag{4.2.1}$$

tiles, where $D = \log 4/\log 3 = 1.262$ is the dimension of the Von Koch curve (see Section 1.4.1). The total area increases for $\varepsilon \to 0$ as:

$$A(\varepsilon) \propto \varepsilon^{2-D_s} \qquad (D_s = 2.262), \tag{4.2.2}$$

that is, the surface is 2.262-dimensional. A variation of this model was used by Bunde et al. (1987) to investigate why the experimentally found surface-fractal dimension depends on the size of the adsorbed probe molecules (as, e.g. what is the cause of the bifractality in Figure 4.5b?).

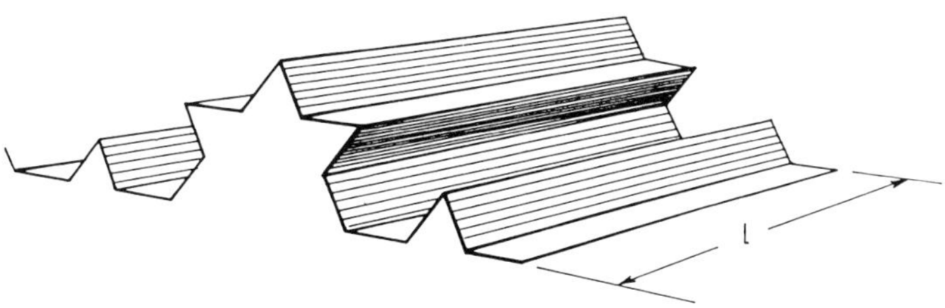

Fig. 4.11. Anisotropic Von Koch surface, $D_s = 2.262$ (from Feder, 1988).

A similar model was used by Liu, Kaplan and Gray (Liu, 1985, 1986; Liu et al., 1986a, b; Gray et al., 1985; Kaplan and Gray, 1985) to study the electrical ac response of fractally rough surfaces. As known since the late 1920's, the impedance $Z(\omega)$ across a rough interface between two materials is proportional to $(j\omega)^{-\eta}$ as the angular frequency tends to zero, where the exponent η lies between 0 and 1. Liu, Kaplan and Gray developed an exactly solvable fractal surface model (Figure 4.12) to theoretically explain this phenomenon, and proved that η is related to the surface fractal dimension as $\eta = 3 - D_s$. The model is based on a Cantor bar; in Figure 4.12b (representing the simplest case) the electrolyte side is shown in black, the electrode is white. The grooves in the electrodes are seen as protrusions on the electrode side. Each groove has the self-similar structure with a magnification ratio a, $a > 2$. The dimension of the Cantor bar is $D_s = \log 2/\log a < 1$, the surface itself has the dimension $D_s = 2 + D$.

In the previous examples, the fractal dimension of the rough surfaces was deter-

Fig. 4.12. Fractal model for the ac response of a rough interface. a. Electron micrograph of the surface of a solid electrolyte (Ag β-alumina).

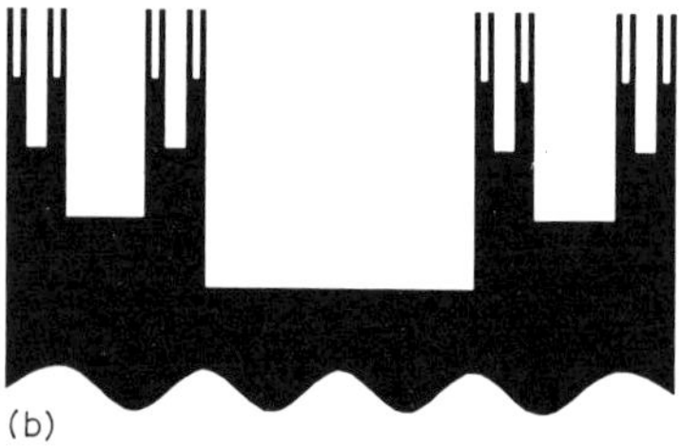

Fig. 4.12 (continued). b. Cantor bar model of the electrolyte/electrode interface.

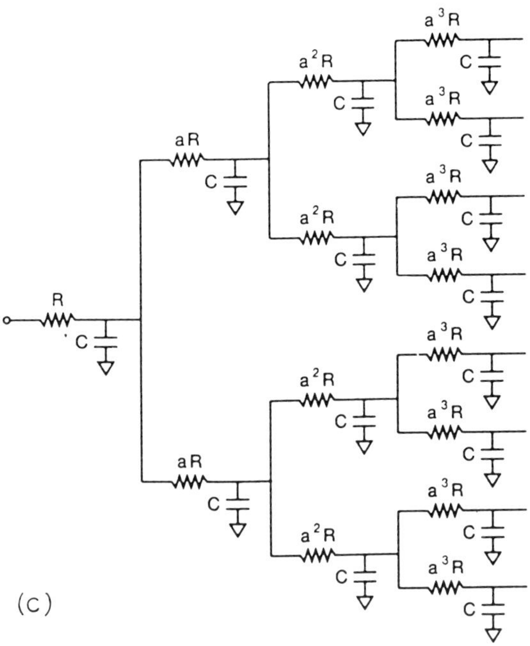

Fig. 4.12 (continued). c. Equivalent circuit of the model interface. (From Liu et al., 1986a.)

mined by covering the surface with small symmetrical objects such as squares [in equations (4.1.15) or (4.2.2)] or spheres [equation (4.1.17c)].

If the total surface area $A(\sigma)$ scaled as:

$$A(\sigma) \propto \sigma^{(2-D_s)/2} \tag{4.2.3}$$

for $\sigma \to 0$, where σ is the area of a small "tile", the surface was considered D_s-dimensional. Although there are techniques (Clarke, 1986; Romeau et al., 1986; Creutzburg and Ivanov, 1989) where the surface fractal dimension is estimated from a coverage with triangles, this approach requires extreme caution (Mandelbrot, 1986). Indeed, as H.A. Schwarz (1890, see Gelbaum and Olmsted, 1964) discovered, the triangulation of even the simplest regular surfaces could lead to paradoxical results: in case of a cylinder of radius R and height H, for example, different triangulations can yield any value from the true area $2\pi RH$ to infinity, and the resulting approximating surface can become arbitrarily corrugated, mimicking a fractal (Mandelbrot, 1986; Feder, 1988). Suppose we use a triangulation as shown in Figure 4.13.

We divide the surface into m bands and n sectors, estimate the surface area as the sum of the areas of all triangles, $A(m,n)$, and make the division finer and finer by letting $m \to \infty$ and $n \to \infty$ in an arbitrary fashion. Because the cylinder has a

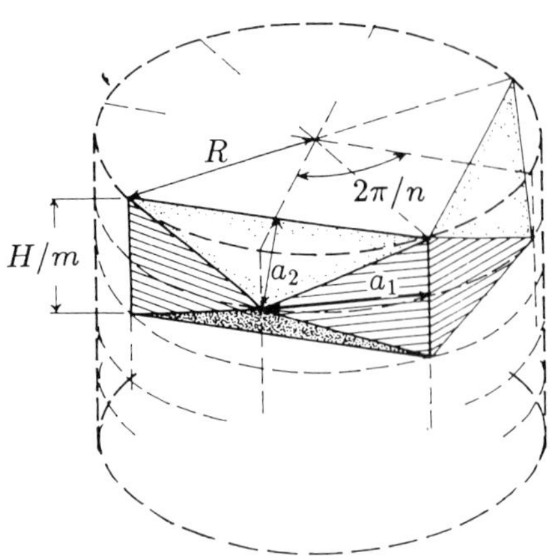

Fig. 4.13. The Schwarz area paradox. Approximation of the vertical surface of a cylinder by triangulation. (From Feder, 1988.)

"smooth" vertical surface of area $A = 2\pi RH$, we naturally expect that:

$$\lim_{\substack{m\to\infty \\ n\to\infty}} A(m,n) = 2\pi RH \tag{4.2.4}$$

Straight trigonometry shows however that this is *not* the case.

Indeed, for $n \gg 1$ we have:

$$A(m,n) \approx \pi RH \left\{ 1 + \left[1 + \left(\frac{R}{H}\right)^2 \left(\frac{\pi^2 m}{n^2}\right)^2 \right]^{1/2} \right\} \tag{4.2.5}$$

where the first term is the contribution of the triangles of the type marked a_1 in Figure 4.13, the second term with the square root is the total area of the a_2-type triangles. We see that:

$$\lim_{\substack{m\to\infty \\ n\to\infty}} A(m,n) = 2\pi RH = A$$

if and only if $(m/n^2) \to 0$. For a triangulation where:

$$m = \lambda n^2 \qquad (\lambda > 0)$$

we get:

$$\underset{m,n\gg 1}{A(m,n)} \approx \text{const} > A$$

As a matter of fact, this area can be made arbitrarily large with a proper choice of $\lambda \gg 1$.

If we choose m, n such that:

$$m = n^{2+\delta} \qquad (\delta > 0) \tag{4.2.6}$$

we find that the total area tends to infinity as:

$$\underset{m,n \gg 1}{A(m,n)} \propto n^{\delta} \tag{4.2.7}$$

(a)

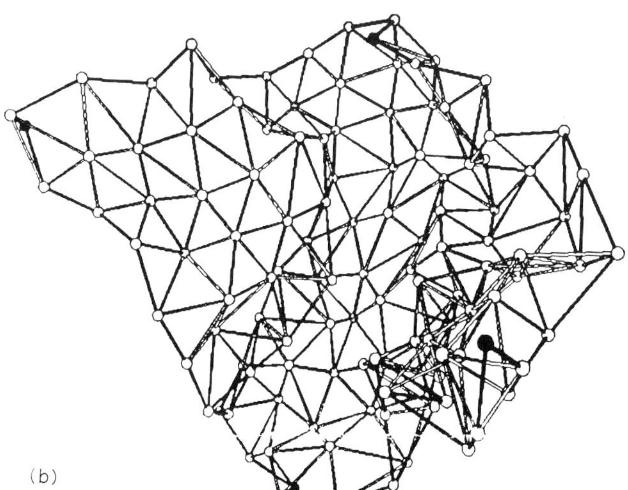

(b)

Fig. 4.14. Examples for triangular and prismatic fractal surfaces. a. Quartz druse partially covering a grain, adjacent grains have large rhombohedral overgrowths (*A*). Coarse grained, poorly sorted sandstone, Toolachee Formation, South Australia. (From Schulz-Rojahn and Phillips, 1989.). b. Triangular net model of a self-avoiding molecular surface (from Kantor et al., 1986).

Je m'imaginais voyager à travers un diamant.

(c) *Voyage au centre de la terre.*

Fig. 4.14 (continued). c. "I thought I'm walking across a diamond" — classical illustration of Jules Verne's "Journey to the Centre of the Earth".

As in this case the area of an individual small triangle will be $\sigma \propto 1/n^3$, equation (4.2.7) can be expressed in terms of σ as:

$$\underset{m,n \gg 1}{A(m,n)} \propto n^{\delta} \propto \left(\frac{1}{n^3}\right)^{-\delta/3} = \sigma^{-\delta/3} \tag{4.2.8}$$

that is, according to equation (4.2.3), the approximation of the surface actually mimics the scaling behaviour of a $D_s = 2 + 2\delta/3$-dimensional fractal if $0 < \delta < 3/2$. The reason for the infinite area is that as the ratio m/n^2 increases, the thin a_2-type triangles tend to be perpendicular to the surface of the cylinder. Observe that no matter how rough these boundary layers are, in the limit their thickness tends to zero. As Mandelbrot (1986) noted, "good airports' runways are as rough as the Himalayas — but their vertical scale is smaller".

Quartz overgrowths, observed under the scanning electron microscope in diagenetically altered sandstones (Figure 4.14a) constitute a fractal surface similar to that in the Schwarz paradox. Kantor et al. (1985) approximated fractal surfaces by hard spheres tethered together by strings into a triangular net (Figure 4.14b); Jingzhong et al. (1989) proposed a 2.3347-dimensional Al–Mn quasicrystal model based on a generalised Von Koch construction with icosahedra (see also Burkov, 1985). The heroes of Jules Verne's "Voyage au centre de la terre" also found diamond-like fractal surfaces in the centre of the Earth (Figure 4.14c).

The most powerful geometrical model of the internal fractal surfaces of sedimentary and igneous rocks is the "pigeon-hole" model of Pape and his group (1981a; 1985a, b; a recent review is Pape et al., 1987a. Pape et al., 1981b, 1982, 1984 apply the pigeon-hole model to relate specific surface of the rock to permeability). Figure 4.15 shows the pigeon-hole model in cross-section, we recognise a variant of the Koch construction, based on semicircles. To reproduce the experimentally found (Pape et al., 1982) $D_s = 2.36$ fractal dimension of the pore wall of the sedimentary rocks, Pape et al. assumed that the self-similarity quotient $v = r_{k+1}/r_k$ is equal to 0.25 and that the number of newly added inwardly concave semispherical pigeon-holes around any pigeon-hole of the previous generation is $N = 26.5$ (detailed

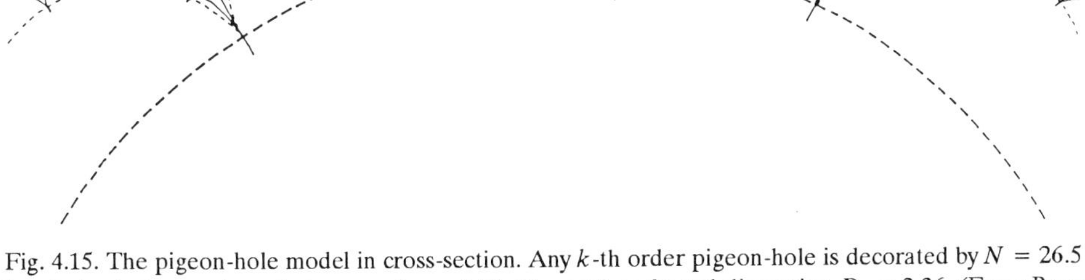

Fig. 4.15. The pigeon-hole model in cross-section. Any k-th order pigeon-hole is decorated by $N = 26.5$ hemispherical new pigeonholes, $r_{k+1}/r_k = 0.25$. The surface fractal dimension $D_s = 2.36$. (From Pape et al., 1987a.)

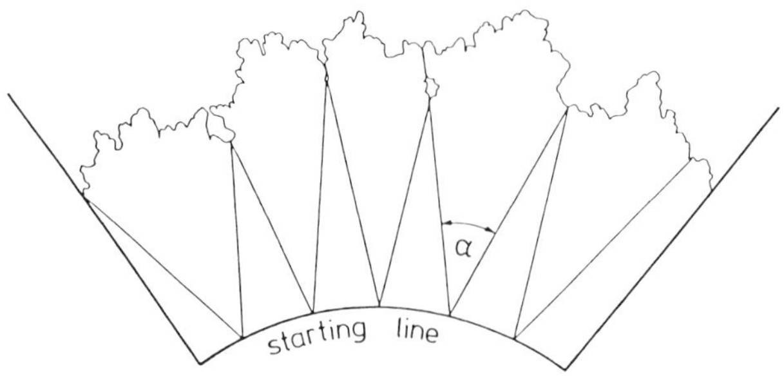

Fig. 4.16. Radial migration of a coloured solution on a sheet of paper, $D = 1.24$ (from Pape et al., 1987a).

calculations are given in Pape et al., 1981a; the non-integer N value takes care of the distortions at the contact of pigeon-holes).

In Pape et al. (1987a) it is shown that the pigeon-hole model is closely related to polar growth (compare Figures 4.15 and 4.16). Figure 4.17a–d shows some geologi-

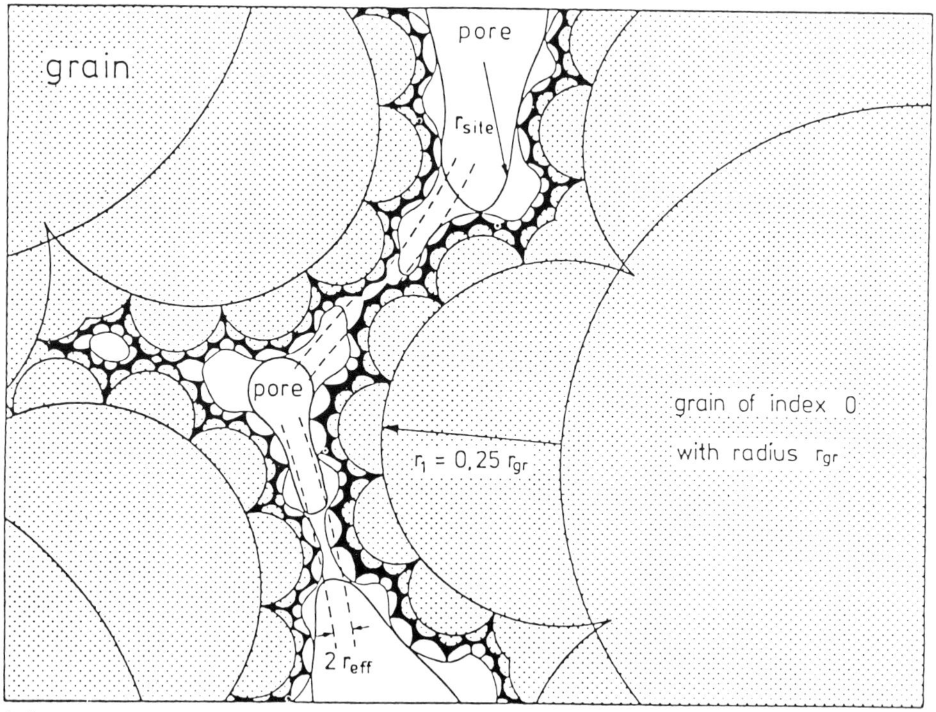

Fig. 4.17. Geological and biological applications of the pigeon-hole model. a. Pore space model of a sandstone. r_{site} = radius of the pore, r_{eff} effective hydraulic radius (from Pape et al., 1987a).

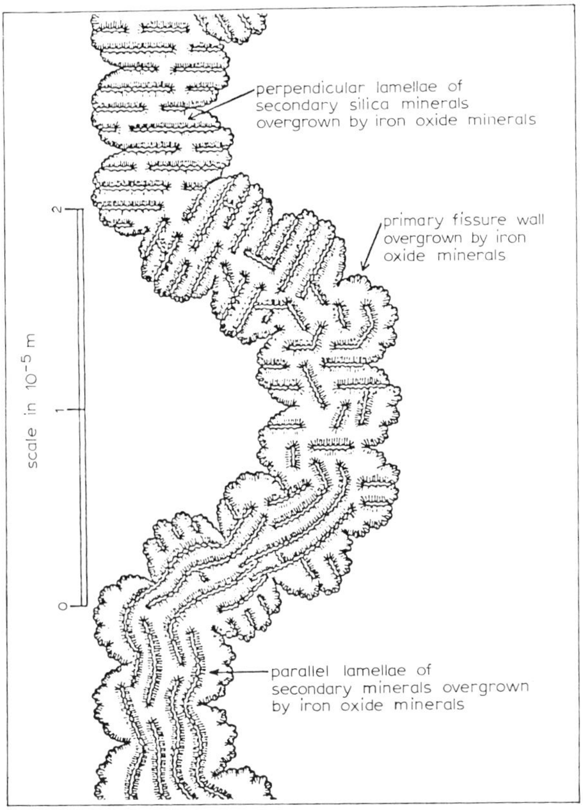

(b)

Fig. 4.17 (continued). b. Model of the microfissure of Falkenberg granite (Bavaria, Germany) (from Pape et al., 1985c).

cal and biological examples: (a) pore space of a sedimentary rock; (b) microfissure in granite; (c) the mouse small intestine (see also Goldberger et al., 1990); and (d) the shape of tree canopies. Pape et al. (1981a) derive from the pigeon-hole model the allometric relation:

$$S \propto V^{2.36/3} = V^{0.79} \tag{4.2.9}$$

between the surfaces and volumes of animal organisms, which is close to the $V^{0.75}$ law (Sernetz et al., 1985; Figure 4.3) discussed in Section 4.1. A pigeon-hole-like model of coastal erosion was discussed in Section 1.4.2 (Figures 1.32 and 1.33); similar features can be observed on artistic representations of chaotic natural phenomena (Figure 4.18) as e.g the traditional representation of flames in Tibetan Thangka painting, or Hokusai's "Wave". The physicist-photographer McGuire (1988) writes: "I can think of no more explicitly fractal image than Hokusai's Great Wave of Kanazawa. The great capital C shape of the wave on the left as it threatens to

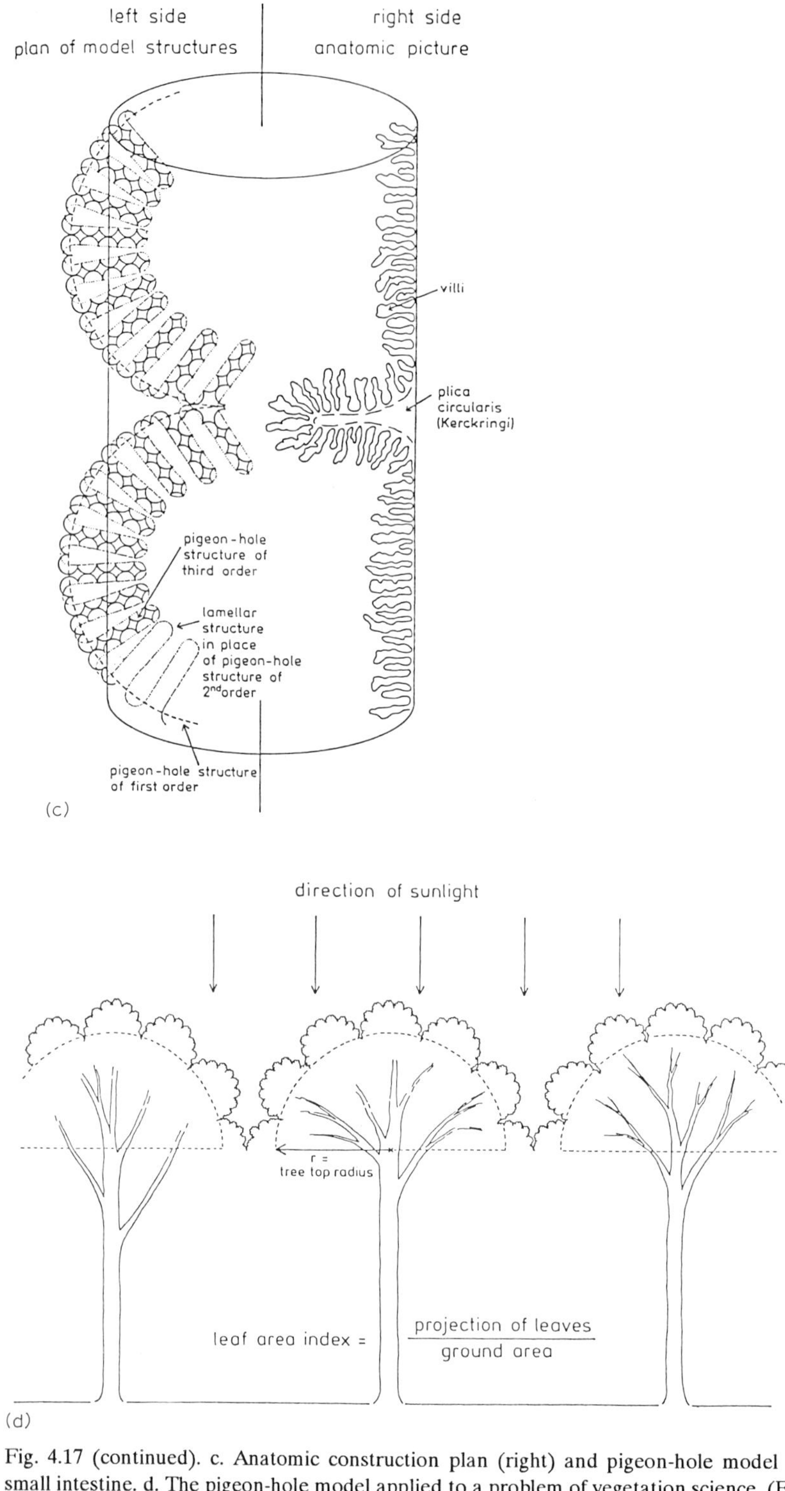

Fig. 4.17 (continued). c. Anatomic construction plan (right) and pigeon-hole model (left) of mouse small intestine. d. The pigeon-hole model applied to a problem of vegetation science. (From Pape et al., 1981a.)

Fig. 4.18. Pigeon-hole like motifs in the representation of chaotic natural phenomena. a. Traditional representation of flames in Tibetan painting (from Jackson and Jackson, 1988). b. Hokusai's "The Wave" (International Society for Educational Information, Tokyo).

engulf the small sampan like boat repeats itself in ever smaller waves and wavelets throughout the picture."

4.3 Analytical treatment of fractal surfaces

Fractal surfaces are rough. If one considers surface roughness as a random variation of height (or depth) values, the realm of statistics offers a multitude of standard measures of data dispersion to describe roughness: root mean square roughness, standard variation, or variance.

If the surface roughness is analysed along some line e at equidistantly spaced points $\{e_i\}$, the rms roughness is generally defined as:

$$\text{rms roughness} = \left\{ \frac{\sum_{i=1}^{n} (S_i - \overline{S}_i)^2}{(n-1)} \right\}^{1/2} \tag{4.3.1}$$

where S_i is the height of the surface $S(x,y)$ at the i-th sampling value e_i; $\overline{S}_i$ is some predicted value, as e.g. the smoothed bathymetry in case of seafloor roughness (Fox and Hayes, 1985).

This very useful measure, however, only characterises the overall strength of roughness but reveals no information on its spatial variation, as e.g. how rapidly does it change along the random line e.

More complex measures of roughness include the autocovariance function:

$$R(i,k) = \langle (S_i - \overline{S}_i)(S_{i+k} - \overline{S}_{i+k}) \rangle \tag{4.3.2}$$

and the variance of increments:

$$V(i,k) = \langle [S_i - S_{i+k}]^2 \rangle \tag{4.3.3}$$

It is generally assumed that the functions R and V do not depend on the direction of the line e, that is the rough surface is *isotropic* (Figure 4.23, however, is example for an *anisotropic* surface) and also, that the statistical properties of S_i do not depend on the position e_i. In the latter case the surface is called *statistically homogeneous*, the sequence of height values $\{S_i\}$ is *stationary*, and R will only depend on k:

$$R(k) = \langle (S_i - \overline{S})(S_{i+k} - \overline{S}) \rangle \tag{4.3.4}$$

By normalising the autocovariance by the sample-variance $R(0)$, we obtain the *autocorrelation function*:

$$\rho(k) = \frac{R(k)}{R(0)} \tag{4.3.5}$$

For isotropic, homogeneous surfaces the variance of increments $V(i,k)$ [defined in equation (4.3.3)] is also a function of k alone. Geostatisticians (Journel and Huijbregts, 1978) generally use the half of $V(k)$, called the *semivariogram*. For statistically homogeneous, isotropic surfaces the semivariogram, σ, and the covariance function are simply related:

$$\sigma(k) = \frac{\langle (S_i - S_{i+k})^2 \rangle}{2} = \frac{\langle ([S_i - \overline{S}] - [S_{i+k} - \overline{S})^2 \rangle}{2}$$
$$= \langle (S_i - \overline{S}) \rangle^2 - \langle (S_i - \overline{S})(S_{i+k} - \overline{S}) \rangle$$
$$= R(0) - R(k) = R(0)[1 - \rho(k)] \quad (4.3.6)$$

The surface $S(x,y)$ can also be characterized by its power spectrum:

$$S(k) = 4 \int_0^\infty R(x) \cos(2\pi k x) \, dx \quad (4.3.7)$$

(see e.g. Bell, 1975).

The rough surfaces encountered in practice are continuous but generally not differentiable. The continuity of $S(x)$ means that $|S(x + \delta) - S(x)| \to 0$ as $\delta \to 0$. The differentiability of $S(x)$ however is not guaranteed by this condition, as it also requires that:

$$\lim_{\delta \to 0} \frac{[S(x + \delta) - S(x)]}{\delta}$$

should exist. Consider a class of functions whose increments:

$$S(x + \delta) - S(x)$$

depend on δ as stated in the Lipschitz condition:

$$|S(x + \delta) - S(x)| \propto \delta^\alpha \qquad (0 < \alpha \leq 1) \quad (4.3.8)$$

or, if $S(x)$ is random:

$$\langle \{S(x + \delta) - S(x)\}^2 \rangle = 2\sigma(\delta) \propto \delta^{2\alpha} \quad (4.3.9)$$

The exponent α is called the Lipschitz exponent (Rothrock and Thorndike, 1980), equation (4.3.9) is often quoted in the form:

$$\langle \{S(x + \delta) - S(x)\}^2 \rangle \propto \delta^{2H} \quad (4.3.9a)$$

where the exponent H is called the Lipschitz-Hölder exponent or Hurst exponent. The derivative of S behaves as $\delta^{\alpha-1}$ (for $\alpha \to 0$), which is only finite for $\alpha = 1$. For $\alpha = 0$ the function $S(x)$ fails to be continuous. Thus, the range $0 < \alpha < 1$ can be used to define functions which are continuous but non-differentiable at different degrees. As α increases through this range, the functions become smoother.

The behaviour of the autocovariance $\langle [S(x)S(x + \delta)] \rangle$ for $\delta \ll 1$ is another indicator of roughness. From equations (4.3.6) and (4.3.9) we have:

$$R(\delta) \approx R(0) - c\delta^{2\alpha} \quad (4.3.10)$$

(see Figure 4.19). It can be shown that for random functions with an autocovariance

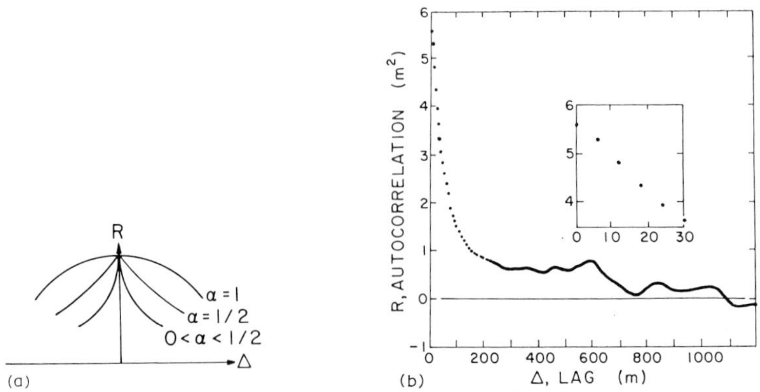

Fig. 4.19. a. Autocovariance of a non-differentiable function for small lags δ. α is the Lipschitz exponent defined in equation (4.3.9). b. Autocorrelation of a sonar profile of the underside of sea ice. The inset shows R for small lags. (From Rothrock and Thorndike, 1980.)

of the form (4.3.10), their power spectrum $S(k)$ falls off for large wave numbers as:

$$S(k) \propto k^{-s} \qquad \text{with } s = 2\alpha + 1 \tag{4.3.11}$$

Let us now assume that the vertical profile $S(x)$ of the surface is a *self-similar fractal curve* with dimension D. Let us consider a finite interval $[x_1,x_2]$, and determine the number $N(\delta)$ of small squares of side δ that are necessary to cover the trace of $S(x)$ in this interval. As the interval $[x_1,x_2]$ was divided to $\propto \delta^{-1}$ small intervals $[x_i,x_i + \delta]$, and by the Lipschitz condition (4.3.8) the increment of the function $S(x)$ over any of these small intervals can be covered by $\delta^{\alpha-1}$ small squares, we need altogether $\propto \delta^{\alpha-2}$ squares of size δ to cover the whole curve between x_1 and x_2. Consequently, the fractal dimension D and the Lipschitz exponent α of the vertical surface profile are related as:

$$D = 2 - \alpha \qquad (1 \leq D < 2) \tag{4.3.12}$$

The fractal dimension of the surface itself is:

$$D_{\text{surf}} = D + 1 \qquad (2 \leq D_{\text{surf}} \leq 3) \tag{4.3.13}$$

Summarising: for a vertical profile $S(x)$ of a D_{surf}-dimensional fractal surface:

– $S(x)$ is a $D = D_{\text{surf}} - 1$- dimensional curve.

– $S(x)$ is continuous but non-differentiable, it satisfies the Lipschitz condition (4.3.8) with the exponent:

$$\alpha = 2 - D \tag{4.3.14}$$

– The autocovariance of $S(x)$, $R(\delta) = \langle S(x)S(x + \delta)\rangle$ behaves for small δ as:

$$R(\delta) \approx R(0)\{1 - \text{const.}\, \delta^{2\alpha}\} = R(0)\{1 - \text{const.}\, \delta^{4-D}\} \tag{4.3.15}$$

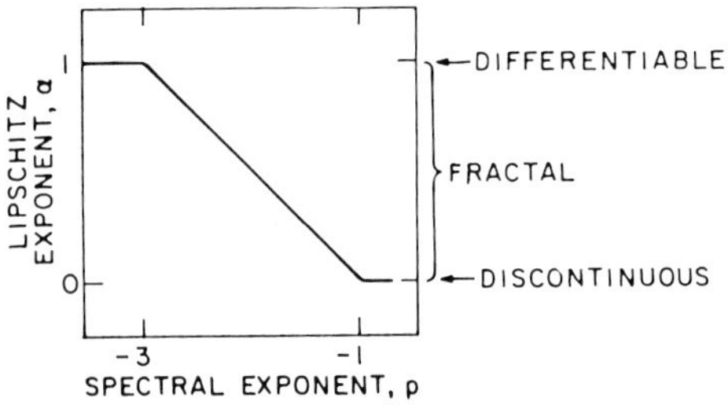

Fig. 4.20. Connection between the Lipschitz exponent and the fall-off of the power spectrum, $S(k) \propto k^{-p}$. In certain cases, the fractal dimension is also related to α: $D = 2 - \alpha$. (From Rothrock and Thorndike, 1980.)

– The power spectrum of $S(x)$ falls off for high wave numbers as:

$$S(k) \propto k^{-2\alpha-1} = k^{-(5-2D)} \tag{4.3.16}$$

The interconnection between the Lipschitz exponent and the spectral exponent s of (4.3.11) is shown in Figure 4.20.

Strictly speaking, there are only two fractal functions, the fractional Brownian function and the Weierstrass-Mandelbrot function where the above properties have been rigorously established (Berry and Lewis, 1980; Mandelbrot and Van Ness, 1968; Voss, 1985a, b, 1988).

However, it can be proved that the equation:

$$2\sigma(\delta) = \langle\{S(x+\delta) - S(x)\}^2\rangle \propto \delta^{2H} \tag{4.3.9a}$$

is characteristic of *statistically self-similar* functions. Indeed, if (4.3.9a) holds, the variations over any scale $r\delta$ are related to the variations over a scale δ by:

$$\sigma(r\delta) = r^{2H}\sigma(\delta) \tag{4.3.17}$$

that is, the variance at any scale can be determined from the variance measured at another scale. Or, in other words, samples of the function S have the same appearance at different scales when subjected to the proper r^H-fold magnification. Conversely, suppose that $S(x)$ is statistically self-similar, that is its variogram σ satisfies:

$$\sigma(r\delta) = f(r)\,\sigma(\delta) \tag{4.3.18}$$

for all positive values r. Then, by Pexider's Theorem (THEOREM 8 in Appendix A1.1), $\sigma(\delta)$ is necessarily of the form:

$$\sigma(\delta) \propto \delta^{\beta} \tag{4.3.19}$$

where the exponent β is positive, because σ is an increasing function of δ, and it can be taken as $\beta = 2H$.

Thus, the relations (4.3.9), (4.3.15) and (4.3.16) are very likely true for most random fractal functions, and are generally used for the estimation of the fractal dimension.

The determination of D from the slope of the log $S(k)$ vs. log k plot is the most widely applied technique of analysing fractal surfaces [Brown and Scholz (1985), Fox and Hayes (1985), Rothrock and Thorndike (1980), Sayles and Thomas (1978a), Turcotte (1987, 1989). Earlier studies, where it has been verified that for different rough surfaces the power spectrum falls off as $S(k) \propto k^{-s}$, with $0 \leq s \leq 4$, but this had not been related to the fractality of the surfaces, include Bell (1975), Berkson (1975), Hibler and Leshack (1972), Jaeger and Schuring (1966)].

A few examples are shown in Figure 4.21, from different fields of Earth Sciences.

The power spectrum decaying as k^{-2} in Figure 4.21b is associated with 2.5-dimensional surfaces, such as the random Brownian surface. This is the most frequently encountered spectrum in fractal surface studies. In a much debated paper Sayles and Thomas (1978a) claimed that all random surfaces occurring in nature

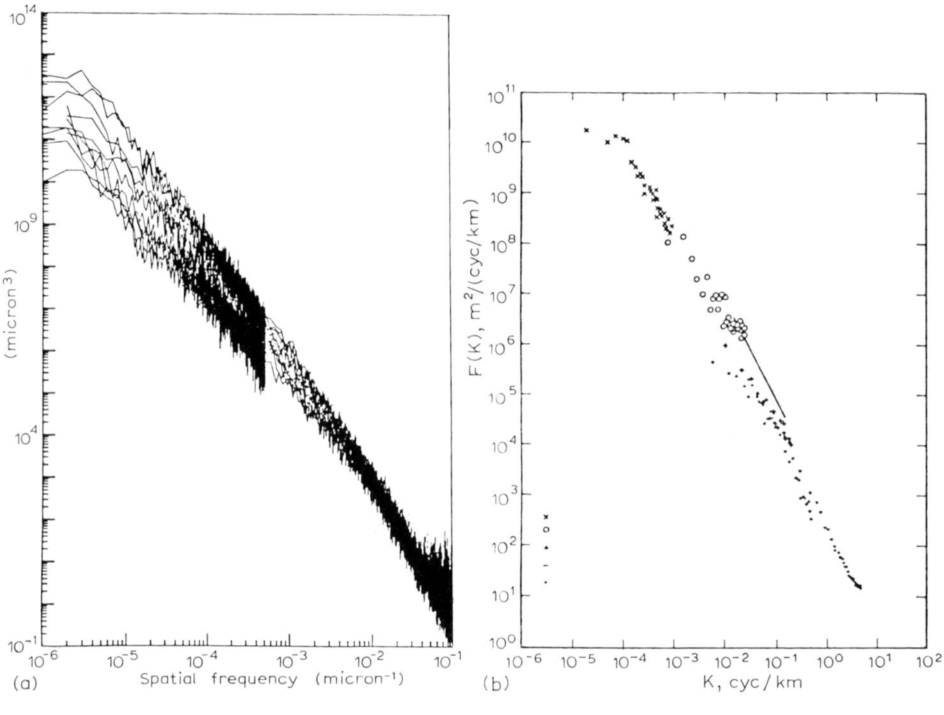

Fig. 4.21. Spectral analysis of fractal surfaces. a. Power spectrum of the topography of natural rock surfaces. $S(k) \propto k^{-s}$ where s ranges between 1.63 and 3.35 (that is, D ranges between 0.82 and 2.68). The topography of these surfaces cannot be described by a single fractal dimension. (From Brown and Scholz, 1985.) b. Spectral estimates for the topography of the Pacific Ocean floor, collected by Bell (1975). $S(k) \propto k^{-2}$, that is, $D \approx 1.5$. (From Fox and Hayes, 1985.)

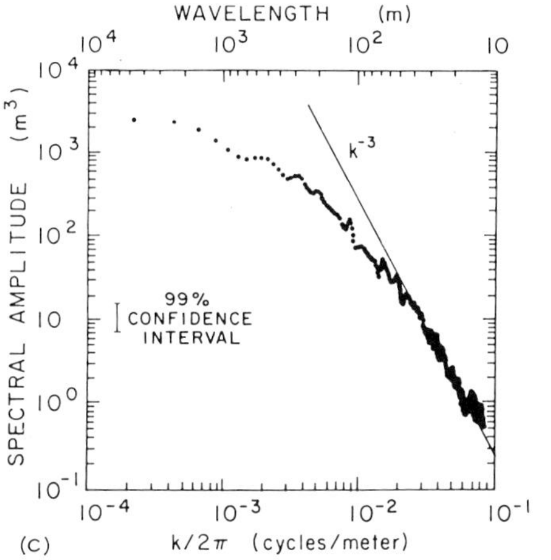

Fig. 4.21 (continued). c. Power spectrum of the underside of sea ice, offshore Beaufort Sea. $S(k) \propto k^{-3}$, that is, $D \approx 1$. (From Rothrock and Thorndike, 1980.)

have this type of spectra, and substantiated this by plotting the spectra of 23 different surfaces, ranging from supertanker hulls to concrete runaways, along a single k^{-2} curve over 8 decades of wavelength. They assumed that the power spectra of these surfaces have the universal form:

$$G(\omega) = \frac{2\pi L}{\omega^2} \tag{4.3.20}$$

and claimed that the single constant L, called "topothesy" (from the Greek $\tau o\pi o$-$\theta\epsilon\sigma\acute{\iota}\alpha$ = description of a region), uniquely defines the statistical geometry of the surface. A discussion between Berry and Hannay (1978) and Sayles and Thomas (1978b) has led to a re-examination of the original data (Figure 4.22), showing that the spectral exponent is not necessarily 2 but ranges between 1 and 3 with a mean value of 2. (The controversy is reviewed in Brown and Scholz, 1985, and Feder, 1988). The generalisation of the concept of "topothesy" for a more general spectral decay of the form $k^{-s} = k^{-(5-2D)}$ is due to Berry (Berry and Hannay, 1978; Berry, 1979a; Berry and Blackwell, 1981), who defined the topothesy as the distance L for which:

$$\langle\{S(x+L) - S(x)\}^2\rangle^{1/2} = L \tag{4.3.21}$$

Because of the relation (4.3.17) we have:

$$\sigma(\delta) = \sigma\left(L\frac{\delta}{L}\right) = \left(\frac{\delta}{L}\right)\sigma(L)$$

$$= L^{2-2\alpha}\,\delta^{2\alpha} \tag{4.3.22a}$$

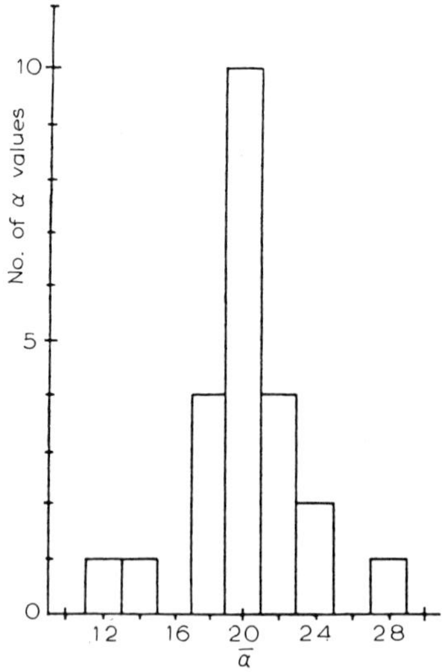

Fig. 4.22. Histogram of the spectral exponents for 23 surfaces (from Sayles and Thomas, 1978b).

or, in terms of D and s:

$$\sigma(\delta) = L^{2D-2}\,\delta^{4-2D} \tag{4.3.22b}$$

$$= L^{3-s}\,\delta^{s-1} \tag{4.3.22c}$$

Three special cases are of particular interest. $D = 1$ ($\alpha = 1, s = 3$; see Figure 4.21c) is the *marginal fractal*: $S(x)$ is an almost smooth function, the topothesy disappears from equations (4.3.22); as it has been often noticed (Jaeger and Schuring, 1966; Nye, 1973; Rothrock and Thorndike, 1980) there is no natural length-scale to the surface. $D = 1.5$ ($\alpha = 0.5, s = 2$; see e.g. Figure 4.21b) corresponds to the *Brownian fractal surface*. $D = 2$ ($\alpha = 0, s = 1$) is the *extreme fractal*, where the graph of $S(x)$ almost fills a finite area. For functions of time it corresponds to the so-called $1/f$ *noise*, to be discussed in Section 5.1.2.

It should be noted that Bell (1975) explained the observed spectrum of the sea-floor topography in terms of a model based on the random distribution of hills. He showed that the inverse-square-law power spectrum corresponds to the case when the energy of formation is uniformly distributed over hill sizes.

In practice, the estimation of power spectra requires extreme care, as the availability of a (software or hardware) FFT is no substitute for the personal responsibility in optimally selecting the profile length, sampling rate, prewhitening, windowing, etc. Familiarity with Blackman and Tukey's (1958) classic "The Measurement of Power Spectra" is a must; some very useful practical details can be found in Fox and Hayes, 1985.

If the surface is anisotropic, the fractal dimensions determined from the power spectra of $S(x,y)$ computed along different directions might be different (Figure 4.23).

In case of large-scale isotropy, it is common practice to expand the topography of

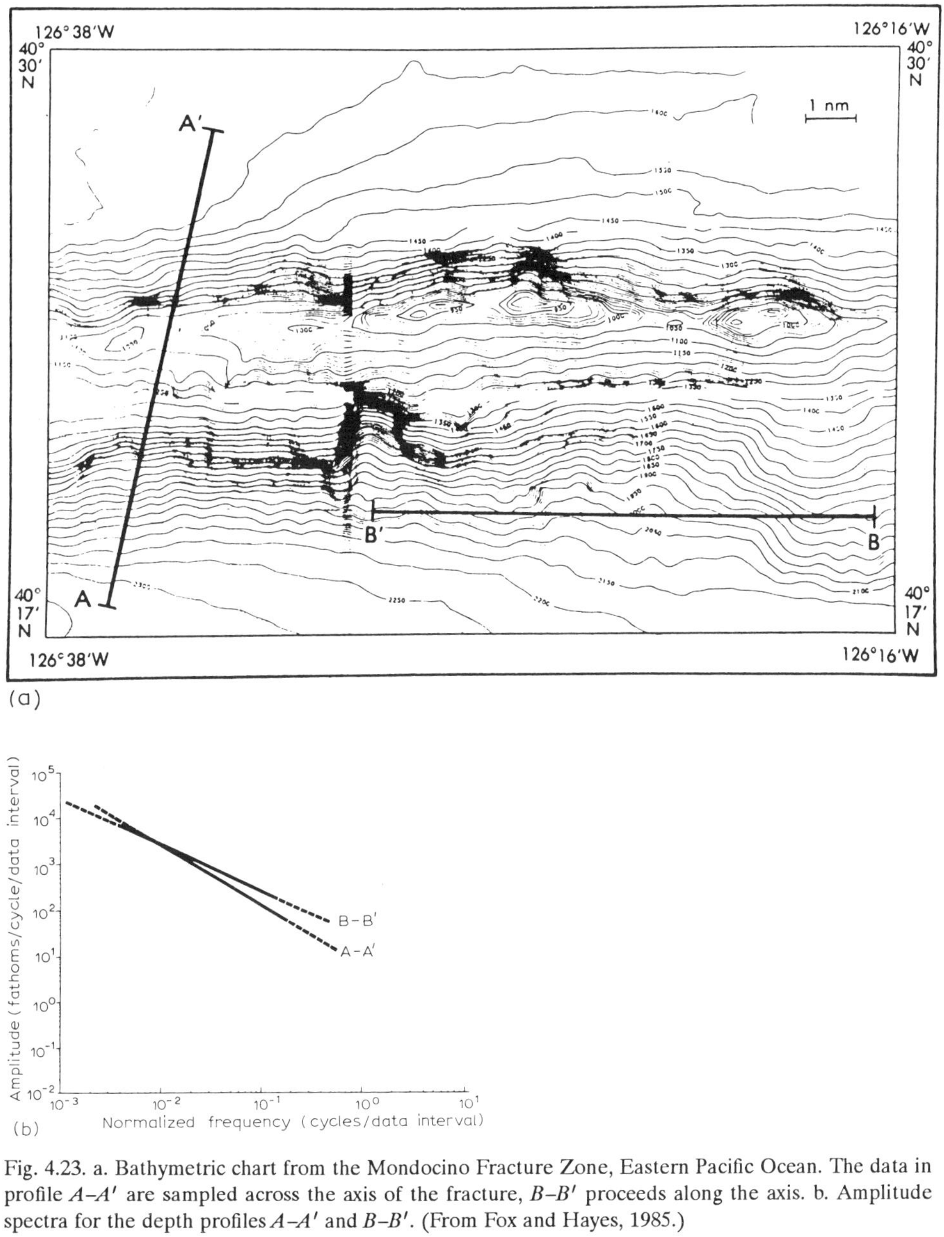

Fig. 4.23. a. Bathymetric chart from the Mondocino Fracture Zone, Eastern Pacific Ocean. The data in profile A–A' are sampled across the axis of the fracture, B–B' proceeds along the axis. b. Amplitude spectra for the depth profiles A–A' and B–B'. (From Fox and Hayes, 1985.)

the Earth (or other planets) in terms of spherical harmonics (Turcotte, 1987, 1989) as:

$$R(\theta, \varphi) = R_0 \left\{ 1 + \sum_{l=1}^{\infty} \sum_{m=0}^{l} [C_{\text{tlm}} \cos(m\varphi) + S_{\text{tlm}} \sin(m\varphi)] P_{\text{lm}}(\sin\theta) \right\} \tag{4.3.23}$$

where R_0 is the reference radius, R is the radius, θ and φ are polar coordinates, C_{tlm} and S_{tlm} are coefficients, P_{lm} are normalized associated Legendre functions. [The coefficients of the harmonic expansion (4.3.23) were published by Balmino et al. (1973) for the Earth, by Bills and Ferrari (1977) for the Moon, by Bills and Kobrick (1985) for Venus, by Bills and Ferrari (1978) for Mars.] The power spectral density $S_{\text{t}}(k_l)$ is given by:

$$S_{\text{t}}(k_l) = 2\pi R_0^3 \sum_{m=0}^{l} (C_{\text{tlm}}^2 + S_{\text{tlm}}^2) \tag{4.3.24}$$

where k_l is the wave number:

$$k_l = \frac{l}{2\pi R_0} \tag{4.3.25}$$

As found by Turcotte (1987; Figure 4.24), for the Earth, Venus, Mars and the Moon the respective spectra can be reasonably well approximated by an inverse square law:

$$S_{\text{t}}(k_l) \propto k_l^{-2} \tag{4.3.26}$$

corresponding to $D = 1.5$, that is to $D_{\text{surf}} = 2.5$-dimensional (Brownian) surfaces.

According to equations (4.3.9) and (4.3.10) the fractal dimension can also be estimated from the behaviour of the (semi-)variogram or auto-covariance function for small lags. Burrough (1981) observed that if $W(x)$ is a fractal random function, its fractal dimension can be found from the logarithmic derivative of its variance of increments:

$$\frac{\mathrm{d}\log\langle[W(x_0) - W(x_0 + x)]^2\rangle}{\mathrm{d}\log x} = 4 - 2D \qquad (\text{if } x \to 0) \tag{4.3.27}$$

As the computation of the variance of increments (or the half thereof, the "semi-variogram") is the first step in the interpolation procedure called "kriging" (Journel and Huijbregts, 1978), which is used in geostatistics for the estimation of mineral resources, contouring ground water levels, etc.; many published data are available in this form and their fractal dimension can be estimated. D can also be determined from covariances and — as Burrough (1981) noted — from the so-called "block variances" as well. (Block variances are used in agriculture and soil science, where

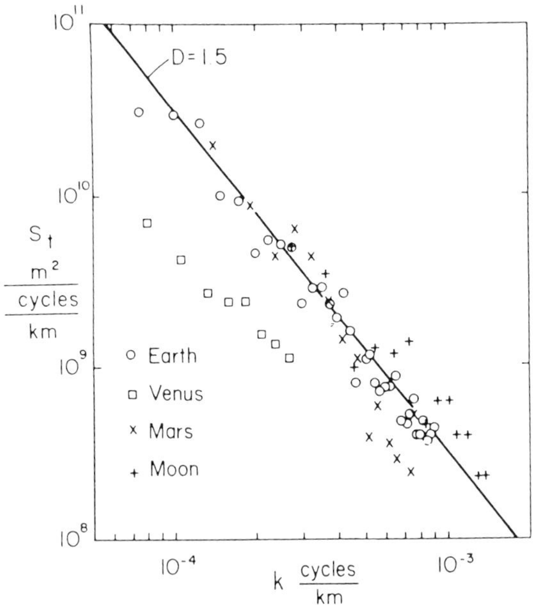

Fig. 4.24. Power spectral density S_t of the topography of different planets as a function of wave number k_l (from Turcotte, 1987).

the variances within blocks of equal size are plotted against block size:

$$s_k^2 = \frac{2}{k(k+1)} \sum_{h=1}^{k} (k - h + 1) V(h) \tag{4.3.28}$$

where s_k^2 is the block variance for block length k, $V(h)$ is the variance increment for lag h.) A few representative results of Burrough are shown below in Table 4.2.

A similar technique was used by Culling and Datko (1987), who estimated the fractal dimension of 15 sites of varied terrains in southern England from the semi-variograms of the profile curves.

As Feder (1988) noted in connection with Burrough's data: "the high values of D reported in the table for some soil and geological data seem to question the wisdom of interpolating mapping in certain instances, and it would seem worthwhile to use D as a guide to how further mapping and interpolation should proceed".

As a matter of fact, the problem has been recognised by geostatisticians (Agterberg, 1982; Figure 4.25). The first results concerning the conditional simulation of reservoir heterogeneities and geological surfaces with fractals have been reported by Hewett (1986), Emanuel et al. (1988), Hewett and Behrens (1988), Yfantis et al. (1988), Marechal (1989), Taggart and Salisch (1991).

TABLE 4.2

Fractal dimensions of environmental data

Location	Property	$D + 1 = D_{surf}$	Technique
Australia	Soil-		Estimated from block variance
	phosphorus level	3.0	
	pH	2.5	
	potassium level	2.6	
	bulk density	2.5	
Worldwide	Crop yields	2.6–2.8	
France	Iron ore in		Estimated from variogram
	chlorite	2.6	
	quartz	2.9	
Wales	Soil sodium content	2.7–2.9	
England	Soil electric resistivity	2.4–2.6	
India	Water table depth	2.6	
France	Topographic heights	2.5	
France	Sea anemones	2.6	Estimated from covariance

Source: Burrough (1981).

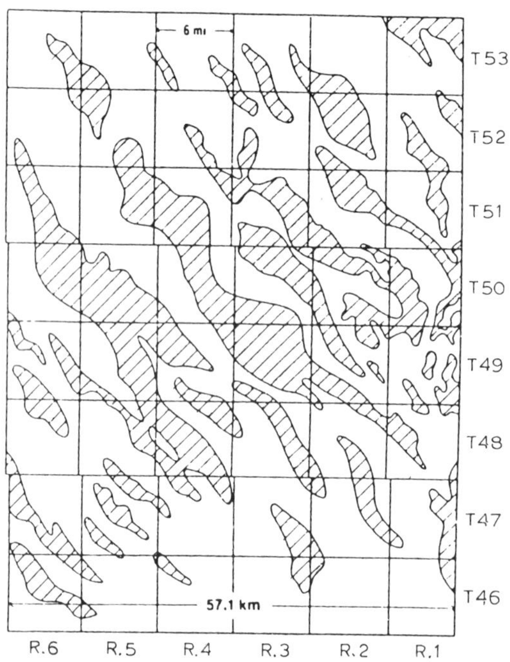

Fig. 4.25. The 30 feet contours of the Sparky Sandstone, Alberta, Canada. Ordinary kriging is not applicable for the statistical treatment of such surfaces. (From Agterberg, 1982, based on Burnett and Adams, 1977.)

4.4 Wave scattering from fractal surfaces

> 15 naive subjects (mostly language researchers) were shown digitized images of eight natural textured surfaces... They were asked 'if you were to draw your finger horizontally along the surface pictured here, how rough or smooth would the surface feel?' — PENTLAND (1984)

Fractal surfaces *look* rough. Pentland's (1984) interesting experiments, carried out with photographs of textures (drawn from Brodatz's, 1966, album) and plots of fractal Brownian curves and surfaces, have shown that there is a nearly perfect (0.91–0.98) correlation between our visual perception of roughness and the surface's fractal dimension.

Our ability to classify and recognise roughness is dependent on the information carried to the eye by light scattered from the object. (Pentland has shown, mathematically, that in this respect the rough surface and its photographic image are equivalent: "a spatially isotropic fractal Brownian surface produces an image whose intensity surface is fractal Brownian and whose fractal dimension is identical to that of the components of the surface normal, given a Lambertian surface reflectance function and constant illumination and albedo".) It is well known that waves acquire random fluctuations in their wave functions ψ and intensities $I = |\psi|^2$ when they encounter a random structure S (Chernov, 1960; Beckmann and Spizzichino, 1963; Tatarski, 1961; Frisch, 1968). Familiar examples are starlight passing through turbulent atmosphere ("twinkling"), sunlight reflected by water waves (Figure 4.26), radio

Fig. 4.26. Paul Klee, 1929: Fleeting appearances in the water (Galerie Beyeler, Basle).

or radar reflected by landscapes, light and sound scattered in materials near phase transition. If the random object S is fractal, that is it has a hierarchical structure down to arbitrarily small scales, the waves passing through it or diffracted by it will interact with its finest details and the observed waves will be rich in information even in the short-wave limit. In other words: *waves interacting with fractals become fractals themselves.*

Good general reviews on scattering by fractal objects are Berry (1979b) and Jakeman (1984, 1986). Waves that have encountered fractals were termed *diffractals* by Berry (1979a, b; Berry and Blackwell, 1981), they constitute a new regime in wave propagation theory, because conventional perturbation theory and geometrical optics do not apply.

In the case of simple deterministic fractals [Cantor bar or "Vicsek fractal" (Allain and Cloitre, 1985, 1986; Figure 4.27); or Menger sponge (Schmidt and Dacai, 1986)], the diffraction pattern is also self-similar and preserves the symmetries of the fractal object. In these examples the scattered intensity $I(k)$ could also be determined analytically, and — apart from the oscillations — its decay was found proportional to k^{-D} for large wave numbers, where D is the fractal dimension of the object.

In the more complicated cases of random fractals we can also observe the power-law fall-off of the scattered intensity:

$$I(k) \propto k^{-\alpha} \tag{4.4.1}$$

with "anomalous" non-integer exponents α. Equation (4.4.1) with non-integer exponents is generally accepted as the "signature of a fractal object" (Schaefer et al., 1987), though there is not always a simple connection between α and D.

A notable exception is the small angle light, X-ray or neutron scattering from fractal objects (Liu, 1986; Schaefer and Keefer, 1986a, b; Schaefer et al., 1987; Thompson et al., 1987; Feder, 1988; Zumofen et al., 1988), where D and the exponent α are related as:

$$\text{for mass fractals of fractal dimension } 1 < D < 3: \ \alpha = D \tag{4.4.2a}$$

[see equation (4.4.15)];

$$\text{for surface fractals of fractal dimension } 2 < D_s < 3: \ \alpha = 6 - D_s \tag{4.4.2b}$$

[see equation (4.4.19)].

For *self-affine* fractal surfaces Po-zen Wong has derived, instead of equation (4.4.1):

$$I(k) \propto \frac{A}{k^4} + \frac{B}{k^{3+\alpha}} \tag{4.4.3}$$

[see equation (4.4.20)], where α is between 0 and 1 (Wong, 1985, 1987; Wong and Lin, 1988).

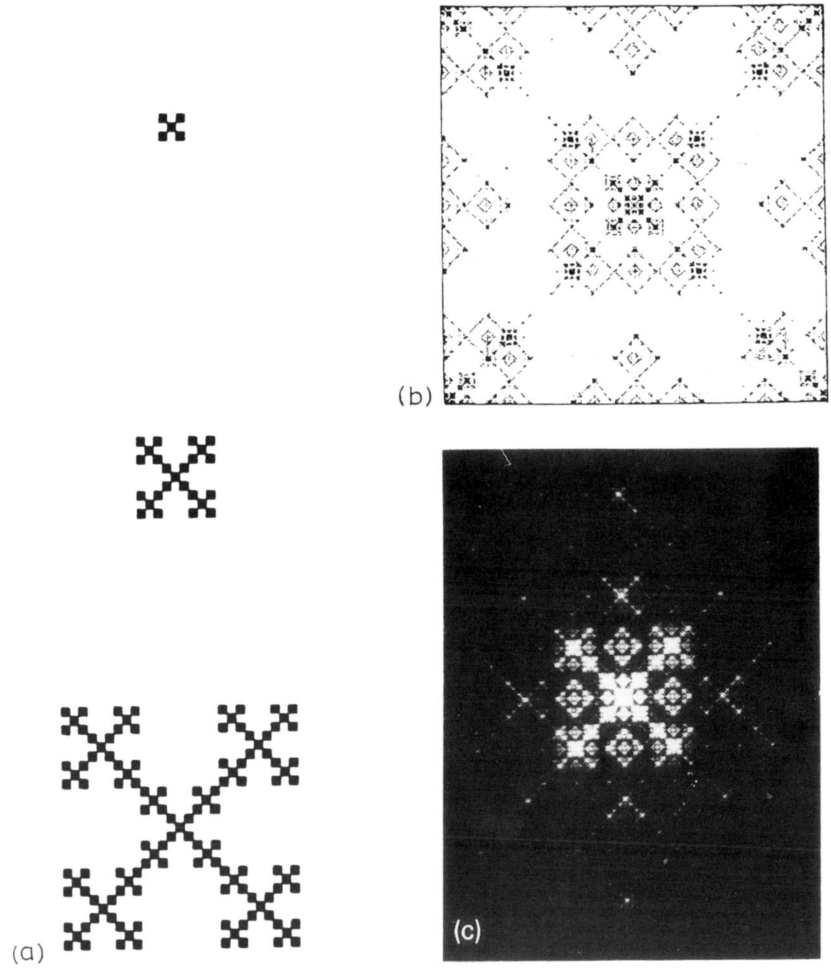

Fig. 4.27. a. The first few generations of the Vicsek fractal (Vicsek, 1983) drawn on a HP 7470 precision graphics plotter. b. Computed structure function of the fifth generation of the Vicsek fractal. c. Optical Fourier transform of b), exhibiting self-similarity. (From Allain and Cloitre, 1986.)

The derivations of equations (4.4.1) and (4.4.3) are based on the first Born approximation of diffraction theory, where multiple scattering is neglected and the scattering cross-section $I(k)$ is simply assumed to be proportional to the Fourier transform of the geometrical correlation function of the diffracting structure.

If the scattering object is a *fractal aggregate* built up from small spherical particles of size a, multiple scattering of the wave inside the aggregate can not be neglected if $ka \gg 1$, especially not for "opaque" aggregates with high fractal dimension $D > 2$. Theoretical (Berry and Percival, 1986) and numerical (Frey et al., 1988; Nelson, 1989) studies of the multiple scattering problem have proved that the rule:

$$I(k) \propto k^{-D} \quad \text{only applies if} \quad L^{-1} < k < a^{-1} \tag{4.4.4}$$

where L is the size of the whole aggregate, a is the size of an elementary particle. For $ka \gg 1$ there is a significant multiple scattering contribution to the scattered intensity, and there seems to be no simple analytical expression relating $I(k)$ to D. [For the application of these ideas to nuclear-winter simulations see Nelson (1989). She proved that the absorptivity of a sooty smoke which is fractal is greater than that of a "Euclidian" distribution of spheres and concluded that if the sooty smoke provides about half the initial optical depth of the post-nuclear-war atmosphere, then "the short-term effect of soot fractality is to make the nuclear winter colder by several (up to 5) degrees Celsius." Another example where multiple scattering on fractals has approximately been taken into account is Wu and Aki (1985), who determined the fractal dimension of the lithospheric inhomogeneities from the coda excitation spectra and from the high-frequency attenuation of *multiply scattered* S waves.]

The small-angle scattering of X-rays (SAXS) or neutrons (SANS), mentioned above, is one of the most important experimental methods of measuring the fractal dimension of the internal surfaces of rocks and other porous objects. [The standard textbook on small angle scattering is still Guinier and Fournet (1955); the case of fractal objects is reviewed in Liu (1986), Mildner and Hall (1986), Schaefer et al. (1987), Thompson (1991), Thompson et al. (1987), Wong and Lin (1988).]

The principle of the method is that the measured intensity is related to the Fourier transform of the charge distribution (in SAXS) or the mass distribution (in SANS) of the object. If the object is fractal, its non-integer fractal dimension will manifest itself in an anomalous fall-off of intensity with wave vector k, according to equations (4.4.1)–(4.4.4).

The experimental setup is shown in Figure 4.28. A beam of photons or neutrons is scattered by the object, and the intensity of the scattered wave depends on the angles θ and φ. The dependence on the azimuthal angle φ is averaged out by the symmetry or the random orientation of the object. The scattering angle and the

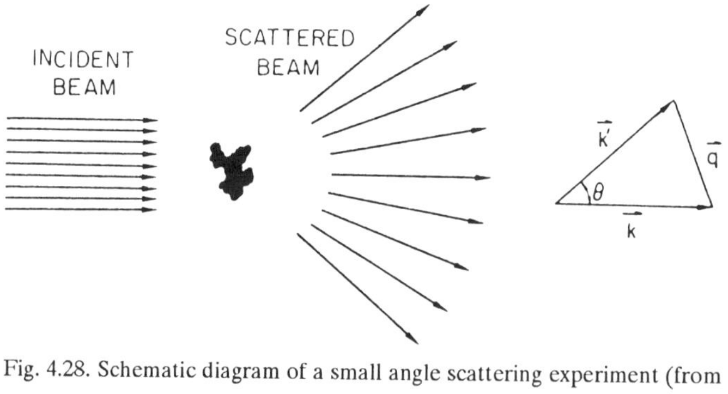

Fig. 4.28. Schematic diagram of a small angle scattering experiment (from Liu, 1986).

momentum transfer q are related by:

$$q = 2k \sin\left(\frac{\theta}{2}\right) \tag{4.4.5}$$

where k is the wave vector of the incident particles. As well known, the scattering problem can only be so solved exactly for perfect spheres, perfect spheroids and infinite cylinders (Van de Hulst, 1957; Kerker, 1969; Bourrely et al., 1986). The case of uniform sphere (see, for example, Liu, 1986) will illustrate the basic concepts. The scattering potential is:

$$\begin{aligned} V(r) &= V_0 \quad \text{for } r < a \\ &= 0 \quad \text{for } r > a \end{aligned} \tag{4.4.6}$$

where a is radius of the sphere. In the first Born approximation the scattering amplitude $f(q)$ is the Fourier transform of $V(r)$:

$$f(q) \propto \int V(r) e^{iqr}\, \mathrm{d}r^3 = \frac{4\pi V_0[\sin(qa) - qa\cos(qa)]}{q^3} \tag{4.4.7}$$

For $qa \ll 1$ the r.h.s. of (4.4.7) is expanded to obtain:

$$f(q) = f(0)\left(1 - \frac{R^2q^2}{6}\right) \tag{4.4.8}$$

where the forward scattering amplitude $f(0) \propto V_0\ (2\pi a^3/3)$ is proportional to the total amount of matter in the sphere, R — defined by $R^2 = 3a^2/5$ — is a measure of the size of the sphere (the "radius of gyration"). For large wave vectors, $qa \gg 1$, from (4.4.5) and (4.4.7):

$$I(k) \propto f(q)^2 \propto \frac{\{\cos[2k\sin(\theta/2)]\}^2}{k^4\sin^4(\theta/2)} = \frac{A(k)}{k^4} \tag{4.4.9}$$

that is, the intensity oscillates with an envelope decaying as k^{-4}.

In the more general case, when the (charge or mass) density is $\rho(\boldsymbol{r})$, the scattering amplitude is:

$$f(q) \propto \int \rho(\boldsymbol{r}) e^{iqr}\, \mathrm{d}^3r \tag{4.4.10}$$

For small values of q such that $q \ll 1/R$ we obtain again:

$$f(q) = f(0)\left(1 - \frac{1}{6}R^2q^2\right) \tag{4.4.11}$$

independently of the nature of the distribution $\rho(\boldsymbol{r})$.

For $q \gg 1/R$ we study the expected value of the intensity:

$$\langle I(q)\rangle = \langle |f(q)|^2\rangle \propto \int \langle \rho(\boldsymbol{r}')\, \rho(\boldsymbol{r} + \boldsymbol{r}')\rangle e^{iqr}\, \mathrm{d}^3r = \int \rho_2(\boldsymbol{r}) e^{iqr}\, \mathrm{d}^3r \tag{4.4.12}$$

where:

$$\rho_2(\boldsymbol{r}) = \langle \rho(\boldsymbol{r}')\, \rho(\boldsymbol{r} + \boldsymbol{r}') \rangle = \int \rho(\boldsymbol{r}')\rho(\boldsymbol{r} + \boldsymbol{r}')\, \mathrm{d}^3 r' \tag{4.4.13}$$

is the density–density correlation function, that is the probability of finding two particles separated in space by a vector $\boldsymbol{r}$. For a *fractal* mass distribution we choose an arbitrary point $P(\boldsymbol{r}')$ and construct a spherical shell of radius r and thickness dr around it (where $r = |\boldsymbol{r}|$). If the mass distribution is fractal with dimension D, the amount of matter enclosed in the shell is proportional to $4\pi r^{D-1}\mathrm{d}r$. As the volume of the shell is $4\pi r^2 \mathrm{d}r$, the probability of finding a particle at a distance r from $\mathrm{P}(\boldsymbol{r}')$ is proportional to:

$$\frac{4\pi r^{D-1}\,\mathrm{d}r}{4\pi r^2\,\mathrm{d}r} \propto r^{D-3}$$

while the probability that there is a particle at point $P(\boldsymbol{r}')$ and one a distance r apart is proportional to $\rho(\boldsymbol{r}')\, r^{D-3}$.

Finally, after carrying out the r'-integration, we get:

$$\rho_2(r) \propto M r^{D-3} \tag{4.4.14}$$

where M is the total mass of the object. Substituting this into equation (4.4.12) and expressing q in terms of k (equation 4.4.5):

$$\langle I(k) \rangle \propto \frac{M\,\Gamma(D-1)\,\sin[\pi(D-1)/2]}{k^D} \tag{4.4.15}$$

where Γ is the gamma-function. In the derivation it has been assumed that the object is a *mass fractal*, that is its mass enclosed in a volume of size R scales as $M(R) \propto R^D$.

Let us consider next the case of surface scattering, first for bodies with arbitrary, smooth surfaces. If the material distribution ρ is uniform inside the body, the value of the correlation function is $\rho_2(0) = \rho^2 V$ for zero shift, while for a displacement $\boldsymbol{r}$ it is proportional to the non-overlapping volume between the object and its identical, displaced copy (Figure 4.29). For $\boldsymbol{r}$ small:

$$\rho_2(\boldsymbol{r}) \approx V - aSr \tag{4.4.16}$$

where V, S are the volume and the surface, respectively, and a is a geometric factor of order one. Substitution into equation (4.4.12) yields:

$$I(k) \propto \frac{\rho^2 a S}{k^4} \tag{4.4.17}$$

for the non-oscillatory term (see e.g. Liu, 1986).

For the case of fractal surfaces of dimension D_s, Bale and Schmidt (1984) estimated the volume of the non-overlapping part in Figure 4.29 by covering it with

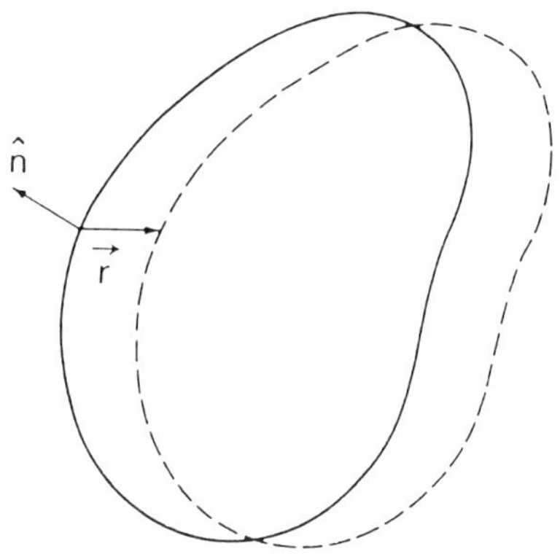

Fig. 4.29. To the calculation of surface scattering (from Liu, 1986).

small cubes of volume r^3 ($r \ll 1$). As the number of cubes scales as Nr^{-D_s} where N is a constant characteristic of the surface, we get instead of (4.4.16):

$$\rho_2(\boldsymbol{r}) \approx V - Nr^{3-D_s} \tag{4.4.18}$$

and the scattered intensity from the fractal surface becomes for large wave vectors k:

$$\langle I(k) \rangle \approx \frac{\pi N\,\Gamma(5 - D_s)\,\sin[\pi(D_s - 1)/2]}{k^{6-D_s}} \tag{4.4.19}$$

Observe that both formulae (4.4.15) and (4.4.19) give a $k^{-\alpha}$ dependence. Since the fractal dimension is less than three in both cases, *the volume scattering part gives a $k^{-\alpha}$ dependence with $\alpha < 3$, the surface scattering part gives a dependence with $\alpha > 3$. Thus we can deduce the nature of the fractal object depending on whether the exponent α is greater than or less than 3.*

Further discussion of volume fractals (sometimes called "mass fractals") and surface fractals can be found in Martin (1986), Schaefer and Keefer (1986b), Dos Santos et al. (1987), Hurd et al. (1987), Kolb and Herrmann (1987; see Figure 4.30).

An heuristic derivation of the k^{-D} and $k^{-(6-D_s)}$ dependences, respectively, based on simple scaling arguments, can be found in Witten (1987a), without the $\sin[\pi(d - 1)/2]$ prefactor, however. The meaning of this factor is that there is no k^{-3} contribution to the scattering amplitude, either for volume fractals or surface fractals with fractal dimension $D = 3$ or $D_s = 3$ [see, however, the discussion of Wong and Bray (1988), with the answer of Pfeifer and Schmidt].

A schematic illustration of the above-discussed scattering laws is shown in Figure 4.31: in the ideal case we may observe up to three distinct regions, separated by the cross-over values $\eta^{-1}, \xi^{-1}, l^{-1}, a_0^{-1}$:

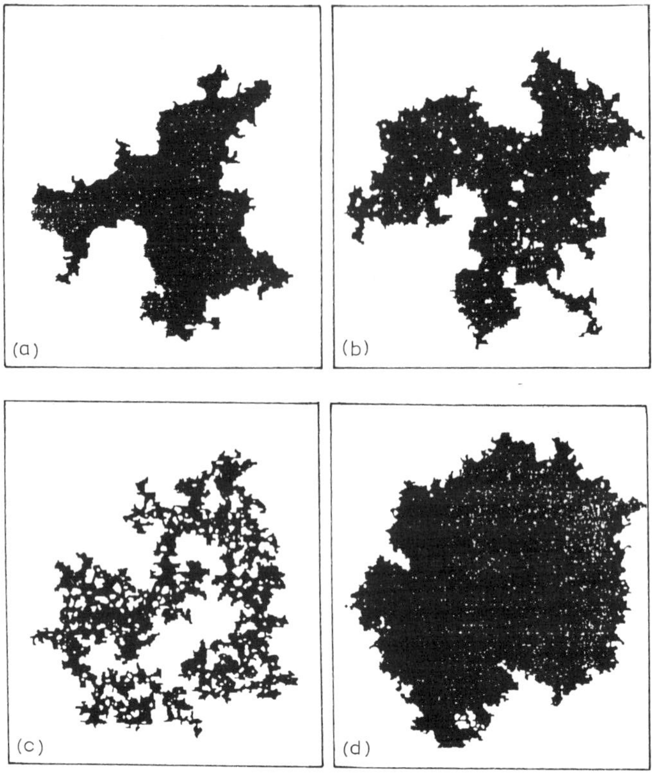

Fig. 4.30. Computer generated clusters illustrating the planar cross-section of (a) a compact surface-fractal; (b) a porous surface-fractal; (c) a volume-fractal; (d) a non-fractal object (from Kolb and Herrmann, 1987).

(1) if $\eta^{-1} < q < \xi^{-1}$ then $I \propto q^{-D}$: we measure the fractal dimension of the mass distribution of the object;

(2) if $\xi^{-1} < q < l^{-1}$ then $I \propto q^{D_s - 6}$: we measure the fractal dimension of the interfacial region;

(3) if $l^{-1} < q < a_0^{-1}$ then $I \propto q^{-4}$: the interface is sensed as smooth, we observe a true Porod scattering;

(4) if $a_0^{-1} < q$ (where a_0 is a lattice constant) we observe more complicated scattering behaviour — related to the atomic structure — in the Bragg regime (Mildner and Hall, 1986; Schaefer et al., 1987).

Wong (1985, 1987; Wong and Lin, 1988; see also Lucido et al., 1988) has recently studied the scattering from *self-affine* surfaces. Self-affinity (Mandelbrot, 1982, 1986; Voss 1985a, b; Feder, 1988; for an elementary introduction see Turcotte, 1991) is a generalization of self-similarity: while a self-similar object looks the same when space is stretched (dilated) uniformly by a factor λ, a self-affine object looks the

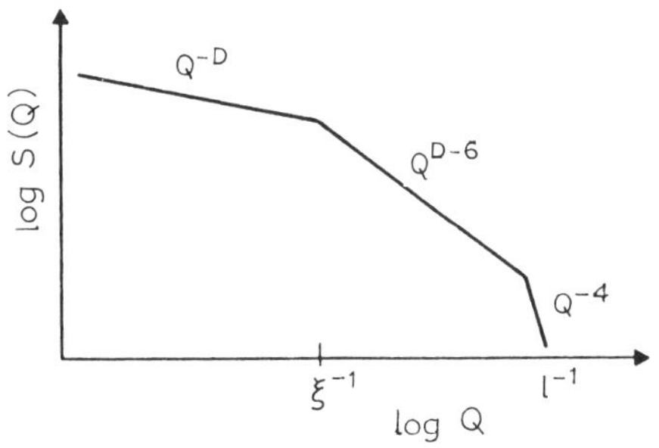

Fig. 4.31. Schematic diagram of the different scattering regimes: ξ^{-1} represents cross-over from volume- to rough-surface scattering; l^{-1} is cross-over from rough-surface to smooth-surface (Porod-)scattering (from Mildner and Hall, 1986).

same after the *affine transformation* $\boldsymbol{x} = (x_1,x_2,x_3) \rightarrow \boldsymbol{x}' = (r_1x_1, r_2x_2, r_3x_3)$, where the scaling ratios r_1, r_2, r_3 are not all equal. (Examples for two-dimensional affine transformation are given in Figure 4.32.) In the more general case, the object will be statistically similar to its original version when stretched by the factors λ^{b_x}, λ^{b_y}, λ^{b_z} in the respective directions (where the exponents b_x, b_y, b_z are not all equal). Well known examples are the fractional Brownian function $z = B_H(t)$ and its counterpart, the fractional Brownian surface $z = B_H(x,y)$, of Hurst exponent H $(0 < H < 1)$. In case of $B_H(t)$, if $B_H(0) = 0$, the random processes $B_H(t)$ and $\lambda^{-H}B_H(\lambda t)$ are statistically identical. If time is "stretched" by a factor λ, z is stretched by λ^H , that is $b_t = 1$ and $b_z = H_z$. A similar relation holds for the fractal surface $z = B_H(x,y)$: if $B_H(0,0) = 0$, the surfaces $B_H(x,y)$ and $\lambda^{-H}B_H(\lambda x, \lambda y)$ are statistically identical: $b_x = b_y = 1$, $b_z = H$. The fractional Brownian surface is *self-similar in x and y, and self-affine in x and z or y and z.*

Fractal surfaces without "overhangs", that is whose height z is a single-valued function $z = f(x,y)$, are frequently self-affine, that is, z will be stretched by the factor λ^α $(0 < \alpha < 1)$ when the horizontal coordinates are stretched by the factor λ. Wong (1985, 1987; Wong and Lin, 1988) has shown that scattering from a self-affine fractal surface is given by:

$$I(q) = \frac{A}{k^4} + \frac{B}{k^{3+\alpha}} \tag{4.4.20}$$

[the physical meaning of the ratio B/A is explained in Wong and Lin (1988); see also equation (4.5.54) in Section 4.5. As noted by Witten (1987a), for macroscopically smooth self-affine surfaces the coefficient B must be much smaller than A.]

Strictly speaking, the "theoretical" k^{-4} scattering is only valid for three-dimensional, smooth scattering centres. The outer portion of the average scattering intensity is proportional to k^{-2} for randomly oriented platelets, and to k^{-1} for randomly

Fig. 4.32. a. Affine transformations of a human face. The distortions in the famous portraits of El Greco (b) and Modigliani (c) were produced by affine transformations. (From Schillinger, 1948.)

oriented rods (Schmidt, 1982). Thus, for all three scattering systems there is a region of wave vectors where the scattered intensity is:

$$I(k) \propto k^{-\alpha} \qquad (\alpha = 1, 2, \text{ or } 4) \tag{4.4.21}$$

In the early 80's the experimentally found exponents α were used to determine whether the scatterers in the sample are three-dimensional, with $\alpha = 4$, or can they be considered as platelets ($\alpha = 2$) or rods ($\alpha = 1$). No definite interpretation had been known for those scattering curves where $I(k)$ behaved as $k^{-\alpha}$ where α was not equal to 4, 2, or 1. In 1982, Paul W. Schmidt proved that a power-law size distribution of d-dimensional randomly oriented scattering particles:

$$P(r \leq \text{particle size} < r + dr) \propto r^{-(2d+1-\alpha)} \tag{4.4.22}$$

leads to the scattered intensity:

$$I \propto k^{-\alpha} \tag{4.4.23}$$

where:

$$\begin{aligned} &0 < \alpha < 4 \qquad \text{for } d = 3 \\ &0 < \alpha < 2 \qquad \text{for } d = 2 \\ &0 < \alpha < 1 \qquad \text{for } d = 1 \end{aligned} \tag{4.4.24}$$

[see also the more recent results of Martin (1986) on the scattering exponents for the power-law distribution of surface- and mass fractals].

When a few years later Bale and Schmidt measured an "anomalous" scattering curve:

$$I(q) \propto q^{-(3.44 \pm 0.03)} \tag{4.4.25}$$

for a Beulah lignite (Figure 4.33), they first remeasured the scattering and when they found it reproducible, ascribed the power-law scattering to a power-law size distribution of the submicroscopic pores (Figure 4.34).

Later (Bale and Schmidt, 1984), when they learned about the adsorption studies of Avnir and his group (see Section 4.1), they proved that the observed result can also be explained by assuming scattering from a D_s-dimensional fractal pore boundary surface [$D_s = 6 - \alpha = 2.56 \pm 0.03$ by equations (4.4.2b)].

SAXS and SANS studies on different kinds of rocks (shales: Mildner and Hall, 1986; Mildner et al., 1986; Wong, 1987; Wong and Lin, 1988; Wong et al., 1986; sandstones: Thompson et al., 1987; Wong, 1987; Wong and Lin, 1988; dolomite, limestone: Wong et al., 1986; Wong, 1987; Wong and Lin, 1988; volcanic rocks: Lucido et al., 1988) have shown that shales and sandstones behave as surface fractals ($I \propto k^{D_s-6}$), dolomite, limestone and the volcanic rocks follow the self-affine scattering law [equations (4.4.3) or (4.4.20)]. Representative examples are shown in Figure 4.35.

The fractal dimensions indicated in Figure 4.35a are very high; the highest, $D_s = 2.96$ corresponds to the Coconino sandstone where the quartz grains are completely covered by fibrous and sheetlike clay (illite: see the SEM image in Wong et al.,

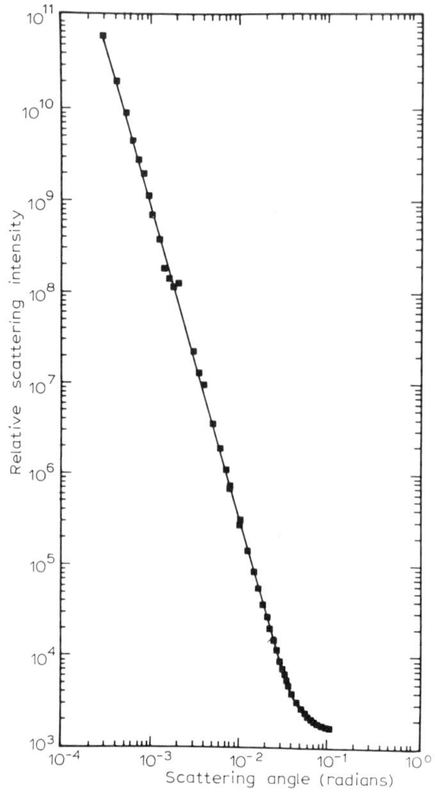

Fig. 4.33. X-ray scattering intensity for Beulah lignite. The curve corresponds to the least squares fit by the function $I(q) = B + Cq^{-\alpha}$ ($\alpha = 3.44 \pm 0.03$). The boundary surface of the pores in this coal is 2.56 ± 0.03 dimensional. (From Bale and Schmidt, 1984.)

1986, figure 2). Wong et al. (1986) have drawn the general conclusion that in sandstones the fractal character of the pore surfaces is due to the presence of diagenetic clay.

The value of α in the fit of equation (4.4.3) to the carbonate data (Figure 4.35b) is close to zero, indicating that these surfaces are randomly rough but not self-affine (Wong et al., 1986).

Though Mildner and Hall (1986, see Figure 4.31) predict a cross-over from k^{-D} to k^{D_s-6} scattering at smaller wave vectors, it seems very likely that rocks are not volume fractals. To date, the k^{-D} type behaviour has only been observed in colloidal aggregates (Courtens and Vacher, 1987; Dos Santos et al., 1987; Schaefer et al., 1984).

The interpretation of small-angle scattering curves in terms of a fractal (surface) dimension presents many difficulties. The fractal dimensions found from scattering are systematically greater than those obtained from chemical adsorption:

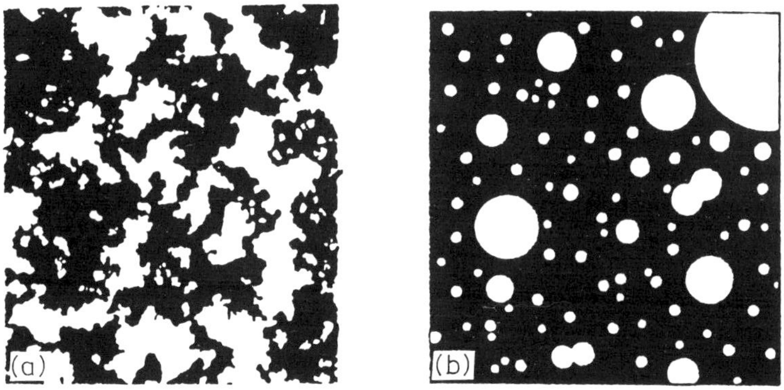

Fig. 4.34. Small angle scattering can not differentiate between fractally rough porous surfaces (a), and a power-law distribution of smooth pores (b) (from Schaefer et al., 1987).

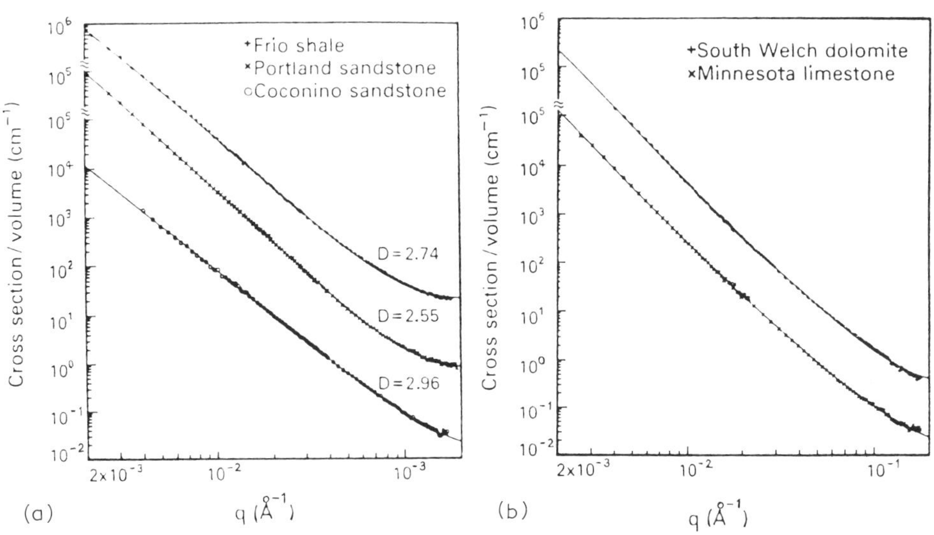

Fig. 4.35. Scattering curves for (a) sandstones and shales, fitted by equation (4.4.2b), and (b) for carbonate rocks, fitted by equation (4.4.3) (from Wong and Lin, 1988).

$$D_{\text{scatt}} > D_{\text{ads}} \tag{4.4.26}$$

(Hall, 1986), also — for partially saturated rocks — :

$$D_{\text{SANS}} > D_{\text{SAXS}} \tag{4.4.27}$$

(Mildner and Hall, 1986; Mildner et al., 1986). Pitfalls in the interpretation of the scattering curve are pointed out in Thompson et al. (1987), Schaefer et al. (1987) and Witten (1987a).

Elastic waves can also develop anomalous properties when propagating in a medium with fractal inhomogeneities. As well known, seismic and ultrasonic waves in rocks are exponentially attenuated with distance, that is the wave intensity decays as $\exp[-\alpha(\omega)x]$, where x is the distance between source and receiver, and $\alpha(\omega)$ is a frequency-dependent attenuation factor. The attenuation factor consists of two parts, the absorption coefficient $\alpha_a(\omega)$ and the scattering attenuation coefficient $\alpha_s(\omega)$ (Bhatia, 1967). The absorption coefficient corresponds to the inelastic energy losses and, in rocks, it is proportional to the first power of frequency:

$$\alpha(\omega) = A\omega \tag{4.4.28}$$

over a wide frequency range of $10^{-3} < f < 10^8$ Hz (Attewell and Ramana, 1966). The scattering part of the attenuation has a more complicated frequency dependence.

Denoting by r_0 the correlation distance of the inhomogeneities, (for polycrystalline materials, r_0 can be taken as the average grain size, in the general case r_0 is the distance beyond which the correlation function of the inhomogeneities becomes less then $1/e$ times its value at zero) and by λ the wavelength, we have:

– for $\lambda \gg r_0$:

$$\alpha(\omega) \propto \omega^4 \tag{4.4.29a}$$

(Rayleigh scattering) for three-dimensional inhomogeneities;

$$\alpha(\omega) \propto \omega^2 \tag{4.4.29b}$$

for one-dimensional inhomogeneities (as e.g. a sequence of sedimentary layers);

– for $\lambda \ll r_0$:

$$\alpha(\omega) \propto \frac{1}{r_0} \tag{4.4.29c}$$

[for the derivation of equations (4.4.29a)–(4.4.29c) see Korvin (1977)].

If the medium contains inhomogeneities at all scales neither the Rayleigh- or the high frequency limits are appropriate (Figure 4.36), and the expression:

$$\alpha(\omega) \propto \omega^\mu \qquad (1 \leq \mu \leq 4) \tag{4.4.30}$$

is generally used to describe the overall frequency-dependence of the attenuation coefficient where μ usually has a non-integer value.

This non-integer exponent suggests, by the analogy with equations (4.4.29a) and (4.4.29b), that the total mass of the scattering imperfections in an object of size R also scales as a non-integer power of R, i.e. the density of scatterers is fractal.

Indeed, West and Shlesinger (1984) proved that the fractional part of the scattering exponent μ is connected to the fractal dimension of the distribution of scatterers. First they proved, both for longitudinal and shear waves, that in the case

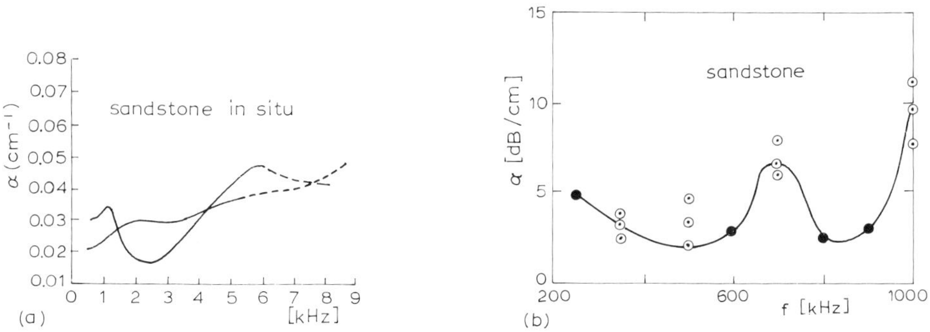

Fig. 4.36. Wide-band measurements of the attenuation coefficient in sandstone: (a) in situ (Militzer and Stoll, 1968); (b) in the laboratory, ⊙ denotes three identical results (Auberger and Rinehart, 1961; from Korvin, 1977.)

of weak scattering — when the correlations between individual scatterers can be neglected — the attenuation coefficient is proportional to the product of the total cross-section of an individual scatterer and the density of scatterers:

$$\alpha(\omega) \propto \sigma(\omega)\rho \tag{4.4.31}$$

where the scattering cross-section has the frequency dependence:

$$\sigma(\omega) \propto \omega^s \qquad \text{with } s = 0, 2 \text{ or } 4 \tag{4.4.32}$$

Because the scatterers occur at all scales, a high frequency wave interacts with more scatterers than does a low-frequency wave. The density of scatterers experienced by the incident wave is frequency-dependent: $\rho = \rho(\omega)$. If the volume of the object is V, a wave of frequency ω_1 is scattered by $N_1 = \rho_1 V$ scatterers, that of frequency ω_2 ($\omega_2 < \omega_1$) is scattered by $N_2 = \rho_2 V < N_1$ scatterers. Thus $\rho_2 < \rho_1$ if $\omega_2 \ll \omega_1$. We write with real constants a and b:

$$\rho(\omega_1) = b\rho(\omega_2) = b\rho\left(\frac{\omega_1}{a}\right) \qquad \text{with } a, b > 1 \tag{4.4.33}$$

The functional equation (4.4.33) has the solution:

$$\rho(\omega) = \text{const}\, \omega^{\log b/\log a} \tag{4.4.34}$$

where the exponent $\log b/\log a$ is reminiscent of a fractal dimension. (Suppose that at a resolution corresponding to frequency ω we can observe individual scattering objects of size R, while at a finer resolution ω/a we discover that each object actually consists of b smaller objects of size R/a, etc. Repeating this process *ad infinitum* leads to a mass distribution with fractal dimension $\log b/\log a$.)

Combining equations (4.4.31), (4.4.32) and (4.4.34) we get that the attenuation coefficient can be written:

$$\alpha(\omega) = \rho(\omega)\,\sigma(\omega) \propto \omega^{\mu} \quad \text{with } \mu = s + \frac{\log b}{\log a} \quad (s = 0, 2 \text{ or } 4) \tag{4.4.35}$$

Koch (1987) considered the petrophysically important case when a compressional sound wave is propagating in a porous solid with fractally rough pore surfaces. At high frequencies ω, the absorption is caused primarily by the viscous dissipation within a boundary layer near the solid surface of thickness:

$$\delta = \left(\frac{\nu}{\rho\omega}\right)^{1/2} \tag{4.4.36}$$

where ν is viscosity and ρ the density of the fluid. As the frequency increases, the boundary layer becomes thinner and — in case of fractal pore walls — its surface area increases as an inverse power of the thickness δ. Koch shows, using an analytically tractable surface model (Figure 4.37), that in this case the absorbtion coefficient will follow an "anomalous" power law:

$$\alpha(\omega) \propto \delta^{2-d_a} \approx \omega^{(-1+d_a/2)} \tag{4.4.37}$$

where the "absorption" fractal dimension d_a however, is in no simple relation with the fractal dimension of the pore wall.

In an important study Wu and Aki (1985) proved that the observed coda excitation spectrum of the earthquakes and the frequency-dependence of the scattering

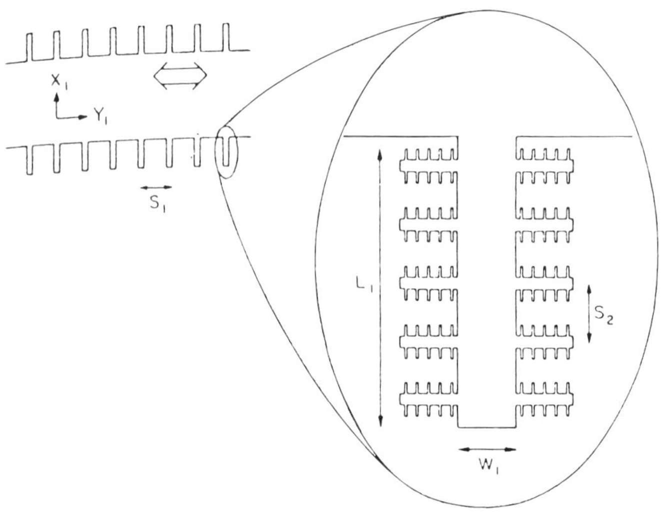

Fig. 4.37. Fractal pore-wall model to study sound wave absorbtion in fluid-filled porous solids (from Koch, 1987).

attenuation of seismic shear waves can be explained in terms of a fractal structure of the lithospheric (shear wave velocity) inhomogeneities. First they proved that the coda excitation spectrum $g(\omega)$ and the scattering attenuation coefficient $\eta(\omega)$ are related to the power spectrum $W(k)$ of the inhomogeneities as:

$$g(\omega) \propto k^2 W(2k) \tag{4.4.38}$$

and:

$$\eta(\omega) \propto k^2 W(k), \tag{4.4.39}$$

respectively, where k is the (shear-)wave number corresponding to the frequency ω. [The derivation of (4.4.38) assumes single back-scattering, while (4.4.39) is based on the forward multiple scattering approximation of Wu (1982).]

From the measurements of coda excitation and the scattering attenuation in Kanto (Japan), Hindu-Kush and other regions, Wu and Aki inferred that the lithospheric shear wave velocity inhomogeneities have a power spectrum $W \propto k^{-p}$, where $p = 1.5$–2 for different regions and depth ranges. The corresponding Hurst number H of the fractal velocity distribution $v = v(x,y,z)$, and the fractal dimension D of the $v(x,y,z)$ function in the four-dimensional space (v,x,y,z) can be found from the relations:

$$p = 2H + 1 \qquad\qquad D = 4 - H$$

that is, $H = 0.25$–0.5, $D = 3.5$–3.75.

Wu and Aki (1985) also noted that if $v(x,y,z)$ is D-dimensional, its "iso-strength" surfaces in the lithosphere (i.e. surfaces where the seismic velocities have the same value) should be $D - 1 = 2.5$–2.75-dimensional, while their intersection with the Earth's surface would be $D - 2 = 1.5$–1.75-dimensional fractal curves — in agreement with the measured fractal dimensions of fault traces (Okubo and Aki, 1987; Aviles et al., 1987; Scholz and Aviles, 1986).

Randomly rough surfaces can also be studied by *pulsed scattering*, as in the case of reflection seismic surveys. Due to the random diffractions and reflections from the surface irregularities, following the directly reflected pulse as an echo, it is not only difficult to map the rough surface itself, but because of the slowly decaying tail of the return pulse it is difficult or even impossible to detect reflections coming from eventual deeper horizons.

The existence of the diffuse echo following the reflection from rough boundaries has been demonstrated by model experiments (Leong et al., 1971; Voskresenskiy, 1962). The temporal behaviour of the echo for Gaussian random surfaces was theoretically investigated in the low-frequency limit by Biot (1957) for non-differentiable surfaces and in the high-frequency limit by Berry and Blackwell (1981), Korvin (1982) for isotropic, differentiable random surfaces. The case of *fractal surfaces*, which is especially complicated since the surface is irregular on all scales so that

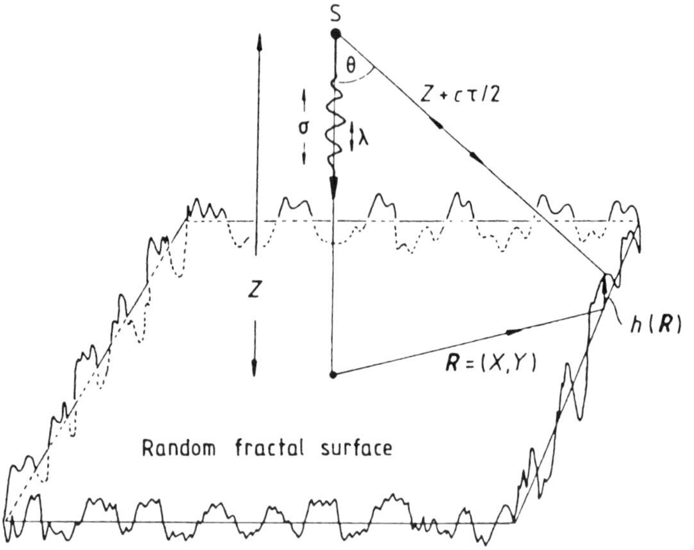

Fig. 4.38. Geometry of the diffractal echo problem. Source and receiver are at the same position S; Z is height of S above the mean surface; $h(R)$ is the statistically isotropic surface; c is propagation velocity above the surface; θ is scattering angle; λ is wavelength; σ is pulse duration; τ is delay time measured from the time $2Z/c$ when the pulse would be received from a plane boundary. (From Berry and Blackwell, 1981.)

neither the low-frequency nor the high-frequency approximation applies, has been solved by Berry and Blackwell (1981) in their paper "Diffractal echoes".

The geometry of the "diffractal" echo problem is shown in Figure 4.38. Figure 4.39 shows the contours of a smooth and rough boundary, respectively, contributing to the different parts of the echo; Figure 4.40 presents a simplified example of how pulses are superimposed.

The results of the above-mentioned studies can be summarized as follows (τ is the delay time measured from the time $T_0 = 2Z/c$ corresponding to the specular reflection):

(a) for an ideal plane boundary we get a single reflected pulse at $\tau = 0$;

(b) for an isotropic, Gaussian random surface the intensity of the reflected pulse train falls off exponentially:

$$I(\tau) \propto \exp\left(-\frac{\tau}{t_0}\right) \tag{4.4.40}$$

Two cases should be distinguished:

In the *low frequency limit* there appears a coherent specular reflection from the mean surface $\langle h(x,y) \rangle$ (Tolstoy and Clay, 1966); this is followed by the exponentially decaying tail. The time constant of the decay, t_0 is frequency dependent: it is proportional to the square of the wave number (Biot, 1957).

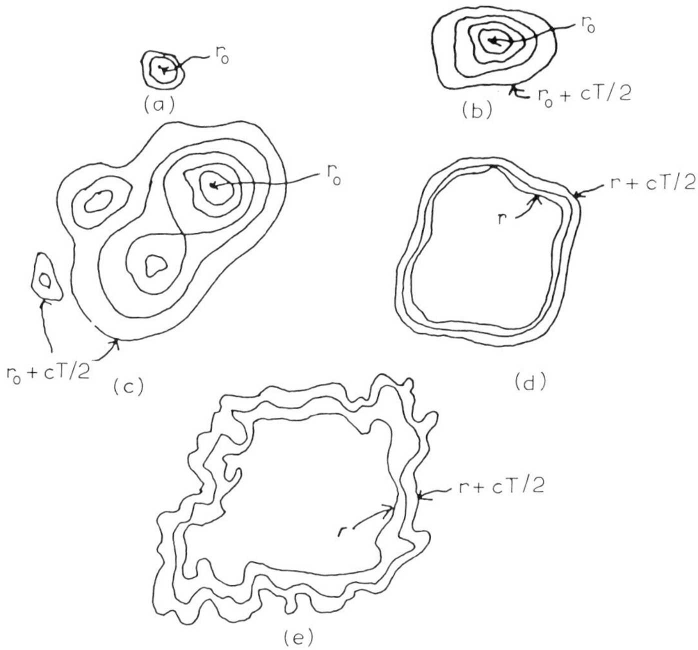

Fig. 4.39. Contours of points at equal distances from the source and contributing to different parts of the echo: (a) shortly after the first return; (b) at time T after the first return (smooth surface); (c) same, for rough surface; (d) in the tail of the echo (smooth surface); (e) same, for rough surface (from Berry, 1972).

In the *high-frequency limit* there is no specular reflection, for large values of τ the intensity of the pulse train decays with a frequency-independent time-constant as:

$$I(\tau) \propto \exp\left(\frac{\tau}{T_0 \beta^2}\right) \tag{4.4.41}$$

where β^2 is the RMS slope of the surface $h(x,y)$ (Berry and Blackwell, 1981). The behaviour of $I(\tau)$ for *finite values* of τ was analysed by Korvin (1982), who showed that prior to the exponential decay, the pulse train starts with zero expected energy at $\tau = 0$, its energy gradually builds up, attains its maximum around $T_{max} = T_0(1 + \beta^2)^{1/2}$; then decreases exponentially and becomes practically negligible around $T_{end} = T_0(1 + 3\beta^2)^{1/2}$. [For the case $c = 4000$ m/s; $f = 40$ Hz; $Z = 4000$ m; and a Gaussian random surface with correlation distance $r_0 = 250$ m and variance $\sigma^2 = 5000$ m^2 we have $\beta^2 = \sigma^2/r_0^2$ by a general formula of Rice (1944, 1945) and $T_0 = 2$ s; $T_{max} \approx 2.078$ s; $T_{end} \approx 2.228$ s, that is, the tail is only about 230 msec long.]

(c) For *fractally rough* surfaces Berry and Blackwell (1981) have shown in the "weak fractal limit" (where the fractal height variations are small and the reflections

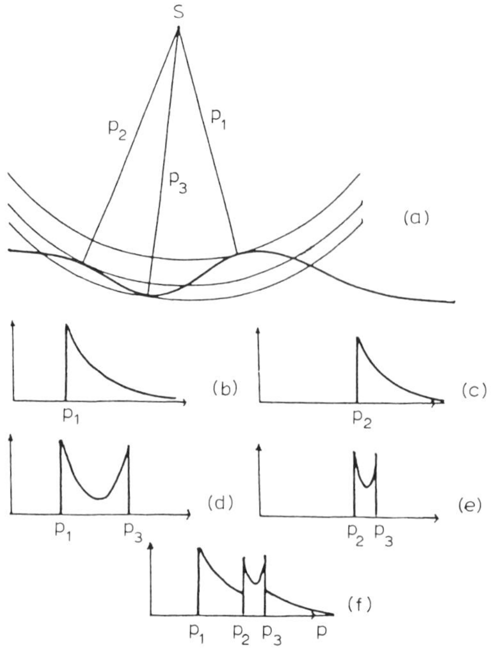

Fig. 4.40. Superposition of pulses from a surface with three reflecting points. The pulses are indexed by the distances from S. (From Berry, 1972.)

are paraxial) that *the pulse-train decays anomalously slowly*, as a power-law:

$$I(\tau) \propto \tau^{-(4-D_s)} \tag{4.4.42}$$

where D_s ($2 \leq D_s \leq 3$) is the fractal dimension of the surface.

4.5 Fractal models of porous rocks

The four Mists of Chaos, the North, the East, the West, and the South, went to visit Chaos himself. He treated them all very kindly and when they were thinking of leaving, they consulted among themselves how they might repay his hospitality. Since they had noticed that he had no holes in his body, as each had (eyes, nose, mouth, ears, etc.), they decided each day to provide him with an opening. At the end of seven days, Kwang-tse tells us, Chaos died.

— From CHUANG-TZU's *Seven Inner Chapters*, as retold by John Cage (1975)

The fractal structure of the pore space of sedimentary rocks has been experimentally verified over a wide length range, with different techniques (the measurements are reviewed, and compared, in Brochard, 1985; Thompson, 1991; Thompson et al., 1987; Feder, 1988; Wong, 1988):

– The range of small scales (from 1 Å to ≈100 Å) was studied by Avnir and his coworkers at the Hebrew University of Jerusalem.

They covered the porous surface with a monolayer of molecules of size r and observed that the number $N(r)$ of molecules adsorbed per unit volume scales as:

$$N(r) \propto r^{-D_S} \tag{4.5.1}$$

where D_S is the surface fractal dimension (see Section 4.1 and references therein).

– The intermediate scales (from 10 Å to ≈1000 Å) have been studied by small angle X-ray or neutron scattering, by the group of A. Thompson at Exon Production Research, Houston; Po-zen Wong and colleagues of Schlumberger-Doll Research, Ridgefield, CT; and others (see Section 4.4). In this technique the scattered intensity falls of with increasing wave vectors k as:

$$I(k) \propto k^{-(6-D_S)} \tag{4.5.2}$$

for surface fractals, and:

$$I(k) \propto k^{-D_m} \tag{4.5.3}$$

for mass fractals. Experiments on many sedimentary rocks indicated a behaviour corresponding to equation (4.5.2), proving that these rocks can be considered surface fractals in the given range.

– The large scales (from 10 Å to ≈100 μm) were investigated by Thompson and co-workers (Katz and Thompson, 1985; Krohn and Thompson, 1986; Thompson et al., 1987; Krohn, 1988a, b; see also Hansen and Skjeltorp, 1988). They analysed thin sections of the rocks by electronic (SEM) or optical microscopy and determined the fractal dimension from the feature size distribution along straight lines across the picture.

The technique is based on the fact that a fractal surface of dimension D_S has a power-law distribution of feature sizes:

$$N(r) \propto r^{-\alpha} \qquad \text{with } \alpha = D_S + 1 \tag{4.5.4}$$

A random straight line will intersect a feature of size r with a probability proportional to r^3, that is, the feature size distribution along the line will be:

$$N_1(r) \propto r^{3-\alpha} = r^{2-D_S} \tag{4.5.5}$$

This technique was first applied by Katz and Thompson (1985), who found $D_S = 2.78$ for a Coconino sandstone (Figure 4.41). Katz and Thompson assumed that the pore volume is also fractally distributed and that the pore surface and the pore volume have the same fractal dimension, D.

This assumption implies a simple scaling relation between the fractal dimension and the porosity of the rock. Suppose that the self-similarity holds in the length

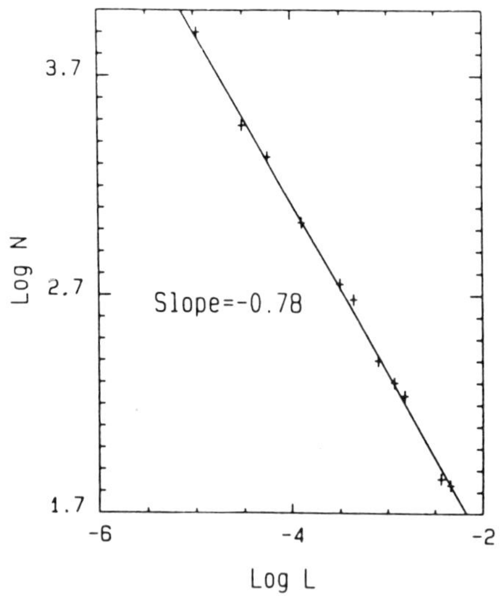

Fig. 4.41. Log–log plot of the number of geometric features of size L per unit length vs. the size L for the Coconino sandstone; L is in centimetres (from Katz and Thompson, 1985).

range $[l_1, l_2]$, where, for example, l_2 is of the order of the grain size and l_1 is the smallest measurable roughness scale of the pore wall. In a volume element of size $L < l_2$ of the rock the fractal pore volume can be covered by $N \propto (L/l)^D$ small cubes of size l ($l_1 < l \ll L < l_2$), that is, the porosity Φ determined for pores greater than l will scale as:

$$\Phi \propto \left(\frac{L}{l}\right)^D \frac{l^3}{L^3} = \left(\frac{l}{L}\right)^{3-D} \tag{4.5.6}$$

Katz and Thompson (1985) used equation (4.5.6) in the form:

TABLE 4.3

Computed and measured sandstone porosities

Sample	Fractal dimension from feature size distribution [Equation (4.5.5)]	l_2 (μm)	Φ(%)	
			calculated	measured
Tight gas sand no. 965	2.57	2.5	4.7	5.3–5.6
Tight gas sand no. 466	2.68	6	7.6	6.9–7.6
Coconino	2.78	98	10	11–12.5
Navajo	2.81	50	15	16.4
St. Peters	2.87	50	27	24–28

Source: Katz and Thompson (1985).

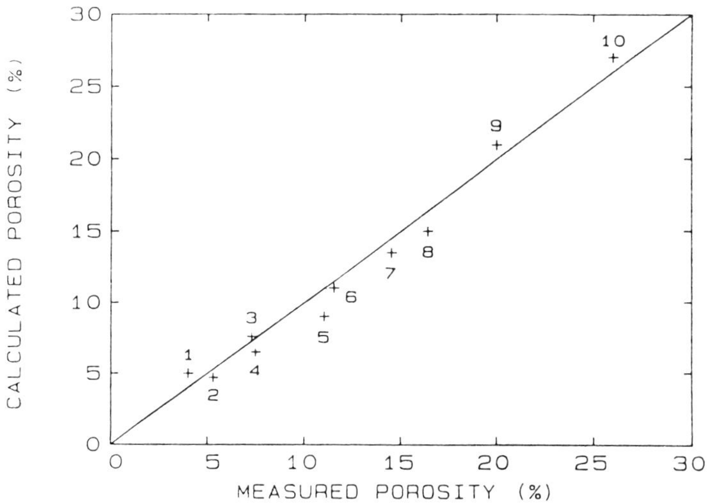

Fig. 4.42. Crossplot of measured and calculated porosities for sandstones of dominantly fractal porosity. Porosities were calculated from equation (4.5.7). (From Thompson et al., 1987.)

$$\Phi = \left(\frac{l_1}{l_2}\right)^{3-D} \tag{4.5.7}$$

for various sandstone samples, to predict their porosity from the fractal dimension D determined from feature size distribution. The lower limit l_1 was selected as a constant of 20 Å for all samples, the upper limit of the self-similar region was determined from optical autocorrelation data. There was a fair agreement between predicted and measured porosities (Table 4.3; Figure 4.42), though the calculated porosities were systematically smaller (see also Krohn, 1988a, b). For scales greater than l_2 the pore volume is strictly Euclidian, as reported by Ben Ohoud et al. (1988) for montmorillonite, sepiolite and attapulgite clay powders (for scales greater than $l_2 \approx 40$ μm) and by Krohn (1988a) for sandstones. In some of Krohn's samples the fractal part constitutes a significant amount of the pore space (Table 4.4). According

TABLE 4.4

Fractal and Euclidian porosities of sandstones

Sample	Core porosity Φ_{tot} (%)	Grain size (μm)	Pore size (μm)	l_2 (μm)	Φ_{fract} (%)	Φ_{Eucl} (%)
Table sandstone	33	370	290	5	3.0	30.0
Coconino sandstone	9.2	180	90	>40	8.0	1.2
Blackhawk foreshore	12	210	140	11	5.4	6.6
Blackhawk shoreface	7.3	60	20	>40	6.5	0.8

Source: Krohn (1988a).

to Krohn the Euclidean part of the pore-space distribution represents the original unaltered porosity, and the ratio of the relative porosities associated with the fractal and Euclidian regimes, respectively, is a measure of the diagenetic alteration of the pore space.

The power-law scaling of porosity, akin to equation (4.5.7), has been known since the late 1930's in civil engineering, where the dense packing of space with objects having a wide distribution of sizes is of great technological interest. In concrete materials the (undesirable) porosity is empirically known to scale as:

$$\Phi \propto \left(\frac{l_1}{l_2}\right)^{1/5} \tag{4.5.8}$$

where l_2 and l_1 are the largest and smallest sizes of the grains, respectively. The empirical relation (4.5.8) is called *Caquot's law* (A. Caquot, 1937, Mem. Soc. Ing. Civ., p. 562), its justification can be found in Baudet et al. (1985), who applied the same scaling argument which has led above to the derivation of equation (4.5.6). Baudet's derivation also suggests that the exponent 1/5 should not be universal, but depends on the fractal dimension as in equation (4.5.7). Omnès (1985) approached the problem by assuming a randomly packed system of spheres of randomly distributed radii. He proved that in a "self-similar" iterative packing, when we use spheres of decreasing radii to fill in the voids of the previous filling, the porosity scales as:

$$\Phi \propto \left(\frac{l_1}{l_2}\right)^{(\log\beta-1)/\log K} \tag{4.5.9}$$

where K is the ratio of the successive radii, $K = r_i/r_{i+1}$ and in each iteration step a fraction β of the remaining void-space is filled. The values l_1 and l_2 are the smallest and largest diameters, respectively.

Pape et al. (1984, 1987a) derived equation (4.5.7) from their pigeon-hole model of porosity (see Figure 4.17), in the form:

$$\Phi = 0.5\left(\frac{r_{\text{grain}}}{r_{\text{eff}}}\right)^{-0.25} \tag{4.5.10}$$

where r_{grain} is grain radius, r_{eff} is effective hydraulic pore radius, and obtained very good agreement with measured porosities for 41 sandstone and greywacke samples from German oil and gas fields [Figure 4.43; the physical properties of the samples are compiled in Pape et al., 1985b (mechanical properties), and 1987b (electrochemical and hydraulic properties)].

Though it is generally accepted that the pore space of sedimentary rocks is fractal in a certain range, Katz and Thompson's (1985) claims, that (a) the pore–grain interface is fractal, (b) the pore volume is also fractal, and (c) the pore interface and volume have the same fractal dimension over a length range of $\approx$10 Å–100 μm

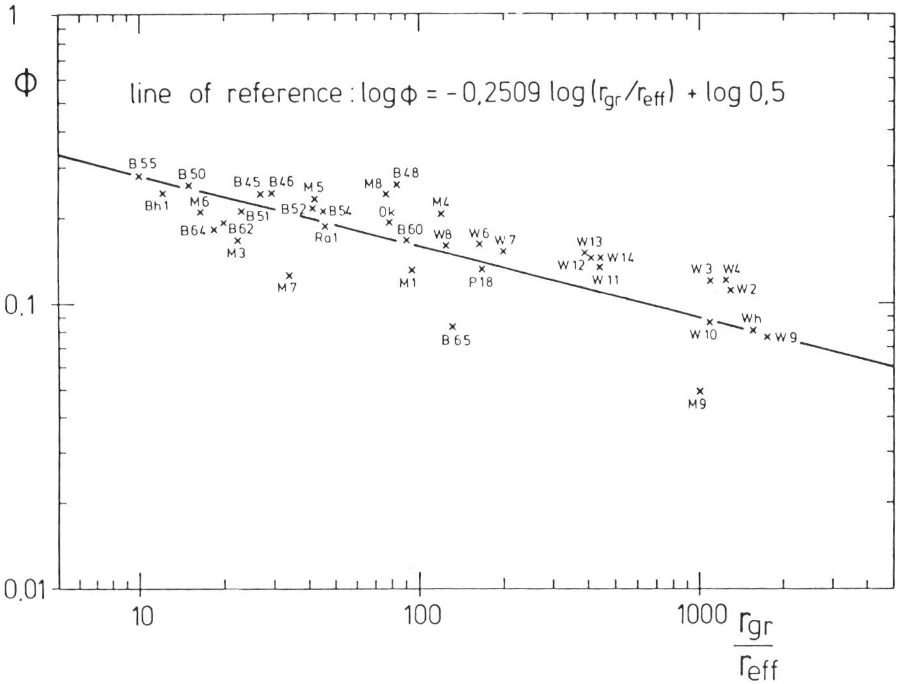

Fig. 4.43. Porosity as a function of the ratio of grain radius to effective pore radius (r_{gr}/r_{eff}) for sandstone and greywacke samples from German oil and gas fields (from Pape et al., 1987a).

have been disputed both on theoretical (Roberts, 1986; Katz and Tompson, 1986) and experimental grounds (optical microscopy of Hansen and Skjeltorp, 1988; and small angle scattering results of Wong et al., 1986).

Roberts' (1986) argument is amusing: he proves that the three claims (a)–(c) of Katz and Thompson, taken together, would imply that the rock's pore space could not be anywhere farther than a few nanometres from the nearest pore/grain interface (as Antonio's blood and flesh in the "Merchant of Venice") — which is obviously *not* the case in Katz and Thompson's sandstone samples containing 100 μm large open pores. Certainly, $D_V = D_S$ can only be true in special cases.

For example, it is easy to construct three-dimensional models of porous media (Menger-sponge models of Ackermann and Spindler, 1988) where the pore surface is fractal and the pore volume is Euclidean. Hansen and Skjeltorp (1988) used optical microscopy to determine the fractal volume and surface dimensions of sandstone samples in the 0.5–200 μm range. They analysed digitised (512 × 512)-size micrographs of sandstones by box-counting, for different resolutions L and determined the scaling of the number of boxes $N_V(L)$ and $N_S(L)$ corresponding to pore space and to the pore/grain interface, respectively. In both cases $N(L)$ scaled with box size

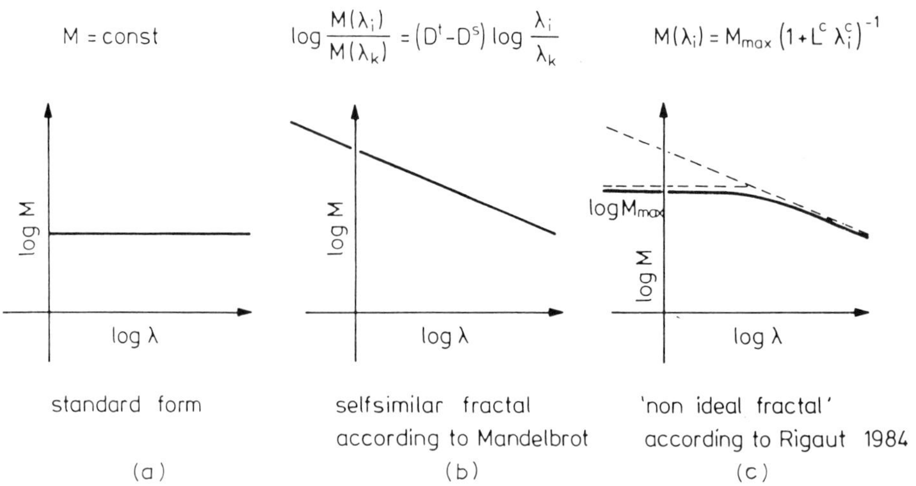

Fig. 4.44. Richardson-Mandelbrot plot for the length of a Euclidian form (a), of an ideal fractal (b), and (c) of a non-ideal fractal according to the definition of Rigout (1984). (From Pape et al., 1987a.)

as a fractal, with different dimensions D_V and D_S, however:

$$N_V(L) \propto L^{-(D_V-1)} \qquad (\langle D_V \rangle = 2.69 \pm 0.05) \tag{4.5.11a}$$

$$N_S(L) \propto L^{-(D_S-1)} \qquad (\langle D_S \rangle = 2.56 \pm 0.07) \tag{4.5.11b}$$

[In equations (4.5.11) the dimensions are averages over 30 samples.]

In 1984 Rigout introduced the concept of "non-ideal fractal" to describe Richardson-Mandelbrot plots of the length of boundary lines $L(\lambda)$ with two asymptotes in a double-logarithmic L vs. λ plot: a horizontal one for small values of $\lambda \to 0$, and an oblique one in the range of the first order structures (Figure 4.44). The "non-ideally fractal" boundary line scales as:

$$L(\lambda) = L_{\max}\left\{1 + \frac{\lambda^c}{L_0^c}\right\}^{-1} \tag{4.5.12}$$

where L_0 (a cross-over length) and the exponent c are constants.

Pape et al. (1987a) claim that the pore surface in sandstone is a "nearly ideal fractal", while the pore volume $\Phi(\lambda)$ measured by resolution λ is "non-ideal fractal" satisfying a generalised Rigout formula:

$$\Phi(\lambda) = \Phi_{\max}\left\{1 + \left(\frac{0.5676\lambda}{r_{\text{pore}}}\right)^{0.6434}\right\}^{-1} \tag{4.5.13}$$

Suppose, for simplicity, that the pore space satisfies the simple scaling law (4.5.6), i.e, with the notations $\delta = 3 - D, l_1 = r, l_2 = R$, we have:

$$\Phi(r) \propto r^{\delta} \tag{4.5.14a}$$

$$\Phi(R) \propto R^{-\delta} \tag{4.5.14b}$$

There are some immediate consequences of formulae (4.5.14), without any further assumptions on rock structure. Mosolov and Dinaryev (1987) considered the change of porosity of dry or gas-filled sedimentary rocks as a function of pressure and proved that it is related to $\delta = 3 - D$ as:

$$\frac{\partial \Phi(P,r)}{\partial P} = -\frac{\delta}{3K} \tag{4.5.15}$$

where $\Phi(P,r)$ is porosity measured at an ambient pressure of P and resolution r; K is compressibility of the porous rock.

It can be assumed that the rock matrix is incompressible when P changes from $P = P_0$ to $P = P_0 + \Delta P$, that is the corresponding change ΔV in the volume of the rock sample is due to a proportional reduction of pore space. By Hooke's law:

$$\Delta P = K\left(\frac{\Delta V}{V}\right) \tag{4.5.16}$$

Denote by $N(r)$ the number of r-cubes of volume r^3 covering the pore space at pressure P_0. Because of the assumed proportional decrease of the pore space it can be assumed that the pore radii r decrease to $r/c(\Delta)$, and the volume change becomes:

$$\Delta V \approx N(r)\left[r^3 - \left(\frac{r}{c(\Delta P)}\right)^3\right] = N(r)r^3[1 - c^{-3}(\Delta P)]$$

and, as $N(r)r^3 \propto \Phi(P_0,r)V$, we have:

$$\frac{\Delta V}{V} \propto \Phi(P_0,r)[1 - c^{-3}(\Delta P)] \tag{4.5.17}$$

By the scaling relation (4.5.14a):

$$\Phi(P_0 + \Delta P,r) \propto \Phi[P_0, r/c(\Delta P)] \propto \Phi(P_0,r)c^{-\delta}(\Delta P) \tag{4.5.18}$$

Combining equations (4.5.17) and (4.5.18) and making use of Hooke's law (4.5.16) we get, for small ΔP:

$$\Phi(P_0 + \Delta P,r) \propto \Phi(P_0,r)\left\{1 - \frac{\Delta P}{K\Phi(P_0,r)}\right\}^{\delta/3} \propto \Phi(P_0,r) - \frac{\delta \Delta P}{3K} \tag{4.5.19}$$

and equation (4.5.15) follows. [Mosolov and Dinaryev do not support (4.5.16) with experimental data; the interested Reader should see Volarovich et al. (1968) on the variation in the porosity, pore diameter and tortuosity of sandstones under pressure.]

The basic empirical transport laws of sedimentary rocks describe the electrical conductivity σ of fluid-filled rocks, and the hydraulic permeability k of the rocks, respectively, as:

$$\sigma \propto \Phi^{m} \tag{4.5.20}$$

(Archie's law), and:

$$k \propto \Phi^{m'} \tag{4.5.21}$$

(a special form of the Kozeny-Carman law; permeability can also be expressed in terms of porosity and specific surface, or porosity and tortuosity, see Section 1.3). The exponents m and m' are interrelated, according to Wong et al. (1984), for example, $m' \approx 2m$. Equations (4.5.20) and (4.5.21) are generally viewed as relations satisfied by a *family of porous rocks with a range of porosities and a common geological history* (Roberts and Schwartz, 1985). The intriguing aspect of these equations is that the exponents m and m' are usually non-integers, suggesting that the value of these exponents might be connected to the fractality of the pore space.

Here we only review those attempts where the Kozeny-Carman relation (4.5.21), or one of its equivalent forms, has been derived from simple scaling considerations. Fluid flow in porous rocks is governed by D'Arcy's law:

$$\boldsymbol{u} = -\frac{k}{\mu}\,\mathrm{grad}\,P \tag{4.5.22}$$

where $\boldsymbol{u}$ is the velocity of the flow, P is pressure, μ viscosity of the fluid, and k is permeability. Suppose that $\boldsymbol{u}$ does not depend on $\boldsymbol{x}$, let A be a cross-section of linear size a perpendicularly to $\boldsymbol{u}$. The discharge Q in one second across A is:

$$Q \propto \rho\Phi \boldsymbol{u} a^2 = \rho\Phi\frac{k}{\mu}a^2|\mathrm{grad}\,P| \tag{4.5.23}$$

Suppose that the porosity is built up of "channels" of characteristic size r, the number of such channels being determined by the distribution $n(r)$. As $n(r)r^2 \propto \Phi(r)$, by (4.5.14a) $n(r) \propto r^{\delta-2}$.

Using this relation, the discharge can also be expressed by Poiseuille's law:

$$Q \propto \rho a^2 \frac{n(r)r^4\,|\mathrm{grad}\,P|}{\mu} \tag{4.5.24}$$

and matching equations (4.5.23) with (4.5.24) yields:

$$k(r) \propto \frac{n(r)\,r^4}{\Phi(r)} \propto r^2 \approx \Phi^{2/\delta} \tag{4.5.25}$$

that is, equation (4.5.21) holds in the form:

$$k \propto \Phi^{m'} = \Phi^{2/\delta} = \Phi^{2/(3-D)} \tag{4.5.26}$$

This result, combined with equation (4.5.19), predicts that permeability would decrease with pressure as:

$$k(P_0 + \Delta P) \propto k(P_0) - \frac{2}{3}\frac{\Delta P}{K\Phi(P_0)}k(P_0) \tag{4.5.27}$$

(Mosolov and Dinaryev, 1987; again, without experimental evidence).

In the more common forms of the Kozeny-Carman relation permeability is expressed in terms of specific surface per unit bulk volume and hydraulic tortuosity τ as:

$$k = \frac{1}{b}\Phi^3 \frac{1}{S_{\text{spec}}^2} \cdot \frac{1}{\tau^2} \tag{4.5.28}$$

where b is a shape factor of order one [Kozeny (1927), Carman (1956); see also equation (1.3.1.1) in Section 1.3.1]. Let us derive equation (4.5.26) from simple scaling considerations. The specific area of granular materials scales with particle size R as:

$$S_{\text{spec}} \propto R^{D_S - 3} \tag{4.5.29}$$

for "compact" particles of surface fractal dimension D_S (see e.g. Van Damme et al., 1988). For the estimation of τ, it is tempting to simply assume that the hydraulic path follows the cross-sectional outline of the pore wall (Figure 4.45). In this case

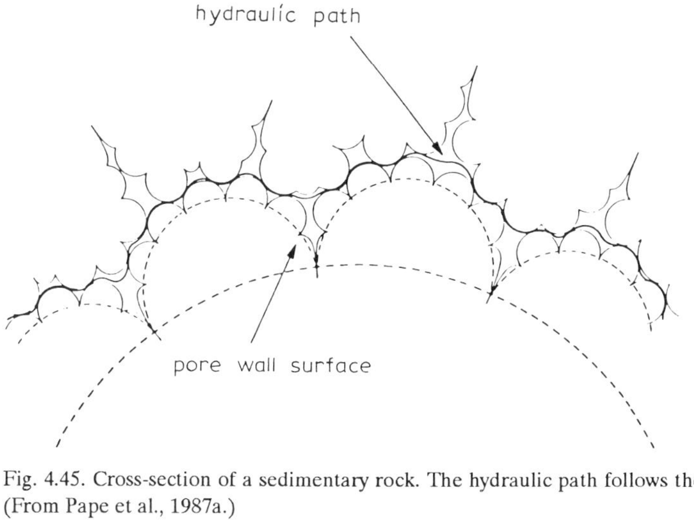

Fig. 4.45. Cross-section of a sedimentary rock. The hydraulic path follows the surface of the pore wall. (From Pape et al., 1987a.)

the hydraulic path $L(R)$ through a domain of size $\propto R$ would scale as $L(R) \propto R^{D_S-1}$ and we would get:

$$\tau = \frac{L}{R} \propto R^{D_S-2} \tag{4.5.30}$$

Combining equations (4.5.14b), (4.5.28), (4.5.29) and (4.5.30), and assuming (with Katz and Thompson, 1985) that $D = D_V = D_S$ we obtain:

$$k \approx \Phi^{(D-1)/(3-D)} \tag{4.5.31}$$

which differs from (4.5.26). The contradiction shows that the scaling of the permeability (and other physical properties of the sediments) can not be estimated in an unambiguous manner from D_V, D_S, r and R alone, without further assumptions on the structure of the rock [as Figure 4.46 shows, for example, the expression for tortuosity, (4.5.30) could only be valid for the *tortuous* capillary of Figure 4.46a, while $\tau = 1$ for the *constricted* capillary of Figure 4.46b, even though D_S is the same for both cases].

There has been very intensive research in different parts of the world in the last ten years to describe sedimentary rocks by fractals, and to predict their petrophysical properties based on these models.

There exist excellent reviews, highlighting the research activities of, and controversies between, the different groups: Thompson et al. (1987: EXON, Houston), Baudet et al. (1985: "Group Poreaux", Paris), Pape et al. (1985b, 1987a: Clausthal University, Germany), Wong (1988: Schlumberger Doll, Ridgefield, Conn.); Thomp-

Fig. 4.46. Tortuous (a) and constricted (b) capillaries. The tortuosity is fractal in (a), Euclidian in (b). (From Pape et al., 1985b.)

son (1991), Mosolov and Dinaryev (1987) is a general overview. The fractal aspects of viscous fingering and invasion percolation in porous media are reviewed in Stanley et al. (1985), Feder (1988), Wong (1988), but the topic is outside the scope of the present book.

The most widely used models of porous media include:

– The *pigeon-hole* model (Figures 4.15 and 4.17a) of Pape et al. [introduced in 1981a; detailed description and proof of optimality: 1982; a review of applications to the scaling of porosity, specific surface, tortuosity, permeability, formation factor, conductivity: 1987a, b. A modified form of the pigeon-hole model, Figure 4.17b, was applied to predict the well-log response of microfissured granites: Pape et al. (1985c)]. Characteristic results are shown in Figures 4.43 and 4.47.

– The *iterative pits* (similar to the pigeon-hole model) and *iterative flocs* model of De Gennes (1985, see Figure 4.48a, b). De Gennes used these models to study the partial filling of the fractal pore space of rocks by a wetting fluid, and found for both models that if the macroscopic capillary radius r lies between the limits of the fractal pore regime, $l_1 < r < l_2$, then the fluid fraction Φ_{fl} scales as:

$$\Phi_{fl} \propto \Phi \left(\frac{l_2}{r}\right)^{3-D_S} \tag{4.5.32a}$$

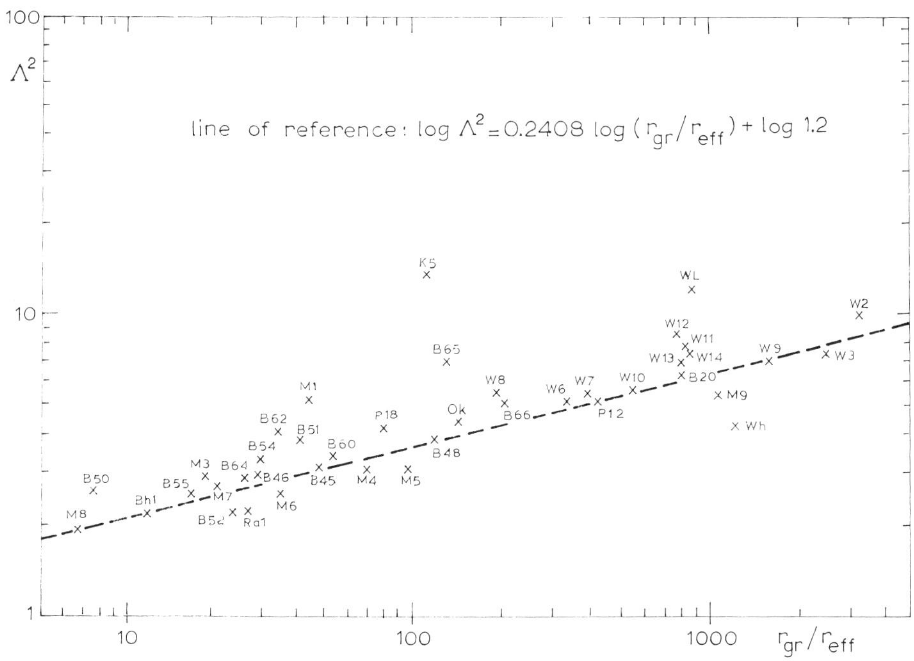

Fig. 4.47. Plot of the logarithm of pore tortuosity vs. $\log(r_{gr}/r_{eff})$ for the same sandstone and graywacke samples as in Figure 4.43 (from Pape et al., 1984).

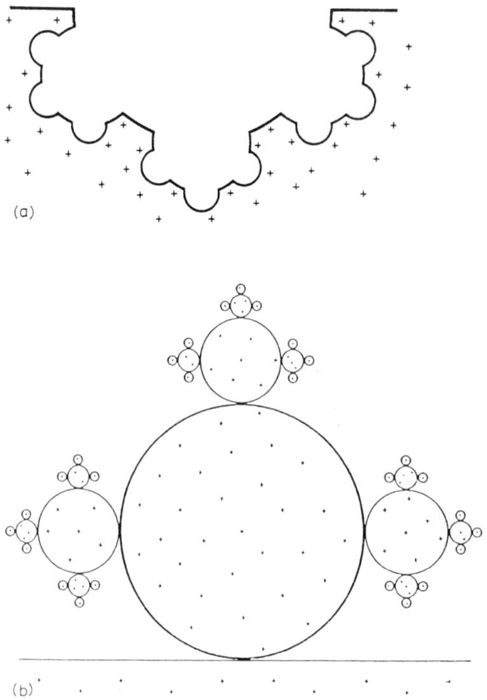

Fig. 4.48. Fractal models of porous materials. Iterative pits (a) and flocs (b) of De Gennes (1985).

while for the non-fractal case:

$$\Phi_{\mathrm{fl}} \propto \Phi \left(\frac{l}{r}\right)^2 \tag{4.5.32b}$$

In equations (4.5.32) Φ is the total porosity, D_S the fractal dimension of the pore surface, l ($r \ll l$) is the grain size.

– *Packing models* [general theory of packing for a random distribution of spheres: Omnès (1985); Apollonian packing and Leibniz packing: Mandelbrot (1982), Adler (1985), Kaye (1989); applications: Pape et al. (1984; Figure 4.48c) to study the electric properties of clayey sandstones; and Hansen and Skjeltorp (1988) to derive the scaling of permeability].

Hansen and Skjeltorp (1988) constructed a porous medium with arbitrary porosity by introducing in the first stage a fraction β of grains of characteristic size l_1. The pore volume is changing from V_0 to $\beta_1 V_0$.

Next, a fraction β_2 of grains with characteristic size l_2 is introduced and the process is repeated until some smallest size l_n is reached. In each filling the total surface will change from S_i to $\gamma_i S_i$.

Hansen and Skjeltorp showed that the volume and surface dimensions, respec-

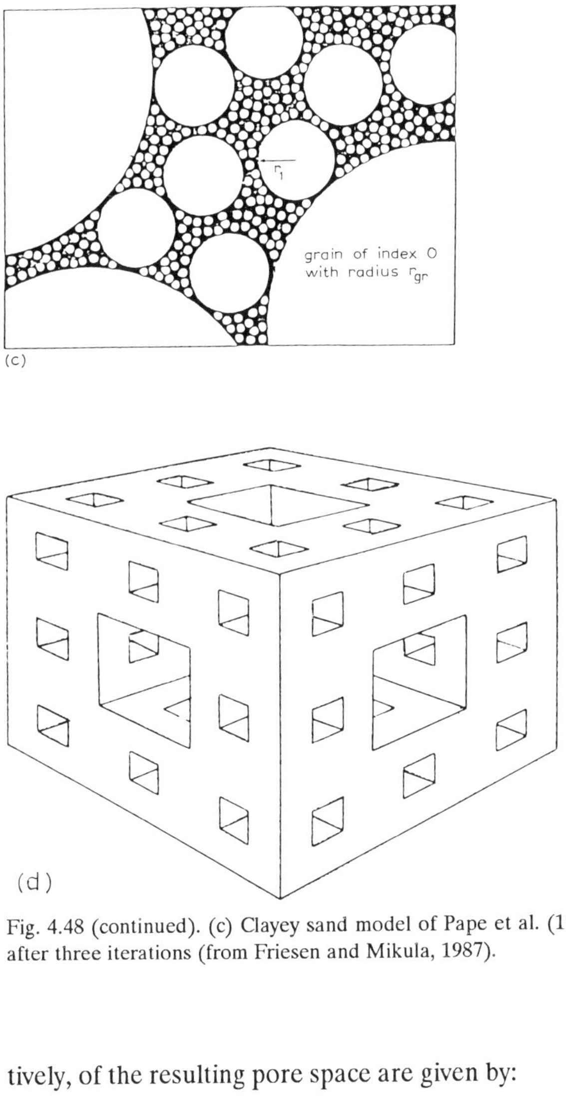

Fig. 4.48 (continued). (c) Clayey sand model of Pape et al. (1984); (d) Menger sponge model, $m = 3$, after three iterations (from Friesen and Mikula, 1987).

tively, of the resulting pore space are given by:

$$D_V \approx E_V + \frac{1}{n}\sum_{i=1}^{n} \frac{\log(\beta_i)}{\log(l_i/l_{i+1})} \tag{4.5.33a}$$

$$D_S \approx E_S + \frac{1}{n}\sum_{i=1}^{n} \frac{\log(\gamma_i)}{\log(l_i/l_{i+1})} \tag{4.5.33b}$$

where $E_V = 3$ and $E_S = 2$ are the Euclidian dimensions.

Assuming constant tortuosity and that the fractions l_i/l_{i+1} are all equal, Hansen and Skjeltorp used equations (4.5.33a) and (4.5.33b) to express the specific surface

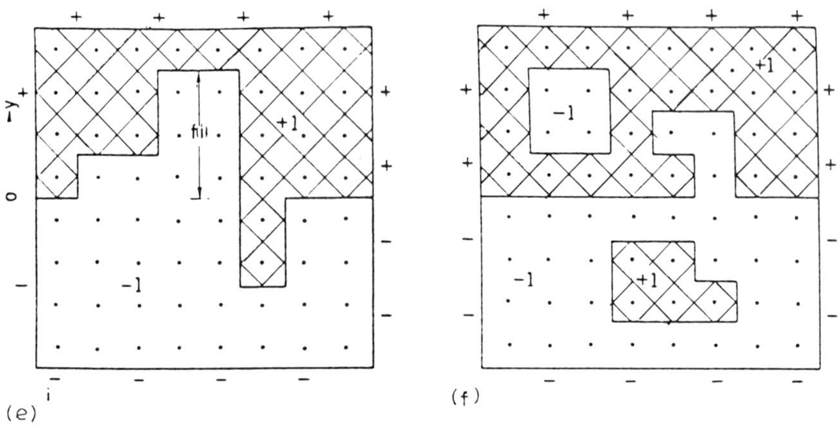

Fig. 4.48 (continued). (e), (f) Interface separating the domains of +1 and −1 Ising spins in the random field Ising model. Note that in the "solid-on-solid model", where we require that $f(i)$ is a single-valued function of i (e), there are no isolated "droplets" or "overhangs" as in (f) (from Grinstein and Ma, 1983).

in the Kozeny-Carman equation (4.5.28) and found the scaling law:

$$k \propto \Phi \left(\frac{l_1}{l_n}\right)^{2[(E_S - D_S)+(D_V - E_V)]} \tag{4.5.34}$$

– Different versions of the *Menger sponge* are ideally suited to study the fractal scaling of the pore volume and pore surface (Friesen and Mikula, 1987; Ackermann and Spindler, 1988). Friesen and Mikula (1987) started out from a unit cube, subdivided each face to m^2 smaller squares and removed n squares subject to the restrictions that they do not have common sides or lie on the face's edges (this requires $n \leq m^2 - 4m + 5$). At each missing square they inserted a cube with the top face missing. Thus, the surface area of an original face increased from 1 to N_1/m^2, where:

$$N_1 = m^2 + 4n \tag{4.5.35}$$

Repeating the procedure iteratively (Figure 4.48d) leads to a fractal porous surface with fractal dimension:

$$D_S = 2 + \frac{\log\{1 + (4n/m^2)\}}{\log m} \tag{4.5.36}$$

It is easy to prove that $D_S \to 2$ if $m \to \infty$ and it attains its maximal value, 2.37, for $m = 7$ and $n = 13$. It should be noted that this value is close to the maximum possible surface dimension, $D_S^{\max} \approx 2.36$, achievable by pigeon-hole models. (Pape et al., 1987b, discuss the "optimality" of $D_S = 2.36$).

Friesen and Mikula (1987) used this model to derive the basic equation of mercury intrusion porosimetry (MIP) in fractally porous materials. As well known (see e.g. Friesen and Mikula, 1987; Thompson et al., 1987), in MIP mercury is forced into the evacuated pore space. To overcome the interfacial tension between the mercury and the solid, a pressure $P(r)$ must be exerted on the mercury before it will fill a pore of radius r. P and r are related by the Washburn equation:

$$P = \frac{2\gamma \cos(\theta)}{r} \tag{4.5.37}$$

where γ (485 dyne/cm) is the surface tension of the mercury, θ ($\approx 130°$) is the mercury–solid contact angle. In MIP we measure the volume of mercury forced into the rock at a given pressure P, which is equal to the total volume of pores with radius $>r$.

The pore volume of the above model can be easily expressed in terms of D_S. In the first step the pore volume on one face is $V_1 = n/m^3$ and there are N squares as given by equation (4.5.35). After two steps, the pore volume becomes:

$$V_2 = V_1 + \frac{nN_1}{(m^3)^2}$$

while after k steps:

$$\begin{aligned} V_k &= V_1 + n\sum_{j=1}^{k-1} \frac{N_j}{(m^3)^{j+1}} = \frac{n}{m^3}\sum_{j=0}^{k-1}\left\{\frac{N_1}{m^3}\right\}^j \\ &= V_{\text{tot}}\left\{1 - \left(\frac{N_1}{m^3}\right)^k\right\} = V_{\text{tot}}\{1 - r_k^{3-D_S}\} \end{aligned} \tag{4.5.38}$$

where $r_k = 1/m^k$ and $V_{\text{tot}} = n/(m^3 - N_1)$ is the total pore volume for $k \to \infty$. By differentiation, the continuum limit of (4.5.38) yields:

$$\log\left[\frac{-\mathrm{d}V(r)}{\mathrm{d}r}\right] \approx (2 - D_S)\log r \tag{4.5.39}$$

that is, by the Washburn equation (4.5.37) we get:

$$\log\left[\frac{\mathrm{d}V}{\mathrm{d}P}\right] \approx (D_S - 4)\log P \tag{4.5.40}$$

The surface fractal dimension of the porous medium can be obtained from the slope of the log–log plot of $\mathrm{d}V/\mathrm{d}P$ *vs.* P. According to Kaye (1989) a similar relation can be derived from an *Apollonian gasket* model. Friesen and Mikula applied this technique to the MIP of low-rank coal samples (Figure 4.49). They identified three pressure regimes: the low-pressure regime ($P < 0.1$ MPa); the intermediate regime ($1 < P < 10$ MPa), and the high-pressure regime ($P > 10$ MPa), each with its

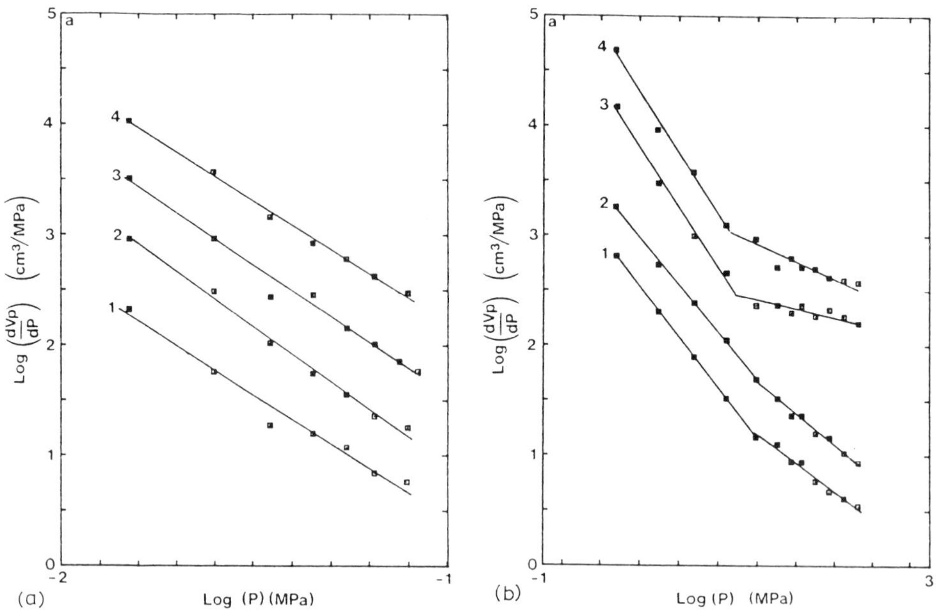

Fig. 4.49. Mercury intrusion porosimetry data in different pressure ranges: (a) 0.01–0.1 MPa; (b) 1–200 MPa. Key: *1* = coal A; *2* = coal A (treated); *3* = coal B; *4* = coal B (treated). (From Friesen and Mikula, 1987.)

characteristic fractal dimension D_{I}, D_{II} and D_{III}, respectively. For the low-pressure regime $1 < D_{\mathrm{I}} < 2$; for the intermediate regime $2 < D_{\mathrm{II}} < 3$; for the high-pressure regime $3 < D_{\mathrm{III}}$ (Table 4.5). According to Friesen and Mikula, D_{I} corresponds to the structure of the interparticle voids, D_{II} is found when the mercury penetrates the pores related to the surface, while the "unphysical" D_{III} describes the compression of the coal.

In the fractal models of the physics of porous rocks, there is a tendency to focus on either the *rock matrix* (to predict elastic properties or heat conduction),

TABLE 4.5

Fractal dimensions obtained from Figure 4.49a, b

Sample	D_{I} ($P < 0.1$ MPa)	D_{II} ($P = 1$–10 MPa)	D_{III} ($P = 10$–200 MPa)
Coal A	1.84	2.84	3.37
Coal A treated	1.54	2.92	3.33
Coal B	1.70	2.64	3.87
Coal B treated	1.82	2.62	3.67

Source: Friesen and Mikula (1987).

or the distribution and connectivity of the *pore space* (to predict transport properties).

The rock matrix is generally considered as Euclidean. Ben Ohoud et al. (1988) have experimentally confirmed that the mass fractal dimension D_m is ≈ 3 in the particle size range of ≈ 50 to ≈ 1500 μm, for powders of glass, cement, different clays (sepiolite, attapulgite, montmorillonite), silica gel, alumina and charcoal. This does not preclude the grains in porous rocks being mass fractals at smaller length scales (as e.g. soot or metallic smokes are: Forrest and Witten, 1979; Nelson, 1989; etc.). Though there has been as yet no experimental evidence for the fractality of the rock matrix in the $\approx < 10$ μm length scale, Pape et al. (1985b) have developed a fractal matrix model, and its possible consequences. In their recent works (Pape et al., 1987a) the rock matrix is considered as a "non-ideal" fractal (in the sense of Rigout, 1984), for example its volume density $(1 - \Phi)$ scales as:

$$(1 - \Phi) = (1 - \Phi)_{\max} \left\{1 + L^c r_{\text{eff}}^c\right\}^{-1} \tag{4.5.41}$$

for sandstones and greywackes. In equation (4.5.41) r_{eff} is effective hydraulic radius, $(1 - \Phi)_{\max} = 1, c = 0.25, L = 0.0625/r_{\text{grain}}$.

Both in the classical statistical models of transfer in porous rocks (Schopper, 1966, 1967; Rink and Schopper,1976) or in the recent percolation approach (Section 1.3.3; Thompson et al., 1987; Wong, 1988) the pore space is considered as a *network*, that is a connected system of grains, pores, grain contacts and throats which has geometrical and *topological* properties. Schopper (1966, 1967) characterised the influence of network topology on the macroscopical permeability or resistivity of such a network through a single network constant:

$$\varepsilon = \frac{M}{N} - \frac{Z}{N} \tag{4.5.42}$$

where M is the number of meshes per unit volume, N is number of branches per unit volume, Z is number of cells per unit volume. In particular, for the pore network and grain network, respectively:

$$\varepsilon_{\text{pore}} = \frac{n_{\text{contact}}}{n_{\text{throat}}} - \frac{n_{\text{grain}}}{n_{\text{throat}}} \tag{4.5.42a}$$

$$\varepsilon_{\text{solid}} = \frac{n_{\text{throat}}}{n_{\text{contact}}} - \frac{n_{\text{pore}}}{n_{\text{contact}}} \tag{4.5.42b}$$

where n_{pore}, n_{grain}, n_{throat} and n_{contact} are the numbers of pores, grains, pore throats and grain contacts per unit volume. In the fractal pigeon-hole model (Pape et al., 1985b, 1987a) the pore throats and grain contacts are distributed in space as fractal "dusts" of the same fractal dimension $D = 2.3566$ (that is, at any resolution there are $\propto R^{2.3566}$ throats and $\propto R^{2.356}$ grain contacts in a region of size R), so that $n_{\text{throat}}/n_{\text{contact}} = $ const ($\approx 4/3$) in equations (4.5.42a) and (4.5.42b). The other ratios scale with the effective hydraulic radius as:

$$\frac{n_{\text{throat}}}{n_{\text{pore}}} \propto \left\{\frac{r_{\text{pore}}}{r_{\text{eff}}}\right\}^{D'} \tag{4.5.43a}$$

$$\frac{n_{\text{contact}}}{n_{\text{grain}}} \propto \left\{\frac{r_{\text{grain}}}{r_{\text{eff}}}\right\}^{D''} \tag{4.5.43b}$$

where $D' = 1.9$, $D'' = 2.08$ for the pigeon-hole model (Pape et al., 1987a). [It would be interesting to compare the compaction laws that follow from (4.5.43) with the findings of Taylor (1950).]

In topology, complicated, multiple connected surfaces are characterised by their *genus*, defined as "the greatest number of distinct continuous non-self-intersecting closed curves which may be drawn on a surface without separating it into distinct regions" (Flegg, 1974; Figure 4.50a, b). The genus of the surface of a sphere is 0, the genus of the surface of a one-fold torus is 1, of a two-fold torus 2, of an n-fold torus, n. In Smalley's (1967; Figure 4.50c) sandstone diagenesis model the genus was used to characterise the advancement of diagenesis.

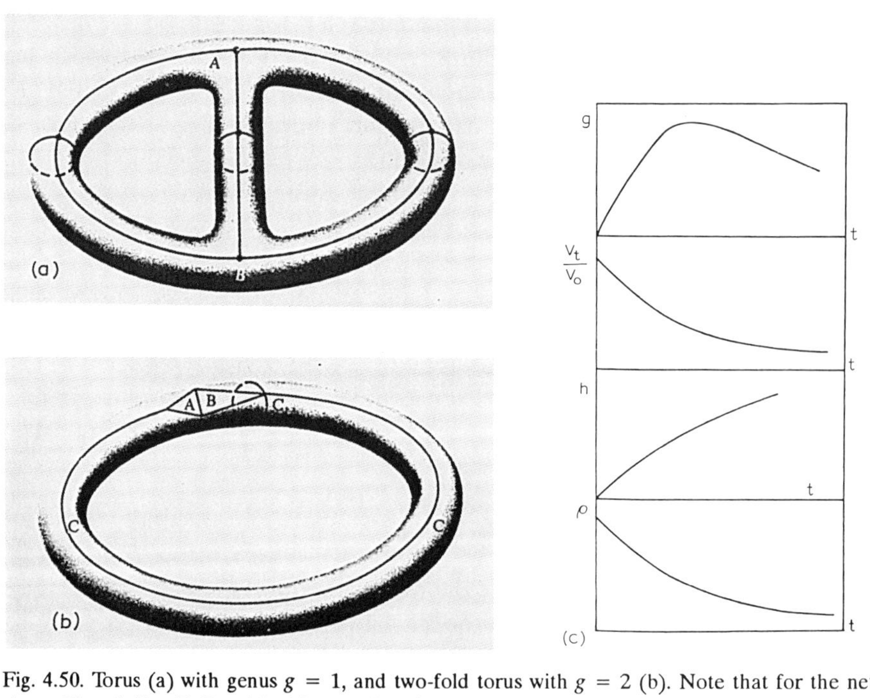

Fig. 4.50. Torus (a) with genus $g = 1$, and two-fold torus with $g = 2$ (b). Note that for the networks shown $V = 4$, $E = 7$, $C = 3$ in (a), and $V = 5$, $E = 9$, $C = 2$ in (b), that is, Euler's relation (4.5.44) is satisfied in both cases (from Flegg, 1974.). (c) Variation of the genus, fractional volume V_t/V_0, depth of burial and density during sandstone diagenesis (from Smalley, 1967).

If two sand particles are joined together, the genus does not change but if a third one is attached to both of these, the configuration becomes topologically equivalent to a torus and the genus rises to one.

If more and more particles are joined together the genus will steadily rise (sand to sandstone transition). In the compaction-diagenetic phase however, pores begin to fill, tangential contacts develop between the grains, and the genus decreases (sandstone to quartzite transition).

As we have seen in Section 2.4.1 [equation (2.4.1.1)], for any network drawn on a surface we have Euler's equation for the numbers of cells (C), edges (E), and vertices (V):

$$C - E + V = \text{const}$$

where the constant, the so-called Euler characteristic, is a topological invariant of the surface. The Euler characteristic can be expressed in terms of the genus g (Flegg, 1974):

$$C - E + V = 2 - 2g \tag{4.5.44}$$

Consequently, if n_{grain}, n_{pore}, n_{contact}, n_{throat} are fractally scaling with $(r_{\text{grain}}/r_{\text{eff}})$ or $(r_{\text{pore}}/r_{\text{eff}})$ as in equations (4.5.43a) and (4.5.43b), the scaling of the genus g can also be predicted.

It has only recently been recognised (Wong, 1987) that the fractal nature of the *pore–solid interface* can also affect the macroscopic transfer properties. As Wong (1987) notes "the notion that surfaces can affect bulk properties may seem unusual" — if one thinks, however, of porous rocks as aggregates of small particles with very high (>2.5) fractal dimension, it is clear that a non-negligible part of the atoms are on the surface. Wong applied the so-called *solid-on-solid* (SOS) model of porous surfaces to describe small angle scattering from, and the electrical transfer properties of, sedimentary rocks (Wong et al., 1986; Wong, 1987; Wong and Bray, 1987; Wong and Lin, 1988). The SOS model is inherently anisotropic as there are only vertical fluctuations above the base plane. The mean-square-width w of the interface $Y(X)$ diverges over a range $(X_0, X_0 + X)$ as:

$$w = \langle Y^2 \rangle^{1/2} \propto X^\alpha \qquad \text{with } 0 < \alpha < 1 \tag{4.5.45}$$

that is, the interface is *self affine*: $Y(X)$ will only look statistically similar to its scaled image after the affine transformation:

$$X \to pX \qquad Y \to p^\alpha Y \tag{4.5.46}$$

If we introduce a characteristic length b defined by the equation:

$$b = w(b) \tag{4.5.47}$$

equation (4.5.45) can be re-written as:

$$w = b\left(\frac{X}{b}\right)^{\alpha} \tag{4.5.48}$$

The length b is a *cross-over length* for the following reasons:

If we measure the length of the trace of $Y(X)$ with a resolution r over a horizontal distance L, then each of the L/r horizontal segments will have a vertical fluctuation of order $w(r)$:

$$l_r \propto \left(\frac{L}{r}\right)[r + w(r)] = L\left[1 + \left(\frac{b}{r}\right)^{1-\alpha}\right] \tag{4.5.49}$$

As $1 - \alpha > 0$, $l_r \approx L$ if $r \gg b$ and $l_r \approx L(b/r)^{1-\alpha}$ if $r \ll b$. In other words: the line is Euclidean, smooth, for $r \gg b$, and behaves as a fractal for $r \ll b$. Recalling that the length of a fractal boundary line diverged as $\propto r^{1-D}$ for $r \to 0$, we have:

$$l_r \propto \left(\frac{1}{r}\right)^{1-\alpha} \propto r^{1-D} \qquad \text{for } r \ll b$$

that is:

$$D = 2 - \alpha \tag{4.5.50}$$

for self-affine curves and, in general:

$$D = d - \alpha \tag{4.5.51}$$

for self-affine surfaces in d-dimensions.

Po-zen Wong and his group (Wong, 1985, 1987; Wong and Lin, 1988) used this self-affine surface model to describe small angle X-ray and neutron scattering from fractal rock surfaces. As mentioned in Section 4.4 [equation (4.4.20)], in this model the scattered intensity is described as:

$$I(k) \propto \frac{A}{k^4} + \frac{B}{k^{3+\alpha}} \tag{4.5.52}$$

By (4.5.51) this becomes:

$$I(k) \propto \frac{A}{k^4} + \frac{B}{k^{6-D}} \tag{4.5.53}$$

which is a mixture of Porod scattering ($\propto k^{-4}$) and scattering from a D-dimensional fractal surface ($\propto k^{-(6-D)}$; see Bale and Schmidt, 1984, and Section 4.4). The relative intensities of the two modes are related to the cross-over length and D as:

$$\frac{B}{A} = b^{1-\alpha} = b^{D-2} \tag{4.5.54}$$

As $D - 2 > 0$, for macroscopically smooth surfaces, where b is small, $B \ll A$ by equation (4.5.54) and the Porod scattering dominates.

Wong (1987) also applied the self-affine surface model to derive the frequency dependence of the ac electric response of brine saturated porous rocks. The rock matrix is insulating, but the ions in the pore water conduct. When the electric field is applied, the ions move along the field but get blocked by the solid surfaces. Assume that the pore wall is a self-affine surface defined by (4.5.45), and the electric field is applied in the vertical direction. The ions first move vertically along the field; when they are blocked by the surface they diffuse horizontally and then drift again along the field vertically. During a small time interval t the horizontal diffusion distance r and the vertical displacement w become $r \propto t^{1/2}$ and $w \propto r^{\alpha}$. As the conductivity σ is proportional to the average velocity w/t, we have, letting n denote the charge density and making use of equation (4.5.51):

$$\sigma \propto \frac{nw}{t} \propto r^{\alpha} t^{-1} \propto t^{[(3-D)/2]-1} \propto \omega^{(D-1)/2} \tag{4.5.55}$$

where ω is the circular frequency. Experimental verification of (4.5.55) is shown in Figure 4.51 for Portland sandstone saturated with 1 Ω-m NaCl solution. The high-frequency conductivity behaviour is consistent with equation (4.5.55), with an exponent 0.68. The fractal surface dimension of the rock was found to be $D = 2.55$ according to small angle neutron scattering (Wong, 1987), that is the predicted exponent $(D - 1)/2 = 0.78$ was reasonably close to the measured one.

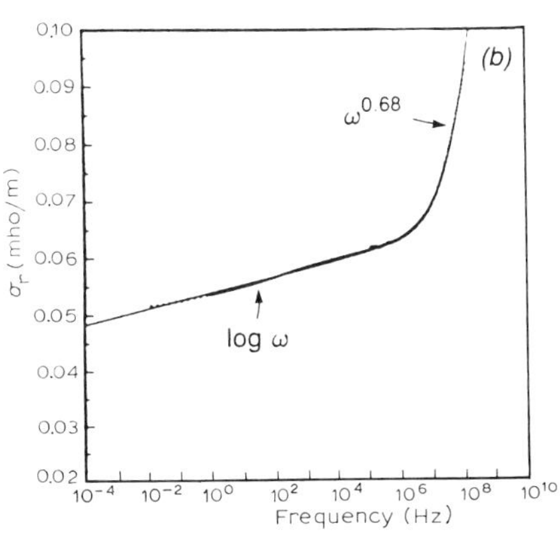

Fig. 4.51. The ac conductivity of a Portland sandstone saturated by 1 Ω-m NaCl solution (from Wong, 1987).

4.5.1 *Why are the pores fractal rather than smooth?*

My brother is a hairy man, and I am a smooth man. — *Genesis*, 27. 11

There is an obvious contradiction between the low fractal dimensions (2.05–2.36) of the reservoir rocks, obtained by the Clausthal school from the pigeon-hole model (Pape at al., 1987a, b) and the high fractal dimensions (2.5–2.9) found from the analysis of micrographs and small angle scattering (see the review of Thompson et al., 1987). A possible explanation is that the lower range corresponds to the original unaltered porosity of the mechanically stable grain–pore configuration (Peleg and Normand, 1985; Pape et al., 1987b) while the higher range is connected to the fractal structure of the diagenetic clay particles grown along the rock/pore interface (Krohn and Thompson, 1986; Wong et al., 1986; Wong, 1987).

The controversy is far from having been settled — there are both mechanical and thermodynamical arguments against the existence of very high fractal-dimensional pore spaces. Peleg and Normand (1985) compiled the published fractal dimensions of solid particle silhouettes and found that D is always limited to the range of 1.05–1.36 — though it is easy to construct arbitrarily "hairy" theoretical Von Koch islands with dimensions 1.6 or greater (see plates 53–55 in Mandelbrot, 1982). Peleg and Normand suggested that particles having silhouettes with $D > \approx 2.36$ are mechanically unstable. Pape et al. (1987b) followed up this suggestion and claimed that the $D = 2.36$ of their pigeon-hole model is the highest possible fractal dimension for a *mechanically stable* pore–grain configuration.

Cohen (1987) considered the morphology of sedimentary rocks from the thermodynamical point of view. Taking into account that deeply buried sediments could have experienced $\approx 200°C$ temperatures and ≈ 1000 atm pressures for as long as 10^8 years, he finds several morphological features (among others the very rough — let alone fractal — nature of the pore–grain interface) incompatible with the theory of sintering, as there had been ample time for sintering to occur.

To explain the observed rock morphologies, Cohen made the hypothesis that the free energy σ_{gf} associated with the interface between grains and the pore filler fluid is much less than the grain-boundary free energy σ_{gb}:

$$\sigma_{gf} \ll \frac{\sigma_{gb}}{2} \tag{4.5.1.1}$$

("antisintering regime"). According to Cohen, for silica grains in water the interfacial free energy is actually *negative*: $\sigma_{gl} \approx -60$ ergs/cm^2.

From equation (4.5.1.1) Cohen found that there exists a critical temperature T_R (the index refers to the "roughening" transition), or, at a fixed temperature, a critical value σ_R where the free energy of the rough surface overtakes that of the smooth one due to the additional entropy associated with the roughness. At this point there is a spontaneous roughening transition: the pore/rock interface becomes rough though not fractal. Further increase of the σ_{gb}/σ_{gf} ratio leads to a cross-over to a fractal pore/rock interface.

Wong et al. (1986; Wong, 1987) also tackled the question "why do the pore surfaces become fractal?". They associated the high-dimensional fractal structure with the negative surface tension of the clay minerals: the lowest free-energy state for the clay minerals is that where the *surface area is maximal.* Wong et al. (1986) attributed the negative surface tension of the clay particles to impurity ions lying along the clay surface: the solid surface tries to bend around to exclude as many impurities as possible, and a negative surface tension develops.

Wong et al. (1986) suggested that the development of fractal pore surfaces is in many ways similar to surface roughening in the self-affine *random field Ising model* (RFIM) of Grinstein and Ma (1983; Figure 4.48e, f; see also the very important Burkov, 1985). The RFIM considers a set of Ising spins placed on a d-dimensional lattice. The Hamiltonian is:

$$H = -\sum_{[i,j]} J\, S_i S_j - \sum_j H_j S_j \tag{4.5.1.2}$$

where S_i ($S_i = +1$ or $S_i = -1$) is the spin variable at the i-th site, J is the exchange energy, H_j is a random magnetic field which points up and down randomly at every site; $\langle H_j \rangle = 0$, $\langle H_j^2 \rangle = H^2$; $\langle H_i H_j \rangle = 0$ if $i \neq j$. The summation in equation (4.5.1.2) is only carried out for sites $[i,j]$ which are nearest neighbours. The first term in equation (4.5.1.2) favours the spins to be parallel, while the random field term favours the spins point up and down randomly, as governed by the magnetic field H_j. Grinstein and Ma (1983) assumed that due to the competition of the two terms, there exists a "spin up" domain, and a "spin down" domain, separated by an interface $y = f(i)$, where the index i labels lattice sites in the $(d-1)$-dimensional horizontal space and y is the vertical coordinate (Figure 4.48e). They also assumed that the exclusion of "droplets" and "overhangs" (Figure 4.48f) does not affect the gross features of the surface $f(i)$. The surface is assumed to be self-affine that is [by equation (4.5.45)] for a horizontal length r and for any i, $f(i)$ diverges as:

$$\langle f(i)^2 \rangle^{1/2} = w \propto r^{\alpha} \tag{4.5.1.3}$$

The exponent α [that is, by equation (4.5.51), the fractal dimension of the surface $f(i)$ as well] can be determined from the condition that the surface is in equilibrium.

The number of sites involved is $r^{d-1}w$. The expected coherent part of the random field over these sites is:

$$E_{\mathrm{H}} = (r^{d-1} w H)^{1/2} \tag{4.5.1.4}$$

that is, the presence of the surface lowers the system's energy by this amount. On the other hand, the surface breaks r^{d-1} horizontal and $\sum_{[i,i']} f(i) - f(i') \propto r^{d-2} w$ vertical exchange bonds, that is, it costs an energy:

$$E_J = J(r^{d-1} + r^{d-2}w) \tag{4.5.1.5}$$

The equilibrium is determined from the condition that the net surface energy $E_S = E_J - E_E$ is minimal: $\partial E_S/\partial w = 0$, wherefrom:

$$w = \left(\frac{H}{4J^2}\right) r^{3-d} \tag{4.5.1.6}$$

Consequently, α is $3 - d$ in (4.5.1.3) and, by (4.5.51), the surface is:

$$D = 2d - 3 \tag{4.5.1.7}$$

-dimensional.

The energy of the equilibrium surface is, by equations (4.5.1.4)–(4.5.1.6):

$$E_S = Jr^{d-1}\left(1 - \frac{H}{4J^2} r^{2-d}\right) \tag{4.5.1.8}$$

The *surface tension* is the surface energy per unit area:

$$T = \frac{E_S}{r^{d-1}} = J\left\{1 - \frac{w(r)}{r}\right\} \tag{4.5.1.9}$$

and is negative for $w > r$; positive for $w < r$.

According to Wong et al. (1986; Wong, 1987) the exchange energy in the RFIM is like the chemical bonds between the atoms in the grains of the porous rocks, the random magnetic vectors $\{H_j\}$ are similar to the impurity atoms that cost extra energy to incorporate into the solid. The analogies, however, are not exact, the three-dimensional RFIM yields a unique dimension $D = 2d - 3 = 3$ and cannot explain the experimentally observed wide range of fractal dimensions in diagenetically altered sandstones ($D = 2.5$–2.96).

Cohen (1987) mentions another mechanism leading to fractal pore walls. From the earliest stages of sedimentation the very small clay particles can freely diffuse within the pore space defined by the larger quartz grains. Thus, there is a possibility that the rough walls are formed by *diffusion-limited aggregation* (DLA, see Meakin, 1988a, b; Feder, 1988), which also leads to self-similar structures.

4.6 Multifractal measures — not for the squeamish

"Do you like card tricks?"
"No, I hate card tricks," I answered.
"Well, I'll just show you this one."
He showed me three. — SOMERSET MAUGHAM: *Mr. Know-All*

Session W101, "Fractals, Statistics, and Chaos in Exploration and Development" of the 1991 Annual AAPG Meeting (Dallas) attracted record audience. In one of the lectures, Barton et al. reported on the fractal analysis on the spatial distribution of producing and showing wells combined, and of dry wells, in 1600 mi^2 portions of the Denver and Powder River basins. The analysis was limited to single stratigraphic horizons. The fractal dimensions of the reservoirs were in the two regions 1.5 and 1.4, respectively; the fractal dimensions of the dry holes were 1.8 and 1.7, respectively; the fractal dimension of the pattern of all wells drilled was 1.8 for both basins; the average grid spacing was about 1/4 mile.

Let us denote the average grid spacing by ε, and determine the probability $P(\varepsilon)$ of finding hydrocarbon in a randomly selected well as $\varepsilon \to 0$. For the Denver basin, say, the number of producing holes and dry holes, respectively, will scale as:

$$N_{\text{res}} = C_{\text{res}}\varepsilon^{-1.5} \qquad N_{\text{dry}} = C_{\text{dry}}\varepsilon^{-1.8}$$

with appropriate constants C_{res} and C_{dry}. For small values of ε:

$$P(\varepsilon) = \frac{N_{\text{res}}}{N_{\text{tot}}} = \frac{C_{\text{res}}\varepsilon^{-1.5}}{C_{\text{res}}\varepsilon^{-1.5} + C_{\text{dry}}\varepsilon^{-1.8}} \approx \frac{C_{\text{res}}\varepsilon-^{1.5}}{C_{\text{dry}}\varepsilon^{-1.8}} \propto$$
$$\propto \varepsilon^{0.3} \to 0 \quad \text{as } \varepsilon \to 0 \tag{4.6.1}$$

This simple example shows that if we measure a *physical field* on the points of a fractal, the probabilities (or, more exactly, "measures") describing the distribution of the functional values on the fractal could show unexpected, scaling behaviour, as the resolution increases. In the previous example the *points of the fractal set* were the wells, the *physical field* was a geochemical quantity or well-log response on the basis of which a given well was classified as producing or dry, the probabilities $P(\varepsilon)$ and $1 - P(\varepsilon)$ were the *measures* associated with the two possible ranges of field values, and they scaled with $\varepsilon \to 0$ according to equation (4.6.1).

A more complicated example of this kind concerns ore deposits (Mandelbrot, 1982, 1989; after De Wijs, 1951, 1953): the areal distribution of all ore bodies is a fractal set wherein the bodies of different grades constitute subsets of different fractal dimensions depending on grade.

Further examples for physical fields measured on fractals include the magnetization M_i measured in a region labelled i of the Ising model (see the previous section), the hydrodynamic force distribution at points on and inside a fractal aggregate moving in a fluid (Meakin and Deutsch, 1987; Witten, 1987b); the local growth rate on the surface of aggregates (Amitrano et al., 1986; Nittman et al., 1987) or the local density of sites:

$$p_i = \frac{N_i}{N_{\text{total}}} \tag{4.6.2}$$

in small regions labelled by i (Hentschel and Procaccia, 1983; Halsey and Jensen, 1986; Halsey et al., 1986). In some cases the local fractal dimension is inhomogeneous: for different points X_i of the fractal the number of points N_i within a neighbourhood of size r scales with different exponents $N_i \propto r^{\delta_i}$ (Figure 4.52).

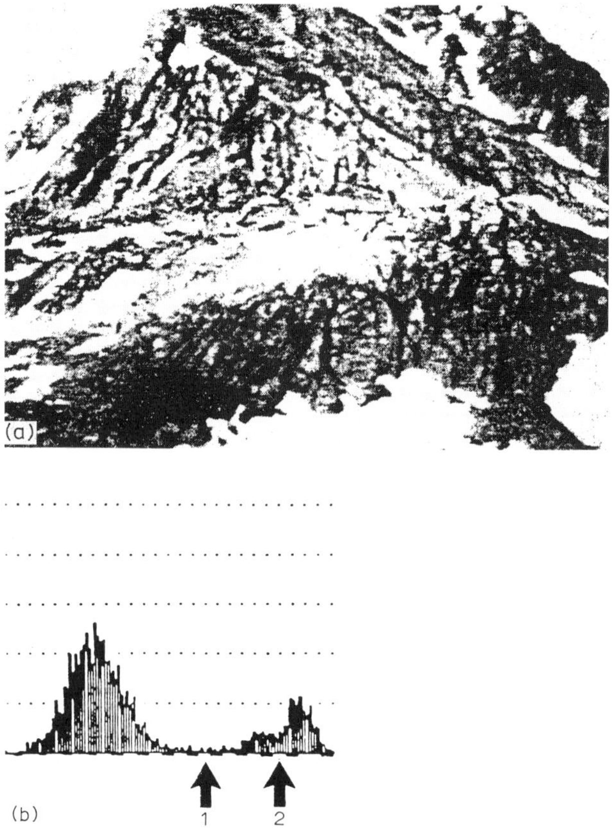

Fig. 4.52. a. Photo of a landscape digitised to 512 x 512 pixels. b. Histogram of the local fractal dimensions of the brightness distribution in (a). The histogram is tri-modal. The first arrow shows the break between land (low dimensions) and snow plus sky; the second arrow shows the sky (highest dimensions) separated from the snow. (From Pentland, 1984.)

Histograms like Figure 4.52, or the other standard tools of statistics, however, can not be used to characterise the distribution of the physical field on a fractal if the sets $\{X_i | y(X_i) = y\}$ have different fractal dimensions $D(y)$ depending on the value y, that is, if:

$$\dim(\{X_i | y(X_i) = y\}) = D(y) \tag{4.6.3}$$

Physical fields with such *multifractal* properties "present a picture of treacherous complexity" (Witten, 1987b). In standard statistics, if we are given a physical field $\{y_1, y_2, \ldots, y_M\}$ distributed over M sites, one way to define a typical field value is to take the median $y(1/2)$. By definition, half of the $\{y_i\}$ values are greater than $y(1/2)$. Similarly, the a-th percentile field $y(a)$ is exceeded on only a fraction aM $(0 < a < 1/2)$ of the sites. Witten (1987b) proved the remarkable theorems that for multifractal physical fields the median $y(1/2)$, and the ratio $y(a)/y(1/2)$ *diverge* with the number of sites as:

$$\log\left[y\left(\frac{1}{2}\right)\right] = \sigma_0 \log M \tag{4.6.4}$$

$$\log\left[\frac{y(a)}{y(1/2)}\right] = \Delta\, \mathrm{erf}^{-1}(1-2a)[\log M]^{1/2} \tag{4.6.5}$$

where $\mathrm{erf}^{-1}(x)$ is the inverse error function, and the exponents σ_0 and Δ can be determined from the function $F(q)$ describing the scaling of the moments:

$$\langle y^q \rangle \propto M^{F(q)} \tag{4.6.6}$$

[see equation (4.6.30)]. For non-multifractal distributions the ratio on the l.h.s. of (4.6.5) is constant.

Witten (1987a, b) proved that the scaling of the statistical distribution of the values $\{y_i\}$ is completely determined by the function $F(q)$. If we normalise the field values to unity:

$$\langle y \rangle = \sum_{i=1}^{M} y_i = \int y\, n(y)\, \mathrm{d}y = 1 \tag{4.6.7}$$

and introduce the variable:

$$\sigma(y) = \frac{\log(y)}{\log(M)} \tag{4.6.8}$$

then, for $M \gg 1$:

$$\frac{\log[y\, n(y)]}{\log(M)} = \Gamma[\sigma(y)] \tag{4.6.9}$$

where the function Γ can be obtained from $F(q)$ [via a Legendre transform, as will be shown later when deriving equations (4.6.24)–(4.6.27)]. The functional form of $F(q)$ and Γ is independent of the number of sites M.

As we see, for a complete characterisation of multifractally distributed fields on fractals we have to know the function $F(q)$ defining the scaling of the moments of the field [equation (4.6.6)] or its Legendre transform Γ.

The physical meaning of $F(q)$ and Γ, and their connection, will be discussed for the simplest case when the physical field is a *measure* defined on the fractal. The treatment will follow Halsey and Jensen (1986) which I have found the most enjoyable introduction. (More detailed treatments are Halsey et al., 1986; Feder, 1988; Stanley and Meakin, 1988; and the authoritative Mandelbrot, 1989.)

We should first recall the definition of measure (Sazonov, 1990): The measure is a notion that generalizes those of the length of segments, area of figures and volume of bodies. Let $X = \{X_i\}$ be a set of the points X_i; denote by $E = \{E_i\}$ the set of all subsets of X, including the whole set X and the empty set $\emptyset$. A non-negative function $P(E_i)$ defined for all subsets $E_i \in E$ is called a measure if:

$$P(\emptyset) = 0 \tag{4.6.10a}$$

$$P\left(\bigcup_{i=1}^{n} E_i\right) = \sum_{i=1}^{n} P(E_i) \text{ whenever } E_i \cap E_j = \emptyset \text{ if } i \neq j \text{ for } n \leq \infty \tag{4.6.10b}$$

(In words, condition 4.6.10b means that the measure is *additive* for an arbitrary number of mutually disjunct sets.) We shall also assume that the measure is normalized, to make it akin to the probability distributions:

$$P(X) = 1 \tag{4.6.10c}$$

Suppose now that X is a finite (fractal or Euclidian) point set in the d-dimensional space, and $P(\cdot)$ is a measure on X. Let us cover X by N small, d-dimensional, non-overlapping cubes of size $l \ll 1$, and assume that the set of points of X inside the i-th box have the measure p_i, $\sum p_i = 1$. The distribution $\{p_i\}$ depends on l, a convenient way to describe this dependence is to compute the moments:

$$M_q = M(q,l) = \sum_{i=1}^{N} p_i^q \tag{4.6.11}$$

Let us assume that the moments scale as:

$$M_q \propto l^\tau \tag{4.6.12}$$

and determine τ, first for Euclidian shapes and fractals with uniformly distributed measures.

Let X be first the one-dimensional unit interval, let us divide it to $N = 1/l$ equal pieces of length $l \ll 1$, and let P be the ordinary measure of length. Then $p_i = l$ and:

$$M_q = \sum_{i=1}^{N} p_i^q = N l^q = l^{-1} l^q = l^{q-1} \tag{4.6.13}$$

that is, $\tau = q - 1$. Similarly, if X is a d-dimensional unit cube and we divide it to small cubes of size l:

$$M_q = \sum_{i=1}^{N} p_i^q = N\{l^d\}^q = l^{-d} l^{dq} = l^{d(q-1)} \tag{4.6.14}$$

that is:

$$\tau = d(q-1) \tag{4.6.15}$$

Suppose now that X is a D-dimensional fractal set in the d-dimensional space. X can be covered by $N = Cl^{-D}$ d-dimensional non-overlapping cubes of size $l \ll 1$, where C is a constant which does not depend on l.

If the measure $P(\cdot)$ is uniformly distributed over the points of the fractal, the measure of the points inside any small cube will be about the same: $p_i = 1/N = C^{-1}l^D$.

The moment M_q will scale for $l \to 0$ as:

$$M_q = Np_i^q \propto (l^{-D})\{l^D\}^q = l^{D(q-1)} \tag{4.6.16}$$

that is:

$$\tau = D(q-1) \tag{4.6.17}$$

In conclusion, we can say that *for uniformly distributed measures* the moments M_q scale for $l \to 0$ as:

$$M_q = \sum p_i^q \propto l^{D(q-1)} \tag{4.6.18}$$

where D is the (Euclidian or fractal) dimension of X.

In an influential paper, Hentschel and Procaccia (1983) have shown that if the measure $P(\cdot)$ is not uniformly distributed on the fractal, the exponent D in (4.6.18) should be substituted by D_q, that is the moments scale as:

$$M_q = \sum p_i^q \propto l^{D_q(q-1)} \tag{4.6.19}$$

where the D_q's are different for each q. Equation (4.6.19) serves as a definition of the numbers D_q:

$$D_q = \lim_{l \to 0} \left\{ \frac{1}{q-1} \frac{\log\left(\sum p_i^q\right)}{\log l} \right\} \tag{4.6.20}$$

Hentschel and Procaccia (1983) called D_q"generalised" dimensions to indicate that D_q $(-\infty < q < \infty)$ describe the distribution of a measure on a fractal, and not the fractal itself.

Hentschel and Procaccia proved that the set of all values $\{D_q\}$ uniquely characterises the measure $P(\cdot)$. Special cases of D_q are the similarity dimension (for $q \to 0$); the information dimension (for $q \to 1$) and the correlation dimension ($q = 2$). D_q is a strictly decreasing function of q if the measure $P(\cdot)$ is non-uniform, while — by (4.6.18) — all D_q are the same for fractals with an homogeneous measure.

An alternative way to characterise the fractal measure has been introduced by Halsey et al. (1986; Halsey and Jensen, 1986). Let us consider the local behaviour of the measure $P(\cdot)$ on the fractal. Let C_i be a small cube of size l and let $p_i = P(X_i \mid X_i \in C_i)$ be the measure of the sites covered by C_i. If l is very small, it can be assumed that the measure $P(\cdot)$ is locally homogeneous inside C_i, that is, p_i scales as:

$$p_i \propto l^{\alpha_i} \tag{4.6.21}$$

where α_i is a local scaling index. A particular scaling index α' can occur in different places of the fractal X. *Suppose that the subset of points where the local scaling index has the same value α' is also fractal, with fractal dimension $f(\alpha')$.* Thus, the number of those cubes C_i where the local scaling index takes on a particular value between α' and $\alpha' + d\alpha'$ is given by:

$$n(\alpha') = d\alpha' \rho(\alpha') l^{-f(\alpha')} \tag{4.6.22}$$

where $\rho(\alpha')$ describes the spectrum of the scaling indices.

In order to relate the α–$f(\alpha)$ picture to the previously discussed q–D_q description, we express $M_q = \sum p_i^q$ in terms of α and $f(\alpha)$. Expressing M_q as an integral and making use of equations (4.6.21) and (4.6.22):

$$M_q = \sum p^q \approx \int \mathrm{d}(\alpha')\rho(\alpha') l^{-f(\alpha')} l^{q\alpha'} \propto l^{-f(\alpha)+q\alpha} \tag{4.6.23}$$

In the last step we made the usual saddle-point approximation: the dominant part of the integral will correspond to the minimum exponent of l for $l \ll 1$ (provided that $\rho(\alpha) \neq 0$). From the minimum condition:

$$\left(\frac{\mathrm{d}[-f(\alpha') + q\alpha']}{\mathrm{d}\alpha'}\right)_{\alpha'=\alpha(q)} = 0$$

and:

$$\left(\frac{\mathrm{d}^2[-f(\alpha) + q\alpha]}{\mathrm{d}(\alpha')^2}\right)_{\alpha'=\alpha(q)} > 0$$

so that:

$$f'[\alpha(q)] = q \tag{4.6.24}$$

$$f''[\alpha(q)] < 0 \tag{4.6.25}$$

Matching equations (4.6.12) and (4.6.23) we have:

$$f[\alpha(q)] = q\alpha(q) - \tau(q) \tag{4.6.26}$$

$$\alpha(q) = \frac{\mathrm{d}\tau(q)}{\mathrm{d}q} \tag{4.6.27}$$

where, by equation (4.6.19), $\tau = (q-1)D_q$. The pair of equations (4.6.26) and (4.6.27) constitute a so-called Legendre transform from the independent variables (τ, q) to the independent variables (f, α) (Halsey and Jensen, 1986; Feder, 1988). As the "generalised dimensions" can be measured according to equation (4.6.20), we can find $\alpha(q)$ from (4.6.27), and then — knowing $\alpha(q)$ — $f[\alpha(q)]$ can be obtained from (4.6.26).

Alternatively, if we know the spectrum of α values and $f(\alpha)$, we obtain $D_q = \tau(q)/(q-1)$ from (4.6.26)

According to equation (4.6.25), if $f(\alpha)$ is continuous and differentiable, it should be convex. As D_0 is equal to the similarity dimension (Hentschel and Procaccia, 1983), by (4.6.24) the maximum value of $f(\alpha)$ is the dimension D of X as a whole. By the definition (4.6.21) the minimum scaling exponent $\alpha_{\min}$ corresponds to the most concentrated region of the measure on the fractal, the maximum exponent $\alpha_{\max}$ corresponds to the most rarefied region of the measure.

The α–$f(\alpha)$ plots or q–D_q plots can be exactly constructed for Cantor bars with non-uniformly distributed measures (see Halsey et al., 1986; Halsey and Jensen, 1986; Feder, 1988). Here, we present the famous "first experimental evidence for the multifractality of the surface measure of a viscous fingering structure" (Nittman et al., 1987; also discussed in Feder, 1988; Stanley and Meakin, 1988).

The very complicated surface of aggregates, viscous fingering structures observed in Hele-Shaw cells (Figure 4.53a) and other fractals can be characterised by the so-called harmonic measure p_i defined on the surface sites i of the fractal. The measure p_i is the probability that a random walker approaching the fractal from the infinity first strikes the fractal at site i (Meakin, 1987). As the "tips" of these structures are more easily accessible than the deep "fjords", and there also are almost inaccessible "screened" parts, it has been suggested (Amitrano et al., 1986) that p_i is a multifractal measure.

For a given cluster shape in two dimensions the harmonic measure can be estimated by computer simulation in which a random walk particle is started from a random position on a circle enclosing the cluster, and then moves on a random trajectory until it contacts the cluster at some site i of the perimeter. If we repeat this for a very large number of times, say N times, the probability p_i can be estimated as $p_i = N_i/N$, where N_i is the number of contacts with the site i. Let us repeat this experiment for clusters of the same surface-fractal dimension D but different characteristic size L. If we denote the *fixed* grid spacing by l, the surface can be covered by $N \propto (L/l)^D$ squares of size l, that is, the moments M_q will scale with L as:

$$M_q = \sum p_i^q \propto N p_i^q \propto N\left(\frac{1}{N}\right)^{D_q} \propto \left(\frac{L}{l}\right)^D \left(\frac{l}{L}\right)^{D_q} \propto L^{D(1-q)} \tag{4.6.28}$$

for uniformly distributed fractal measures $\{p_i\}$, and as:

$$M_q \propto L^{D_q(1-q)} \propto L^{-\tau(q)} \tag{4.6.29}$$

for multifractal measures.

If the exponents $\tau(q)$ have been obtained from the simulations, the scaling indices α, and the fractal dimension $f(\alpha)$ of the set of all points where the scaling index has the value α, can be determined from the Legendre transform (4.6.26) and (4.6.27).

Nittman et al. (1987) used this procedure for the harmonic measure of viscous fingering patterns observed in Hele-Shaw cells, and found the $\tau(q)$ curve and $f(\alpha)$ curve shown in Figure 4.53b, c. Observe that $\tau(q)$ consists of two nearly linear sections, suggesting a "bifractal" distribution of the harmonic measure (corresponding to the "fjords" and the "tips", respectively). The $f(\alpha)$ curve shows the above-discussed general properties, its maximal value however (≈ 1.15), is much lower than the expected $f_{\max} = D_{\text{perimeter}}$ value. (See Feder, 1988, for further discussion.)

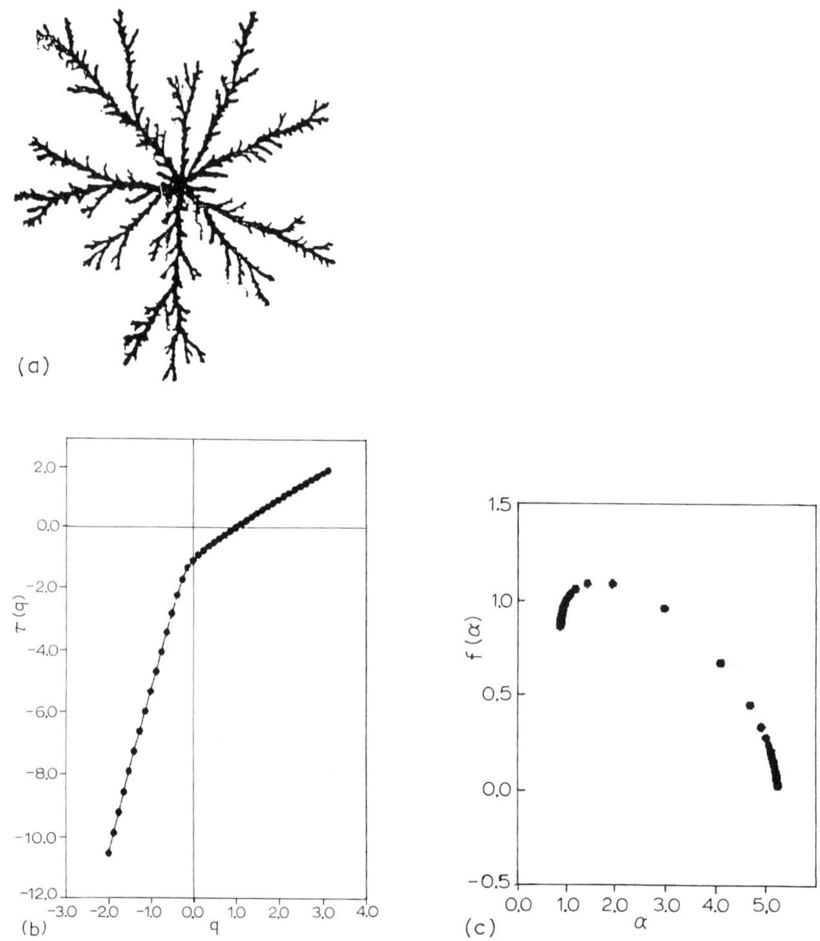

Fig. 4.53. a. A typical viscous fingering structure in a circular Hele-Shaw cell: water displacing a non-Newtonian high-viscosity fluid. The structure is fractal, with $D \approx 1.7$ (from Daccord et al., 1986). b. The hierarchy of scaling exponents $\tau(q)$, and c. the $f(\alpha)$ function for the harmonic measure on experimentally observed two-dimensional viscous fingers (from Nittman et al., 1987).

In order to reconcile Witten's equations (4.6.4)–(4.6.9) describing the scaling of the statistics of multifractal physical fields with the α–$f(\alpha)$ picture of Halsey et al., observe that Witten's σ and Γ relate to α and $f(\alpha)$ as:

$$\sigma = \frac{\alpha}{D} \qquad \Gamma(\sigma) = \left(\frac{1}{D}\right) f(\alpha) \tag{4.6.30}$$

There is also another convention: Readers of Feder (1988) are warned that his definition of the exponents $\tau(q)$ (cf. his equation 6.33) differs by a sign from the one used here.

Witten (1987b) represents the $\Gamma(\sigma)$ function near its peak as:

$$\Gamma(\sigma) \approx 1 - \frac{(\sigma - \sigma_O)^2}{\Delta^2} + O[(\sigma - \sigma_O)^3] \tag{4.6.31}$$

The peak value σ_O and the width Δ control the scaling of the percentile fields, as seen in equations (4.6.4) and (4.6.5).

References

Ackermannn, W.-G. and Spindler, H. (1988). Die Fraktale Dimension als Katalysatorkenngrösse. 1. Eine mathematische Einführung. Z. Phys. Chem., 269, 5: 1000–1010.

Adler, P.M. (1985). Transport processes in fractals. I. Conductivity and permeability of a Leibniz packing in the lubrication limit. Int. J. Multiph. Flow, 11, 1: 91–108.

Agterberg, F.P. (1982). Recent developments in geomathematics. Geo-Processing, 2: 1–32.

Allain, C. and Cloitre, M. (1985). Diffraction on fractals. *In*: E. Pynn and A. Skjeltorp (Editors), Scaling Phenomena in Disordered Systems. Plenum Press, New York, pp. 193–195.

Allain, C. and Cloitre, M. (1986). Optical diffraction on fractals. Phys. Rev., B, 33, 5: 3566–3569.

Amitrano, C., Coniglio, A. and Di Liberto, F. (1986). Growth probability distribution in kinetic aggregation processes. Phys. Rev. Lett., 57: 1016–1019.

Attewell, P.B. and Ramana, Y.W. (1966). Wave attenuation and internal friction as functions of frequency in rocks. Geophys., 31, 6: 1049–1056.

Auberger, M. and Rinehart, J.S. (1961). Ultrasonic velocity and attenuation of longitudinal waves in rocks. J. Geophys. Res., 66: 191–199.

Aviles, C.A., Scholz, C.H., and Boatwright, J. (1987). Fractal analysis applied to characteristic segments of the San Andreas fault. J. Geophys. Res., 92, B1: 331–344.

Avnir, D. (1986). Fractal aspects of surface science — an interim report. *In*: C.J. Brinker (Editor), Better Ceramics Through Chemistry, Material. Res. Soc., 1–10.

Avnir, D. and Pfeifer, P. (1983). Fractal dimension in chemistry. An intensive characteristic of surface irregularity. Nouv. J. Chim., 7, 2: 71–72.

Avnir, D., Farin, D. and Pfeifer, P. (1983). Chemistry in noninteger dimensions between two and three. II. Fractal surfaces of adsorbents. J. Chem. Phys. 79, 7: 3566–3571.

Avnir, D., Farin, D. and Pfeifer, P. (1984). Molecular fractal surfaces. Nature, 308, 5956: 261–263.

Avnir, D., Farin, D. and Pfeifer, P. (1985). Surface geometric irregularity of particulate materials: The fractal approach. J. Colloid Interf. Sci., 103, 1: 112–123.

Bale, H.D. and Schmidt, P.W. (1984). Small-angle X-ray-scattering investigation of submicroscopic porosity with fractal properties. Phys. Rev. Lett., 53, 6: 596–599.

Balmino, G., Lambeck, K. and Kaula, W.M. (1973). A spherical harmonic analysis of the Earth's topography. J. Geophys. Res., 78: 478–481.

Barenblatt, G.I., Zhivago, A.V., Neprochnov, Yu.P. and Ostrovskiy, A.A. (1984). Fractal dimension: A quantitative characteristic of the ocean bottom geomorphological evolution processes. Okeonologiya, 24, 6: 924–928 (in Russian).

Barton, C.C., Scholz, C.H., Schutter, T.A., Herring, P.R. and Thomas, W.J. (1991). Fractal nature of hydrocarbon deposits, 2: Spatial distribution. Ann. AAPG Mtg. (Dallas), Abstracts, p. 74.

Baudet, C., Charlaix, E., Clement, E., Gyron, E., Hulin, J.-P. and Leroy, C. (1985). Scaling concepts in porous media. *In*: Pynn, R. and Skjeltorp, A. (Editors), Scaling Phenomena in Disordered Systems. Plenum Press, New York, pp. 399–422.

Beckmann, P. and Spizzichino, A. (1963). The Scattering of Electromagnetic Waves from Rough Surfaces. Pergamon Press, New York.

Bell, T.H., Jr. (1975). Statistical features of sea-floor topography. Deep-Sea Res., 22: 883–892.

Ben Ohoud, M., Obrecht, F., Gatineau, L., Levitz, P. and Van Damme, H. (1988). Surface area, mass fractal dimension, and apparent density of powders. J. Colloid Interf. Sci., 124, 1: 156–161.

Berkson, J.M. (1975). Statistical properties of the ocean bottom. J. Acoust. Soc. Am., 58, Suppl. No. 1, Abstr. No. PP12.

Berry, M.V. (1972). On deducing the form of surfaces from their diffracted echoes. J. Phys. A, Gen. Phys., 5: 272–291.

Berry, M.V. (1979a). Diffractals. J. Phys. A, Math. Gen., 12, 6: 781–797.

Berry, M.V. (1979b). Catastrophe and fractal regimes in random waves. *In*: W. Güttinger and H. Eikemeier (Editors), Structural Stability in Physics. Springer, Heidelberg–New York, pp. 43–50.

Berry, M.V. and Blackwell, T.M. (1981). Diffractal echoes. J. Phys. A, Math. Gen., 14: 3101–3110.

Berry, M.V. and Hannay, J.H. (1978). Topography of random surfaces. Nature, 273: 573.

Berry, M.V. and Lewis, Z.V. (1980). On the Weierstrass-Mandelbrot fractal function. Proc. R. Soc. Lond., A, 370: 459–484.

Berry, M.V. and Percival, I.C. (1986). Optics of fractal clusters such as smoke. Optica Acta, 33, 5: 577–591.

Bhatia, A.B. (1967). Ultrasonic Absorption. Oxford Univ. Press, Oxford.

Bills, B.G. and Ferrari, A.J. (1977). A harmonic analysis of lunar topography. Icarus, 31: 244–259.

Bills, B.G. and Ferrari, A.J. (1978). Mars topography harmonics and geophysical implications. J. Geophys. Res., 83: 3497–3508.

Bills, B.G. and Kobrick, M. (1985). Venus topography: A harmonic analysis. J. Geophys. Res., 90: 827–836.

Biot, M.A. (1957). Reflection on a rough surface from an acoustic point source. J. Acoust. Soc. Am., 29, 11: 1193–1200.

Blackman, R.B. and Tukey, J.W. (1958). The Measurement of Power Spectra. Dover Publ., New York.

Bourrely, C., Chiapetta, P. and Torresani, B. (1986). Light scattering by particles of arbitrary shape: a fractal approach. J. Opt. Soc. Am., A3, 2: 250–255.

Brochard, F. (1985). Some surface effects in porous fractals. J. Phys., 46: 2117–2123.

Brodatz, P. (1966). Textures. A Photographic Album for Artists and Designers. Dover Publ., New York.

Brown, S.R. and Scholz, C.H. (1985). Broad bandwidth study of the topography of natural rock surfaces. J. Geophys. Res., 90, B14: 12575–12582.

Bunde, A., Gouyet, J.-F. and Rosso, M. (1987). On the problem of measuring fractal dimensions of random interfaces. J. Phys. A, Math. Gen., 20: 6127–6132.

Burkov, S.E. (1985). Fractal structure of the equilibrium crystal shape. J. Physique, 46: 317–327.

Burnett, A.I. and Adams, K.C. (1977). A geological, engineering and economic study of a portion of the Lloydminster Sparky Pool, Lloydminster, Alberta. Bull. Can. Pet. Geol., 25, 2: 341–366.

Burrough, P.A. (1981). Fractal dimensions of landscapes and other environmental data. Nature, 294: 240–242.

Cage, J. (1975). A Year from Monday. Marion Boyars, London.

Carman, P.C. (1956). Flow of Gases through Porous Media. Butterworth Sci. Publ., London.

Chernov, L.A. (1960). Wave Propagation in Random Media. McGraw Hill, New York.

Clarke, K.C. (1986). Computation of the fractal dimension of topographic surfaces using the triangular prism surface area method. Comp. Geosci., 12, 5: 713–722.

Cohen, M.H. (1987). The morphology of porous sedimentary rocks. *In*: Banavar, J.R. et al. (Editors), Physics and Chemistry of Porous Media II.. AIP Conf.Proc. No. 154, Am. Inst. Phys., New York, pp. 3–16.

Comroe, J.H., Jr. (1966). The lung. Sci. Am., (February), pp. 57–68.

Courtens, E. and Vacher, R. (1987). Structure and dynamics of fractal aerogels. Z. Phys. B, Cond. Matt., 68: 355–361.

Creutzburg, R. and Ivanov, E. (1989). Fast algorithm for computing fractal dimensions of image segments. *In*: V. Cantoni et al. (Editors), Recent Issues in Pattern Analysis and Recognition, Lecture Notes in Computer Science No.399, Springer-Verlag, Heidelberg–New York, pp. 42–51.

Culling, W.E.H. and Datko, M. (1987). The fractal geometry of the soil-covered landscape. Earth Surf. Proc. Landforms, 12: 369–385.

Daccord, G., Nittmann, J. and Stanley, H.E. (1986). Fractal viscous fingers: Experimental results. *In*: H.E. Stanley and N. Ostrowsky (Editors), On Growth and Form, Martinus Nijhoff, Dordrecht, pp. 203–210.

De Gennes, P.-G. (1984). Adsorption de polymeres linéaires flexibles sur une surface fractale. C.R. Acad. Sci. Paris, 299, Ser. II, 14: 913–915.

De Gennes, P.-G. (1985). Partial filling of a fractal structure by a wetting fluid. *In*: Adler D. et al. (Editors), Physics of Disordered Materials. Plenum Publ., New York, pp. 227–241.

De Wijs, H.J. (1951). Statistics of ore distribution. I. Geol. Mijnbouw, 13: 365–375.

De Wijs, H.J. (1953). Statistics of ore distribution. II. Geol. Mijnbouw, 15: 12–24.

Dos Santos, D.I., Aegerter, M.A., Craievich, A.F., Lours, T. and Zarzycki, J. (1987). Structural studies of fractal silica aerogels. J. Non-Cryst. Solids, 95 and 96: 1143–1150.

Elias, H. and Schwartz, D. (1969). Surface areas of the cerebral cortex of mammals determined by stereological methods. Science, 166: 110–113.

Emanuel, A.S., Alameda, G.K., Behrens, R.A. and Hewett, T.A. (1987). Reservoir performance prediction methods based on a fractal geostatistics. 62nd Ann. SPE Tech. Conf. (Dallas), Pap. SPE 16971.

Farin, D., Peleg, S., Yavin, D. and Avnir, D. (1985). Applications and limitations of boundary-line fractal analysis of irregular surfaces: Proteins, aggregates, and porous materials. Langmuir, 1: 399–407.

Feder, J. (1988). Fractals. Plenum Press, New York and London.

Flegg, H.G. (1974). From Geometry to Topology. The English Universities Press Ltd., London.

Forrest, S.R. and Witten, T.A., Jr. (1979). Long-range correlations in smoke-particle aggregates. J. Phys. A, Math. Gen., 12, 5: L109–L117.

Fox, C.G. and Hayes, D.E. (1985). Quantitative methods for analysing the roughness of the sea-floor. Rev. Geophys., 23, 1: 1–48.

Frey, J., Pinvidic, J.J., Botet, R. and Jullien, R. (1988). Light scattering by fractal aggregates: a numerical investigation. J. Phys., 49: 1969–1976.

Friesen, W.I. and Mikula, R.J. (1987). Fractal dimension of coal particles. J. Colloid Interf. Sci., 120, 1: 263–271.

Frisch, U. (1968). Wave propagation in random media. *In*: A.T. Bharucha-Reid (Editor), Probabilistic Methods in Applied Mathematics. I. Academic Press, New York–London.

Gelbaum, B.R. and Olmsted, J.M.H. (1964). Counterexamples in Analysis. Holden-Day, San Francisco–London–Amsterdam.

Goldberger, A.L. and West, B.J. (1987). Fractals in physiology and medicine. Yale J. Biol. Med., 60: 421–435.

Goldberger, A.L., Rigney, D.R. and West, B.J. (1990). Chaos and fractals in human physiology. Sci. Am. (February), pp. 34–41.

Gomes, M.A.F. (1987a). Fractal geometry in crumpled paper balls. Am. J. Phys., 55: 649–650.

Gomes, M.A.F. (1987b). Paper crushes fractally. J. Phys. A, Math. Gen., L283–L284.

Gray, L.J., Liu, S.H. and Kaplan, T. (1985). A fractal model for charge distribution across a rough interface. *In*: R. Pynn and A. Skjeltorp (Editors), Scaling Phenomena in Disordered Systems. Plenum, New York, pp. 79–83.

Gregg, S.J. and Sing, K.S.W. (1967). Adsorption, Surface Area and Porosity. Academic Press, London–New York.

Grinstein, G. and Ma, Sh.-K. (1983). Surface tension, roughening, and lower critical dimension in the random field Ising model. Phys. Rev. B, 28: 2588–2601.

Guinier, A. and Fournet, G. (1955). Small-Angle Scattering of X-Rays. John Wiley and Sons, New York.

Hall, P.J. (1986). The fractal properties of microporous solids as determined by X-ray diffraction and adsorption techniques. Chem. Phys. Lett., 124, 5: 467–469.

Halsey, T.C. and Jensen, M.H. (1986). Spectra of scaling indices for fractal measures: Theory and experiment. Physica, 23D: 112–117.

Halsey, T.C., Jensen, M.H., Kadanoff, L.P., Procaccia, I. and Shraiman, B. (1986). Fractal measures and their singularities: The characterization of strange sets. Phys. Rev. A, 33, 2: 1141–1151.

Hansen, J.P. and Skjeltorp, A.T. (1988). Fractal pore space and rock permeability implications. Phys. Rev. B, 38, 4: 2635–2638.

Hentschel, H.G.E. and Procaccia, I. (1983). The infinite number of generalized dimensions of fractals and strange attractors. Physica, 8D: 435–444.

Hewett, T.A. (1986). Fractal distributions of reservoir heterogeneity and their influence on fluid transport. 61st Ann. SPE Tech. Conf. (New Orleans), Pap. SPE 15386.

Hewett, T.A. and Behrens, R.A. (1988). Conditional simulation of reservoir heterogeneity with fractals. 63rd Ann. SPE Tech.Conf. (Houston), Pap. SPE 18326.

Hibler, W.D., III and Leshack, L.A. (1972). Power spectrum analysis of undersea and surface sea-ice profiles. J. Glaciol., 11, 63: 345–356.

Hurd, A.J., Schaefer, D.W. and Martin, J.E. (1987). Surface and mass fractals in vapor-phase aggregates. Phys. Rev. A, 35, 5: 2361–2364.

Jackson, D.P. and Jackson, J.A. (1988). Tibetan Thangka Painting. Methods and Materials. 2nd (revised) Edition. Serindia Publishers, London.

Jaeger, R.M. and Schuring, D.J. (1966). Spectrum analysis of terrain of Mare Cognitum. J. Geophys. Res., 71, 8: 2023–2028.

Jakeman, E. (1984). Scattering by fractal objects. Nature, 307: 110.

Jakeman, E. (1986); Scattering by fractals. *In*: L. Pietronero and E. Tosatti (Editors), Fractals in Physics. Elsevier, Amsterdam, pp. 55–60.

Jensen, R.A. (1966). High Density Living. Leonard Hill, London.

Jingzhong, C., Boren, D., Jinchen, Y, and Zhaolu, P. (1989). The Hausdorff dimension of the CP model for quasi-crystalls. Earth Sci. J. China Univ. Geosci., 14: 254–258 (in Chinese).

Journel, A.G. and Huijbregts, Ch. (1978). Mining Geostatistics. Academic Press, New York.

Kantor, Y., Kardar, M. and Nelson, D.R. (1986). Statistical mechanics of tethered surfaces. Phys. Rev. Lett., 57, 7: 791–794.

Kaplan, T. and Gray, L.J. (1985). Effect of disorder on a fractal model for the ac response of a rough interface. Phys. Rev. B, 32, 11: 7360–7366.

Katz, A.J. and Thompson, A, H. (1985). Fractal sandstone pores: Implications for conductivity and pore formation. Phys. Rev. Lett., 54, 12: 1325–1328.

Katz, A.J. and Thompson, A.H. (1986). Katz and Thompson reply. Phys. Rev. Lett., 56, 19: 2112.

Kaye, B.H. (1989). A Random Walk Through Fractal Dimensions. VCH Verlagsgesellschaft, Weinheim, Germany.

Kerker, M. (1969). The Scattering of Light. Academic Press, New York.

Koch, D.L. (1987). Attenuation of a compressional sound wave in the presence of a fractal boundary. Phys. Fluids, 30, 10: 2922–2927.

Kolb, M. and Hermann, H.J. (1987). Surface fractals in irreversible aggregation. Phys. Rev. Lett., 59, 4: 454–457.

Korvin, G. (1977). Certain problems of seismic and ultrasonic wave propagation in a medium with inhomogeneities of random distribution. II. Wave attenuation and scattering on random inhomogeneities. Geophys. Trans. 24, Suppl. 2: 3–36.

Korvin, G. (1982). Certain problems of seismic and ultrasonic wave propagation in a medium with inhomogeneities of random distribution. III. Statistics of the diffuse reflection shadow following a rough reflecting boundary. Geophys. Trans. 28, 1: 5–19.

Kozeny, J. (1927). Über die kapillare Leitung des Wassers im Boden (Aufstieg, Versickerung und Anwendung auf die Bewässerung). Sitzungsber. Akad. Wiss. Wien, Math. Naturwiss. (Abt. 2a), 136a: 271–306.

Krohn, C.E. (1988a). Sandstone fractal and Euclidian pore volume distribution. J. Geophys. Res., 93, B4: 3286–3296.

Krohn, C.E. (1988b). Fractal measurements of sandstones, shales, and carbonates. J. Geophys. Res. 93, B4: 3297–3305.

Krohn, C.E. and Thompson, A.H. (1986). Fractal sandstone pores: Automated measurements using SEM images. Phys. Rev. B, 33: 6366–6374.

Leong, W.K., Kan, T.K. and Clay, C.S. (1971). Use of acoustic scattering theory to interpret marine geophysical data. Univ. Wisconsin Geophys. Polar. Res. Cent., Res. Rep., No. 71-7.

Lewis, M. and Rees, D.C. (1985). Fractal surfaces of proteins. Science, 230: 1163–1165.

Liu, S.H. (1985). Fractal model for the ac response of a rough interface. Phys. Rev. Lett., 55, 5: 529–532.

Liu, S.H. (1986). Fractals and their applications in condensed matter physics. Solid State Phys. 39: 207–273.

Liu, S.H., Kaplan, T. and Gray, L.J. (1986a). Theory of the ac response of rough surfaces. *In*: L. Pietronero and E. Tosatti (Editors), Fractals in Physics. Elsevier, Amsterdam–New York, pp. 383–389.

Liu, S.H., Kaplan, T. and Gray, L.J. (1986b). Ac response of fractal interfaces. Solid State Ionics, 18 and 19: 65–71.

Lucido, G., Triolo, R. and Caponetti, E. (1988). Fractal approach in petrology: Small-angle neutron scattering experiments with volcanic rocks. Phys. Rev. B, 38, 13: 9031–9034.

Mandelbrot, B.M. (1967). How long is the coast of Britain? Statistical self-similarity and fractional dimension. Science, 155: 636–638.

Mandelbrot, B.M. (1982). The Fractal Geometry of Nature. Freeman and Co., New York.

Mandelbrot, B.M. (1986). Self-affine fractal sets. I: The basic fractal dimensions; II: Length and surface dimensions; III: Hausdorff dimension anomalies and their implications. *In*: L. Pietronero and E. Tosatti (Editors), Fractals in Physics. Elsevier, Amsterdam–New York, pp. 3–16; 17–20; 21–28.

Mandelbrot, B.M. (1989). Multifractal measures, especially for the geophysicist. PAGEOPH, 131, 1/2: 5–42.

Mandelbrot, B, M., Passoja, D.E., and Paullay, A.J. (1984). Fractal character of fracture surfaces of metals. Nature, 308: 721–722.

Mandelbrot, B.M. and Van Ness, J.W. (1968). Fractional Brownian motion, fractional noises, and applications. SIAM Rev., 10: 422–437.

Marechal, J.-C. (1989). Fractal reconstruction of sea-floor topography. PAGEOPH, 131, 1/2: 197–219.

Martin, J.E. (1986). Scattering exponents for polydisperse surface and mass fractals. J. Appl. Crystallogr., 19: 25–27.

Mcguire, M. (1988). An eye for fractals. *In*: H.-O Peitgen and D. Saupe (Editors), The Science of Fractal Images. Springer, Heidelberg–New York, pp. 261–271.

Meakin, P. (1987). Scaling properties for the growth probability measure and harmonic measure of fractal structures. Phys. Rev. A, 35: 2234–2245.

Meakin, P. (1988a). The growth of fractal aggregates and their measures. *In*: C. Domb and J.L. Lebowitz (Editors), Phase Transitions and Critical Phenomena, Vol. 12. Academic Press, London–New York, pp. 335–489.

Meakin, P. (1988b). Fractal aggregates. Adv. Colloid Interf. Sci., 28: 249–331.

Meakin, P. and Deutsch, J.M. (1987). Properties of the fractal measure describing the hydrodynamic force distributions for fractal aggregates moving in a quiescent fluid. J. Chem. Phys., 86, 8: 4648–4656.

Mendes France, M. (1984). Folding paper and thermodynamics. Phys. Rep., 103, 1/4: 161–172.

Mildner, D.F.R. and Hall, P.L. (1986). Small-angle scattering from porous solids with fractal geometry. J. Phys. D, Appl. Phys., 19: 1535–1545.

Mildner, D.F.R., Rezvani, R., Hall, P.L. and Borst, R.L. (1986). Small-angle scattering of shaly rocks with fractal pore interfaces. Appl. Phys. Lett., 48, 19: 1314–1316.

Militzer, H. and Stoll, R. (1968). Vibratorseismische Messungen *in situ* zur Bestimmung eines frequenzabhängigen Absorptionskoeffizienten. Bergakademie, 20, 9: 518–521.

Morse, D.R., Lawton, J.H., Dodson, M.M. and Williamson, M.H. (1985). Fractal dimension of vegetation and the distribution of arthropod body lengths. Nature, 314: 731–733.

Mosolov, A.B. and Dinaryev, O.Yu. (1987). Fractal models of porous media. J. Technic. Phys., 57, 9: 1679–1685 (in Russian).

Nelson, J. (1989). Fractality of sooty smoke: implications for the severity of nuclear winter. Nature, 339: 611–613.

Nittmann, J., Stanley, H.E., Touboul, E. and Daccord, G. (1987). Experimental evidence for multifractality. Phys. Rev. Lett., 58, 6: 619.

Nye, J.F. (1973). A note on the power spectra of sea-ice profiles. AIDJEX Bull., 21: 20–21.

Okubo, P.G. and Aki, K. (1987). Fractal geometry in the San Andreas Fault System. J. Geophys. Res., B1: 345–355.

Omnés, R. (1985). Sur l'empilement désordonné d'un système de sphères de rayons différents. J. Phys., 46: 139–147.

Orford, J.D. and Whalley, W.B. (1983). The use of fractal dimension to quantify the morphology of irregular-shaped particles. Sedimentology, 30: 655–668.

Pande, C.S., Richards, L.R. and Smith, S. (1987a). Fractal characteristics of fractured surfaces. J. Mat. Sci. Lett., 6: 295–297.

Pande, C.S., Richards, L.E., Louat, N., Dempsey, B.D. and Schwoeble, A.J. (1987b). Fractal characterization of fractured surfaces. Acta Metall., 35, 7: 1633–1637.

Pape, H., Riepe, L. and Schopper, J.R. (1981a). A solution attempt to the "Coastline of Britain Problem". Stereol. Iugosl., 3, Suppl. 1: 331–336.

Pape, H., Riepe, L. and Schopper, J.R. (1981b). Calculating permeability from surface area measurements. 7th Eur. Logging Symp. (Paris), Trans. Pap. No.17.

Pape, H., Riepe, L., and Schopper, J.R. (1982). A pigeon-hole model for relating permeability to specific surface. Log Anal., 23, 1: 5–13 (Errata: *ibid.*, 23, 2: 50).

Pape, H., Riepe, L. and Schopper, J.R. (1984). The role of fractal quantities, as specific surface and tortuosities, for physical properties of porous media. Part. Charact., 1: 66–73.

Pape, H., Riepe, L. and Schopper, J.R. (1985a). Permeability of porous rocks derived from internal surface effects. 47th EAEG Meet. (Budapest), Pap. C-44.

Pape, H., Riepe, L. and Schopper, J.R. (1985b). Shape description of agglomerates by physically relevant geometrical properties in terms of the theory of fractal dimensions. 4th Int. Symp. on Agglomeration (Toronto), pp. 683–695.

Pape, H., Riepe, L. and Schopper, J.R. (1985c). Petrophysical detection of microfissures in granites. Trans. SPWLA 26th Annu. Logging Symp. (Dallas), pp. 1–16.

Pape, H., Riepe, L. and Schopper, J.R. (1987a). Theory of self-similar network structures in sedimentary and igneous rocks and their investigation with microscopical and physical methods. J. Microsc., 148, Pt. 2: 121–147.

Pape, H., Riepe, L. and Schopper, J.R. (1987b). Interlayer conductivity of rocks — A fractal model of interface irregularities for calculating interlayer conductivity of natural porous mineral systems. Colloids Surf., 27: 97–122.

Paumgartner, D., Losa, G. and Weibel, E.R. (1981). Resolution effect on the stereological estimation of surface and volume and its interpretation in terms of fractal dimensions. J. Microsc., 121, Pt. 1: 51–63.

Peleg, M. and Normand, M.D. (1985). Mechanical stability as the limit to the fractal dimension of solid particle silhouettes. Powder Technol., 43: 187–188.

Pentland, A.P. (1984). Fractal based description of natural scenes. IEEE Trans. Pattern Anal. Machine Intelligence, PAMI-6, 6: 661–674.

Peters, R.H. (1983). The Ecological Implications of Body Size. Cambridge Univ. Press, Cambridge.

Pfeifer, P., Avnir, D. and Farin, D. (1983). Ideally irregular surfaces, of dimension greater than two, in theory and practice. Surf. Sci., 126: 569–572.

Pfeifer, P., Avnir, D. and Farin, D. (1984). Scaling behavior of surface irregularity in the molecular domain: From adsorption studies to fractal catalysts. J. Stat. Phys., 36, 5/6: 699–716.

Pfeifer, P. and Schmidt, P.W. (1988). Pfeifer and Schmidt reply. Phys. Rev. Lett., 60, 13: 1345.

Ragette, F. (1980). Architecture in Lebanon. Caravan Books, Delmar, New York.

Rice, S.O. (1944 and 1945). The mathematical analysis of random noise. Bell Syst. Technol. J., 23, No. 3; and 24, No. 1.

Rigout, J.P. (1984). An empirical formulation relating boundary lengths to resolution in specimens showing "non-ideally fractal" dimensions. J. Microsc., 133, Pt. 1: 41–54.

Rink, M. and Schopper, J.R. (1976). Pore structure and physical properties of porous sedimentary rocks. PAGEOPH, 114: 273–284.

Roberts, J.N. (1986). Comment about fractal sandstone pores. Phys. Rev. Lett., 5, 19: 2111.

Roberts, J.N. and Schwartz, L.M. (1985). Grain consolidation and electrical conductivity in porous media. Phys. Rev. B, 31, 9: 5990–5997.

Romeau, D., Gomez, A., Perez-Ramirez, J.G., Silva, O., Perez, O.L., Gonzalez, A.E. and José-Yacaman, M. (1986). Surface fractal dimension of small metallic particles. Phys. Rev. Lett., 57, 20: 2552–2555.

Rothrock, D.A. and Thorndike, A.S. (1980). Geometric properties of the underside of sea ice. J. Geophys. Res., 85, C7: 3955–3963.

Sayles, R.S. and Thomas, T.R. (1978a). Surface topography as a nonstationary random process. Nature, 271: 431–434.

Sayles, R.S. and Thomas, T.R. (1978b). Sayles and Thomas reply. Nature, 273: 573.

Sazonov, V.V. (1990). Measure. *In*: Enc. Math. VI. Kluwer Acad. Publ., Dordrecht–Boston–London, pp. 181–187.

Schaefer, D.W. and Keefer, K.D. (1986a). Structure of random silicates: Polymers, colloids, and porous solids. *In*: L. Pietronero and E. Tosatti (Editors), Fractals in Physics. Elsevier, Amsterdam–New York, pp. 39–45.

Schaefer, D.W. and Keefer, K.D. (1986b). Structure of random porous materials: Silica aerogel. Phys. Rev. Lett., 56, 20: 2199–2202.

Schaefer, D.W., Martin, J.E., Wiltzius, P. and Cannell, D.S. (1984). Fractal geometry of colloidal aggregates. Phys. Rev. Lett., 52, 26: 2371–2374.

Schaefer, D.W., Wilcoxon, J.P., Keefer, K.D., Bunker, B.C., Pearson, R.K., Thomas, I.M. and Miller, D.E. (1987). Origin of porosity in synthetic materials. *In*: J.R., Banvar, J. Koplik and K.W. Winkler (Editors), Physics and Chemistry of Porous Media. AIP Conf. Proc., AIP, New York, 154: 63–80.

Schillinger, J. (1948). The Mathematical Basis of the Arts. Philosophical Library, New York.

Schmidt, P.W. (1982). Interpretation of small-angle scattering curves proportional to a negative power of the scattering vector. J. Appl. Crystallogr., 15: 567–569.

Schmidt, P.W. and Dacai, X. (1986). Calculation of the small-angle X-ray and neutron scattering from nonrandom (regular) fractals. Phys. Rev., A, 33, 1: 560–566.

Scholz, C.H. and Aviles, C.A. (1986). The fractal gerometry of faults and faulting. *In*: Das, S., Boatwright, J. and Scholz, C.H. (Editors), Earthquake Source Mechanics. Geophys. Monogr. 37, Maurice Ewing Vol.6, AGU, Washington, D.C., pp. 147–155.

Schopper, J.R. (1966). A theoretical investigation on the formation factor/permeability/porosity relationship using a network model. Geophys. Prospect., 14, 3: 301–341.

Schopper, J.R. (1967). A theoretical study on the reduction of statistical pore system parameters to measurable quantities. Geophys. Prospect., 15: 262–287.

Schulz-Rojahn, J.P. and Phillips, S.E. (1989). Diagenetic alteration of Permian reservoir sandstones in the Nappameri Trough and adjacent areas, southern Cooper Basin. *In*: B.J. O'Neil (Editor), The Cooper and Eromanga Basins, Australia. Petr. Expl. Soc. Austr., Adelaide, South Australia, pp. 629–645.

Sernetz, M., Gelleri, B. and Hofmann, J. (1985). The organism as a bioreactor. Interpretation of the reduction law of metabolism in terms of heterogeneous catalysis and fractal structure. J. Theor. Biol., 117: 209–230.

Smalley, I.J. (1967). A simple model of a diagenetic system. Sedimentology, 8: 27–33.

Stanley, H.E. and Meakin, P. (1988). Multifractal phenomena in physics and chemistry. Nature, 335: 405–409.

Stanley, H.E., Daccord, G., Herrmann, H.J. and Nittmann, J. (1985). Applications of scaling and disorderly growth phenomena to oil recovery. *In*: R. Pynn and A. Skjeltorp (Editors), Scaling Phenomena in Disordered Systems. Plenum Press, New York–London, pp. 85–97.

Taggart, I.J. and Salisch, H.A. (1991). Fractal geometry, reservoir characterisation and oil recovery. APEA Journal, 1991: 377–385.

Tatarski, V.I. (1961). Wave Propagation in a Turbulent Medium. McGraw Hill, New York.

Taylor, J.M. (1950). Pore space reduction of sandstone. Bull. AAPG, 34, 4: 701–716.

Thompson, A.H. (1991). Fractals in rock physics. Annu. Rev. Earth Planet. Sci., 19: 237–262.

Thompson, A.H., Katz, A.J. and Krohn, C.E. (1987). The microgeometry and transport properties of sedimentary rock. Adv. Phys., 36, 5: 625–694.

Tolstoy, I. and Clay, C.S. (1966). Ocean Acoustics: Theory and Experiment in Underwater Sound. McGraw Hill, New York.

Turcotte, D.L. (1987). A fractal interpretation of topography and geoid spectra on the Earth, Moon, Venus, and Mars. J. Geophys. Res., 92, B4: E597–E601.

Turcotte, D.L. (1989). Fractals in geology and geophysics. PAGEOPH, 131, 1/2: 171–196.

Turcotte, D.L. (1991). Fractals in geology: What they are and what are they good for? GSA Today, 1, 1: 1–4.

Twidale, C.R. (1982). The evolution of Bornhardts. Am. Sci., 70: 268–276.

Van Damme, H. and Fripiat, J.J. (1985). A fractal analysis of adsorption processes by pillared swelling clays. J. Chem. Phys., 82, 6: 2785–2789.

Van Damme, H., Levitz, P., Gatineau, L., Alcovere, J.F. and Fripiat, J.J. (1988). On the determination of the surface fractal dimension of powders by granulometric analysis. J. Colloid Interf. Sci., 122, 1: 1–8.

Van de Hulst, H.C. (1957). Light Scattering by Small Particles. Wiley, New York.

Varnier, F., Dussert, C., Rasigni, G., Rasigni, M. and Llebaria, A. (1988). Defining the fractal nature of thin-film top surfaces from microdensitometer analysis of electron micrographs of surface replicas. J. Vac. Sci. Technol. A 6, 3: 1627–1630.

Vicsek, T. (1983). Fractal models for diffusion controlled aggregation. J. Phys. A, Math. Gen., 16: L647–L652.

Volarovich, M.P., Marmorshteyn, L.M. and Mekler, Yu.B. (1968). Variation in the structure of the pore space in sandstones under pressure. Izv. Earth Phys., 6: 346–349.

Voskresenskiy, Yu.N. (1962). A three-dimensional model study of seismic wave reflection from non-mirrorlike boundaries. Izv. ANSSSR Ser. Geofiz., 5: 620–629 (in Russian).

Voss, R. (1985a). Random fractals: Characterization and measurement. *In*: R. Pynn and A. Skjeltorp (Editors), Scaling Phenomena in Disordered Systems. Plenum Press, New York, pp. 1–11.

Voss, R. (1985b). Random fractal forgeries. *In*: R.A. Earnshaw (Editor), Fundamental Algorithms for Computer Graphics. NATO ASI Ser., Vol. 17. Springer Verlag, Berlin–Heidelberg, pp. 805–835.

Voss, R.F. (1988). Fractals in Nature: From characterization to simulation. *In*: H.-O. Peitgen and D. Saupe (Editors), The Science of Fractal Images. Springer Verlag, Berlin–Heidelberg, pp. 21–70.

West, B.J. and Shlesinger, M.F. (1984). The fractal interpretation of the weak scattering of elastic waves. J. Stat. Phys., 36, 5/6: 779–786.

Witten, T.A. (1987a). Fractal aggregates and other tenuous structures. *In*: J. Souletie, J. Vannimenus and R. Stora (Editors), Le Hasard et la Matière/Chance and Matter. Elsevier, Amsterdam–New York, pp. 159–210.

Witten, T.A. (1987b). Flow and diffusion near a fractal: Partial saturation phenomena. Phys. Rev. Lett., 59, 8: 900–903.

Wong, P.-Z. (1985). Scattering by inhomogeneous systems with rough internal surfaces: Porous solids and random-field Ising systems. Phys. Rev. B, 32, 11: 7417–7424.

Wong, P.-Z. (1987). Fractal surfaces in porous media. *In*: J.R., Banavar, J. Koplik and K.W. Winkler (Editors), Physics and Chemistry of Porous Media, II. AIP Conf. Proc., 154, AIP, New York, pp. 304–317.

Wong, P.-Z. (1988). The statistical physics of sedimentary rock. Phys. Today (December), pp. 24–32.

Wong, P.-Z. and Bray, A.J. (1987). Fractal dimension of rough surfaces in the solid-on-solid model. Phys. Rev. Lett., 59, 9: 1057.

Wong, P.-Z. and Bray, A.J. (1988). Porod scattering from fractal surfaces. Phys. Rev. Lett., 60, 13: 1344.

Wong, P.-Z. and Lin, J.-S. (1988). Studying fractal geometry on submicron length scales by small-angle scattering. Math. Geol., 20, 6: 655–665.

Wong, P.-Z., Koplik, J. and Tomanic, J.P. (1984). Conductivity and permeability of rocks. Phys. Rev. B, 30, 11: 6606–6614.

Wong, P.-Z., Howard, J. and Lin, J.-S. (1986). Surface roughening and the fractal nature of rocks. Phys. Rev. Lett., 57, 5: 637–640.

Wu, R.-S. (1982). Attenuation of short period seismic waves due to scattering. Geophys. Res. Lett., 9: 9–12.

Wu, R.-S. and Aki, K. (1985). The fractal nature of the inhomogeneities in the lithosphere evidenced from seismic wave scattering. PAGEOPH, 123: 805–818.

Yfantis, E.A., Flatman, G.T. and Englund, E.J. (1988). Simulation of geological surfaces using fractals. Math. Geol., 20, 6: 667–672.

Zel'dovich, Ya.B. and Sokolov, D.D. (1985). Fractals, similarity, intermediate asymptotics. Sov. Phys. Usp., 28, 7: 608–615.

Zumofen, G., Blumen, A. and, Klafter, J. (1988). Fractal patterns in chemistry. *In*: A. Amann, L. Cederbaum and W. Gans (Editors), Fractals, Quasicrystals, Chaos, Knots and Algebraic Quantum Mechanics. Kluwer Academic Publ., Dordrecht–Boston–London, pp. 1–20.

5 Of time and chance

I returned, and saw under the sun, that the race is not to the swift, nor the battle to the strong, neither yet bread to the wise, nor yet riches to men of understanding, nor yet favour to men of skill; but *time and chance* happeneth to them all. — *Ecclesiastes*, 11.9

5.1 Paradoxes of time

The following problem has conventionally been attributed to Jean Buridan (c. 1295–1356), philosopher at the University of Paris.

Suppose a donkey is halfway between two identical bales of hay. What will be its decision? In the solution attributed to Buridan the donkey starves in indecision. (For the history of the problem, see the fascinating essay of Rescher, 1959/60. Needless to say, "Buridan's ass" is nowhere to be met with in Buridan's writings.)

The conventional solution of statistical mechanics (Arecchi, 1987) is that the donkey as a material point is coupled to the rest of the Universe which provides for the random fluctuations to immediately (that is, in a short time τ) remove it from the unstable fixed point between the two choices.

The use of the lots for a random choice is well known from ancient and biblical history [e.g. the selection of the successor of the apostle Judas, *Acts* 1.23–26; as a modern example Rescher (*op. cit.*) mentions that when Hawaii was admitted as the 50th state of the U.S., and two new senators were elected, coin-toss and card drawing were applied to decide which one would have seniority and which would serve the longer term]. Random "thermal baths" are also used in the powerful optimisation technique called "simulated annealing", where the multivariate function which is to be globally minimised is randomly perturbed in each step to help it get out from the local minima (Kirkpatrick et al., 1983; geophysical applications: Rothman, 1985, 1986; Schneider and Whitman, 1990).

Random selections in these and other examples are based on two popular dogmas of science:

DOGMA I: *Natural (physical, economical, etc.) systems have short time memory*: the after-effects of random perturbations acting upon a system during a limited time interval τ would exponentially decorrelate as $\exp[-t/\tau]$; the power spectrum of the induced noise will behave like:

$$G(\omega) \propto \frac{1/\tau}{\omega^2 + (1/\tau)^2} \tag{5.1.1}$$

that is, it will fall off for ω ($\gg 1/\tau$) like ω^{-2} (as in the case of a "Brownian noise"), while for low frequencies it will be white, expressing the lack of long-time correlations.

DOGMA II: *Small random perturbations lead to predictable, or at least negligibly small, changes in the future behaviour of the system.*

We shall present examples from various fields of Earth Sciences where these dogmas are not valid: the observed phenomena display unexpected long-term persistence (the *Hurst phenomenon*) or an anomalous sensitivity to the initial conditions (*chaotic dynamics*).

As we shall see, *fractals* will naturally arise in both cases.

5.1.1 *The puzzle called the Hurst phenomenon*

> The Hurst phenomenon has continued to haunt statisticians and hydrologists. To some it has become a puzzle to be explained, to others a feature to be reproduced by their models, and to others still, a ghost to be conjured away.
>
> — V. KLEMES: *The Hurst Phenomenon: A Puzzle?* (1974)

The concept of the *Hurst effect* was engendered by H.E.Hurst (1880–1978), an English physicist working in Cairo, who spent a lifetime studying the Nile and problems related to water storage, including his involvement with the planning of the Aswan Dam (see Mandelbrot, 1982, for bibliographic details; Klemes, 1974 and Boes, 1988 about the very persistent afterlife of the Hurst effect in hydrology and statistics).

The problem of design of an ideal reservoir based upon records of observed discharges from a river can be formulated as follows (Boes, 1988; Feder, 1988):

Let X_t denote the net inflow of water into the reservoir in a given year t, $t = 1, 2, \ldots, N$. The average influx over a period of N years is $\overline{X}_N = (1/N) \sum_{t=1}^{N} X_t$. Suppose that every year we release from the reservoir the same volume, which is equal to the mean influx $\overline{X}_N$ for the period under discussion. In a given year t the reservoir will contain:

$$X(t,N) = \sum_{i=1}^{t} \{\overline{X}_i - X_N\} = \left\{ \sum_{i=1}^{t} X_i \right\} - tX_N \tag{5.1.1.1}$$

volume of water. Obviously, at the end of the period, $X(N,N) = 0$.

A reservoir designed for a period of N years should never overflow or go dry, that is its storage capacity must be larger than the difference $R(N)$ between the maximum and minimum amounts of water contained in the reservoir:

$$R(N) = \max_{1 \le t \le N} X(t,N) - \min_{1 \le t \le N} X(t,N) \tag{5.1.1.2}$$

$R(N)$ can be computed for any sequence of random variables, it is called the range of the sequence $\{X_t\}$. Clearly, the range depends on N and on the standard deviation of the data, $S(N)$. In order to exclude the effect of the standard deviation, Hurst (1951) introduced the dimensionless *rescaled range*:

$$R^*(N) = \frac{R(N)}{S(N)} \tag{5.1.1.3}$$

where $S(N)$ is the standard deviation:

$$S(N) = \left\{ \frac{1}{N} \sum_{t=1}^{N} [X_t - \overline{X}_N]^2 \right\}^{1/2} \tag{5.1.1.4}$$

Hurst (1951) first considered a simple theoretical case of tossing k coins N times, where the random variable X_t is defined as the difference between the number of heads and number of tails in the t-th experiment. It is easy to prove that for this process:

$$R = \left(\frac{\pi}{2}kN\right)^{1/2} - 1 \qquad S = k^{1/2}$$

that is:

$$R^*(N) = \frac{R}{S} \propto \left(\frac{\pi}{2}N\right)^{1/2} \tag{5.1.1.5}$$

[Feller (1951) proved that equation (5.1.1.5) is true for any independent random process X_t with finite variance.]

Hurst (1951) actually carried out the above experiment by tossing 10 sixpences 1000 times, and by cutting cards from a pack of 52 "probability cards" containing thirteen "+1"-s and "−1"-s each, eight "+3"-s and "−3"-s; four "+5"-s and "−5"-s; one "+7" and one "−7" (the numbers of each kind are proportional to the Gaussian frequency curve: Hurst, 1951, 1957; Feder, 1988). In these "Monte Carlo" experiments Hurst found again:

$$\frac{R}{S} \propto \left(\frac{\pi}{2}N\right)^{1/2} \tag{5.1.1.6}$$

Next, he analysed a large number [75; in later works (Hurst et al., 1965) 120] of geophysical and statistical time series (river discharges, lake and water levels, rainfall, temperature, pressure, tree rings, varves, sunspot numbers, and wheat prices) with the R/S technique and found that in case of natural phenomena $R^* = R/S$ is very well approximated by the formula:

$$R^* = \frac{R}{S} \propto N^H \tag{5.1.1.7}$$

where H, later called the "Hurst exponent" by Mandelbrot, was $H = 0.73 \pm 0.09$, that is, *it significantly differed from the theoretically expected $H=1/2$ figuring in equations (5.1.1.5)–(5.1.1.6).* Hurst could not prove this finding mathematically; the only model he could propose which had apparently lead to $H \approx 0.7$ was an ingenious card experiment with a randomly biased hand of 26 "probability" cards and a joker (Hurst, 1957; see Figure 5.2a).

Lloyd (1967) summarised the impact of Hurst's discoveries as follows: "We are, then, in one of those situations, so salutary to theoreticians, in which empirical discoveries stubbornly refuse to accord with theory. All of the researches... have led to the conclusion that in the long run $\langle R_N \rangle$ should increase like $N^{1/2}$, whereas Hurst's extraordinarily well-documented empirical law shows an increase like N^H, where H is about 0.7."

Boes (1988) reviewed the several possible explanations for the Hurst effect:

(1) the $\{X_t\}$ have heavy-tailed marginal distributions;

(2) the $\{X_t\}$ are non-stationary;

(3) the N^H behaviour is only transient (pre-asymptotic), with an ultimate asymptotic growth of $\propto N^{1/2}$;

(4) the $\{X_t\}$, though stationary, possess a long-time correlation.

For the case when the $\{X_t\}$ are independent standard normal variables, the rescaled range can be exactly calculated (Anis and Lloyd, 1976):

$$R_N^* = \left\langle \frac{R}{S} \right\rangle = \frac{\Gamma[(N-1)/2]}{\Gamma(N/2)\pi^{1/2}} \sum_{i=1}^{N-1} \left[\frac{(N-i)}{i} \right]^{1/2} \tag{5.1.1.8}$$

$$R_N^* \propto \left(\frac{\pi}{2} \right)^{1/2} N^{1/2} \tag{5.1.1.9}$$

Figure 5.1 shows the behaviour of the "local" Hurst exponent H_N which is computed from expression (5.1.1.8) as:

$$H_N = \frac{\log R_{N+1}^* - \log R_N^*}{\log(N+1) - \log N} \tag{5.1.1.10}$$

Equation (5.1.1.8) holds for independent variables $\{X_t\}$. The "biased card" experiment of Hurst produced a *correlated* sequence, but the dependence was *limited in time* as on the average the joker appears after the 27th cut, and a new biased hand is generated (Hurst, 1957).

When Feder (1988) simulated by a computer Hurst's card experiment to about 100,000 cuts, he found that the estimated H slowly decreased with N and approached the "theoretical" $H \approx 0.5$, as in Figure 5.1.

Thus, for the class of stationary processes, only those having auto-correlations falling off slowly (as a power of the time lag) are capable of preserving the Hurst

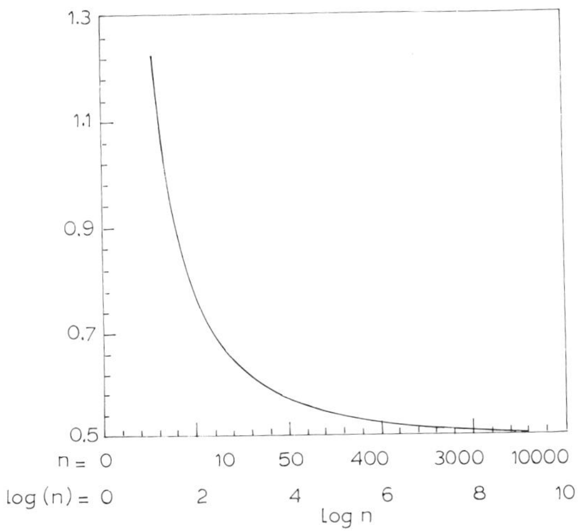

Fig. 5.1. H_N (local slope of R_N^* versus log as function of N, for independent standard Gaussian variables (from Boes, 1988).

property for $N \to \infty$. The most successful of such models is the fractional Brownian noise (fBn), proposed by Mandelbrot and co-workers (Mandelbrot and Van Ness, 1968; Mandelbrot and Wallis, 1968, 1969). The mathematical proof that the fBn satisfies Hurst's law was given in Mandelbrot and Van Ness (1968), Mandelbrot (1975). Here we shall follow the heuristic reasoning of Mandelbrot and McCamy (1970), who used two-dimensional R/S analysis to study the secular component of the Earth's pole motion.

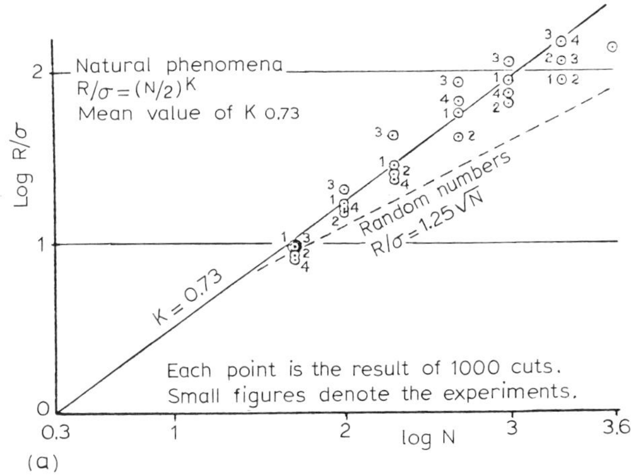

Fig. 5.2. Classical and recent examples for the Hurst effect. a. Hurst's (1951) experiment with a randomly biased hand of twenty-six cards plus joker (from Mandelbrot and Wallis, 1969).

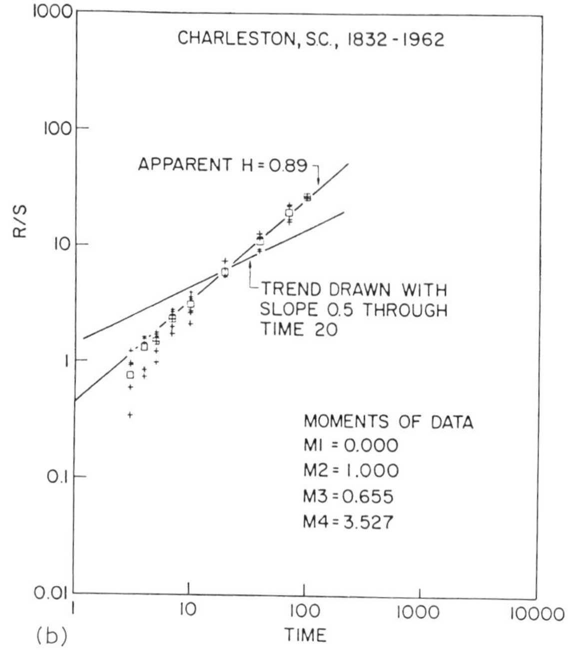

Fig. 5.2 (continued). b. *R/S* analysis of the precipitation data of Charleston, S.C., 1832–1962. (From Mandelbrot and Wallis, 1969.)

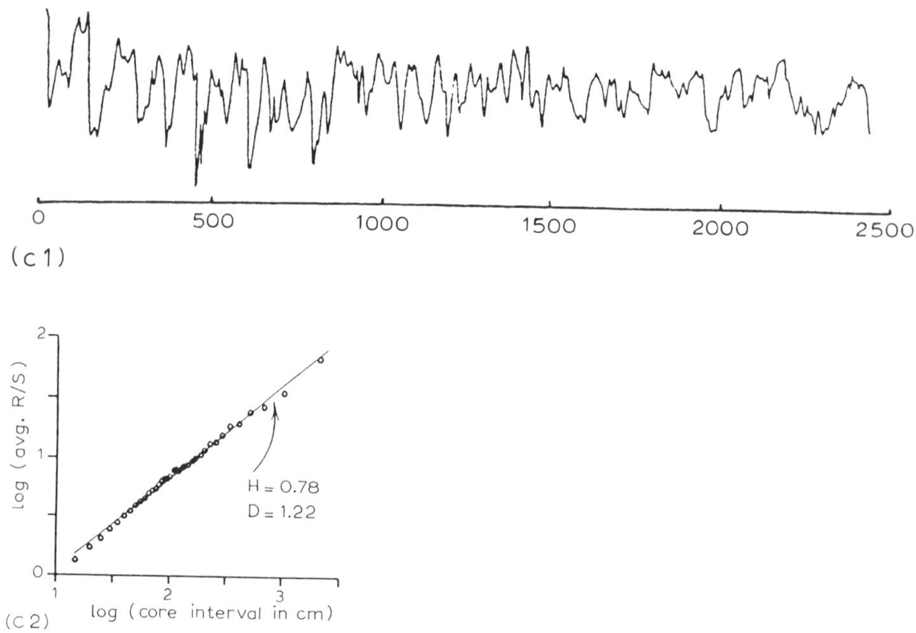

Fig. 5.2 (continued). c. (1) ^{18}O variations (arbitrary scale) of the Pacific core V28-239. Horizontal scale is 1000 yr BP (from Grassberger, 1986); and (2) its *R/S* analysis. Sedimentation rate is 1 cm/1000 yr (from Fluegeman and Snow, 1989).

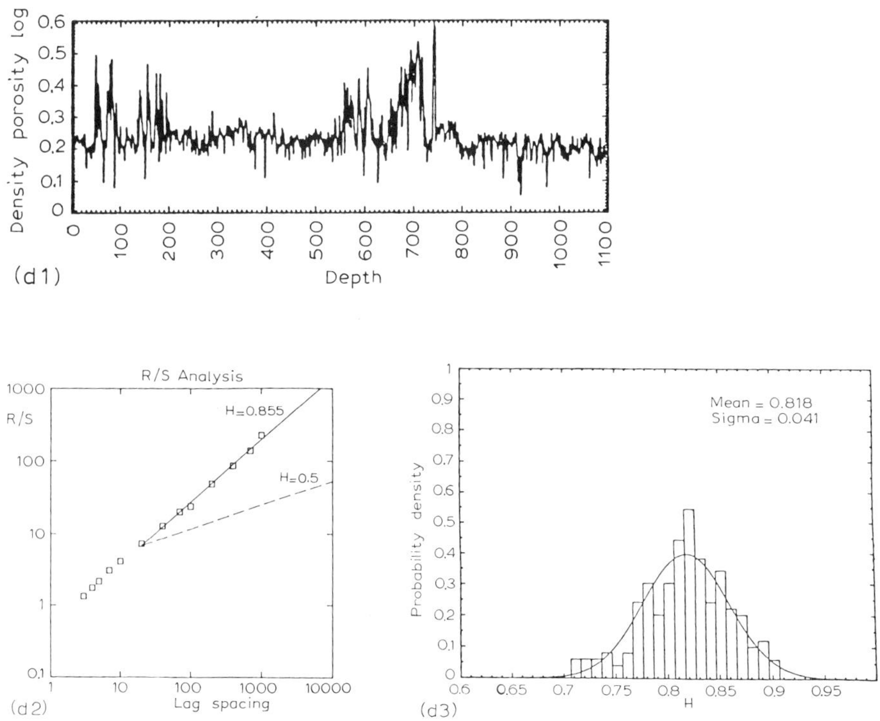

Fig. 5.2 (continued). d. (1) Density porosity log through a sandstone formation; (2) its R/S analysis (from Hewett, 1986); (3) distribution of the H values determined from more than 200 wells (from Emanuel et al., 1987).

Let $X^*(t)$ denote the partial sum $\sum_{i=1}^{t} X_i$. Then the range (5.1.1.2) can be written as:

$$R(N) = \left\{ \max_{0 \leq t \leq N} - \min_{0 \leq t \leq N} \right\} \left[X^*(t) - \frac{t}{N} X^*(N) \right] \tag{5.1.1.11}$$

We are going to show that if $\{X_t\}$ is a fractional Brownian noise (fBn) of exponent H, then $R(N)/S(N) \propto N^H$ for $N \to \infty$.

We define first *fractional Brownian motions*. A continuous-time process $B_H(\tau)$ is a *fBm with exponent H* if:

(1) the increments of B_H are stationary;

(2) $B_H(\tau)$ is Gaussian with:

$$\langle B_H(\tau) \rangle = 0 \qquad \left\langle [B_H(1)]^2 \right\rangle = \sigma^2 \tag{5.1.1.12}$$

(3) $B_H(\tau)$ is statistically invariant with respect to the affine transformation:

$$\tau \to r\tau \qquad B_H \to r^H B_H \tag{5.1.1.13}$$

where the exponent H is between zero and one. (The case $H = 1/2$ corresponds to the ordinary Brownian motion.)

It is easy to prove that:

$$\left\langle [B_H(\tau + T) - B_H(\tau)]^2 \right\rangle \propto \sigma^2 T^{2H} \tag{5.1.1.14}$$

Let us also define the increments $X_t^{(H)}$ of $B_H(\tau)$ as:

$$X_t^{(H)} = B_H(t) - B_H(t-1) \tag{5.1.1.15}$$

The process $X_t^{(H)}$ is called a *fractional Brownian noise* (fBn) with exponent H (Mandelbrot and Van Ness, 1968), it is stationary, of zero mean and variance σ^2. Simple algebra yields from equations (5.1.1.13) and (5.1.1.14) that the correlation between $X_1^{(H)}$ and $X_K^{(H)}$ is:

$$\left\langle X_1^{(H)} X_K^{(H)} \right\rangle \propto \sigma^2 H(2H-1)K^{2H-2} \quad \text{if } K \gg 1 \text{ and } H \neq \frac{1}{2} \tag{5.1.1.16a}$$

$$= 0 \quad \text{for all } K \text{ if } H = \frac{1}{2} \tag{5.1.1.16b}$$

Observe that the correlations are all positive and fall off as a power of K. The case $H = 1/2$ corresponds to the uncorrelated white noise.

After these preliminaries, it is easy to prove that the Hurst law (5.1.1.7) holds for the process $\{X_t\} = \{X_t^{(H)}\}, t = 0, 1, 2, \ldots$. Because of the definition (5.1.1.15) we have $X^*(t) = B_H(t)$. By equation (5.1.1.13), if we change the variable t to $\tau = t/N$ in equation (5.1.1.11), we obtain:

$$R(N) = \left\{ \max_{0 \leq \tau \leq 1} - \min_{0 \leq \tau \leq 1} \right\} \left[N^H B_H(\tau) - \tau N^H B_H(1) \right] = \text{const}\, N^H$$

and, as $S(N) \propto \sigma$ for the fBn $\{X_t\}$, we have Hurst's rule:

$$\frac{R(N)}{S(N)} \propto N^H \tag{5.1.1.17}$$

where the (Hurst) exponent H is simply related to the exponent figuring in the power-law autocorrelation function of the sequence $\{X_t^H\}$, see equation (5.1.1.16). [Even though the fBn process is a most elegant model for the Hurst effect, the puzzle still remains: Why should natural phenomena show the Hurst statistics? As Klemes (1974) noticed, " the success of the modern school has a rather Pyrrhic flavor; a solution of one mathematical puzzle has been paid for by the introduction of several physical riddles". Several such riddles are discussed in Klemes' paper; see also Press (1978).]

Figure 5.2b–d presents several examples for R/S analysis. Figure 5.2b is taken from Mandelbrot and Wallis' classical study (1969), the precipitation data for

Charleston, S.C., 1832–1962, show a Hurst effect with $H = 0.89$. Fluegemann and Snow (1989) applied the R/S technique to analyse the global climatic change over the last 2 million years as expressed by the oxygen isotopic curve of the Pacific core V28-239 (Figure 5.2c-1, c-2). [The classic reference on the use of oxygen isotope ratios to determine Pleistocene palaeotemperatures is Emiliani (1955).] They found $H = 0.78$ over time spans from *a few thousand years to two million years*, and noted that this value fairly agrees with the Hurst exponents obtained by Mandelbrot and Wallis (1969) for the precipitation records over *less than hundred and fifty years* for nine major cities in the eastern U.S. (ranging from 0.64 to 0.89, with a mean of $H = 0.74$; cf. Figure 5.2b).

Hewett's work (Figure 5.2d-1, d-2, from Hewett, 1986; see also Emanuel et al., 1987; Hewett and Behrens, 1988) is especially important, as this was one of the first applications of fractal concepts in reservoir performance prediction.

The simulation of fluid flow in reservoirs with heterogeneous porosity and permeability is based on the concept of "effective dispersivity". It is generally assumed that the effective dispersivity of the medium becomes constant when the fluid path L is large compared with the correlation length of the permeability fluctuations, λ:

$$\alpha = C_k^2 \lambda \tag{5.1.1.18}$$

where C_k^2 is the coefficient of variation of the permeability k:

$$C_k^2 = \frac{\langle (k - \langle k \rangle)^2 \rangle^{1/2}}{\langle k \rangle}$$

Field measurements however indicated a power-law relation:

$$\alpha \propto L^{0.75} \tag{5.1.1.19}$$

and it has been suggested (Arya et al., 1985) that this scaling behaviour may be due to the fractal structure of the permeability distribution. Hewett (1986) has proved that this is indeed the case: the R/S analysis of a density porosity log resulted in $H = 0.855$, implying that the porosity fluctuation $\Phi(h)$ is a fractal function of depth, of dimension $D = 2 - H = 1.145$ (Figure 5.2d-1, d-2). (Permeability cannot be directly determined from well logs. It can be computed from porosity as $k \propto \Phi^m$ where m is an empirical exponent, dependent on lithology.)

As an independent check, Hewett computed the power spectrum of the log, and found an $\propto f^{-\beta}$-like fall-off with frequency, with $\beta \approx 2H - 1 = 0.71$, as theoretically expected for a fBn process (Voss, 1985).

In a later study (Emanuel et al., 1987) Hurst exponents were calculated by R/S analysis on about 200 hundred density and sonic porosity logs for a large sandstone reservoir (Figure 5.2d-3). The distribution of H is quite tight and it ranges from 0.71 to 0.91, the mean being $\overline{H} = 0.82$.

As already mentioned in Section 4.3, Agterberg (1982) analysed the 30 foot net sand contours of the Lloydminster Sparky Oil Pool in Alberta by means of Richardson-Mandelbrot plots, and found a fractal dimension $D = 1.3$ (i.e. $H = 2 - D = 0.7$). He mentioned that the conventional techniques of geostatistics can not be applied for the interpolation and extrapolation of such fractally distributed properties.

Hewett (1986; Emanuel et al., 1987; see also Taggart and Salisch, 1991) suggested two methods for interpolating fractal functions of time or space: a "smooth" one, based on kriging, and a stochastic interpolation using fractals.

As well known (Journel and Huijbregts, 1978) kriging provides the best linear unbiased estimator of the property studied. The linear estimator of the value of a field $Z(x)$ in terms of the measured values $Z(x_i)$ is sought for in the form:

$$Z^*(x) = \sum_{i=1}^{n} \lambda_i Z(x_i) \tag{5.1.1.20}$$

where the weights λ_i are calculated to ensure that the estimator is unbiased and the estimation error variance is minimum. This leads to the system of equations:

$$\sum_{j=1}^{1} \lambda_j \gamma(x_i, x_j) + \mu = \gamma(x_i, x) \qquad (i = 1, 2, \ldots, n) \tag{5.1.1.21a}$$

$$\sum_{j=1} \lambda_j = 1 \tag{5.1.1.21b}$$

where μ is a Lagrange parameter, and γ is the semivariogram:

$$\gamma(x_i, x_j) = \frac{1}{2} \left\langle \left[Z(x_i) - Z(X_j)\right]^2 \right\rangle \tag{5.1.1.22}$$

Suppose that $Z(x)$ is a fractional Brownian motion (fBm) function with exponent H, which is only known at the two ends of an interval $[0, L]$. Equations (5.1.1.21) reduce to:

$$\begin{aligned} &\lambda_2 \gamma(0, L) + \mu = \gamma(0, x) \\ &\lambda_1 \gamma(0, L) + \mu = \gamma(L, x) \\ &\lambda_1 + \lambda_2 = 1 \end{aligned} \tag{5.1.1.23}$$

Assuming that $Z(0) = 0$ and recalling that for an fBm we have:

$$\gamma(x_1, x_2) = \frac{1}{2} \sigma^2 \left|x_i - x_j\right|^{2H}$$

equation (5.1.1.23) can be easily solved and the interpolation–extrapolation formula becomes:

$$Z^*(x) = Z(L) \left\{ \frac{1}{2} \left[1 - |1 - \xi|^{2H} + \xi^{2H}\right] \right\} = Z(L)\, Q(\xi) \tag{5.1.1.24}$$

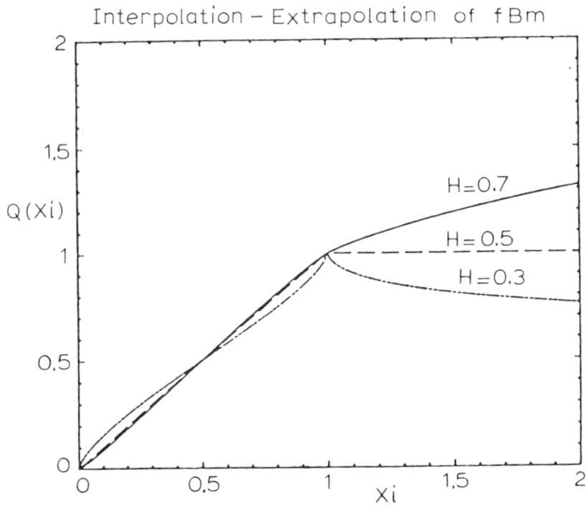

Fig. 5.3. The interpolation-extrapolation function $Q(\xi)$ for fBm [equation (5.1.1.24)] (from Hewett, 1986).

where $\xi = x/L$. [The formula was first derived in Mandelbrot and Van Ness (1968), obviously, without reference to "kriging". The geo-statistical formulation is due to Hewett (1986).] The interpolation–extrapolation function $Q(\xi)$ is shown in Figure 5.3 for different values of H. The difference between the various ranges of H becomes evident outside the interval $[0, L]$ (i.e. for $\xi > 1$).

For $H = 1/2$ the successive changes become independent and the best prediction is the last measured value; for $H > 1/2$ the local trend over the interval will be continued ("persistence"), for $H < 1/2$ the local trend will be reversed and the predicted value tends to the mean value over the interval ("antipersistence").

Stochastic interpolation between wells with fBm-like correlation structure can be achieved by several methods. Hewett (1986) and Emanuel et al. (1987) recommend Successive Random Additions, introduced by Voss (1985) for his famous "fractal forgeries". [An alternative approach is based on one-dimensional (Hewett, 1986; Crane and Tubman, 1990), or two-dimensional (Taggart and Gray, 1991) spectral analysis.]

In this method, values are first "smoothly" interpolated to the midpoint by linear interpolation or kriging, and then a random component is added to all values, with a variance which decreases with each level of resolution.

If the wells have the correlation structure of a fBm with exponent H, one can start with some overall variance σ_0^2 and add to the interpolated values a Gaussian random number with zero mean and variance $\sigma_1 = \sigma_0/2^{2H}$. The procedure is recursively repeated until the desired resolution is achieved. At the n-th stage, the random component added to each interpolated value has a variance $\sigma_n = \sigma_0/2^{2nH}$.

An example for fractal interpolation between three neutron porosity wells in a

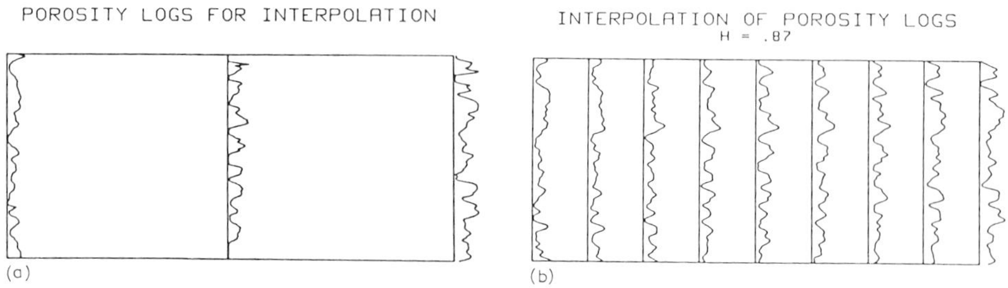

Fig. 5.4. a. Three porosity logs for fractal interpolation. $H = 0.87$ from R/S analysis. b. Pseudo-well logs created by fractal interpolation. (From Hewett, 1986.)

carbonate formation is shown in Figure 5.4a, b (from Hewett, 1986). R/S analysis of the logs gave $H \approx 0.87$, the spacing between the wells was 900 ft (300m), the depth range 145 ft (44 m).

As Hurst's studies were meant to help regularise the river Nile (he was the greatest Nilologist of all times, nicknamed Abu Nil), and the Nile was the site of the biblical story of Joseph who interpreted the Pharaoh's dream as:

> ...there come seven years of great plenty throughout all the land of Egypt. And there shall arise after them seven years of famine; and all the plenty shall be forgotten...
> — *Genesis*, 41. 29–30

Mandelbrot and Wallis (1968) aptly designated by "Joseph effect" Hurst's observation that periods of unusual (high or law) precipitations can be extremely long. Since then, the term has often been used to designate persistent processes with $H > 1/2$, as for example in a recent paper of Todoeschuck and Jensen (1988), "Joseph geology and seismic deconvolution".

Mandelbrot and Wallis (1968) discovered another biblical analogy in the hydrological record, the "Noah effect" (after Genesis, 7.11–12), relating to the very rapid and large changes. Much later, this observation has served as basis for Lovejoy and Mandelbrot (1985), to describe the temporal-spatial behaviour of *rain fields* as a "Fractal Sum of Pulses" process (see also Lovejoy and Schertzer, 1983, 1986; and the criticism of Zawadzki, 1987).

Fractional Brownian noise and fractional Brownian motion processes are widely applied in fractal interpolation, and simulation of natural scenes (Voss, 1985, 1988). The fractal dimension of the trajectories of the real molecular motion however is still controversial.

Powles and Quirke (1984) found that "the connection between these mathematical models and the Brownian motion of molecules in real or at least realistic liquids is not very clear".

They analysed the lengths of the molecular trajectories in a theoretical Lennard-Jones liquid model consisting of 108 molecules, using Richardson-Mandelbrot plots, and obtained — instead of the theoretically expected $D = 2$ value (Mandelbrot, 1982) — $D = 1.4$ after 400 steps; $D = 1.65$ after 5,000 steps, 1.75 for 100,000 steps. Powles (1985) concluded that D apparently tends to 2 but in order to get $D = 1.95$ about 10^{18} steps are needed and this would take 10^{10} years on the VAX 11/750. Rapaport (1985) used a hard-sphere fluid model of 1372 molecules and obtained $D = 2$ for trajectories longer than 25 times the mean free path length. Tsurumi and Takayashu (1986) reanalysed the results of Powles and Rapaport and claimed that the path length $L(r)$ scales with the resolution r as:

$$L(r) \propto \left(1 + \frac{r}{d}\right)^{-1} \tag{5.1.1.25}$$

where d is the mean free path,rather than according to the Richardson-Mandelbrot law:

$$L(r) \propto r^{1-D} \tag{5.1.1.26}$$

5.1.2 *Paradoxes of the 1/f noise*

> The main difference between the astronomer's outlook and that of other people in other fields seems to be that astronomers refer to these stochastic fluctuations as "signal", while in other fields the most common terminology is "noise".
> — PRESS (1978)

The light curves of quasars (Figure 5.5) owe their "exotic attractiveness" (Press, 1978) to the fact that their details only vary by order of unity when observed on different timescales. Curves like this are prototypes of statistically self-similar, rather than self-affine, processes.

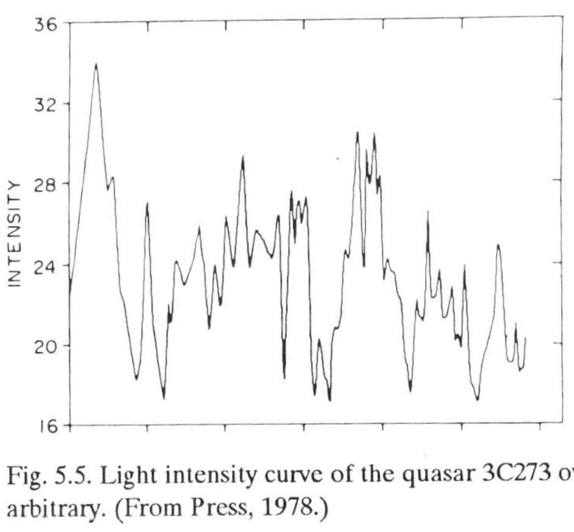

Fig. 5.5. Light intensity curve of the quasar 3C273 over a period of 80 years, 1887–1967. The ordinate is arbitrary. (From Press, 1978.)

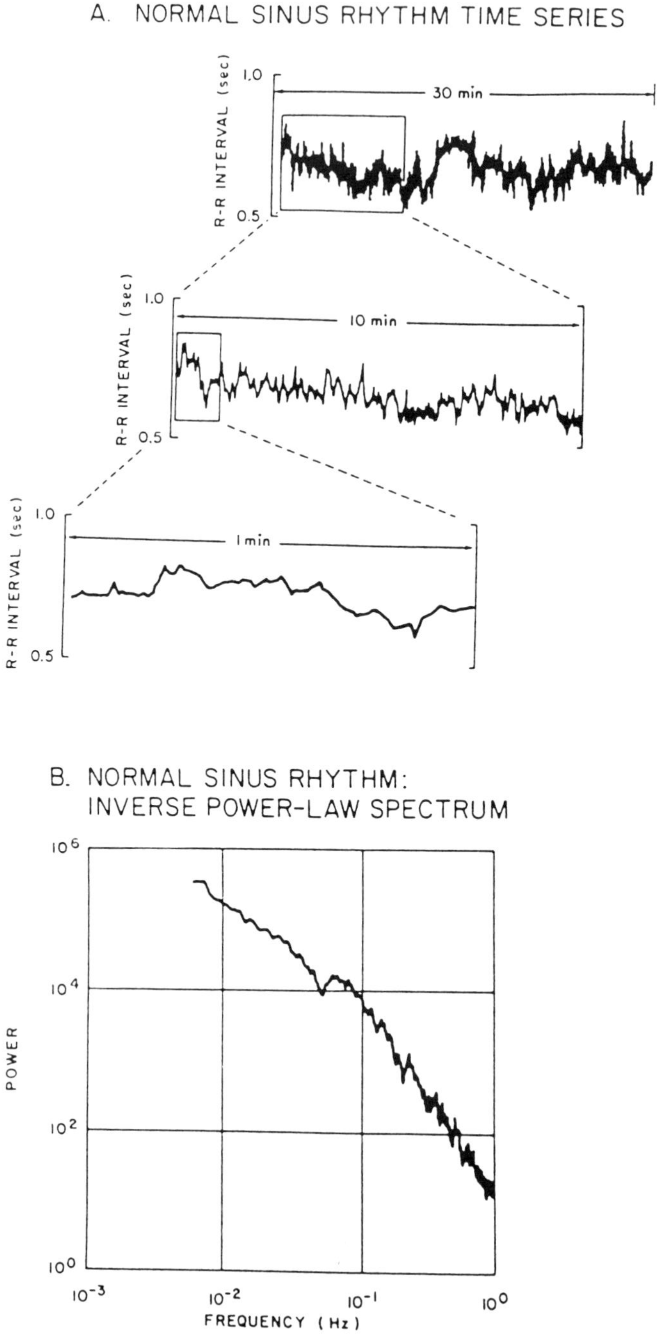

Fig. 5.6. A. Normal sinus rhythm of a healthy individual. B. The corresponding broad-band $1/f$-like power spectrum. (Kobayashi and Musha, 1982; from Goldberger and West, 1987.)

As defined in the last section [equation (5.1.1.14)], a random process $X(t)$ is self-affine if its statistical properties are invariant under the transformation:

$$X(rt) \rightarrow r^{H}X(t) \tag{5.1.2.1}$$

It follows from equation (5.1.1.16) that a fractional Brownian noise with exponent H has a power spectrum:

$$G(f) \propto f^{-2H+1} \tag{5.1.2.2}$$

For *self-similar* processes, (5.1.2.1) is satisfied with $H = 1$ and the expression for $G(f)$ reduces to:

$$G_{1/f}(f) \propto f^{-1} \tag{5.1.2.3}$$

Figure 5.6 shows the self-similarity of the normal sinus rhythm of a healthy, 61-year-old woman, and the corresponding $1/f$-like power spectrum. (The superimposed peak around 0.2 Hz corresponds to the respiratory rate).

Random processes with $1/f$-like power spectrum are called *1/f noise or flicker noise.* In variability, the $1/f$ noise is in a sense "in between" the well known white noise (with $G_{\text{white}}(f) \propto f^{0}$); and the Brownian noise which is by definition the integral of a white noise. As integration brings in a division by f in the amplitude of the Fourier spectrum, and the power spectrum is the square of the amplitude spectrum, for the Brownian noise:

$$G_{\text{Brown}}(f) \propto f^{-2} \tag{5.1.2.4}$$

Comparison of the three noise types can be found in Mandelbrot (1967, 1982), Voss and Clarke (1975), Gardner (1978), Press (1978), Feder (1988), Voss (1988). Figures 5.7 and 5.8 are unconventional examples. Figure 5.7 shows three fractal velocity distributions used in a seismic refraction model study. Figure 5.8 (Voss, 1988) shows samples of "stochastically composed" fractal music (the rules of the composition are explained in popular form in Gardner, 1978). In both figures, the white noise process looks too random, the Brownian too correlated, the $1/f$ process is "in between", "the most natural", "the most pleasing", "just right".

As a matter of fact, according to the studies of Voss and Clarke (1975), Voss (1988), the measured loudness- and pitch fluctuations of the output of different radio stations, and the pitch fluctuations of music from different cultures and various styles, have $1/f$-like power spectra over almost five decades of frequency (Figure 5.9). As yet, there has been no explanation for the surprisingly precise agreement with $1/f$. [Press (1978) playfully suggests a hierarchical model for fractal music: "there are three notes to a phrase, say, and three phrases to a bar, and three bars to a theme, and three repetitions of a theme in a movement, and three move-

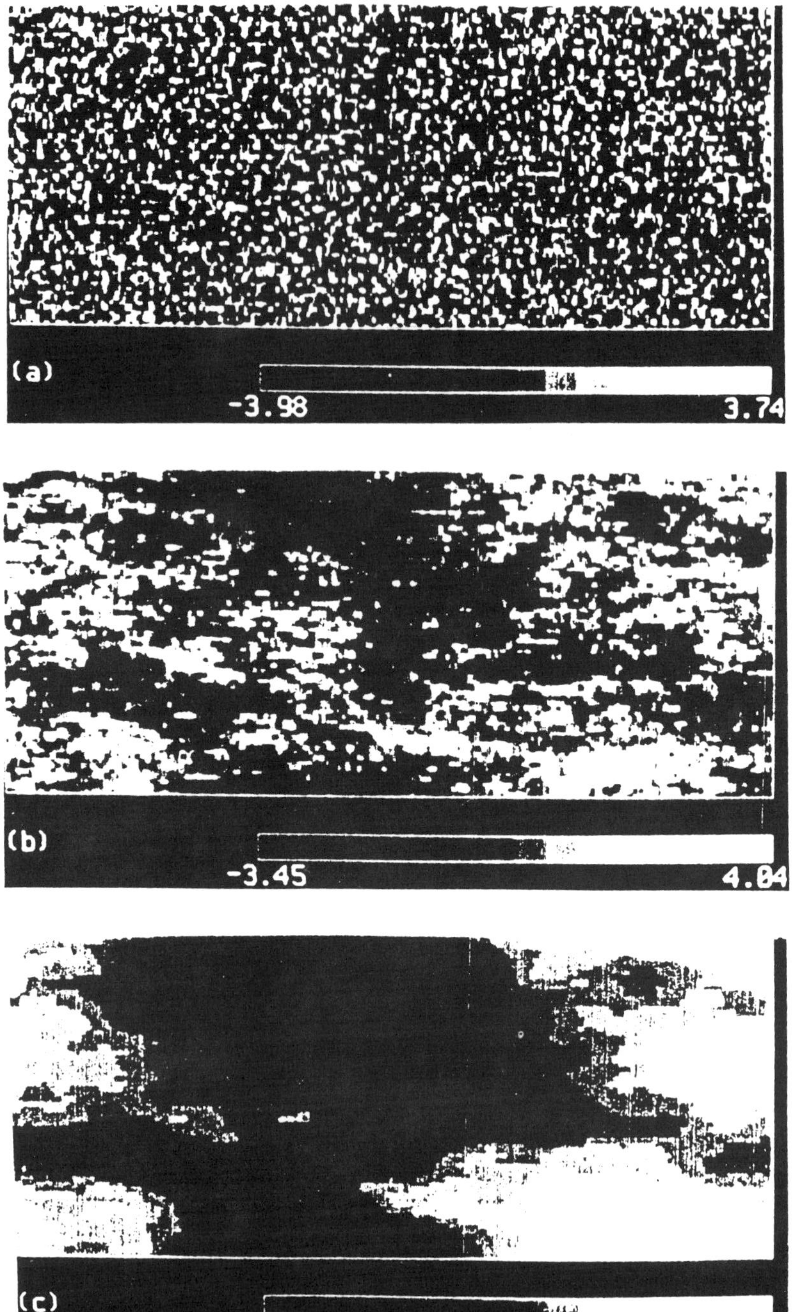

Fig. 5.7. Fractal velocity distribution in a two-dimensional seismic refraction modeling study. (a) White noise; (b) $1/f$ noise; c) Brownian noise. (From Crossley and Jensen, 1989.)

Fig. 5.8. Stochastically composed fractal music based on different types of noise: (a) white noise; (b) $1/f$ noise; (c) Brownian noise. (From Voss, 1988.)

ments in a concerto, and perhaps three concertos in a radio broadcast"; see also Mandelbrot (1982, p. 374a). Schillinger (1948) already observed that works of different composers can be transformed into each other by a proper affine transformation.]

In fractal music even paradoxes are paradoxical! Suppose that we record a piece of music and then play the tape faster or slower than usual. One expects the character of what is heard to change considerably. For example, if it is played back at twice the tape speed, it is expected to sound *one octave higher*. For fractal noises however — as Mandelbrot (1982, pp. 249b–250a) predicted — we only have to adjust the volume to make it sound *exactly as before* [see equation (5.1.2.1)].

However, human perception has its own paradoxes. In 1985, M. Schroeder of the Göttingen University actually carried out this experiment using a tone complex comprising 11 pure tones spanning out the auditory range whose adjacent frequencies were related in the constant ratio $1:2^{13/12}$ (10 Hz, 21 Hz, . . . , 428 Hz, 907 Hz,

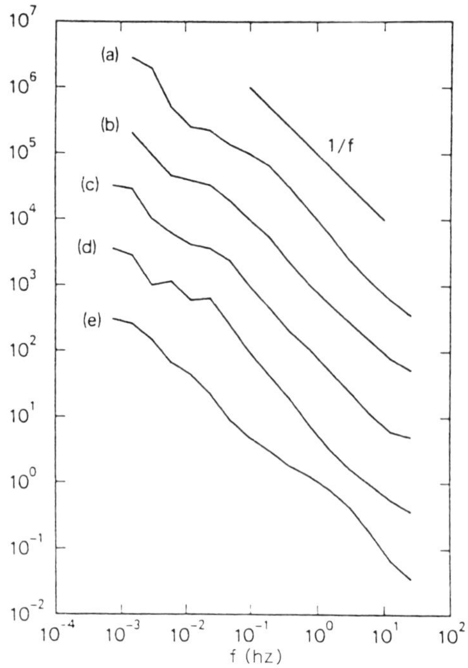

Fig. 5.9. Power spectra of the pitch fluctuations for different styles of western music. (*a*) Early medieval music (to 1300); (*b*) Beethoven, 3rd Symphony; (*c*) Debussy, piano works; (*d*) R. Strauss, "Ein Heldenleben"; (*e*) The Beatles, "Sgt. Pepper". (From Voss, 1988.)

..., 18284 Hz), see Figure 5.10. The waveform, when played back at twice the tape speed, *sounded a semitone lower* (rather than one octave higher than, or exactly as, before) (Schroeder, 1986, 1987).

$1/f$ noise appears very persistently in all fields of science — usually defying explanation (see however Mandelbrot, 1967; Press, 1978; Montroll and Shlesinger, 1983; Procaccia and Schuster, 1983; Arecchi, 1987, on the possible causes of the wide occurrence of $1/f$ in different systems). According to Press, "the explanation of electrical flicker noise is in about the same state of certainty as the explanation of quasar energy sources". There are several examples for $1/f$-like spectra in Earth Sciences: velocity of the undersea ocean currents, solar normal modes, sunspot numbers (reviewed in Press, 1978); and a whole variety of meteorological and climatic time series (analysed by Mandelbrot and Wallis, 1969): annual thickness of glacial varves from Sweden; weakly earthquake frequencies; annual maxima and minima of the Nile between 622–1469 ($\propto f^{-0.68}$ and $\propto f^{-0.82}$, respectively). To have one less mystery, the water level at Bermuda — minus the tidal components — varies as a Gaussian Brown noise ($G \propto 1/f^2$: Wunsch, 1972).

The Earth's pole motion (Munk and MacDonald, 1960) has three large components: an annual term caused by seasonal mass transport, the Chandler wobble with

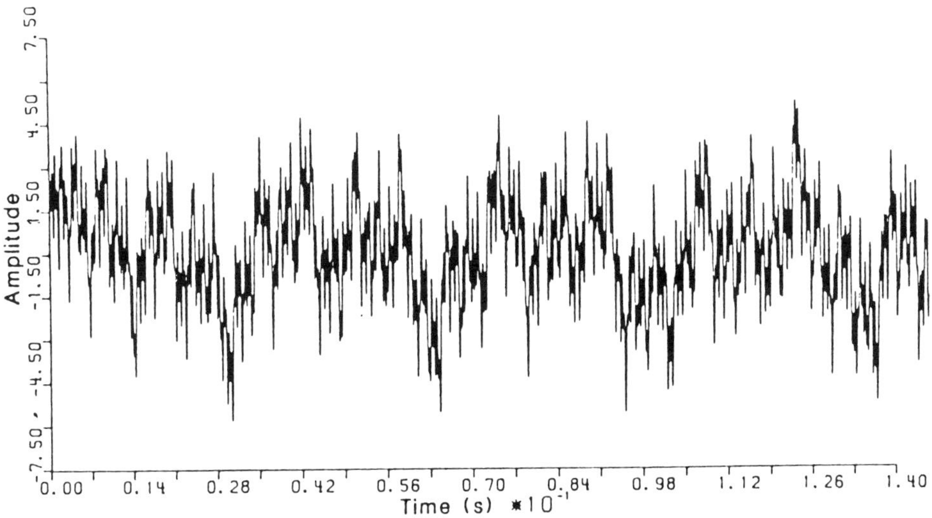

Fig. 5.10. Fractal waveform leading to auditory paradox (from Schroeder, 1987).

$\approx$ 1/3–4/3 cycles/year frequency due to the damped oscillations of the Earth, and an erratic low-frequency secular motion.

Mandelbrot and McCamy (1970) proved by R/S analysis that the behaviour of the secular motion of the Earth's pole is indistinguishable from a two-dimensional $1/f$ noise. They also observed another interesting property of the secular motion: *averaging had practically no effect in reducing its randomness*. This a general property of $1/f$-like processes: because their spectrum is very "red", that is rich in low frequencies, the processes are invariant under averaging.

A striking example is shown in Figure 5.11. The fractional accuracy of different frequency standards can not be arbitrarily increased. As the averaging time is increased, accuracy first improves, then it becomes independent of the length of the interval. The effect is due to the electrical flicker noise generated in the electronic components of the clock. [Press (1978) notes that according to Schick and Verveen's (1974) findings, the flow pattern of grains through an hour-glass also has an $1/f$-like density fluctuation; which rules out the otherwise clever idea of using huge hour-glasses for precise measurement of long time intervals.] The electronic flicker noise raises another intriguing puzzle: "How does a resistor remember over a period of weeks and months that it is in an 'up' fluctuation?" asks Press (1978). Klemes (1974) has recognised the same puzzle in the Hurst phenomenon: "By what sort of physical mechanism can the influence of, say, the mean temperature of this year at a particular geographic location be transmitted over decades and centuries?" — and this question has become even more intriguing since the realisation that climate is chaotic [see the chapter "The Butterfly Effect"in Gleick's (1987) "Chaos" and ref-

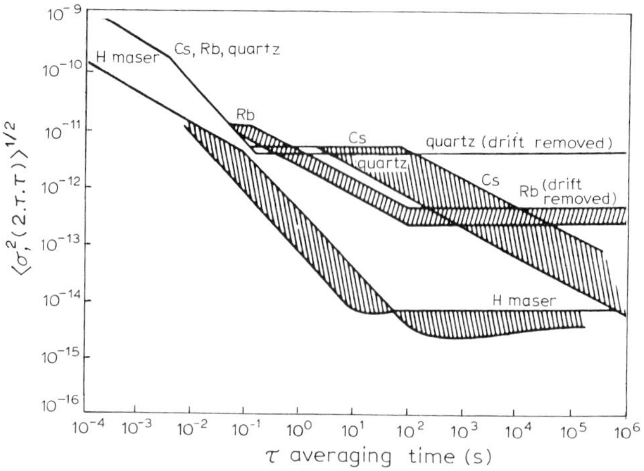

Fig. 5.11. Long-term stability of cesium, rubidium, quartz, and hydrogen frequency standards. The ordinate shows essentially the fractional accuracy to which an interval τ can be measured. (From Press, 1978.)

erences therein; Nicolis and Nicolis (1984), Fraedrich (1986), Grassberger (1986), Henderson and Wells (1988)].

Flicker noise has found its way into seismic data processing. In reflection seismics the standard representation of a seismic trace is (Rietsch, 1982):

$$s(t) = w(t) * r(t) + n(t) \tag{5.1.2.5}$$

where $s(t)$ is the observed seismic trace; $w(t)$ is a wavelet of short duration whose actual form depends on the seismic energy source; "$*$" is the usual designation of the convolution integral; $r(t)$ is the sequence of reflection coefficients; and $n(t)$ a white noise.

If there is a reflecting boundary at a depth h, corresponding to the two-way travel-time t, and $v(t)$, $\rho(t)$ are the longitudinal sound wave velocities, and densities, respectively, at this depth, then the reflection coefficient is given by:

$$r(t) = \frac{Z(t+\delta) - Z(t-\delta)}{Z(t+\delta) + Z(t-\delta)} \tag{5.1.2.6}$$

where $Z(t) = v(t)\rho(t)$ is the acoustic impedance as function of reflection time, δ is a small time interval. As shown by Peterson et al. (1955), the r.h.s. of equation (5.1.2.6) is well approximated by the expression:

$$r(t) \propto \frac{1}{2}\frac{\mathrm{d}[\log Z(t)]}{\mathrm{d}t} \tag{5.1.2.7}$$

The task of *seismic deconvolution* (see e.g. Yilmaz, 1987) is to design an optimum convolutional filter $f(t)$ (in the Wiener sense) such that $f(t) * s(t)$ is the best estimation of the reflectivity function $r(t)$:

$$f(t) * s(t) = f(t) * w(t) * r(t) + f(t) * n(t) \approx r(t)$$

or, more exactly:

$$\left\langle [f(t) * w(t) * r(t) + f(t) * n(t) - r(t)]^2 \right\rangle = \min \tag{5.1.2.8}$$

In practice, we use the observed trace $s(t)$ in its sampled version, $\{s_0, s_1, \ldots, s_k, \ldots\}$, the optimal filter $f(t)$ is also represented by a finite number of filter coefficients $\{f_0, f_1, \ldots, f_M\}$.

In general, the seismic wavelet $w(t)$ and the noise $n(t)$ in (5.1.2.8) are not known. To make the optimisation problem meaningful, further assumptions are needed:

(1) The sequences $\{s_k\}$, $\{r_k\}$, $\{n_k\}$ are stationary.

(2) The sequence of reflection coefficients is an uncorrelated white noise:

$$\langle r_i r_j \rangle = 0 \qquad \text{if } i \neq j \tag{5.1.2.9a}$$

(3) The noise is also white:

$$\langle n_i n_j \rangle = 0 \qquad \text{if } i \neq j \tag{5.1.2.9b}$$

(4) The sequences $\{w_k\}$, $\{r_k\}$, $\{n_k\}$ are uncorrelated:

$$\langle w_i r_j \rangle = \langle w_i n_j \rangle = \langle r_i n_j \rangle = 0 \qquad \text{for any } i, j. \tag{5.1.2.9c}$$

Under these conditions the coefficients of the optimal "deconvolution" filter can be determined from the Wiener-Levinson equation (Levinson, 1946):

$$\begin{bmatrix} R(0) & R(1) & R(2) & \ldots & R(M-1) & R(M) \\ R(1) & R(0) & R(1) & \ldots & R(M-2) & R(M-1) \\ R(2) & R(1) & R(0) & \ldots & R(M-3) & R(M-2) \\ \vdots & & & & & \\ R(M) & R(M-1) & R(M-2) & \ldots & R(1) & R(0) \end{bmatrix} \begin{bmatrix} f_0 \\ f_1 \\ f_2 \\ \vdots \\ f_M \end{bmatrix} = \begin{bmatrix} 1 \\ 0 \\ 0 \\ \vdots \\ 0 \end{bmatrix} \tag{5.1.2.10}$$

where $R(k)$ is the autocorrelation:

$$R(k) = \langle s_i s_{i+k} \rangle \tag{5.1.2.11}$$

It has long been recognised that the basic assumptions of deconvolution are not met, as, among others, the sequence of reflection coefficients is *not* white (O'Doherty and Anstey, 1971; Korvin, 1972; Walden and Hosken, 1986; Todoeschuck and Jensen, 1988, 1989; Todoeschuck et al., 1990). Figure 5.12 shows power spectra computed from the sonic logs of different wells in Canada.

If the spectrum of the reflection sequence is not white, the conventional ways of optimum filter design must be modified (Rietsch, 1982; Todoeschuck and Jensen,

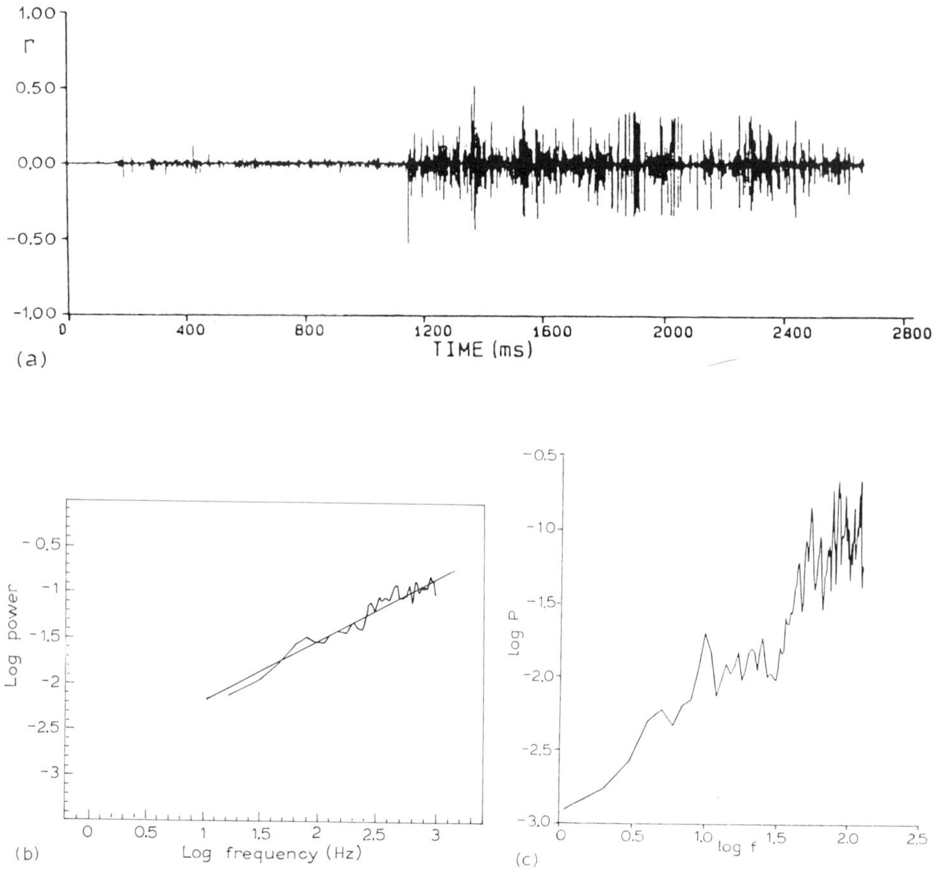

Fig. 5.12. Power spectrum of reflection coefficient sequences. a. A typical reflection coefficient sequence from the Amoco 10E A-1 Puffin B-9 well, New Foundland. Its power spectrum grows as $\propto f^{0.84}$. b. Power spectrum of the reflection coefficients for the Wickham No.1 well, Quebec ($\propto f^{1.02}$). c. Average power spectrum of the reflection coefficients, calculated for 18 wells from the Peace River district, Alberta. (a, b from Todoeschuck and Jensen, 1989; c from Todoeschuck et al., 1990.)

1988, 1989). Here we only comment on the geological reasons of the f^{α}-like reflectivity spectra.

If we assume that in a given depth range the acoustic impedance function consists of a constant part and a randomly fluctuating part:

$$Z(t) = Z_0 + \zeta(t) \tag{5.1.2.12a}$$

where:

$$\langle \zeta(t) \rangle = 0 \qquad \left\langle [\zeta(t)]^2 \right\rangle \ll Z_0^2 \tag{5.1.2.12b}$$

then it can be shown, by Fourier-transforming equation (5.1.2.7) after some approximations, that the power spectra of the fluctuating part of the impedance and of the reflection sequence, respectively, are related as:

$$G_\varsigma(f) \propto \frac{Z_0^2}{f^2} G_r(f) \tag{5.1.2.13}$$

(Korvin, 1972; Todoeschuck and Jensen, 1988). Consequently, the observed power spectra $G_r(f) \propto f^\alpha$ imply that the *acoustic impedance fluctuations are fractal noises*, with:

$$G_\varsigma(f) \propto f^{\alpha-2} = \frac{1}{f^{2-\alpha}} = \frac{1}{f^\beta} \tag{5.1.2.14}$$

According to Todoeschuck et al. (1990), α ranges between 0.11 (almost white noise) and 1.2. Thus, the acoustic impedance fluctuations have $1/f^\beta$-like spectra, with β between 0.8 and 1.89, that is they are in between the $1/f$ noise and the Brownian noise. Todoeschuck and Jensen (1988) have aptly referred to this as *Joseph geology*. (Seismic wave propagation in a "Joseph medium" has recently been studied by Crossley and Jensen, 1989).

The role of internal multiples created in the multi-layer system has been neglected in the previous analysis. They also lead to self-similarities. In a delightful essay on seismic autocorrelogram analysis, much before the advent of fractals, Anstey and Newman (1966) wrote:

"*On the auto-correlation function corresponding to a multi-layered system each secondary event (after the first) is itself the centre of a symmetrically disposed group of tertiary events representing reverberation at levels above that of the secondary*. Big fleas have little fleas..."

5.2 On growth and form

> The waves of the sea, the little ripples on the shore, the sweeping curve of the sandy bay between the headlands, the outline of the hills, the shape of the clouds, all these are so many riddles of form, so many problems of morphology.
>
> — D'ARCY THOMPSON: *On Growth and Form* (1942)

Jonathan Leverkühn, Adrian's father in Thomas Mann's "Doctor Faustus", was a twentieth century alchemist. He found pleasure in ice crystals, "and on winter days when the little peasant windows of the farmhouse were frosted, he would be absorbed in their structure for half an hour, looking at them both with the naked eye and with his magnifying glass... Did, he inquired, these phantasmagorias prefigure the forms of the vegetable world, or did they imitate them? Neither one, nor the other, he answered himself; they were parallel phenomena. Creatively dreaming Nature dreamed here and there the same dream: if there could be a thought of imitation, then surely it was reciprocal." He also collected shells, and "figured stones", some of them showing resemblance to ancient Oriental writings, and tried to decipher their meaning.

These "figured stones" (*Lapides figurati* or *Lapides idiomorphi*) — some of which resembled plants or animals, hieroglyphics or landscapes (Figures 1.44; 5.13i) — seemed to fascinate scientists throughout the Middle Ages and the Renaissance (Adams, 1938). Athanasius Kircher (1602–1680), a member of the Jesuit Order, archaeologist and alchemist, gave a detailed account of the different letters, trees, castles and animals which he found in rocks (in the "Mundus Subterraneus", Amsterdam, 1678; see Godwin, 1979). Figure 5.13i shows a slab of rock with dendritic growth from Kundmann's "Rariora Naturae et Artis" (Breslau, 1737), the animal and the house reflect the imagination of the copyist. A modern representation of the structure of the Rhaetic Cotham Marble ("Landscape Marble") of southern England (Figure 5.13ii, iii) — especially the similarity of the "trees" (ii) with the "hedge" (iii) — suggests that dendritic growth is fractal.

In 1981 Witten and Sander proposed a simple model for fractal growth, called *diffusion-limited aggregation* (DLA). The rules of the model are straightforward: first we define an initial seed particle at a lattice point. Another particle is allowed to walk at random on the lattice (i.e. to "diffuse") from far away until it arrives to a lattice site adjacent to the occupied site. Then it is stopped and a new particle is launched and halted when adjacent to any of the two occupied sites, and so forth. The model can be generalised for diffusion-limited aggregation on a line or surface [Meakin (1983); his related work: "A new model for biological pattern formation" (1986) is especially relevant to stromatolite growth]. In this case one starts with a whole straight line (or surface) of occupied sites as "seed". The statistical properties of this process are usually studied by Monte Carlo simulation. A typical result expresses the connection between the root-mean-square thickness (T) of the deposit grown on a line and the number of particles: $T \propto N^{\varepsilon}$ for $N \gg 1$, where $\varepsilon = 1.30 \pm 0.075$ (Meakin, 1983). Vicsek (1983) proposed a regular fractal model to describe aggregation on a line (Figure 5.13iv), where the scaling properties of the cluster distribution can be exactly calculated.

There are many reviews and research monographs on fractal aggregation: Herrmann (1986), Jullien and Botet (1987), Liu (1986), Meakin (1988a, b), Sander (1985, 1986, 1987), Smirnov (1986), Stanley and Ostrowsky (editors, 1985), Vicsek (1991), Vicsek and Vicsek (1991 — software and manual), Witten (1987). Recent geological applications of fractal growth models include Fowler (1990), who explained the growth of the disequilibrium textures seen in certain igneous rocks (komatites, spherulitic pillow basalts, harrisites, etc.) in terms of the DLA model; Garcia-Ruiz and co-workers' (1990) fractal model for the origin of the ammonite sutures; and a dendritic growth model of river networks (Takayasu and Nishikawa, 1986).

The Witten-Sander model of DLA can be modified. In the original model the cluster is assembled by adding individual particles to it (*particle–cluster* aggregation).

Fig. 5.13. Dendritic growth. (i) Slab of rock showing dendritic growth (from Kundmann's "Rariora Naturae et Artis", Breslau, 1737; in Adams, 1938).

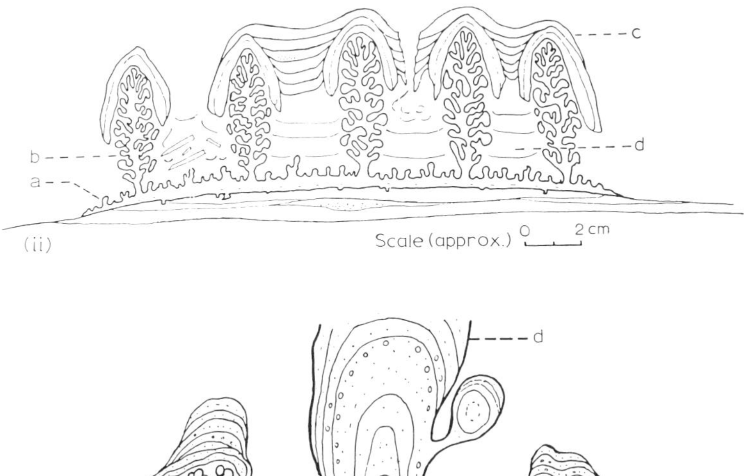

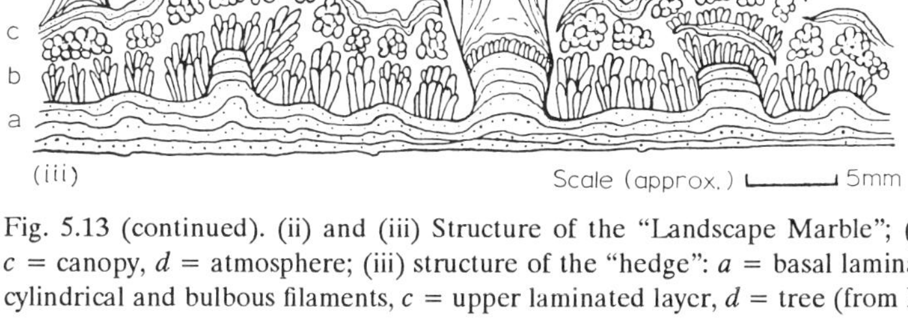

Fig. 5.13 (continued). (ii) and (iii) Structure of the "Landscape Marble"; (ii) a = hedge, b = tree, c = canopy, d = atmosphere; (iii) structure of the "hedge": a = basal laminated horizon, b = zone of cylindrical and bulbous filaments, c = upper laminated layer, d = tree (from Hamilton, 1961).

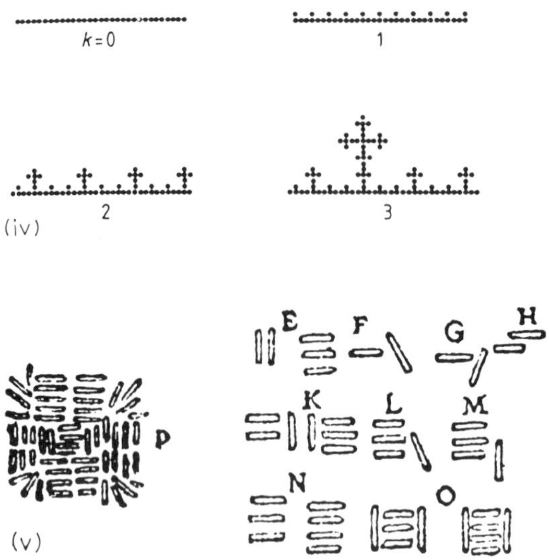

Fig. 5.13 (continued). (iv) Fractal construction simulating dendritic growth on a line; $D = 1.465$ (from Vicsek, 1983). (v) Descartes's model for the aggregation of saline particles, from the "Principia Philosophia" (1644) (from Smith, 1981).

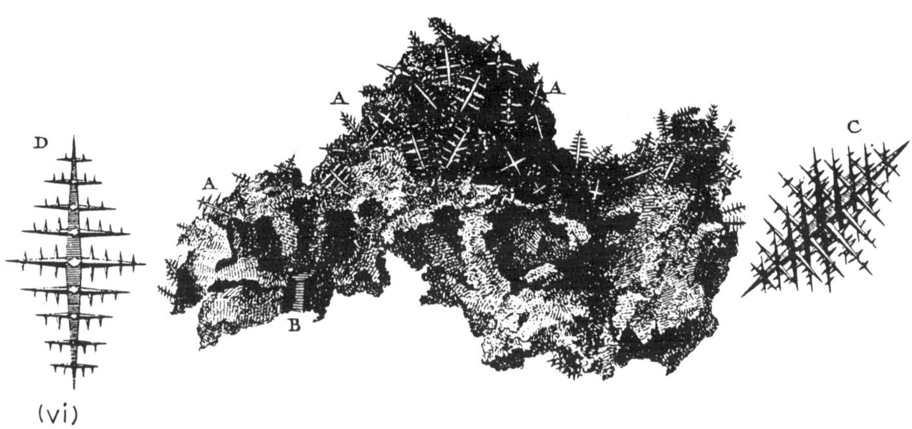

Fig. 5.13 (continued). (vi) Dendrites formed in a shrinkage cavity in a large block of iron that was slowly solidified; compare the small structures designated C and D with the Vicsek fractals in (iv) (from: Grignon's "Mémoire...", 1775; in Smith, 1981).

An alternate possibility is that a given number of particles is introduced at an initial instant of time, and the particles coalesce when they collide with one another. This will initially result in a large number of small clusters, and subsequent collisions will lead to their aggregation (*cluster–cluster* aggregation). In both versions the particles (or clusters) either execute a Brownian motion, or they can travel along straight

TABLE 5.1

Fractal dimension of clusters formed by aggregation

Aggregation model	Dimension of the space	
	2	3
1. Straight line trajectory, particle–cluster	2	3
2. Brownian motion, particle–cluster	1.68 ± 0.02	2.46 ± 0.05
3. Straight line trajectory, cluster–cluster	1.54 ± 0.03	1.94 ± 0.08
4. Brownian motion, cluster–cluster	1.44 ± 0.04	1.77 ± 0.03

Source: Smirnov (1986).

lines, for example drop vertically. The mean values and scatter of the published fractal dimensions for these models are given in Table 5.1.

Note that particle–cluster aggregation leads to higher fractal-dimensional (more dense) structures than cluster–cluster aggregation; the straight line trajectory results in higher dimension than Brownian motion. In the three-dimensional space particle–cluster aggregates are optically opaque (as their dimension is greater than 2); cluster–cluster aggregates are optically transparent, that is, it makes an enormous difference to how we model the smoke aggregates in "fractal winter" studies (see "New Scientist", 1 July, 1989, p. 15; and Nelson, 1989).

Fractals, as their name suggests, are also useful in *fractography* (Mandelbrot et al., 1984; Underwood, 1992; Underwood and Banerji, 1986; Cahn, 1989). Fractography is a fascinating science with a myriad of weird and wonderful terms to describe the features observed (Roulin-Moloney, 1989, lists laceration, shish-kebab, cusps, welts, mist, and shards). Fractures are observed at all scales: microcracks in quartz grains or continental megafaults (Allègre et al., 1982), cracks in a concrete floor (Walker, 1986), — or fractures of a toilet seat (Hertzberg et al., 1983) share the same geometrical features.

Paraphrasing the title of Katz's (1985) paper ("How straight do axons grow?") we can ask: How straight do cracks grow? Lung (1986) and Thomas (1986) considered crack propagation as random walk. In Lung's (1986) model, the crack propagates along zig-zag lines. In the intergranular fracture case, the zig-zag path leads along grain boundaries. In a smaller scale, the crack would propagate along smaller zig-zag subgrain boundaries; in a larger scale it follows some larger weak passage in the general direction of crack propagation, etc. This self-similar hierarchy of zig-zags can be approximated by a random Von Koch curve, which — in case of intergranular brittle fracture — is $D = 1.26$-dimensional.

Thomas (1986) modelled crack propagation as a restricted fractional Brownian motion, where the change in the direction between successive steps is a Fischer-distributed random variable with mean value depending on the elastic properties of the medium. The step length is power-law distributed. Figure 5.14 shows a simulated, $D = 1.1$-dimensional tension crack. [The embryonic frog and chick axons

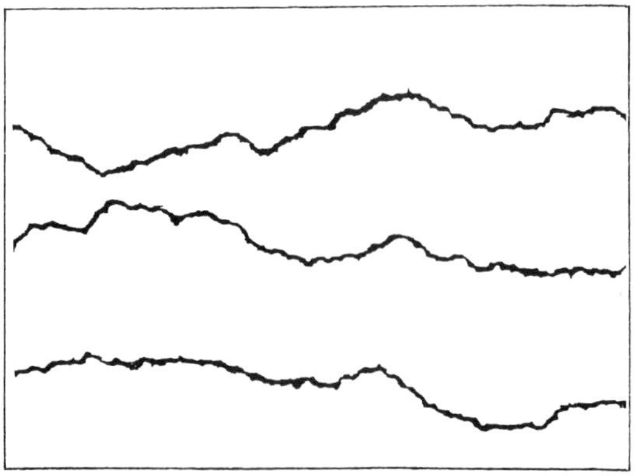

Fig. 5.14. Tensional crack simulated by a fractional random walk process, $D = 1.1$ (from Thomas, 1986).

grown in Katz's (1985) experiments were also low dimensional fractals, with dimensions $D \approx 1.28$ for the frog axons, $D \approx 1.31$ for chick axons. The grow paths resemble Thomas's restricted Brownian motion curve. Both in Thomas's and Katz's cases the low fractal dimension expresses that the growth path is basically straight.]

Because of the similarity of the systems of cracks at all scales at the point of failure, their scaling properties at the percolation transition can be described by renormalization group (RNG) methods (Allègre et al, 1982; Madden, 1983; Smalley et al., 1985; Solla, 1986; Turcotte et al., 1985; Turcotte, 1986; see Section 3.2.2).

To find the temporal evolution of systems of cracks in a material stressed to failure, one has to solve numerically a lattice approximation of the continuum theory of elasticity (Louis et al., 1986; Louis and Guinea, 1987; Takayasu, 1985, 1986; Termonia and Meakin, 1986). Figure 5.15 shows the network model used by Termonia and Meakin (1986). Nodes represent the strong parts of the material, these are joined by elastic bonds with force constants K_x and K_y.

Takayasu (1985, 1986) used a similar network, Louis et al. (1986), Louis and Guinea (1987) applied a triangular lattice (Figure 5.16B.). The distinguishing feature of the models is the rule according to which the bonds are broken. In each model the breaking probability is random and its distribution depends on the bond's elastic constants and the stress accumulated on it. In the Termonia-Meakin model the bond i breaks with the probability:

$$p_i = \exp \frac{\{(-U_i + E_i)/kT\}}{p_{\max}} \tag{5.2.1}$$

where U_i is activation energy, E_i the elastic energy stored in the bond, k is Boltzmann's constant, T is absolute temperature, and the normalising factor $p_{\max}$ ensures that the most stressed bond in the array shall break with probability one. E_i depends on the elongation Δl_i of the bond as:

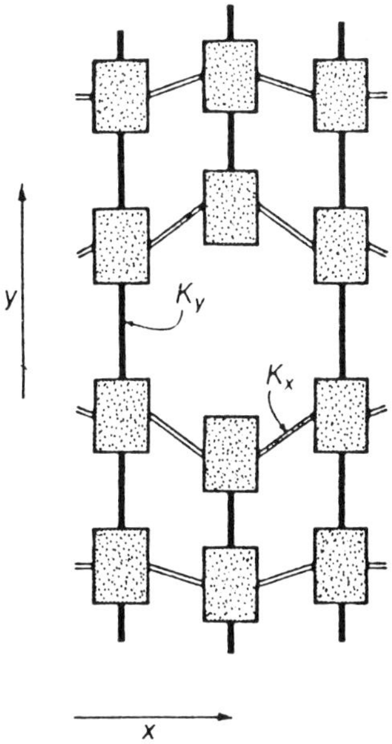

Fig. 5.15. Two-dimensional network model for crack growth. K_x and K_y are elastic constants for the bonds. (From Termonia and Meakin, 1986.)

$$E_i = \frac{1}{2} K_i (\Delta l_i)^2 \tag{5.2.2}$$

In the first step a small number of bonds are randomly broken, then the network is relaxed so as to make the total residual force acting on the net zero; the elastic energies and the probabilities p_i corresponding to the displaced position of the nodes are computed, and so forth, until we get a percolation path across the model (Figure 5.16A, B). All published models resulted in self-similar crack patterns at the point of failure. Termonia and Meakin found $D \approx 1.27$ for a wide range of elastic constants; Louis et al. (1986) obtained $D \approx 1.55$ for uniform dilation, and $\approx$1.60 for shear; Takayasu reported $D \approx 1.65$, and has found that the process is also fractal in time: the number of broken bonds grows with time chain-reaction-like as:

$$N(T) \propto T^{2.4} \tag{5.2.3}$$

Note that the fractal dimensions found by Louis et al. (1986), Louis and Guinea (1987) and Takayasu (1985, 1986) are close to that of the DLA (1.68 ± 0.02 according to Table 5.1); the connection with fractal aggregation has been extensively studied (Meakin, 1991; Van Damme and Ben Ohoud, 1990; Van Damme and Lemaire, 1990).

In all fields of Earth Sciences we encounter *chaotic phenomena*, and wherever there is chaos, we invariably find fractals.

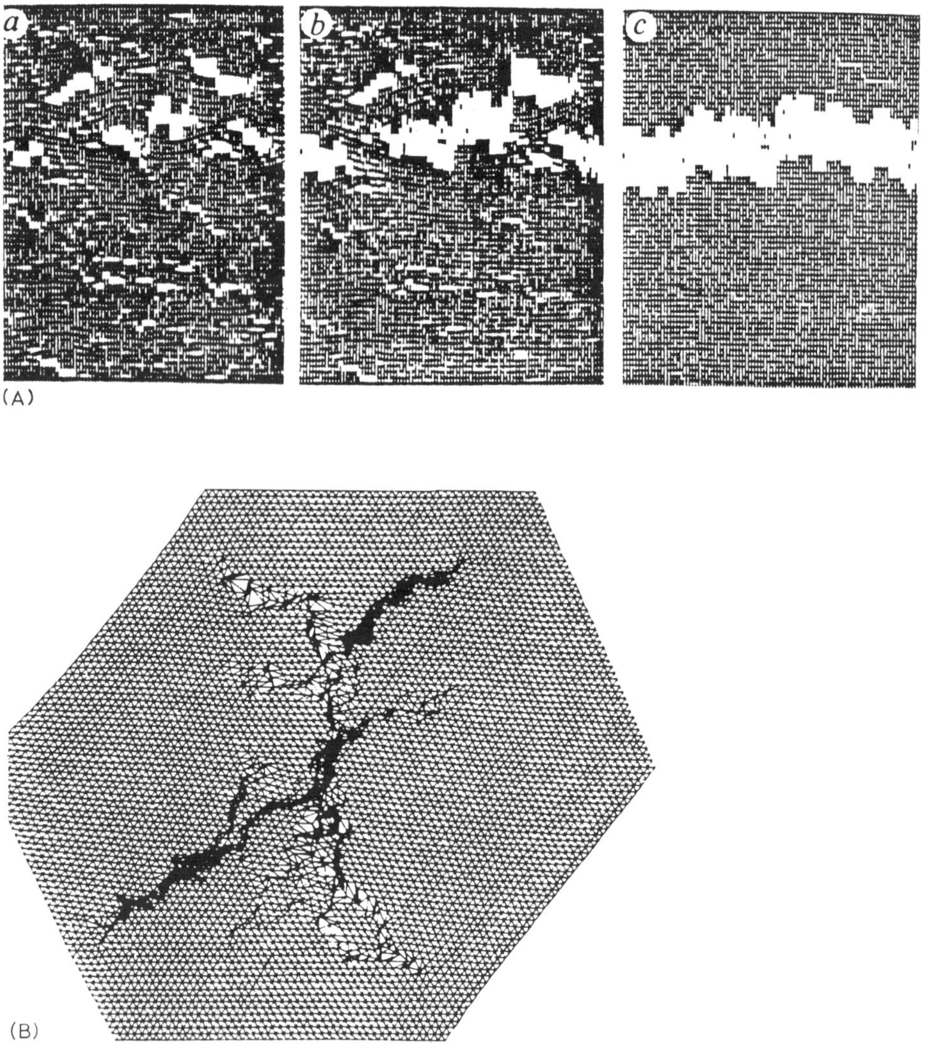

Fig. 5.16. A. Fractal growth of cracks in the network model of Figure 5.15; (a) long before failure; (b) close to failure; (c) at failure. $D \approx 1.27$. (From Termonia and Meakin, 1986.) B. Displacement of the nodes in a triangular lattice at the point of failure under shear stress. The set of broken bonds is ≈ 1.6-dimensional. (From Louis and Guinea, 1987.)

As well known, dissipative physical systems subject to a weak driving force approach states of equilibrium that are independent of the initial conditions. As the driving force is increased, these systems undergo a sequence of transitions and arrive finally to a state which is chaotic or turbulent. As the systems are described by deterministic equations, it has often been stated that there is something paradoxical about well-posed equations having unpredictable, chaotic solutions. There is, of course, no real paradox; the solutions are (in theory) uniquely determined by the initial conditions. However, the effect of small changes in the initial conditions are amplified by the equations of motion, so that if we only have information of finite precision about the initial condition, or we use finite precision arithmetic in the calculations, we cannot predict the state of the system at much later times. There is a relatively low dimensional set — called *attractor* — in the state space, to which all orbits converge, and all we can predict is that the solution will be *somewhere* on this attractor. Generally, the attractor has a fractional dimension ("strange attractor"), and as time proceeds, the orbit shall visit the points of this fractal in a seemingly haphazard manner.

The most famous example of chaotic behaviour was discovered by Lorenz (1963) who studied the short-term predictability of weather using a simple set of non-linear differential equations. In the now standard form, these equations are (Lanford, 1982; West, 1985):

$$\frac{\mathrm{d}x}{\mathrm{d}t} = -\sigma x \tag{5.2.4a}$$

$$\frac{\mathrm{d}y}{\mathrm{d}t} = rx - y - xz \tag{5.2.4b}$$

$$\frac{\mathrm{d}z}{\mathrm{d}t} = -bz + xy \tag{5.2.4c}$$

where t is dimensionless time, x is proportional to the intensity of the convective motion, y is proportional to the temperature difference between the ascending and descending currents and z describes the deviation of the vertical profile from linearity; r, b and σ are constants.

Lorenz (1963) discovered that for certain values of the parameters (for example, for $b = 8/3$, $\sigma = 10$, $r = 28$; see Figure 5.17) the solutions are *chaotic*; and it has been proved later (see e.g. Grassberger and Procaccia, 1983a, b) that the behaviour of these chaotic solutions is governed by a $\approx$2.05 dimensional fractal set in the state space, the so-called *Lorenz attractor*. (Much later, in 1979, Lorenz addressed an annual meeting of the American Association of the Advancement of Science with a talk "Predictability: Does the Flap of a Butterfly's Wings in Brazil Set Off a Tornado in Texas?")

There are many good reviews on chaos; Readers of this book will find West (1985), Crutchfield et al. (1986), Devaney (1988), Lüscher (1988), Ford (1989) and Kaye (1992) especially useful. Henderson and Wells (1988) and Theiler (1990) review the numerical techniques of determining the fractal dimension of chaotic attractors from the observed time series.

Is the natural record periodic, random or chaotic? Shaw (1987) is a general introduction to this controversial topic, with interesting examples from his own work on volcanic episodicity. Scholz (1989) reviews a recent conference "Earthquakes:

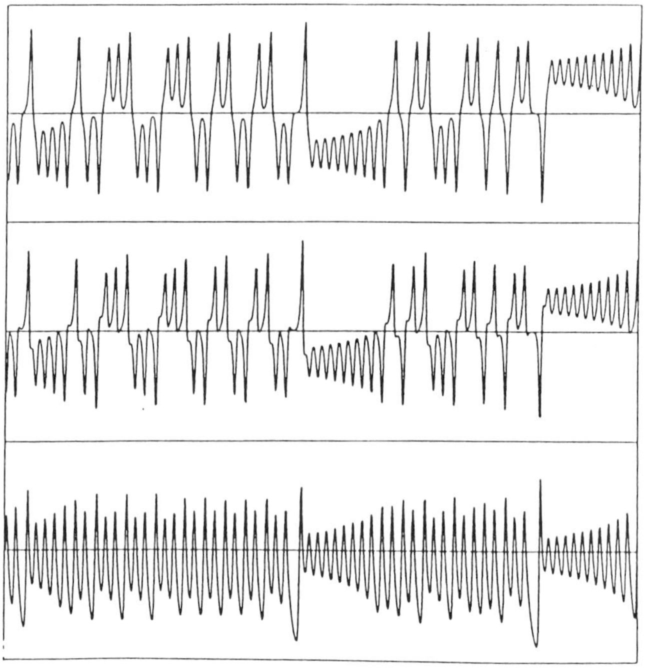

Fig. 5.17. A typical solution of the Lorenz system. From top to bottom: x, y and z against t. (From Lanford, 1982.)

chaotic or deterministic?". Vörös (1990, 1991; see also Burlaga and Klein, 1986) studied the chaotic nature of geomagnetic storms.

Magnetic reversals are briefly discussed in Jacobs (1986); Tritton (1989) refers to his earlier work (1986a, b; 1988) on the chaotic spherical pendulum and suggests a similar chaotic dynamo model for magnetic reversals. Nicolis and Nicolis (1984) and Fraedrich (1986) analysed the oxygen isotope record from oceanic cores and concluded that the *palaeotemperature variation is chaotic* with a low-dimensional attractor ($D \approx 3.1$ according to Nicolis and Nicolis; $D \approx 4.4$ according to Fraedrich). Grassberger (1986) re-analysed the data of Nicolis and Nicolis (1984) and claimed that the variation is not chaotic, but purely random. (The controversy is reviewed in Henderson and Wells, 1988.)

Chaotic phenomena are generally thought of as rapidly varying, fast processes. The geologist, and the poet, recognise chaos in the slow changes:

> The hills are shadows, and they flow
> From form to form, and nothing stands;
> They melt like mist, the solid lands,
> Like clouds they shape themselves and go.
>
> — TENNYSON: *In Memoriam A.H.H.*

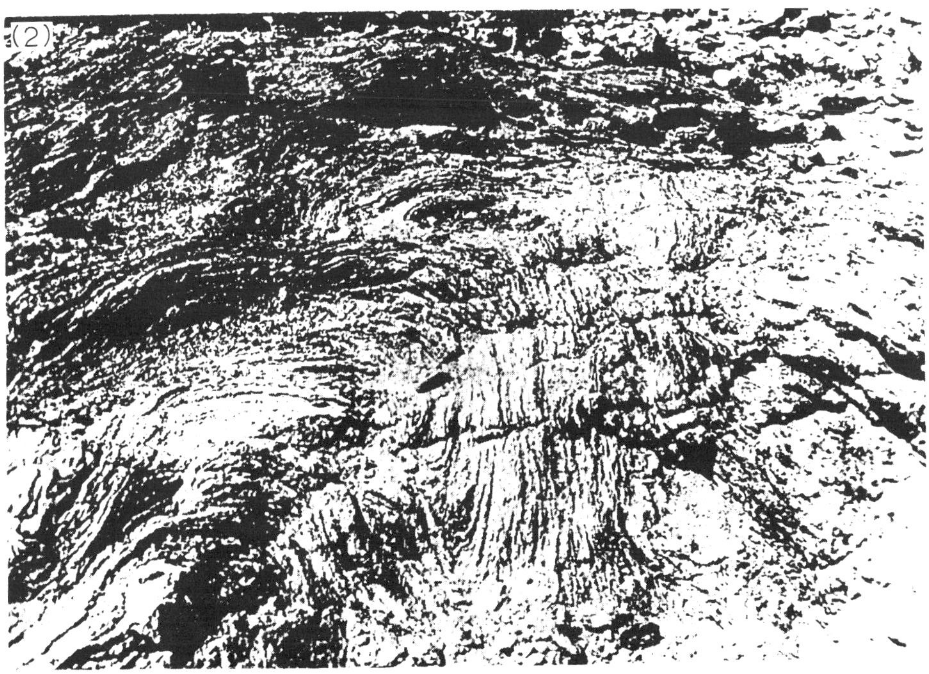

Fig. 5.18. a. (1) Hydrodynamic turbulence: the Maelstroem near to the island of Moshen, Norway (from Herbinus "Dissertationes Admirandis Mundi Cataractis"; in Adams, 1938); (2) turbulent eddy in a magmatic mass (from Lucido, 1984).

The observed geologic section is a "frozen" image of the eternally changing flow. A careful study (as the recent work of Smith and Shaw, 1990) reveals analogies with the different kinds of chaotic phenomena: there are examples for turbulence (Figure 5.18a-1, a-2) or Saffman-Taylor instability (viscous fingering: Figure 5.18b-1, b-2). The well known "baker" (or "Poincaré") transformation (Figure 5.18c-1) — the highly mathematical construction leading to chaos — can also be recognised in the successive folding, flattening and shearing creating chaotic mud injections (Figure 5.18c-2; see also Figure 4.1). As Figure 5.19 suggests, for the case of turbulence, planar cross-sections of a D-dimensional fractal embedded into the three-dimensional space is a $D-1$-dimensional fractal; the intersection of the fractal with a straight line gives a $D-2$-dimensional "Cantor set". Similarly, geological sections, seismic sections, or maps (gravity, magnetic, sand thickness, etc.) of complicated structures are planar fractals (Agterberg, 1982); the level crossings of well logs or the intersection of a vertical line with the boundaries of the sedimentary layers constitute a Cantor set (Hewett, 1986; Plotnick, 1986).

The fractal properties of viscous fingering (Figure 5.18b-1) have been intensively studied recently, partly because of the importance of the phenomenon in secondary oil recovery. Saffman and Taylor (1958) observed that if we push water through a fluid of higher viscosity in a place between two transparent plates (Hele-Shaw cell), there arises a *fluid instability* phenomenon termed "viscous fingering" in which the lower viscosity fluid forms fingers of characteristic shapes, deeply extending into the high viscosity fluid (Paterson, 1981, 1985). It has been recently established that viscous fingers are fractals (of fractal dimension $D \approx 1.7$ in boundary-free radial cells, and $D \approx 1.4$ when the growth is limited by the boundaries of the cell), and their development can be simulated by a modified DLA process (Nittmann, 1986; Nittmann et al., 1985; Feder, 1988; Langer, 1988).

In geology, there are many fascinating examples of viscous fingering [see the recent paper of Bébien et al. (1987) on the associations of acid and basic magmas (Figure 5.18b-2); Brodzikowski and Haluszczak (1987) on the flame structures in

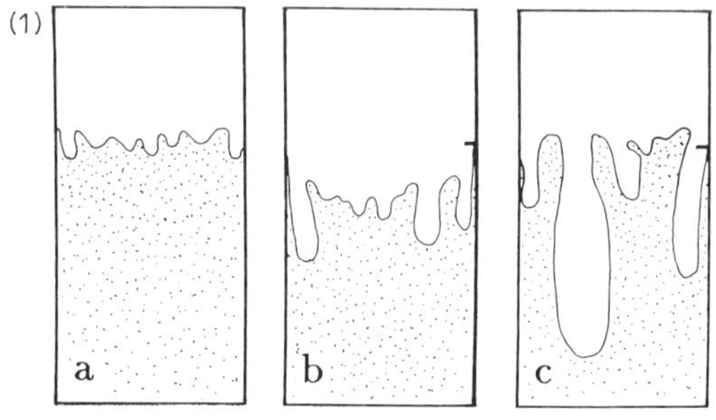

Fig. 5.18 (continued). b. (1) Viscous fingering in a vertical cell where air displaces glycerol from the top and downwards (from Saffman and Taylor, 1958);

(2)

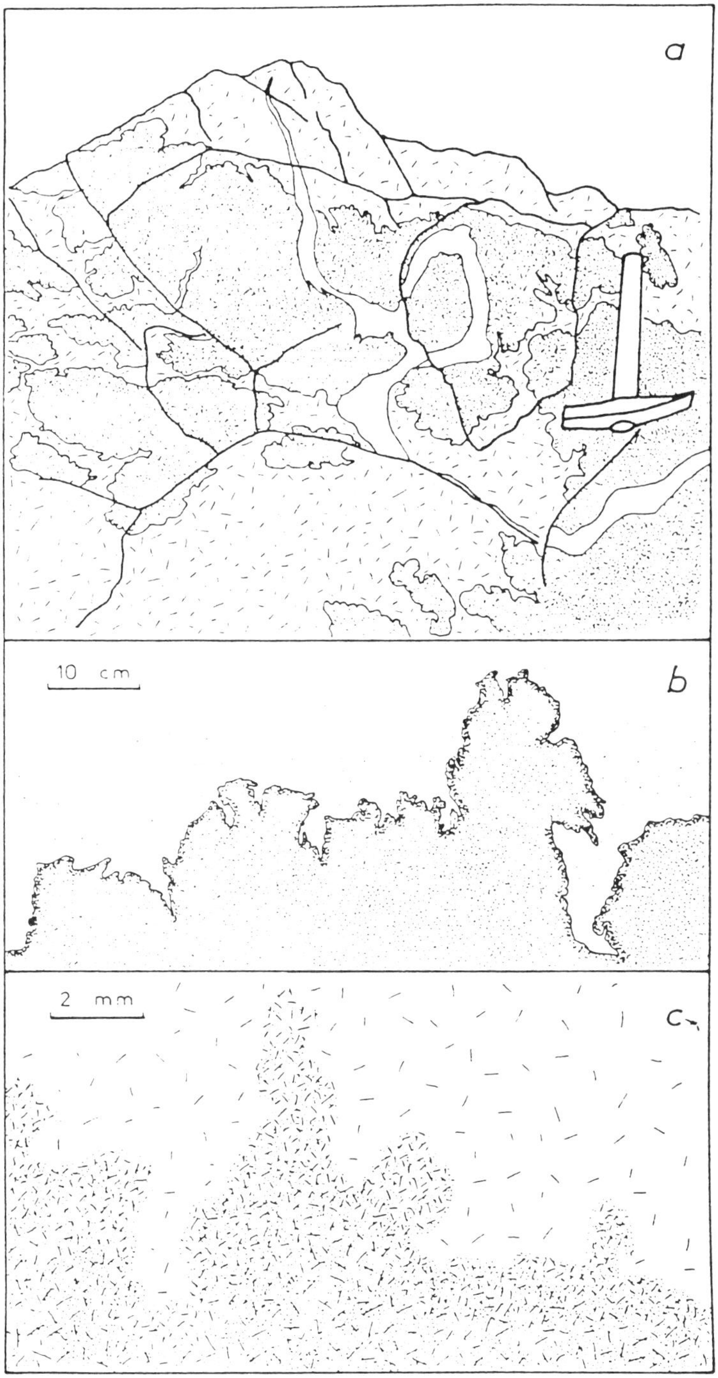

Fig. 5.18 (continued). b. (2) association of dolerites and granodiorites at different scales (from Bébien et al., 1987).

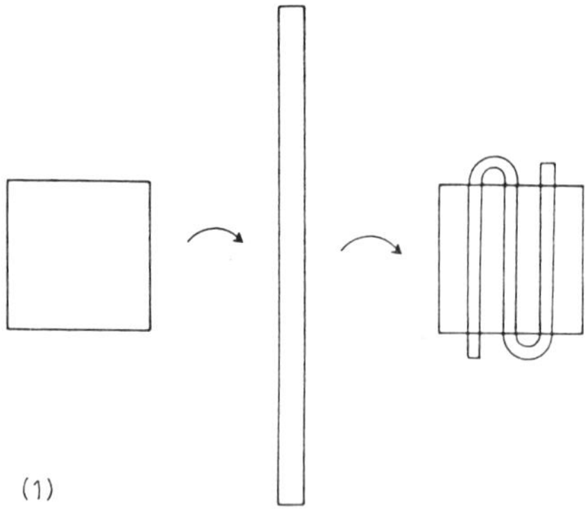

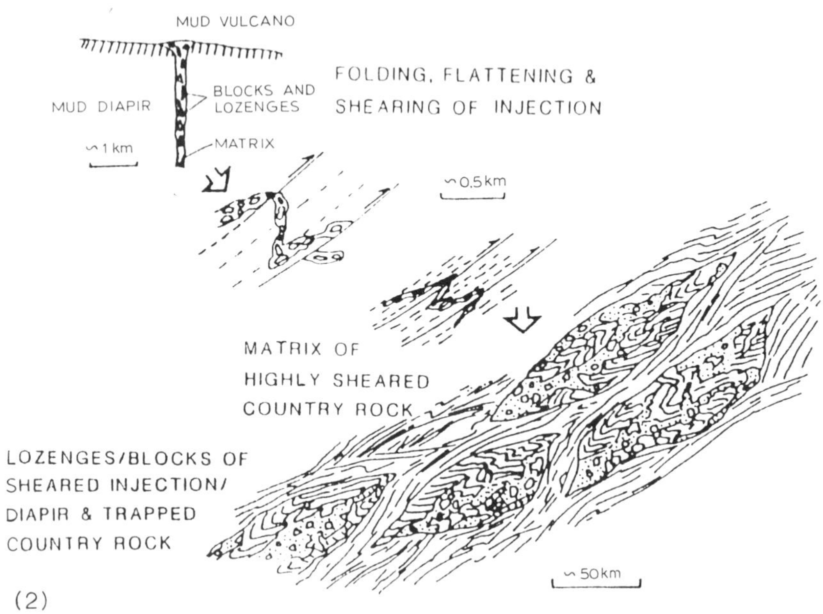

Fig. 5.18 (continued). c. (1) "Kneading" transformation to create chaotic attractors (from Young, 1982); (2) development of a chaotic mud injection (from Pickering et al., 1988).

glaciolacustrine and glaciodeltaic deposits; Brodzikowski et al. (1987) on the large scale gravity-induced sediment deformations].

In a very interesting study, Garcia-Ruiz and co-workers (1990) of the Universidad de Granada, Spain, have proved that the ammonite sutures are fractal and suggested that they are the result of a morphogenetic process that contains a Saffmann-Taylor type instability.

Garcia-Ruiz et al. selected ten ontogenetic sequences in sutures of Jurassic and

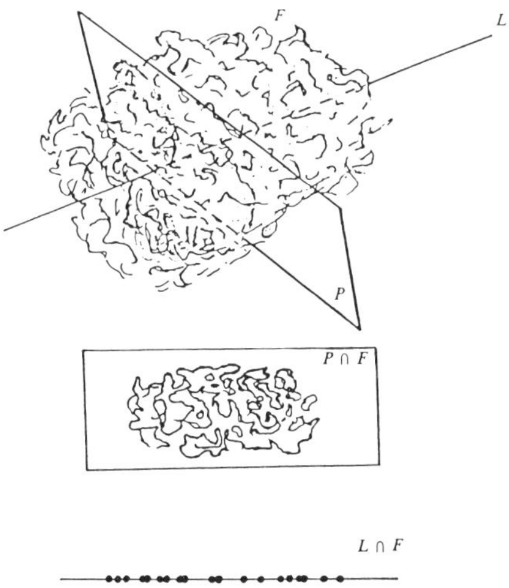

Fig. 5.19. Fractal object F in the three-dimensional space and its intersections with a plane P and a line L (from Sreenivasan and Meneveau, 1986).

Cretaceous ammonites, selecting those that represent a wide range of shell diversity. Figure 5.20 (from Wiedmann and Kullmann, 1981) presents one of the sequences analysed, it shows the suture development in the *Sowerbyceras (Holcophylloceras) calypso*, from the primary suture (a) to the adult suture (i). The fractal dimension of the sutures was determined by box counting at a fixed magnification, using different pixel sizes E. The number of pixels covering the suture line analysed scales as $N(E) \propto E^{-D}$. The most interesting finding of the study has been that for all ontogenetic sequences considered D increased for later sutures (Figure 5.21).

The fractal dimensions of the adult sutures are similar for the different ammonite genera (Table 5.2), that is they can not be used for their classification. The mean value of the fractal dimensions in the table is $D \approx 1.42$.

The fractal nature of the ammonite sutures has been first observed by Damiani (1986), who compared them with the triadic Von Koch curves. The dimension of the

TABLE 5.2

Fractal dimension for the adult sutures belonging to different ammonite genera

No.	Genus	D	No.	Genus	D
1	*Spitidiscus alcoyensis*	1.38	6	*Spiroceras bifurcati*	1.25
2[a]	*Sowerbyceras calypso*	1.44	7	*Prodactylioceras davoei*	1.46
3	*Haploceras grasianum*	1.52	8	*Staufenia staufensis*	1.36
4	*Gaudriceras tenuiliratum*	1.56	9	*Sphaeroceras brongniarti*	1.33
5	*Euphylloceras ponticuli*	1.46	10	*Roloboceras hambrovi*	1.43

Source: Garcia-Ruiz et al.(1990). [a] See Figure 5.20.

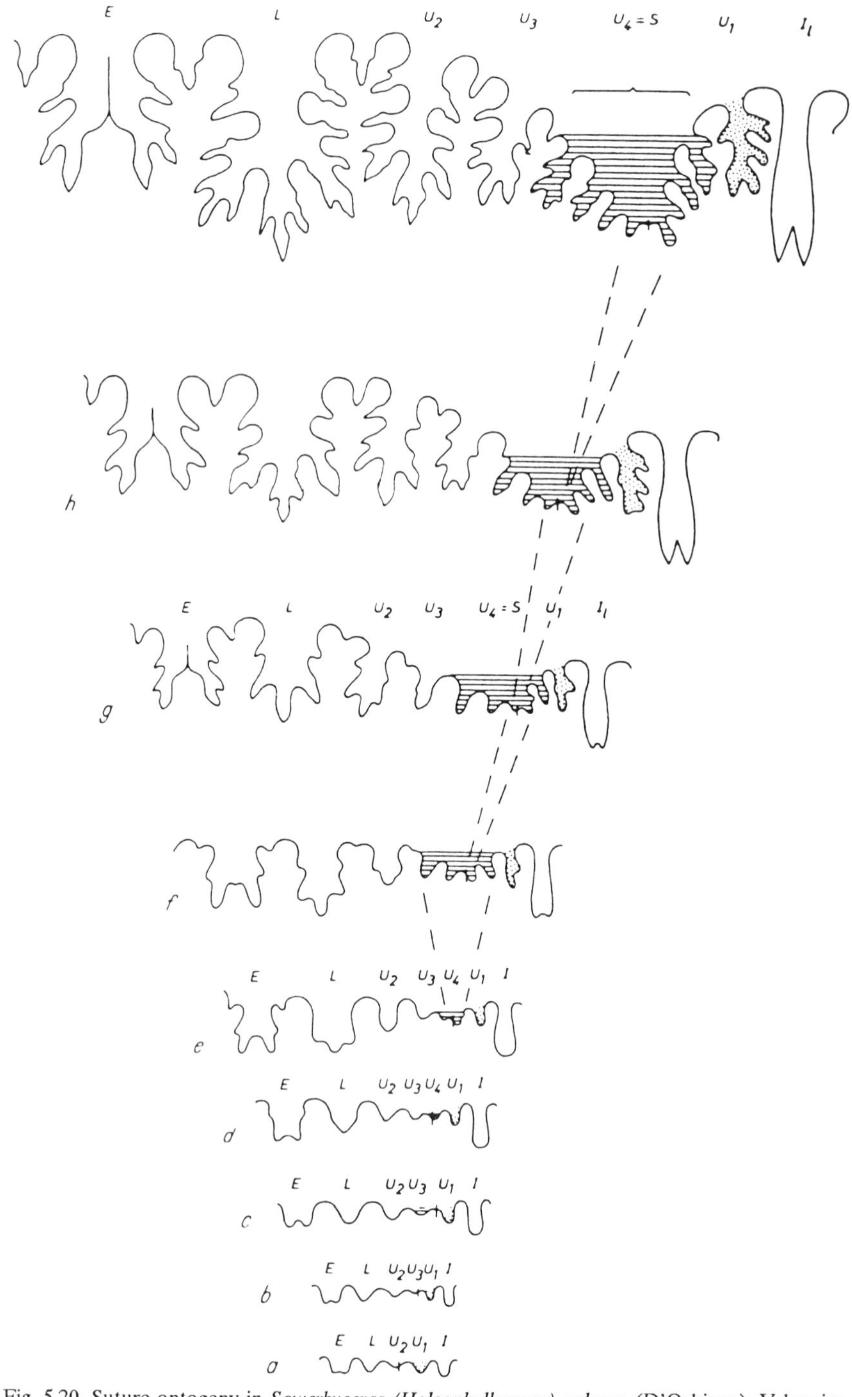

Fig. 5.20. Suture ontogeny in *Sowerbyceras (Holcophylloceras) calypso* (D'Orbigny), Valangian, France. a = Primary suture, i = adult suture. (From Wiedmann and Kullmann, 1981.)

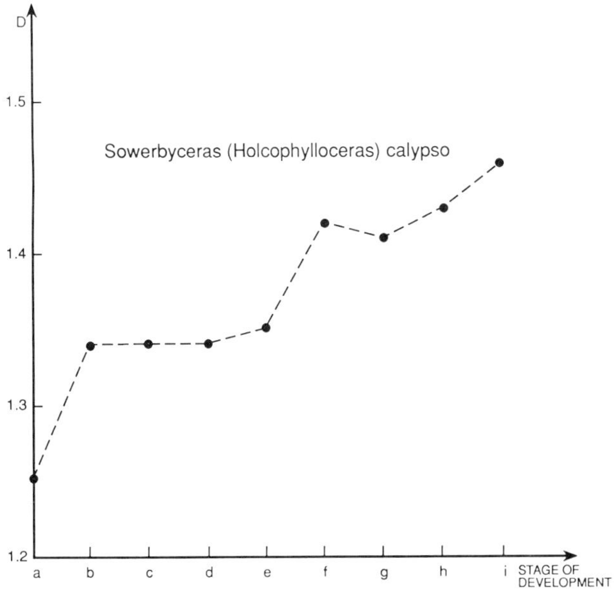

Fig. 5.21. Increase of the fractal dimension of the sutures of *Sowerbyceras (Holcophylloceras) calypso*, shown in Figure 5.20 (after table 1 in Garcia-Ruiz et al., 1990).

regular Von Koch curve ($D = \log 3/\log 4 = 1.26$), however, is lower than the values in Table 5.2.

Garcia-Ruiz et al. (1990) interpreted the value $D \approx 1.42$ in terms of a viscous finger model in a bounded Hele-Shaw cell, where $D \approx 1.40$ (Nittmann et al., 1985; Nittmann, 1986). [The pigeon-hole model of Pape et al. (1987), corresponding to optimal radial growth with $D \approx 1.36$, is a viable alternative.]

TABLE 5.3

Mean values of the parameters of ball lightning

1. Diameter (cm)	28
2. Lifetime (s)	10
3. Speed (m s^{-1})	4
4. Energy (J)	$\approx 10^{4.3}$
5. Energy density (J cm^{-3})	$\approx 10^{1.2}$
6. Luminous flux	≈1400
7. Colour (% of observation)	white (24), yellow (24), red (18), orange (14), blue and violet (12)
8. Observed in stormy weather	in 70% of cases
9. Disintegrated	in 50% of cases
10. Probability of appearance (km^{-2} min^{-1})	$10^{-8.5}$

Source: Smirnov (1986).

Fractal clusters are everywhere and they are also of relevance to *ball lightning.* One of the models (Aleksandrov et al., 1982) capable of explaining the constant shape and size of the balls in the course of their evolution is based on the hypothesis that the active material of ball lightning is a cluster of filamentary aerosols. The submicron filaments are built up by the action of the electric field from charged particles and droplets. Smirnov (1986) compiled the observed properties of ball lightning (Table 5.3), and proved that most of the properties in the table can be understood if one assumes that the *active material of ball lightning is a fractal cluster.*

* * *

Having studied Smirnov's arguments, I decided to re-examine in my old-time favourite, Lewis Carroll's "The Hunting of the Snark", the mysterious disappearance of the Baker. (As you will recall, he had been warned by his uncle that if he would ever meet a Snark which is a Boojum, he would "softly and suddenly vanish away/ and never be met with again". When he finally met one, he uttered the ominous words "Its a Boo... ", and vanished.)

"It is quite possible" — I thought — "that the Boojum is a FRAC... "

References

Adams, F.D. (1938). The Birth and Development of the Geological Sciences. Bailliere, Tindall and Cox, London.

Agterberg, F.P. (1982). Recent developments in geomathematics. Geo-Processing, 2: 1–32.

Aleksandrov, V.Ya., Golubev, E.M. and Padmoskenskii, I.V. (1982). Aerosol mode of ball lightning. Sov. Phys. Techn. Phys., 27, 10: 1221–1224.

Allègre, C.J., Le Mouel, J.L. and Provost, A. (1982). Scaling rules in rock fracture and possible implications for earthquake prediction. Nature, 297: 47–49.

Anis, A.A. and Lloyd, E.H. (1976). The expected value of the adjusted rescaled Hurst range of independent normal summands. Biometrika, 63: 111–116.

Anstey, N.A. and Newman, P. (1966). The sectional auto-correlogram and the sectional retro-correlogram. Geophys. Prospect., 14, 4: 389–426.

Arecchi, F.T. (1987). Hyperchaos and $1/f$ spectra in nonlinear dynamics. *In*: E.R. Pike and L.A. Lugiato (Editors), Chaos, Noise and Fractals. Malvern Physics Series, Adam Hilger, Bristol, pp. 1–14.

Arya, A., Hewett, T.A., Larson, R. and Lake, L.W. (1985). Dispersion and reservoir heterogeneity. 60th Annu. SPE Tech. Conf. (Las Vegas), Pap. SPE 14364.

Bébien, J., Gagny, C. and Tanani, S.S. (1987). Les associations de magmas acides et basiques: des objets fractals? C.R. Acad. Sci. Paris, Ser. II, 305: 277–280.

Boes, D.C. (1988). Schemes exhibiting Hurst behavior. *In*: J.N. Srivastava (Editor), Probability and Statistics. Essays in Honor of F.A.Graybill. Elsevier Science Publishers, Amsterdam–New York, pp. 21–42.

Brodzikowski, K. and Haluszczak, A. (1987). Flame structures and associated deformations in Quaternary glaciolacustrine and glaciodeltaic deposits: examples from central Poland. *In*: M.E. Jones and R.M.F. Preston (Editors), Deformation of Sediments and Sedimentary Rocks. Geol. Soc. Spec. Publ. No. 29: 279–286.

Brodzikowski, K., Haluszczak, A., Krzyszkowski, D. and Van Loon, A. (1987). Genesis and diagnostic value of large-scale gravity induced penecontemporaneous deformation horizons in Quaternary sediments of the Kleszczow Graben, Poland. *Ibid.*, pp. 287–298.

Burlaga, L.F. and Klein, L.W. (1986). Fractal structure of the interplanetary magnetic field. J. Geophys. Res., 91A: 347–350.

Cahn, R.W. (1989). Fractal dimension and fracture. Nature, 338: 201–202.

Carroll, L. (Dodgson, C.L) (1916). The Hunting of the Snark. An Agony in Eight Fits. Macmillan and Co., Ltd., London.

Crane, S.D. and Tubman, K.M. (1990). Reservoir variability and modelling with fractals. 65th Annu. SPE Tech. Conf. (New Orleans), Pap. SPE 20606.

Crossley, D.J. and Jensen, O.G. (1989). Fractal velocity models in refraction seismology. PAGEOPH, 131, Nos. 1–2: 61–76.

Crutchfield, J.P., Farmer, J.D., Packard, N.H. and Shaw, R.S. (1986). Chaos. Sci. Am., 255: 46.

Damiani, G. (1986). Significato funzionalle dell'evoluzione dei setti e delle linee di sutura dei nautiloidi e degli ammonoidi. *In*: G. Pallini (Editor), Fossili, Evoluzione. Ambiente Tecnoscienza, Rome, pp. 123–130.

Devaney, R.L. (1988). Fractal patterns arising in chaotic dynamical systems. *In*: H.-O. Peitgen and D. Saupe (Editors), The Science of Fractal Images. Springer Verlag, Berlin–Heidelberg, pp. 137–167.

Emaliani, C. (1955). Pleistocene temperatures. J. Geol., 63, 6: 538–578.

Emanuel, A.S., Alameda, G.K., Behrens, R.A. and Hewett, T.A. (1987). Reservoir performance prediction methods based on fractal geostatistics. 62nd Annu. SPE Tech. Conf. (Dallas), Pap. SPE 16971.

Feder, J. (1988). Fractals. Plenum Press, New York–London.

Feller, W. (1951). The asymptotic distribution of the range of sums of independent random variables. Ann. Math. Stat., 22: 427–432.

Fluegeman, R.H., Jr. and Snow, R.S. (1989). Fractal analysis of long-range paleoclimatic data: Oxygen isotope record of Pacific core V28-239. PAGEOPH, 131, Nos. 1–2: 307–313.

Ford, J. (1989). What is chaos, that we should be mindful of it? *In*: P. Davies (Editor), The New Physics. Cambridge Univ. Press, Cambridge, pp. 348–371.

Fowler, A.D. (1990). Self-organized mineral textures of igneous rocks: the fractal approach. Earth Sci. Rev., 29: 47–55.

Fraedrich, K. (1986). Estimating the dimensions of weather and climate attractors. J. Atm. Sci., 43, 5: 419–432.

Garcia-Ruiz, J.M., Checa, A. and Rivas, P. (1990). On the origin of the ammonite sutures. Paleobiology, 16, 3: 349–354.

Gardner, M. (1978). White and brown music, fractal curves and one-over-*f* fluctuations. Sci. Am. (April), pp. 16–32.

Gleick, J. (1987). Chaos. Making a New Science. Sphere Books Ltd., London.

Godwin, J. (1979). Athanasius Kircher. Renaissance Man and the Quest for Lost Knowledge. Thames and Hudson, London.

Goldberger, A.L. and West, B.J. (1987). Fractals in physiology and medicine. Yale J. Biol. Med., 60: 421–435.

Grassberger, P. (1986). Do climatic attractors exist? Nature, 323: 609–612.

Grassberger, P. and Procaccia, I. (1983a). Measuring the strangeness of strange attractors. Physica, 9D: 189–208.

Grassberger, P. and Procaccia, I. (1983b). Characterization of strange attractors. Phys. Rev. Lett., 50, 5: 346–349.

Hamilton, D. (1961). Algal growth in the Rhaetic Cotham Marble of southern England. Palaeontology, 4, Pt. 3: 324–333.

Henderson, H.W. and Wells, R. (1988). Obtaining attractor dimensions from meteorological time series. Adv. Geophys., 30: 205–237.

Herrmann, H.J. (1986). Growth: an introduction. *In*: H.E. Stanley and N. Ostrowsky (Editors), On Growth and Form. Martinus Nijhoff Publ., Dordrecht, pp. 3–20.

Hertzberg, R.W., Hahn, M.T., Rimnac, J.A. and Manson, J.A. (1983). A laboratory analysis of a lavatory failure. Int. J. Fracture, 23: R57–R60.

Hewett, T.A. (1986). Fractal distributions of reservoir heterogeneity and their influence on fluid transport. 61st Annu. SPE Tech. Conf. (New Orleans), Pap. SPE 15386.

Hewett, T.A. and Behrens, R.A. (1988). Conditional simulation of reservoir heterogeneity with fractals. 63rd Annu. SPE Tech. Conf. (Houston), Pap. SPE 18326.

Hurst, H.E. (1951). Long-term storage capacity of reservoirs. Trans. Am. Soc. Civ. Eng., 116: 770–808.

Hurst, H.E. (1957). A suggested statistical model of some time series which occur in Nature. Nature, 180: 494.

Hurst, H.E., Black, R.P. and Simaika, Y.M. (1965). Long-Term Storage: An Experimental Study. Constable, London.

Jacobs, J.A. (1986). From the core or the skies? Nature, 323: 296–297.

Journel, A.G. and Huijbregts, C. (1978). Mining Geostatistics. Academic Press, New York.

Jullien, R. and Botet, R. (1987). Aggregation and Fractal Aggregates. World Scientific, Singapore.

Katz, M.J. (1985). How straight do axons grow? J. Neurosci., 5, 3: 589–595.

Kaye, B.H. (1992). Discovering the Surprising Patterns of Chaos and Complexity. VCH Publishers, Weinheim, Germany.

Kirkpatrick, S., Gelatt, C.D., Jr., and Vecchi, M.P. (1983). Optimization by simulated annealing. Science, 220, No. 4598: 671–680.

Klemes, V. (1974). The Hurst phenomenon: A puzzle? Water Resour. Res., 10, 4: 675–688.

Kobayashi, M. and Musha, T. (1982). $1/f$ fluctuation of heartbeat period. IEEE Trans. Biomed. Eng., 29: 456–457.

Korvin, G. (1972). Certain problems of seismic and ultrasonic wave propagation in a medium with inhomogeneities of random distribution. Pt.I. Geophys. Trans. 21, Nos. 1–4: 5–34.

Lanford, O.E.III (1982). The strange attractor theory of turbulence. Ann. Rev. Fluid Mech., 14: 347–364.

Langer, J.S. (1988). Dendrites and viscous fingers. *In*: R.H. Dalitz and R.B. Stinchcombe (Editors), The Breadth of Physics. World Scientific, Singapore, pp. 45–60.

Levinson, N. (1946). The Wiener RMS (root mean square) error criterion in filter design and prediction. J. Math. Phys., 25, 1: 261–278.

Liu, S.H. (1986). Fractals and their applications in condensed matter physics. Solid State Phys., 39: 207–273.

Lloyd, E.H. (1967). Stochastic reservoir theory. Adv. Hydrosci., 4: 281–339.

Lorenz, E.N. (1963). Deterministic nonperiodic flow. J. Atm. Sci., 20: 130–141.

Louis, E. and Guinea, F. (1987). The fractal nature of fracture. Europhys. Lett., 3, 8: 871–877.

Louis, E., Guinea, F. and Flores, F. (1986). The fractal nature of fracture. *In*: L. Pietronero and E. Tosatti (Editors), Fractals in Physics. Elsevier, Amsterdam–New York, pp. 177–180.

Lovejoy, S. and Mandelbrot, B.B. (1985). Fractal properties of rain, and a fractal model. Tellus, 37A: 209–232.

Lovejoy, S. and Schertzer, D. (1983). Rainfronts, fractals and rainfall simulations. *In*: Hydrological Applications of Remote Sensing and Remote Data Transmission. Proc. Hamburg Symp., August, 1983. IAHS Publ.No. 145: 323–334.

Lovejoy, S. and Schertzer, D. (1986). Scale invariance, symmetries, fractals, and stochastic simulations of atmospheric phenomena. Bull. Am. Meteorol. Soc., 67, 1: 21–32.

Lucido, G. (1984). Felsic vortices of the Taylor-Proudman type discovered in basaltic magma of Sicily. Mineral. Petrogr. Acta, 28: 59–70.

Lüscher, E. (1988). Order and chaos. *In*: S.S. Chan and P.G. Debrunner (Editors), Proc. Int. Symp. Frontiers of Science. AIP Conf. Proc. No.180, AIP, New York, pp. 140–163.

Lung, C.W. (1986). Fractals and the fracture of cracked metals. *In*: L. Pietronero and E, Tosatti (Editors), Fractals in Physics. Elsevier, Amsterdam–New York, pp. 189–192.

Madden, T.R. (1983). Microcrack connectivity in rocks: a renormalization group approach to the critical phenomena of conduction and failure in crystalline rocks. Geophys. Res., 88B: 585–592.

Mandelbrot, B.B. (1967). Some noises with $1/f$ spectrum, a bridge between direct current and white noise. IEEE Trans. Info. Theor. IT-13, 2: 289–298.

Mandelbrot, B.B. (1975). Limit theorems on the self-normalized range for weakly and strongly dependent processes. Z. Wahrsch. Verw. Gebiete, 31: 271–285.

Mandelbrot, B.B. (1982). The Fractal Geometry of Nature. Freeman and Co., New York.

Mandelbrot, B.B. and Mccamy, K. (1970). On the secular pole motion and the Chandler wobble. Geophys. J. R. Astrol. Soc., 21: 217–232.

Mandelbrot, B.B. and Van Ness, J.W. (1968). Fractional Brownian motions, fractional noises and applications. SIAM Rev., 10, 4: 422–437.

Mandelbrot, B.B. and Wallis, J.R. (1968). Noah, Joseph, and operational hydrology. Water Resour. Res. 4, 5: 909–918.

Mandelbrot, B.B. and Wallis, J.R. (1969). Some long-run properties of geophysical records. Water Resour. Res. 5, 2: 321–340.

Mandelbrot, B.B., Passoja, D.E. and Paullay, A.J. (1984). Fractal character of fracture surfaces of metals. Nature, 308: 721–722.

Mann, Thomas (1949). Doctor Faustus. The Life of the German Composer Adrian Leverkühn as Told by a Friend. (Translated from the German by H.T. Lowe-Porter.) Secker and Warburg, London.

Meakin, P. (1983). Diffusion-controlled deposition on fibers and surfaces. Phys. Rev. A, 27, 5: 2616–2623.

Meakin, P. (1986). A new model for biological pattern formation. J. Theor. Biol., 118: 101–113.

Meakin, P. (1988a). The growth of fractal aggregates and their fractal measures. *In*: C. Domb and J.L. Lebowitz (Editors), Phase Transitions and Critical Phenomena. Vol. 12. Academic Press, London–New York, pp. 335–489.

Meakin, P. (1988b). Fractal aggregates. Adv. Colloid Interf. Sci., 28: 249–331.

Meakin, P. (1991). Models for material failure and deformation. Science, 252: 226–234.

Montroll, E.W. and Shlesinger, M.F. (1983). Maximum entropy formalism, fractals, scaling phenomena, and $1/f$ noise: a tale of tails. J. Stat. Phys., 32, 2: 209–230.

Munk, W.H. and Macdonald, G.J.F. (1960). The Rotation of the Earth, a Geophysical Discussion. Cambridge University Press, Cambridge.

Nelson, J. (1989). Fractality of sooty smoke: implications for the severity of nuclear winter. Nature, 339: 611–613.

Nicolis, C. and Nicolis, G. (1984). Is there a climatic attractor? Nature, 311: 529–532.

Nittmann, J. (1986). Fractal viscous fingering: experiments and models. Physica, 140A: 124–133.

Nittmann, J., Daccord, G. and Stanley, H.E. (1985). Fractal growth of viscous fingers: Quantitative characterization of a fluid instability phenomenon. Nature, 314, 6007: 141–144.

O'Doherty, R.F. and Anstey, N.A. (1971). Reflections on amplitudes. Geophys. Prospect., 19, 3: 430–458.

Pape, H., Riepe, L. and Schopper, J.R. (1987). Theory of self-similar network structures in sedimentary and igneous rocks and their investigation with microscopical and physical methods. J. Microsc., 148, Pt. 2: 121–147.

Paterson, L. (1981). Radial fingering in a Hele-Shaw cell. J. Fluid Mech., 113: 513–529.

Paterson, L. (1985). Fingering with miscible fluids in a Hele-Shaw cell. Phys. Fluids, 28, 1: 26–30.

Peterson, R.A., Filippone, W.R. and Coker, F.B. (1955). The synthesis of seismograms from well log data. Geophysics, 20, 3: 516–538.

Pickering, K.T., Agar, S.M. and Ogawa, Y. (1988). Genesis and deformation of mud injections containing chaotic basalt–limestone–chert associations: Examples from the southwest Japan forearc. Geology, 16: 881–885.

Plotnick, R.E. (1986). A fractal model for the distribution of stratigraphic hiatuses. J. Geol. 94: 885–890.

Powles, J.G. (1985). Fractal analysis of molecular motion in liquids. Phys. Lett., 107A, 8: 403–405.

Powles, J.G. and Quirke, N. (1984). Fractal geometry and Brownian motion: a new parameter to describe molecular motion. Phys. Rev. Lett., 52, 18: 1571–1574.

Press, W.H. (1978). Flicker noises in astronomy and elswhere. Comments Astrophys., 7, 4: 103–119.

Procaccia, I. and Schuster, H. (1983). Functional renormalization-group theory of universal $1/f$ noise in dynamical systems. Phys. Rev., A28, 2: 1210–1212.

Rapaport, D.C. (1985). The fractal nature of molecular trajectories in fluids. J. Stat. Phys. 40, Nos. 5–6: 751–758.

Rescher, N. (1959/1960). Choice without preference. Kant-Studien, 51, 1: 142–175.

Rietsch, E. (1982). Deconvolution in the presence of noise and a higher than Nyquist signal rate. Geoexploration, 20: 61–73.

Rothman, D.H. (1985). Nonlinear inversion, statistical mechanics, and residual statics estimation. Geophysics, 50, 12: 2784–2796.

Rothman, D.H. (1986). Automatic estimation of large residual statics corrections. Geophysics, 51, 2: 332–346.

Roulin-Moloney, A.C., Editor (1989). Preface to "Fractography and Failure Mechanisms of Polymers and Composites". Elsevier Applied Science, London–New York, pp. I–VIII.

Saffman, P.G. and Taylor, G.I. (1958). The penetration of a fluid into a medium or Hele-Shaw cell containing a more viscous liquid. Proc. R. Soc. Lond., 245: 312–329.

Sander, L.M. (1985). Growth by particle aggregation. *In*: R. Pynn and A. Skjeltorp (Editors), Scaling Phenomena in Disordered Systems. Plenum Press, New York–London, pp. 31–47.

Sander, L.M. (1986). Fractal growth processes. Nature, 322: 789–793.

Sander, L.M. (1987). Fractal growth. Sci. Am., 256, 1: 82–88.

Schick, K.L. and Verveen, A.A. (1974). $1/f$ noise with a low frequency white noise limit. Nature, 251: 599–601.

Schillinger, J. (1948). The Mathematical Basis of the Arts. Philosophical Library, New York.

Schneider, W.A. and Whitman, W.W. (1990). Dipmeter analysis by a Monte Carlo technique. Geophysics, 55, 3: 320–326.

Scholz, C.H. (1989). Global perspectives of chaos. Nature, 338: 459–460.

Schroeder, M.R. (1986). Auditory paradox based on fractal waveform. J. Acoust. Soc. Am., 79, 1: 186–189.

Schroeder, M.R. (1987). Is there such a thing as fractal music? Nature, 325: 765–766.

Shaw, H.R. (1987). The periodic structure of the natural record and nonlinear dynamics. Eos, 68, 50: 1651–1665.

Smalley, R.F., Jr., Turcotte, D.L. and Solla, S.A. (1985). A renormalization group approach to the stick-slip behavior of faults. J. Geophys. Res., B90: 1894–1900.

Smirnov, B.M. (1986). Fractal clusters. Sov. Phys. Usp., 29, 6: 481–505.

Smith, C.S. (1981). A Search for Structure. Selected Essays on Science, Art, and History. The MIT Press, Cambridge, Mass.

Smith, D.K. and Shaw, P.R. (1990). Seafloor topography: A record of a chaotic dynamical system? Geophys. Res. Lett., 17: 1541–1544.

Solla, S.A. (1986). Crack propagation and onset of failure. *In*: H.E. Stanley and N. Ostrowsky (Editors), On Growth and Form. Martinus Nijhoff Publishers, Dordrecht, pp. 260–262.

Sreenivasan, K.R. and Meneveau, C. (1986). The fractal facets of turbulence. J. Fluid Mech., 173: 357–386.

Stanley, H.E. and Ostrowsky, N., Editors (1985). On Growth and Form. Fractal and Non-Fractal Patterns in Physics. Martinus Nijhoff Publishers, Dordrecht.

Taggart, I.J. and Gray, T.A. (1991). Application of fractals to characterise porous media heterogeneities. CTAC-91 Conf. (Adelaide, July 1991). Preprint.

Taggart, I.J. and Salisch, H.A. (1991). Fractal geometry, reservoir characterisation and oil recovery. APEA Journal, 1991: 377–385.

Takayasu, H. (1985). A deterministic model of fracture. Prog. Theor. Phys., 74, 6: 1343–1345.

Takayasu, H. (1986). Pattern formation of dendritic fractals in fracture and electric breakdown. *In*: L. Pietronero and E. Tosatti (Editors), Fractals in Physics. Elsevier, Amsterdam–New York, pp. 181–184.

Takayasu, H. and Nishikawa, I. (1986). Directed dendritic fractals. *In*: S. Ishizaka (General Editor), Science on Form. Proc. First Int. Symp. for Science on Form. KTK Scientific Publishers, Tokyo, pp. 15–22.

Termonia, Y. and Meakin, P. (1986). Formation of fractal cracks in a kinetic fracture model. Nature, 320: 429–431.

Theiler, J. (1990). Estimating the fractal dimension of chaotic time series. Lincoln Lab. J., 3, 1: 63–85.

Thomas, A. (1986). Bases d'un modèle fractal de propagation des fractures dans un milieu rocheux hétérogène. C.R. Acad. Sci. Paris, Ser.II, 303, 3: 225–228.

Thompson, D'Arcy W. (1942). On Growth and Form. Cambridge University Press, Cambridge.

Todoeschuck, J.P. and Jensen, O.G. (1988). Joseph geology and seismic deconvolution. Geophysics, 53, 11: 1410–1414.

Todoeschuck, J.P. and Jensen, O.G. (1989). Scaling geology and seismic deconvolution. PAGEOPH, 131, Nos. 1–2: 273–287.

Todoeschuck, J.P., Jensen, O.G. and Labonte, S. (1990). Gaussian scaling noise model of seismic reflection sequences: Evidence from well logs. Geophysics, 55, 4: 480–484.

Tritton, D.J. (1986a). Ordered and chaotic motion of a forced spherical pendulum. Eur. J. Phys., 7: 162–169.

Tritton, D.J. (1986b). Chaos in the swing of a pendulum. New Scientist (24 July), pp. 37–40.

Tritton, D.J. (1988). Physical Fluid Dynamics, 2nd Edition, Oxford University Press, Oxford.

Tritton, D.J. (1989). Deterministic chaos, geomagnetic reversals, and the spherical pendulum. *In*: F.J. Lowes et al. (Editors), Geomagnetism and Paleomagnetism. Kluwer Academic Publishers, pp. 215–226.

Tsurumi, S. and Takayasu, H. (1986). The fractal dimension in computer-simulated random walks. Phys. Lett., 113A, 9: 449–450.

Turcotte, D.L. (1986). Fractals and fragmentation. J. Geophys. Res., B91: 1921–1926.

Turcotte, D.L., Smalley, R.F., Jr. and Solla, S.A. (1985). Collapse of loaded fractal trees. Nature, 313: 671–672.

Underwood, E.E. (1992). Fractals in fractography. *In*: Encyclopedia of Science and Technology 1992 Year Book. McGraw Hill, N.Y., in press.

Underwood, E.E. and Banerji, K. (1986). Fractals in fractography. Mater. Sci. Eng., 80: 1–14.

Van Damme, H. and Ben Ohoud, M. (1990). From flow to fracture and fragmentation in colloidal media. 2: Local order and fragmentation geometry. *In*: J.C. Charmet et al. (Editors), Disorder and Fracture. Plenum Press, N.Y., pp. 105–116.

Van Damme, H. and Lemaire, E. (1990). From flow to fracture and fragmentation in colloidal media.

1. Non-Newtonian fingering and visco-elastic fracturing. *In*: J.C. Charmet et al. (Editors), Disorder and Fracture. Plenum Press, N.Y., pp. 83–104.

Vicsek, T. (1983). Fractal models for diffusion controlled aggregation. J. Phys. A, Math. Gen., 16: L647–L652.

Vicsek, T. (1991). Fractal Growth Phenomena. 2nd ed. World Scientific, Singapore.

Vicsek, M. and Vicsek, T. (1991). Fractal Growth (Software + Manual). World Scientific, Singapore.

Vörös, Z. (1990). Fractal analysis applied to some geomagnetic storms observed at the Hurbanovo geomagnetic observatory. Ann. Geophys., 8: 191–194.

Vörös, Z. (1991). Synergetic aspects of geomagnetic indices. Geophys. Trans., 36: 347–353.

Voss, R. (1985). Random fractal forgeries. *In*: R.A. Earnshaw (Editor), Fundamental Algorithms for Computer Graphics. NATO ASI Series, Vol. 17. Springer Verlag, Berlin–Heidelberg, pp. 805–835.

Voss, R.F. (1988). Fractals in nature: From characterization to simulation. *In*: H.-O. Peitgen and O. Saupe (Editors), The Science of Fractal Images. Springer Verlag, Berlin–Heidelberg, pp. 21–70.

Voss, R.F. and Clarke, J. (1975). "$1/f$" noise in music and speech. Nature, 258: 317–318.

Walden, A.T. and Hosken, J.W.J. (1986). The nature of the non-Gaussianity of primary reflection coefficients and its sigificance for deconvolution. Geophys. Prospect., 34: 1038–1066.

Walker, J. (1986). Cracks in a surface look intricately random but actually develop rather systematically. Sci. Am., 255, No.4: 178–183.

West, B.J. (1985). An Essay on the Importance of Being Nonlinear. Lecture Notes in Biomathematics, No. 62. Springer Verlag, Berlin–Heidelberg.

Wiedmann, J. and Kullmann, J. (1981). Ammonoid sutures in ontogeny and phylogeny. *In*: M.R. House and J.R. Senior (Editors), The Ammonoidea. Syst. Assoc. Spec. Vol. No. 18. Academic Press, London–New York, pp. 215–255.

Witten, T.A. (1987). Fractal aggregates and other tenuous structures. *In*: J. Souletie et al. (Editors), Le Hasard et la Matière/Chance and Matter. North Holland, Amsterdam, pp. 161–210.

Witten, T.A. and Sander, L.M. (1981). Diffusion limited aggregation, a kinetic critical phenomenon. Phys. Rev. Lett., 47, 19: 1400–1403.

Wunsch, C. (1972). Bermuda sea level in relation to tides, weather, and baroclinic fluctuations. Rev. Geophys. Space Phys., 10, 1: 1–49.

Yilmaz, O. (1987). Seismic Data Processing. Investigations in Geophysics, Vol. 2. Society of Exploration Geophysicists, Oklahoma.

Young, L.-S. (1982). Dimension, entropy and Lyapunov exponents. Ergodic Theor. Dyn. Syst., 2: 109–124.

Zawadzki, I. (1987). Fractal structure and exponential decorrelation in rain. J. Geophys. Res., 92, D8: 9586–9590.

Author index

Subject index